PLEASE STAMP DATE DUE, BOTH BELOW AND ON CARD

DATE DUE	DATE DUE	DATE DUE	DATE DUE
OCT 1 2 2011			

GL-15

Series Editor
John M. Walker
School of Life Sciences
University of Hertfordshire
Hatfield, Hertfordshire, AL10 9AB, UK

For other titles published in this series, go to
www.springer.com/series/7651

Biomedical Nanotechnology

Methods and Protocols

Edited by

Sarah J. Hurst

Center for Nanoscale Materials, Argonne National Laboratory, Argonne, IL, USA

Editor
Sarah J. Hurst, Ph.D.
Center for Nanoscale Materials
Argonne National Laboratory
9700 S. Cass Avenue
Argonne, Illinois 60439
USA
shurst@anl.gov

ISSN 1064-3745   e-ISSN 1940-6029
ISBN 978-1-61779-051-5   e-ISBN 978-1-61779-052-2
DOI 10.1007/978-1-61779-052-2
Springer New York Dordrecht Heidelberg London

Library of Congress Control Number: 2011922747

Preface

Nanoscience and technology focuses on synthesizing structures that have at least one dimension on the sub-100 nm length scale. It deals with investigating the fundamental properties of such structures, which usually differ significantly from that of the bulk material, and taking advantage of these qualities in constructing novel materials and devices or developing unique applications. Owing to widespread interest and investment, biomedical nanotechnology, or the use of nanostructures in medicinal applications, is an area of intense research that is growing and progressing at a rapid pace. This rapid development is driven by the fact that nanomaterials often offer superior capabilities when compared with conventionally used materials for the detection, diagnosis, and treatment of disease. Further, they have the potential to enable real-time disease detection and therapy and to advance point-of-care systems.

The goal of this volume is to provide an overview of biomedical nanotechnology, from the conception of novel materials in the laboratory to the application of such structures in the clinic. After a brief introductory chapter, the first section consists of protocol chapters which provide practical information on the synthesis of a variety of solution-phase and surface-bound nanomaterials and their application in sensing, imaging, and/or therapeutics; most chapters provide step-by-step instructions and insight into overcoming possible pitfalls and challenges. The chapters are written by leading researchers in biology, chemistry, physics, engineering, and medicine from academia, industry, and the national laboratories. The second section consists of a series of case study/review chapters that discuss the toxicology of nanomaterials, the regulatory pathways to US Food and Drug Administration (FDA) approval of these materials, their patenting, marketing, and commercialization, and the legal and ethical issues surrounding their use. These are written by experts in the science, social science, business, law, and ethics communities. Nanotechnology looks not only to revolutionize medical care but the fundamental property differences associated with nanomaterials and the potential for their use as multicomponent/multifunctional structures are also transforming the aspects of these fields that take part in mediating their introduction to the public.

This volume is a useful reference for scientists and researchers at all levels who are interested in working in a new area of nanoscience and technology or in expanding their knowledge base in their current field. The case study/review chapters are meant to inform scientists of routes they can take in moving their research beyond the bench, so they can design their systems with real-world considerations in mind. In turn, this volume also will be of interest to the social scientist, lawyer, or businessperson who wants to learn about the salient points of nanotechnology as they are applied to their field.

I would like to thank Prof. John M. Walker, series editor for *Methods in Molecular Biology*, for his guidance and the authors for sharing their expertise and producing such high-quality chapters. It was a pleasure to work with them on this project. I would also like to acknowledge Dr. Haley D. Hill for her support and helpful discussions.

Argonne, IL   *Sarah J. Hurst*

Contents

Contributors

GHANASHYAM ACHARYA • *Departments of Biomedical Engineering and Pharmaceutics, Purdue University, West Lafayette, IN, USA*

ARCHIE A. ALEXANDER • *Adjunct Instructor of Health Administration (Health Law and Ethics)*

HANENE ALI-BOUCETTA • *Centre for Drug Delivery Research, The School of Pharmacy, University of London, London, UK*

KHULOUD T. AL-JAMAL • *Centre for Drug Delivery Research, The School of Pharmacy, University of London, London, UK*

DARREN J. ANDERSON • *Vive Nano, Inc., Toronto, ON, Canada*

GEORGE P. ANDERSON • *Center for Bio/Molecular Science and Engineering, U.S. Naval Research Laboratory, Washington, DC, USA*

ITAI BENHAR • *Department of Molecular Microbiology and Biotechnology, Tel Aviv University, Ramat Aviv, Israel*

DIANA M. BOWMAN • *Melbourne School for Population Health, University of Melbourne, Victoria, Australia; Department of International and European Law, KU Leuven, Leuven, Belgium*

JERRY C. CHANG • *Department of Chemistry, Vanderbilt University, Nashville, TN, USA*

NADA M. DIMITRIJEVIC • *Center of Nanoscale Materials, Argonne National Laboratory, Argonne, IL, USA*

JOSE AMADO DINGLASAN • *Vive Nano, Inc., Toronto, ON, Canada*

AARON C. EIFLER • *Feinberg School of Medicine, Northwestern University, Chicago, IL, USA*

RUTLEDGE G. ELLIS-BEHNKE • *Department of Brain and Cognitive Sciences, Massachusetts Institute of Technology, Cambridge, MA, USA*

ERIK FISHER • *School of Politics and Global Studies, Consortium for Science, Policy & Outcomes, Center for Nanotechnology in Society, Arizona State University, Tempe, AZ, USA*

STEFAN FRANZEN • *Department of Chemistry, North Carolina State University, Raleigh, NC, USA*

DI GAO • *Department of Chemical and Petroleum Engineering, University of Pittsburgh, Pittsburg, PA, USA*

JERED B. HAUN • *Center for Systems Biology, Massachusetts General Hospital, Boston, MA, USA*

SARAH J. HURST • *Center for Nanoscale Materials, Argonne National Laboratory, Argonne, IL, USA*

ROSHAN JAMES • *Department of Orthopaedic Surgery, Department of Chemical, Materials and Biomolecular Engineering, University of Connecticut Health Center, Farmington, CT, USA*

FABRICE JOTTERAND • *Division of Ethics and Health Policy, Department of Clinical Sciences and Psychiatry, UT Southwestern Medical Center, Dallas, TX, USA*

RAOUL KOPELMAN • *Department of Chemistry, University of Michigan, Ann Arbor, MI, USA*

KOSTAS KOSTARELOS • *Centre for Drug Delivery Research, The School of Pharmacy, University of London, London, UK*

QIN KUANG • *School of Materials Science and Engineering, Georgia Institute of Technology, Altanta, GA, USA*

SANGAMESH G. KUMBAR • *Department of Orthopaedic Surgery, Department of Chemical, Materials and Biomolecular Engineering, University of Connecticut Health Center, Farmington, CT, USA*

CATO T. LAURENCIN • *Department of Orthopaedic Surgery, Department of Chemical, Materials and Biomolecular Engineering, University of Connecticut Health Center, Farmington, CT, USA*

HAKHO LEE • *Center for Systems Biology, Massachusetts General Hospital, Boston, MA, USA*

JAE-SEUNG LEE • *Department of Materials Science and Engineering, Korea University, Seoul, Republic of Korea*

ONE-SUN LEE • *Department of Chemistry, Northwestern University, Evanston, IL, USA*

YONG-EUN KOO LEE • *Department of Chemistry, University of Michigan, Ann Arbor, MI, USA*

HONGWEI LIAO • *Department of Chemistry and Biochemistry, University of Maryland, College Park, MD, USA*

FRANCES S. LIGLER • *Center for Bio/Molecular Science and Engineering, U.S. Naval Research Laboratory, Washington, DC, USA*

DUSTIN LOCKNEY • *Department of Chemistry, North Carolina State University, Raleigh, NC, USA*

STEVEN LOMMEL • *Department of Plant Pathology, North Carolina State University, Raleigh, NC, USA*

MATTHEW MCDERMOTT • *Akina Inc., West Lafayette, IN, USA*

IGOR L. MEDINTZ • *Center for Bio/Molecular Science and Engineering, U.S. Naval Research Laboratory, Washington, DC, USA*

GWEN E. OWENS • *Department of Molecular and Medical Pharmacology, Crump Institute for Molecular Imaging, Institute for Molecular Medicine, University of California, Los Angeles, Los Angeles, CA, USA*

BHAVNA PARATALA • *Department of Biomedical Engineering, State University of New York at Stony Brook, Stony Brook, NY, USA*

HAESUN PARK • *Departments of Biomedical Engineering and Pharmaceutics, Purdue University, West Lafayette, IN, USA*

KINAM PARK • *Departments of Biomedical Engineering and Pharmaceutics, Purdue University, West Lafayette, IN, USA; Akina Inc., West Lafayette, IN, USA*

ROBIN PHELPS • *School of Public Affairs, University of Colorado, Denver, Denver, CO, USA*

DUANE E. PRASUHN • *Center for Bio/Molecular Science and Engineering, U.S. Naval Research Laboratory, Washington, DC, USA*

TIJANA RAJH • *Center for Nanoscale Materials, Argonne National Laboratory, Argonne, IL, USA*

KENNETH L. REED • *DuPont Haskell Global Centers for Health and Environmental Sciences, Newark, DE, USA*

SANDRA J. ROSENTHAL • *Department of Chemistry, Department of Pharmacology, Department of Chemical and Biomolecular Engineering, Department of Physics and Astronomy, Vanderbilt University, Nashville, TN, USA*

ELENA A. ROZHKOVA • *Center for Nanoscale Materials, Argonne National Laboratory, Argonne, IL, USA*

CHRISTIE M. SAYES • *Department of Veterinary Physiology & Pharmacology, Texas A&M University, College Station, TX, USA*

GEORGE C. SCHATZ • *Department of Chemistry, Northwestern University, Evanston, IL, USA*

GERALD E. SCHNEIDER • *Department of Brain and Cognitive Sciences, Massachusetts Institute of Technology, Cambridge, MA, USA*

SOO JUNG SHIN • *Akina Inc., West Lafayette, IN, USA*

LISA C. SHRIVER-LAKE • *Center for Bio/Molecular Science and Engineering, U.S. Naval Research Laboratory, Washington, DC, USA*

BALAJI SITHARAMAN • *Department of Biomedical Engineering, State University of New York at Stony Brook, Stony Brook, NY, USA*

KIMIHIRO SUSUMU • *Division of Optical Sciences, U.S. Naval Research Laboratory, Washington, DC, USA*

DOUGLAS J. SYLVESTER • *Sandra Day O'Connor College of Law, Center for the Study of Law, Science and Technology, Arizona State University, Tempe, AZ, USA*

CHRIS R. TAITT • *Center for Bio/Molecular Science and Engineering, U.S. Naval Research Laboratory, Washington, DC, USA*

C. SHAD THAXTON • *Department of Urology, Feinberg School of Medicine, Institute for BioNanotechnology and Medicine, and the International Institute for Nanotechnology, Northwestern University, Chicago, IL, USA*

KEITH THOMAS • *Vive Nano, Inc., Toronto, ON, Canada*

MICHAEL S. TOMCZYK • *The Wharton School, The University of Pennsylvania, Philadelphia, PA, USA*

UDAYA S. TOTI • *Department of Orthopaedic Surgery, Department of Chemical, Materials and Biomolecular Engineering, University of Connecticut Health Center, Farmington, CT, USA*

HSIAN-RONG TSENG • *Department of Molecular and Medical Pharmacology, Crump Institute for Molecular Imaging, Institute for Molecular Medicine, University of California, Los Angeles, Los Angeles, CA, USA*

LILACH VAKS • *Department of Molecular Microbiology and Biotechnology, Tel Aviv University, Ramat Aviv, Israel*

SHUTAO WANG • *Department of Molecular and Medical Pharmacology, Crump Institute for Molecular Imaging, Institute for Molecular Medicine, University of California, Los Angeles, Los Angeles, CA, USA*

YUHUANG WANG • *Department of Chemistry and Biochemistry, University of Maryland, College Park, MD, USA*

ZHONG LIN WANG • *School of Materials Science and Engineering, Georgia Institute of Technology, Altanta, GA, USA*

David B. Warheit • *DuPont Haskell Global Centers for Health and Environmental Sciences, Newark, DE, USA*

Yaguang Wei • *School of Materials Science and Engineering, Georgia Institute of Technology, Altanta, GA, USA*

Ralph Weissleder • *Center for Systems Biology, Department of Systems Biology, Massachusetts General Hospital, Harvard Medical School, Boston, MA, USA*

Ashish Yeri • *Department of Chemical and Petroleum Engineering, University of Pittsburgh, Pittsburg, PA, USA*

Tae-Jong Yoon • *Center for Systems Biology, Massachusetts General Hospital, Boston, MA, USA*

Jun Zhou • *School of Materials Science and Engineering, Georgia Institute of Technology, Altanta, GA, USA*

Chapter 1

Biomedical Nanotechnology

Sarah J. Hurst

Abstract

This chapter summarizes the roles of nanomaterials in biomedical applications, focusing on those highlighted in this volume. A brief history of nanoscience and technology and a general introduction to the field are presented. Then, the chemical and physical properties of nanostructures that make them ideal for use in biomedical applications are highlighted. Examples of common applications, including sensing, imaging, and therapeutics, are given. Finally, the challenges associated with translating this field from the research laboratory to the clinic setting, in terms of the larger societal implications, are discussed.

Key words: Nanoparticles, Nanodevices, Biomedical nanotechnology, Biodetection, Nanotherapeutics, Implant materials

1. Introduction

Nanoscience and technology is a field that focuses on developing new synthetic and analytical tools for building and studying structures with submicrometer, and more typically sub-100 nm dimensions (1 nm $= 1 \times 10^{-9}$ m) (see Fig. 1) (1, 2). Moreover, it is concerned with the study of the chemical and physical properties of such structures and how these properties change as the size of a material is scaled down from the bulk to a collection of several atoms. Finally, nanoscience and technology centers on utilizing the capabilities and the fundamental property differences associated with such highly miniaturized structures to construct novel functional materials and devices and to develop ground-breaking applications. The nanoscale is a length scale that falls between that of traditional chemistry and physics, which deals with the manipulation of atomic bonds ($\sim 10^{-10}$ m) and that of biology, where most structures are on the order of $\sim 10^{-6}$ to 10^{-7} m in

Sarah J. Hurst (ed.), *Biomedical Nanotechnology: Methods and Protocols*, Methods in Molecular Biology, vol. 726,
DOI 10.1007/978-1-61779-052-2_1, © Springer Science+Business Media, LLC 2011

1

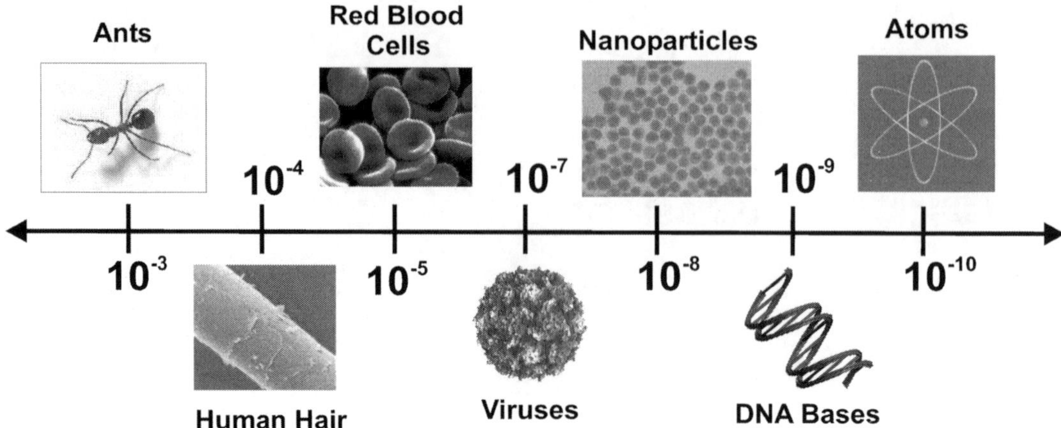

Fig. 1. Length scales.

diameter (e.g., cells, viruses, and bacteria), and as a consequence the field is highly interdisciplinary, encompassing aspects of physics, chemistry, biology, engineering, and medicine.

Although a boom in nanoscience and technology research occurred only recently, nanomaterials, in particular metallic nanoparticles, have been used for centuries in art (i.e., the Lycurgus cup), architecture (i.e., stained glass windows), photography (i.e., the developing process), and medicinal remedies (3, 4). In the late 1800s, Michael Faraday discovered and developed reliable syntheses for pseudo-spherical gold nanoparticles and later (in 1908) a theoretical framework for understanding of their optical properties was put in place by Gustav Mie. In the mid-1900s, enabling technologies for the imaging and manipulation of atoms and nanostructures were pioneered as atomic force, and electron microscopes were designed and put to use. It was also during this time that biologists were unraveling cellular structure and discovering a multitude of biological species (e.g., DNA and proteins) that have nanoscale dimensions. Further, the advent of novel electronic components such as transistors was ushering in the age of high-powered computing. Such findings led prominent scientists such as Richard Feynman and others to speculate that nanoscience and technology would be a revolutionary new research direction for many fields of science (5).

Today, one of the main thrusts of nanoscience research is the synthesis of novel nanoparticle materials and devices. Solution-based syntheses exist for making monodisperse samples of both isotropic and anisotropic metallic, semiconducting, and polymeric nanoparticles of a variety of shapes (6) including spheres (7–10), disks (11–13), prisms (14, 15), cubes (16–18), wires (19), rods

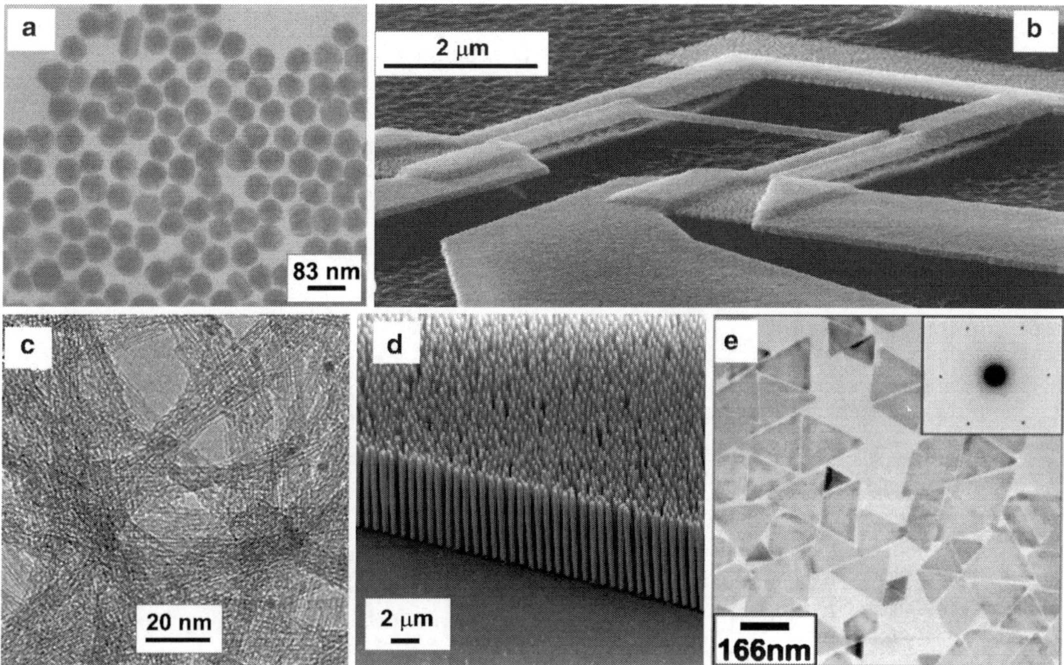

Fig. 2. Electron microscopy images of some common nanostructures. (**b**) Adapted with permission from *Nano Lett.* (2009), **9**, 3116. Copyright 2009 American Chemical Society. (**c**) Adapted with permission from *J. Phys. Chem. B* (2007), **111**, 1249. Copyright 2007 American Chemical Society. (**d**) Adapted with permission from *J. Am. Chem. Soc.* (2008), **130**, 14958. Copyright 2008 American Chemical Society. (**e**) Adapted with permission from *J. Am. Chem. Soc.* (2005), **127**, 5312. Copyright 2005 American Chemical Society.

(20–23), and branched (24–26) structures (see Fig. 2). Techniques such as arc discharge, laser ablation, and chemical vapor deposition (CVD) are being used to create carbon-based nanomaterials (i.e., fullerenes and carbon nanotubes) (27, 28) (see Chapter 15). Surface-based techniques (e.g., electrospinning (29), lithography (30–32), and templation (20)) are being utilized to make surface-bound nanostructures (33) (see Chapters 8 and 10), nanostructured scaffolding materials (see Chapters 15–17) (34–36), nanoelectromechanical devices (NEMS) (see Chapter 9) (37, 38), and nanofluidic devices (39). The nanomaterials produced using these processes are being applied in electronics (40), catalysis (41, 42), energy storage and generation (43, 44), environmental remediation (45, 46), security (47), and especially in biology and medicine (48–50).

In particular, this volume focuses on the application of the above types of nanomaterials in biomedicine. This chapter provides a brief introduction to this field. It highlights the common properties of such structures and the main advantages that they can offer compared to conventionally used materials. It then discusses the key applications in which these structures are used, namely, sensing, imaging, and therapeutics. Finally, it gives a future perspective of this field, not only in terms of new research directions, but also in terms of the larger societal implications of and challenges associated with transitioning nanoscience and technology products and strategies from the research laboratory to the clinical setting.

2. Nanomaterials in Medicine

The types of nanoparticles and nanodevices that are utilized in biomedical applications are chemically and physically diverse, but despite this diversity, they share several commonalities that make them advantageous compared to conventionally used structures.

First, nanomaterials are small in size, having at least one dimension (e.g., particle diameter or feature size) between 1 and 100 nm (1). Due to their small size, these nanostructures have high surface-to-volume ratios and hence are very reactive both during and after their synthesis. This property in part makes them highly tailorable and since their chemical and physical properties depend on their size, shape, and composition, important parameters including their charge, hydrophobicity, solubility, and stability can be easily tuned. For example, a given nanostructure could be designed to be either structurally robust (51) or easily biodegradable (52) over a certain period of time in a biological environment. Also, their small size allows nanostructures to readily interact with biological entities, which have similar dimension. Nanoparticles have been shown to be taken into cells through the pores of their membranes (53) and even localized in particular areas of the cell (54). They also have been known to cross the blood–brain-barrier through tight biological junctions unlike larger macrosized objects (55).

In addition, the nanomaterials employed in biomedical applications usually are multicomponent in nature (20, 49). Often, a nanoparticle or nanostructure is conjugated to one or more types of chemical and/or biological species such as oligonucleotides (short, synthetic DNA strands), proteins, drugs, or lipids through techniques including chemical conjugation, encapsulation, infusion, or adsorption (56, 57) (see Chapters 6 and 7). For instance, the gold nanoparticles discussed in Chapter 2 are functionalized with oligonucleotides of more than one sequence (58) while the bacteriophages discussed in Chapter 13 are modified with antibodies as well as the hydrophobic drug, chloramphenicol (59). Further, the nanomaterial itself is often composed of two or more inorganic components or a combination of inorganic and organic components in an alloy, core-shell, multishell, or dumb-bell arrangement (60, 61). In Chapter 3, for instance, the utilized nanomaterial is a particle with an iron-oxide core and a cross-linked dextran shell (62, 63). These multicomponent nanostructures retain the chemical and physical properties of their individual parts but also are imparted with a synergistic set of properties resulting from their combination.

These primary considerations make nanomaterials advantageous compared to conventional materials in many cases. As a result of their unique nanoscale properties, quantum dots, for example, have higher quantum yields, narrower emission wavelengths, and better photostability than conventional fluorescence dyes; their emission also can be easily tuned by varying their composition (64). As a result, detection systems employing quantum dots can be more sensitive than those utilizing conventional dyes, pushing them toward single molecule level detection (see Chapter 4). In another example, the high density of oligonucleotides on the surface of a metal nanoparticle in part allows them to more readily enter cells (53, 65), be more resistant to degradation (66), have higher target binding properties (67), and produce a weaker immune response compared to systems using traditional gene delivery agents where DNA is not concentrated on a nanostructured surface (68) (see Chapter 2). The details underpinning such differences are being investigated computationally (see Chapter 18) as well as experimentally.

3. Applications of Nanomaterials in Medicine

The main applications of nanomaterials in biology are in the areas of sensing, imaging, and therapeutics. Because of their multicomponent structure, many nanoparticles can carry out more than one of these tasks simultaneously (i.e., multifunctionality) (see Subheading 3.3).

3.1. Sensing and Imaging

Most of the sensing and imaging applications of nanomaterials in biomedicine are based on a two-part process: (1) a self-assembly event and (2) a readout event (see Fig. 3a) (48). Typically, the first step is a chemical or biological recognition event which takes place between the molecule attached to the nanomaterial and a target species in a sample of interest. Some common interactions that are exploited are oligonucleotide base pairing, protein–protein, and peptide–protein interactions as well as the interactions of oligos, proteins, and peptides with small molecules (69). For example, Chapter 4 describes a protocol in which proteins of interest in a cell sample are labeled with antibodies (against proteins including glycine receptors, nerve growth factors, kinesin motors, or GABA receptors, for instance) possessing a biotin tag; these are then exposed to quantum dots modified with streptavidin moieties. The biotin and streptavidin units participate in a

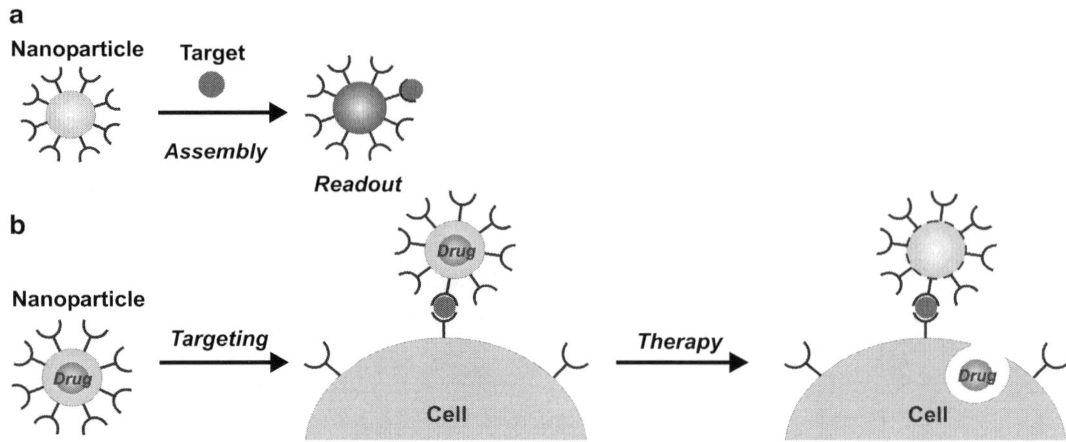

Fig. 3. General scheme for nanoparticle-based (**a**) biodetection and (**b**) therapeutics.

highly specific binding interaction and the quantum dot label is bound to the protein of interest. In another case, silicon substrates possessing nanopillar features were modified with epithelial cell adhesion antibodies (see Chapter 10) (70). When solutions of cells then were exposed to these substrates, the antibodies interact with proteins on the surface of the cell, trapping them on the substrate where they can be further analyzed.

Once the target species has been recognized, the next step in the application is typically a readout event, which often involves the nanomaterial. The readout can be accomplished using colorimetry (71, 72), fluorescence (73), light scattering (74), radiochemistry- (75) or Raman-based approaches (76, 77), magnetic resonance (78), or electrical signal (79) to name a few. The detection of the surface plasmon resonance (SPR) or fluorescence emission signal of a metal or semiconducting particle probe, respectively, could be an indicator of the presence of a given target in a sample. A change in the electronic properties (e.g., conductance) of a fabricated microelectrode linked with a semiconducting nanowire could be indicative of target binding at its surface (see Chapter 8) (79). The difference in magnetic relaxitivity of superparamagnetic particles could be used to detect whether they are dispersed (and not bound to the target molecule) and aggregated (bound to the target molecule of interest) (see Chapter 3).

In these detection strategies, selectivity and sensitivity are important considerations (48, 80). Nonspecific binding should be minimized and care should be taken to ensure that the molecules are oriented on the nanomaterial in such a way that they maintain their biological function (i.e., the ability to participate in subsequent binding interactions). The level of sensitivity attainable in a readout method varies; some methods are more sensitive than others. Many nanomaterial-based biodetection strategies have been shown to rival or even surpass the current limits of detection of conventional techniques [e.g., enzyme-linked immunoabsorbant assay (ELISA), polymerase chain reaction (PCR)]. In addition to advantages in terms of sensitivity and selectivity, nanomaterial-based detection schemes can be easier, faster, and cheaper than conventional strategies. Also, they have the potential to be more readily portable, making them useful in field deployable or point-of-care systems.

3.2. Therapeutics

Nanotherapeutic strategies are being used to treat diseases ranging from genetic disorders to cancer (50, 81). Similar to the detection strategies, these schemes often involve two major steps (see Fig. 3b). In the first step, the nanomaterial is targeted to specific cells or tissues. This targeting step is vital to ensure that the treatment is not delivered to healthy cells, causing unwanted side effects. Targeting can be accomplished using passive or active methods (50). Passive methodology involves tuning the physical and chemical properties (i.e., size and charge) of the material to trap the nanomaterials inside tumor sites, for instance, by exploiting the physiological differences between tumor cells and healthy cells (e.g., vasculature and pH). In active targeting schemes, the material is functionalized with species capable of participating in chemical and biological recognition events with the cells being targeted This volume presents one example of active targeting in Chapter 14, where the role of surface-bound peptides in targeting of viruses is discussed.

After the material is targeted to the cells or tissues of interest, a therapeutic payload can be released or a non-drug-based therapy can be initiated. Small molecule drugs, doped inside polymer particles can be gradually released as the particle degrades, for example (82) (see Chapter 12). The inherent assembly/disassembly properties of a plant virus nanoparticle capsid can be exploited to load and then unload drug molecules at targeted locations (see Chapter 14). Nanoparticle-based schemes are also replacing traditional gene therapy approaches and, in these cases, the payload is plasmid DNA, oligonucleotides, or siRNA (83). Other methods use light or heat (e.g., photodynamic or photothermal therapy) (84, 85) to initiate cytotoxic effects. In Chapter 5, titanium dioxide and semi-conducting nanoparticles (~5 nm in diameter), modified with cancer-targeting proteins, are discussed (86). These particles are capable of generating cytotoxic reactive oxygen species (ROS) when excited by light of a particular wavelength (87).

In another kind of therapeutic scheme, nanostructured implant materials are utilized in the regeneration of organ and nervous system tissue (e.g., brain and spinal cord) (88, 89). Here, the nanomaterial can deliver a drug cargo but primarily provides a scaffold for the mitigation of cell growth and development. These materials can be synthesized and then surgically placed at the site of injured tissue or administered as nanoscale precursor materials, capable of assembling into scaffolds in vivo.

Three examples of nanostructured scaffolds are presented in this volume: carbon nanotube-reinforced polymer nanocomposites, polymeric, nanofiberous scaffolds, synthesized using electrospinning, and porous biodegradable scaffolds, synthesized via the self-assembly of peptide amphiphiles (see Chapters 15–17, respectively).

3.3. Combined Sensing/Imaging and Therapy

Current research is focusing more and more on the development of multicomponent nanomaterials that perform both sensing and/or imaging and therapy concomitantly (90, 91). These nanomaterials can potentially contain multiple types of targeting moieties, drugs, and dyes or contrast agents all on one surface (see Fig. 4). Several examples of the use of such particles are presented in Chapter 11, which highlights photodynamic therapies. Inorganic, photon-upconverting particles made of $NaYF_4:Yb^{3+}$, Er^{3+} and coated with the polymer polyethyleneimine (PEI) can be loaded with photosensitizers (92). Not only did these particles demonstrate the destruction of specific cells with near-IR excitation but they also produced red/green emission allowing them to be imaged both in vitro and in vivo. Using these types of systems, one could imagine designing structures that are capable of providing real-time feedback and diagnostic information as a treatment which is being administered. In this way, treatments and dosages could be tailored on a patient-by-patient basis, moving towards personalized medicine.

As these applications are being further developed and beginning to enter and navigate the US Food and Drug Administration (FDA) approval process (see Chapter 21), it is becoming important to assess not only their limits of detection or therapeutic effectiveness but also the toxicology profiles of the novel nanomaterials used (64, 93, 94). Due to their structural complexity, they must be assessed on a case-by-case basis. The same particle functionalized with different ligands could have different toxicity profiles, for example, or be metabolized differently in the body;

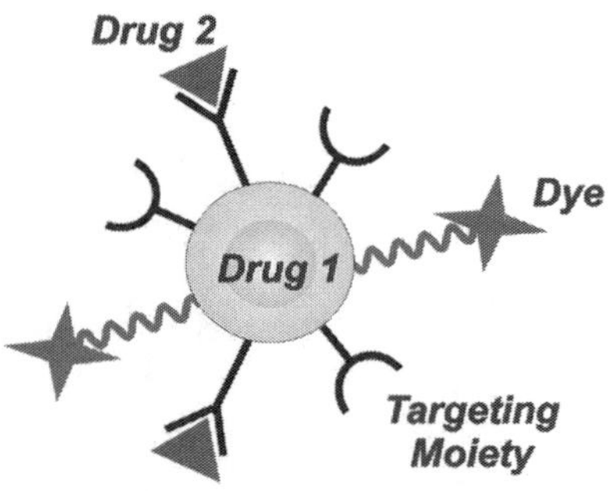

Fig. 4. General schematic of a multicomponent nanoparticle showing targeting moieties, dyes, and drugs on one surface.

the components of the degradation of these complex materials also warrant study. In addition, it is being discovered that nanomaterials can interfere with components of traditional assays that are used to assess nanoparticle toxicology and safety, necessitating the development of novel or modified methods of analysis. Strides are being made in this direction (see Chapters 19 and 20). Chapter 19 assesses cell injury caused by exposure to carbon nanotubes. Chapter 20 introduces a protocol for assessing the pulmonary hazards of nanoparticles (both fine and ultrafine, pigmentary and nano-sized) using biochemical and histopathological evaluations.

4. Future Perspectives

The field of biomedical nanotechnology is growing at an unprecedented rate (4, 95, 96). More research dollars are being spent on this industry than ever before both in the USA and around the world. It is the source of numerous statewide and global collaborations. The patenting of intellectual property in this area is immense. Government and university programs are being developed and user facilities geared toward nanotechnology research have been opened in national laboratory settings in the USA. Nanotechnology startups are being spun out of university research laboratories and larger companies are adopting nanomaterials into their everyday processes. Nanomaterials are already used in products ranging from sunscreen to golf clubs, and the first wave of nanotechnology-based drugs is being approved by the FDA.

The rapid growth of research in nanoscience and technology not only looks to revolutionize the scientific community, but also the business, social science, and the legislative and legal communities, responsible in part for disseminating nanoscience knowledge, products, and applications into society. The multicomponent and multifunctional nature of the nanomaterials that are used in biomedicine leads to key advantages; however, this complexity also leads to complications for decision-makers in these fields. Business people are wrangling with how to most beneficially expand companies around (see Chapter 24) and market (see Chapter 25) nanotechnology products. Politicians are determining how to legislate them such that ground-breaking technology will be quickly moved from the laboratory to the public arena but also in a manner that is responsible and beneficial to both the general public and the environment (see Chapter 22). Lawyers are working to protect the intellectual property of nanoscience researchers (see Chapter 23) and ethicists struggle with the potential moral issues that these novel types of medicines bring up (see Chapter 26).

The continued growth of the field of nanoscience and technology relies almost as much on the actions of these decision-makers as on the success of the research projects carried out by scientists in the laboratory. These types of decisions are partially responsible for shaping public opinion and, as mentioned repeatedly in this volume, a single negative incident has in the past been sufficient to cause a general public mistrust of a particular field of science. Therefore, this volume incorporates a section of chapters written by experts in the above mentioned fields highlighting some of the major challenges that they are facing, in order to foster the translation of knowledge between researchers from the natural science and social science fields. In this way, this volume intends to present a complete picture of the current state of biomedical nanoscience and technology from the laboratory to the clinic.

Acknowledgements

S.J.H. would like to thank Dr. H. Christopher Fry for his careful editing of this chapter and Dr. Tijana Rajh for being an outstanding mentor and advisor. S.J.H also acknowledges Argonne National Laboratory for a Director's Postdoctoral Fellowship. Work at the Center for Nanoscale Materials was supported by the US Department of Energy, Office of Science, Office of Basic Energy Sciences, under contract no. DE-AC02-06CH11357.

References

1. Mirkin, C. A. (2005) The Beginning of a Small Revolution. *Small* 1, 14–16.

2. Mirkin, C. A. (1999) Tweezers for the Nanotool Kit. *Science* 286, 2095–2096.

3. Daniel, M.-C. and Astruc, D. (2004) Gold Nanoparticles: Assembly, Supermolecular Chemistry, Quantum-Size-Related Properties, and Applications Toward Biology, Catalysis, and Nanotechnology. *Chem. Rev.* 104, 293–346.

4. Ratner, M. and Ratner, D. (2003) *Nanotechnology: A Gentle Introduction to the Next Big Idea.* Prentice Hall, Upper Saddle River, NJ.

5. Feynman, R. (1999) *The Pleasure of Finding Things Out.* Perseus Books, New York, NY.

6. Burda, C., Chen, X., Narayanan, R., and El-Sayed, M. (2005) Chemistry and Properties of Nanocrystals of Different Shapes. *Chem. Rev.* 105, 1025–1102.

7. Park, J., Joo, J., Kwon, S. G., Jang, Y., and Hyeon, T. (2007) Synthesis of Monodisperse Spherical Nanocrystals. *Angew. Chem. Int. Ed.* 46, 4630–4660.

8. Smith, D., Pentzer, E., and Nguyen, S. T. (2007) Bioactive and Therapeutic ROMP Polymers. *Polym. Rev.* 47, 419–459.

9. Cushing, B. L., Kolesnichenko, V. L., and O'Connor, C. J. (2004) Recent Advances in the Liquid-Phase Syntheses of Inorganic Nanoparticles. *Chem. Rev.* 104, 3893–3946.

10. Oh, M. and Mirkin, C. A. (2005) Chemically Tailorable Colloidal Particle from Infinite Coordination Polymers. *Nature* 438, 651–654.

11. Maillard, M., Giorgio, S., and Pileni, M.-P. (2002) Silver Nanodisks. *Adv. Mater.* 14, 1084–1086.

12. Hao, E., Kelly, K. L., Hupp, J. T., and Schatz, G. C. (2002) Synthesis of Silver Nanodisks Using Polystyrene Mesospheres as Templates. *J. Am. Chem. Soc.* 124, 15182–15183.

13. Chen, S. and Carroll, D. L. (2002) Synthesis and Characterization of Truncated Triangular Silver Nanoplates. *Nano Lett.* 2, 1003–1007.

14. Hulteen, J. C. and Van Duyne, R. P. (1995) Nanosphere Lithography: A Materials General Fabrication Process for Periodic Particle Array Surfaces. *J. Vac. Sci. Technol. A* **13**, 1553–1558.

15. Millstone, J. E., Hurst, S. J., Metraux, G. S., Cutler, J. I., and Mirkin, C. A. (2009) Colloidal Gold and Silver Nanoprisms. *Small* **5**, 646–664.

16. Ahmadi, T. S., Wang, Z. L., Green, T. C., Henglein, A., and El-Sayed, M. (1996) Shape-Controlled Synthesis of Colloidal Platinum Nanoparticles. *Science* **272**, 1924–1926.

17. Sun, Y. and Xia, Y. (2002) Shape-Controlled Synthesis of Gold and Silver Nanoparticles. *Science* **298**, 2176–2179.

18. Yu, D. and Yam, V. W.-W. (2004) Controlled Synthesis of Monodisperse Silver Nanocubes in Water. *J. Am. Chem. Soc.* **126**, 13200–13201.

19. Hu, J., Odom, T. W., and Lieber, C. M. (1999) Chemistry and Physics in One Dimension: Synthesis and Properties of Nanowires and Nanotubes. *Acc. Chem. Res.* **32**, 435–445.

20. Hurst, S. J., Payne, E. K., Qin, L., and Mirkin, C. A. (2006) Multisegmented One-Dimensional Nanorods. *Angew. Chem. Int. Ed.* **45**, 2672–2692.

21. Peng, X., Manna, L., Yang, W., Wickham, J., Scher, E., Kadavanich, A., et al. (2000) Shape Control of CdSe Nanocrystals. *Nature* **404**, 59–61.

22. Jana, N. R., Gearheart, L., and Murphy, C. J. (2001) Wet Chemical Synthesis of High Aspect Ratio Cylindrical Gold Nanorods. *J. Phys. Chem. B* **105**, 4065–4067.

23. Kim, F., Song, J. H., and Yang, P. (2002) Photochemical Synthesis of Gold Nanorods. *J. Am. Chem. Soc.* **124**, 14316–14317.

24. Manna, L., Milliron, D. J., Meisel, A., Scher, E. C., and Alivisatos, A. P. (2003) Controlled Growth of Tetrapod-Branched Inorganic Nanocrystals. *Nat. Mater.* **2**, 382–385.

25. Chen, S., Wang, Z. L., Ballato, J., Foulger, S. H., and Carroll, D. L. (2003) Monopod, Bipod, Tripod, and Tetrapod Gold Nanocrystals. *J. Am. Chem. Soc.* **125**, 16186–16187.

26. Hao, E., Bailey, R. C., Schatz, G. C., Hupp, J. T., and Li, S. (2004) Synthesis and Optical Properties of "Branched" Gold Nanocrystals. *Nano Lett.* **4**, 327–330.

27. Zhou, W., Bai, X., Wang, E., and Xie, S. (2009) Synthesis, Structure, and Properties of Single-Walled Carbon Nanotubes. *Adv. Mater.* **21**, 4565–4583.

28. Dai, H. J. (2002) Carbon Nanotubes: Synthesis, Integration and Properties. *Acc. Chem. Res.* **35**, 1035–1044.

29. Lu, X., Wang, C., and Wei, Y. (2009) One-Dimensional Composite Nanomaterials: Synthesis by Electrospinning and Their Applications. *Small* **5**, 2349–2370.

30. Gates, B. D., Xu, Q., Stewart, M., Ryan, D., Willson, C. G., and Whitesides, G. M. (2005) New Approaches to Nanofabrication: Molding, Printing, and Other Techniques. *Chem. Rev.* **105**, 1171–1196.

31. Ginger, D. S., Zhang, H., and Mirkin, C. A. (2004) The Evolution of Dip-Pen Nanolithography. *Angew. Chem. Int. Ed.* **43**, 30–45.

32. Euliss, L. E., DuPont, J. A., Gratton, S., and DeSimone, J. (2006) Imparting Size, Shape, and Composition Control of Materials for Nanomedicine. *Chem. Soc. Rev.* **35**, 1095–1104.

33. Hayes, C. L. and Van Duyne, R. P. (2001) Nanosphere Lithography: A Versitile Nanofabrication Tool for Studies of Size-Dependent Nanoparticle Optics. *J. Phys. Chem. B* **105**, 5599–5611.

34. Tang, Z., Wang, Y., Podsiadlo, P., and Kotov, N. A. (2006) Biomedical Applications of Layer-by-Layer Assembly: From Biomimetics to Tissue Engineering. *Adv. Mater.* **18**, 3203–3224.

35. Wei, G. and Ma, P. X. (2009) Nanostructured Biomaterials for Regeneration. *Adv. Funct. Mater.* **18**, 3568–3582.

36. Yoo, P. J., Nam, K. T., Qi, J. F., Lee. S.-K., Park, J., Belcher, A. M., et al. (2006) Spontaneous Assembly of Viruses on Multilayered Polymer Surfaces. *Nat. Mater.* **5**, 234–240.

37. Craighead, H. G. (2000) Nanoelectromechanical Systems. *Science* **290**, 1532–1535.

38. Li, M., Tang, H. X., and Roukes, M. L. (2007) Ultra-Sensitive NEMS-Based Cantilevers for Sensing Scanned Probe and Very High-Frequency Applications. *Nat. Nanotechnol.* **2**, 114–120.

39. Sparreboom, W., van den Berg, A., and Eijkel, J. C. T. (2009) Principles and Applications of Nanofluidic Transport. *Nat. Nanotechnol.* **4**, 713–720.

40. Lu, W. and Lieber, C. M. (2007) Nanoelectronics from the Bottom Up. *Nature* **6**, 841–850.

41. Bonnemann, H. and Richards, R. M. (2001) Nanoscopic Metal Particles – Synthetic Methods and Potential Applications. *Eur. J. Inorg. Chem.* **2001**, 2455–2480.

42. Bell, A. T. (2003) The Impact of Nanoscience on Heterogeneous Catalysis. *Science* **299**, 1688–1691.

43. Zhong, C.-J., Luo, J., Fang, B., Wanjala, B. N., Njoki, P. N., Rameshwori, L., et al. (2010) Nanostructured Catalysts in Fuel Cells. *Nanotechnology* **21**, 062001.

44. Talapin, D. V., Lee, J.-S., Kovalenko, M. V., and Shevchenko, E. V. (2010) Prospects of Colloidal Nanocrystals for Electronic and Optoelectronic Applications. *Chem. Rev.* **110**, 389–458.

45. Savage, N. and Diallo, M. S. (2005) Nanomaterials and Water Purification: Opportunities and Challenges. *J. Nanopart. Res.* **7**, 331–342.

46. Pradeep, T. and Anshup (2009) Nobel Metal Nanoparticles for Water Purification: A Critical Review. *Thin Solid Films* **517**, 6441–6478.

47. Golightly, R. S., Goering, W. E., and Natan, M. J. (2009) Surface-Enhanced Raman Spectroscopy and Homeland Security: A Perfect Match? *ACS Nano* **3**, 2859–2869.

48. Rosi, N. L. and Mirkin, C. A. (2005) Nanostructures in Biodiagnostics. *Chem. Rev.* **105**, 1547–1562.

49. Gao, J., Gu, H., and Xu, B. (2009) Multifunctional Magnetic Nanoparticles: Design, Synthesis and Biomedical Applications. *Acc. Chem. Res.* **42**, 1097–1107.

50. Peer, D., Karp, J. M., Hong, S., Farokhzad, O. C., Margalit, R., and Langer, R. (2007) Nanocarriers as an Emerging Platform for Cancer Therapy. *Nat. Nanotechnol.* **2**, 751–760.

51. Wu, H., Zhu, H., Zhuang, J., Yang, S., Liu, C., and Cao, Y. C. (2008) Water-Soluble Nanocrystals with Dual-Interaction Ligands. *Angew. Chem. Int. Ed.* **47**, 3730–3734.

52. Pridgen, E. M., Langer, R., and Farokhzad, O. C. (2007) Biodegradable, Polymeric Nanoparticle Delivery Systems for Cancer Therapy. *Nanomedicine* **2**, 669–680.

53. Rosi, N. L., Giljohann, D. A., Thaxton, C. S., Lytton-Jean, A. K. R., Han, M. S., and Mirkin, C. A. (2006) Oligonucleotide-Modified Gold Nanoparticles for Intracellular Gene Regulation. *Science* **312**, 1027–1030.

54. Patel, P. C., Giljohann, D. A., Seferos, D. S., and Mirkin, C. A. (2008) Peptide Antisense Nanoparticles. *Proc. Natl. Acad. Sci. U.S.A* **105**, 17222–17226.

55. Silva, G. A. (2006) Neuroscience Nanotechnology: Progress, Opportunities, and Challenges. *Nat. Rev. Neurosci.* **7**, 65–74.

56. Katz, E. and Willner, I. (2004) Integrated Nanoparticle-Biomolecule Hybrid Systems: Synthesis, Properties, and Applications. *Angew. Chem. Int. Ed.* **43**, 6042–6108.

57. Hermanson, G. T. (2008) *Bioconjugate Techniques (2nd ed.)*. Academic Press, New York, NY.

58. Stoeva, S. I., Lee, J.-S., Thaxton, C. S., and Mirkin, C. A. (2006) Multiplexed DNA Detection with Biobarcoded Nanoparticle Probes. *Angew. Chem. Int. Ed.* **45**, 3303–3306.

59. Yacoby, I., Bar, H., and Benhar, I. (2007) Targeted Drug Carrying Bacteriophages as Antibacterial Nanomedicines. *Antimicrob. Agents. Chemother.* **51**, 2156–2163.

60. Cozzoli, P. D., Pellegrino, T., and Manna, L. (2006) Synthesis, Properties and Perspectives of Hybrid Nanocrystal Structures. *Chem. Soc. Rev.* **35**, 1195–1208.

61. Zou, H., Wu, S., and Shen, J. (2008) Polymer/Silica Nanocomposites: Preparation, Characterization, Properties and Applications. *Chem. Rev.* **108**, 3893–3957.

62. Josephson, L., Tung, C. H., Moore, A., et al. (1999) High-Efficiency Intracellular Magnetic Labeling with Novel Superparamagnetic-Tat Peptide Conjugates. *Bioconjug. Chem.* **10**, 186–191.

63. Harisinghani, M. G., Barentsz, J., Hahn, P. F., Deserno, W. M., Tabatabaei, S., Hulsbergen van de Kaa, C., et al. (2003) Noninvasive Detection of Clinically Occult Lymph-Node Metastases in Prostate Cancer. *N. Engl. J. Med.* **348**, 2491–2499.

64. Smith, A. M., Duan, H., Mohs, A. M., and Nie, S. (2008) Bioconjugated Quantum Dots for *In Vivo* Molecular and Cellular Imaging. *Adv. Drug Deliv. Rev.* **60**, 1226–1240.

65. Giljohann, D. A., Seferos, D. S., Patel, P. C., Millstone, J. E., Rosi, N. L., and Mirkin, C. A. (2007) Oligonucleotide Loading Determines Cellular Uptake of DNA-Modified Gold Nanoparticles. *Nano Lett.* **7**, 3818–3821.

66. Seferos, D. S., Prigodich, A. E., Giljohann, D. A., Patel, P. C., and Mirkin, C. A. (2009) Polyvalent DNA-Nanoparticle Conjugates Stabilize Nucleic Acids. *Nano Lett.* **9**, 308–311.

67. Lytton-Jean, A. K. R. and Mirkin, C. A. (2005) A Thermodynamic Investigation into the Binding Properties of DNA Functionalized Gold Nanoparticle Probes and Molecular Fluorophore Probes. *J. Am. Chem. Soc.* **127**, 12754–12755.

68. Massich, M. D., Giljohann, D. A., Seferos, D. S., Ludlow, L. E., Horvath, C. M., and Mirkin, C. A. (2009) Regulating Immune Response Using Polyvalent Nucleic Acid-Gold Nanoparticle Conjugates. *Mol. Pharm.* **6**, 1934–1940.

69. Bloomfield, V. A., Crothers, D. M., and Tinoco, Jr., I. (2000) *Nucleic Acids: Structures, Properties, and Functions.* University Science Books, Sausalito, CA.

70. Wang, S., Wang, H., Jiao, J., Chen, K.-J., Owens, G. E., Kamei, K.-I., et al. (2009) Three-Dimensional Nanostructured Substrates Toward Efficient Capture of Circulating Tumor Cells. *Angew. Chem. Int. Ed.* **48**, 8970–8973.

71. Mirkin, C. A., Letsinger, R. L., Mucic, R. C., and Storhoff, J. J. (1996) A DNA-Based Method for Rationally Assembling Nanoparticles into Macroscopic Materials. *Nature* **382**, 607–609.

72. Storhoff, J. J., Elghanian, R., Mucic, R. C., Mirkin, C. A., and Letsinger, R. L. (1998) One-Pot Colorimetric Differentiation of Polynucleotides with Single Base Imperfections Using Gold Nanoparticle Probes. *J. Am. Chem. Soc.* **120**, 1959–1964.

73. Zhong, W. (2009) Nanomaterials in Fluorescence-Based Biosensing. *Anal. Bioanal. Chem.* **394**, 47–59.

74. Taton, T. A., Mirkin, C. A., and Letsinger, R. L. (2000) Scanometric DNA Array Detection with Nanoparticle Probes. *Science* **289**, 1757–1760.

75. Schipper, M. L., Cheng, Z., Lee, S.-W., Bentolila, L. A., Iyer, G., Rao, J., et al. (2007) Micro-PET-Based Biodistribution of Quantum Dots in Living Mice. *J. Nucl. Med.* **48**, 1511–1518.

76. Cao, Y. C., Jin, R., and Mirkin, C. A. (2002) Nanoparticles with Raman Spectroscopic Fingerprints for DNA and RNA Detection. *Science* **297**, 1536–1540.

77. Schlucker, S. (2009) SERS Microscopy: Nanoparticle Probes and Biomedical Applications. *Chemphyschem* **10**, 1344–1354.

78. Na, H. B., Song, I. C., and Hyeon, T. (2009) Inorganic Particles for MRI Contrast Agents. *Adv. Mater.* **21**, 2133–2148.

79. Cui, Y., Wei, Q. Q., Park, H. K., and Lieber, C. M. (2001) Nanowire Sensors for Highly Sensitive and Selective Detection of Biological and Chemical Species. *Science* **293**, 1289–1292.

80. Giljohann, D. A. and Mirkin, C. A. (2009) Drivers of Biodiagnostic Development. *Nature* **462**, 461–464.

81. Fang, X. and Tan, W. (2010) Aptamers Generated from Cell-SELEX for Molecular Medicine: A Chemical Biology Approach. *Acc. Chem. Res.* **43**, 48–57.

82. Acharya, G., Shin, C. S., McDermott, M., Mishra, H., Park, H., Kwon, I.-C., et al. (2010) The Hydrogel Template Method for Fabrication of Homogeneous Nano/ Microparticles. *J. Control. Release* **141**, 314–319.

83. Ditto, A. J., Shah, P. N., and Yun, Y. H. (2009) Non Viral Gene Delivery Using Nanoparticles. *Exp. Opin. Drug Deliv.* **6**, 1149–1160.

84. Koo, Y.-E., Reddy, G. R., Bhojani, M., Schneider, R., Philbert, M. A., Rehemtulla, A., et al. (2006) Brain Cancer Diagnosis and Therapy with Nanoplatforms. *Adv. Drug Deliv. Rev.* **58**, 1556–1577.

85. Jain, P. K., Huang, X., El-Sayed, I. H., and El-Sayed, M. A. (2008) Nobel Metals on the Nanoscale: Optical and Photothermal Properties and Some Applications in Imaging, Sensing, Biology and Medicine. *Acc. Chem. Res.* **41**, 1578–1586.

86. Rozhkova, E. A., Ulasov, I., Lai, B., Dimitrijevic, N. M., Lesniak, M. S., and Rajh, T. (2009) A High-Performance Nanobio Photocatalyst for Targeted Brain Cancer Therapy. *Nano Lett.* **9**, 3337–3342.

87. Dimitrijevic, N. M., Rozhkova, E., and Rajh, T. (2009) Dynamics of Localized Charges in Dopamine-Modified TiO and their Effect on the Formation of Reactive Oxygen Species. *J. Am. Chem. Soc.* **131**, 2893–2899.

88. Hartgerink, J. D., Beniash, E., and Stupp, S. I. (2001) Self-Assembly and Mineralization of Peptide-Amphiphile Nanofibers. *Science* **294**, 1684–1688.

89. Zhang, L. and Webster, T. J. (2009) Nanotechnology and Nanomaterials: Promises for Improved Tissue Regeneration. *Nano Today* **4**, 66–80.

90. Cheon, J. and Lee, J.-H. (2008) Synergistically Integrated Nanoparticles as Multimodal Probes for Nanobiotechnology. *Acc. Chem. Res.* **41**, 1630–1640.

91. Dhar, S., Daniel, W. L., Giljohann, D. A., Mirkin, C. A., and Lippard, S. J. (2009) Polyvalent Oligonucleotide Gold Nanoparticle Conjugates as Delivery Vehicles for Platinum (IV) Warheads. *J. Am. Chem. Soc.* **131**, 14652–14653.

92. Zhang, P., Steelant, W., Kumar, M., and Scholfield, M. (2007) Versatile Photosensitizers for Photodynamics Therapy at Infrared Excitation. *J. Am. Chem. Soc.* **129**, 4526-4527.

93. Lewinski, N., Colvin, V., and Drezek, R. (2008) Cytotoxicity of Nanoparticles. *Small* **4**, 26–49.

94. Maurer-Jones, M. A., Bantz, K. C., Love, S. A., Marquis, B. J., and Haynes, C. L. (2009) Toxicity of Therapeutic Nanoparticles. *Nanomedicine* **4**, 219–241.

95. http://www.clinicaltrials.gov (2010).

96. http://www.nano.gov (2010).

Part I

Using Nanomaterials in Sensing, Imaging, and Therapeutics

Chapter 2

Multiplexed Detection of Oligonucleotides with Biobarcoded Gold Nanoparticle Probes

Jae-Seung Lee

Abstract

Applications in a variety of fields rely on the high-throughput ultrasensitive and multiplexed detection of oligonucleotides. However, the conventional microarray-based techniques that employ fluorescent dyes are hampered by several limitations; they require target amplification, fluorophore labeling, and complicated instrumentation, while the fluorophore-labeled species themselves exhibit slow binding kinetics, photo-bleaching effects, and overlapping spectral profiles. Among the emerging nanomaterials that are being used to solve these problems, oligonucleotide–gold nanoparticle conjugates (Oligo-AuNPs) have recently been highlighted due to their unique chemical and physical properties. In this chapter, a detection scheme for oligonucleotides that utilize Oligo-AuNPs is evaluated with multiple oligonucleotide targets. This scheme takes advantage of the sharp melting transitions, intense optical properties, catalytic properties, enhanced binding properties, and the programmable assembly/disassembly of Oligo-AuNPs.

Key words: Oligonucleotide, Multiplexing, Gold nanoparticle, Detection, Sensing, Biomolecule, DNA, RNA

1. Introduction

The dramatic development of genomics and proteomics in the last several decades has attracted biologists to investigate how the two fields merge together at the molecular level. For example, the desire to understand how proteins are expressed and what genes cause the expression has strongly motivated researchers to investigate proteins or genes in vivo or in vitro under different experimental conditions. To accomplish this task, they have taken advantage of current molecular biology techniques such as mass spectrometry (1), tap-tags (2), ELISAs (3), microarrays (4), and blottings associated with gel electrophoresis. However, the need to develop alternative assays that are rapid, sensitive, and selective

Sarah J. Hurst (ed.), *Biomedical Nanotechnology: Methods and Protocols*, Methods in Molecular Biology, vol. 726,
DOI 10.1007/978-1-61779-052-2_2, © Springer Science+Business Media, LLC 2011

is growing significantly since the progress of the current research is hampered by the low-throughput, high cost, slowness, and, in certain cases, limited reproducibility of the conventional methods.

Emerging nanotechnologies are considered to be promising to solve these problems. Due to their unique and characteristic properties that have not been observed in macroscopic worlds previously and that excel the conventional materials in efficiency, nanomaterials have been intensively investigated and utilized in a variety of applications in diagnostics and therapeutics (5–7). Among the large number of nanomaterials that have been synthesized, oligonucleotide–gold nanoparticle conjugates (Oligo-AuNPs) have been used as probes in numerous detection schemes for biomolecule targets (8). The unique chemical and physical properties of Oligo-AuNPs that stem from their "inorganic nanoparticle core and densely packed oligonucleotide shell" structure offer high sensitivity and selectivity in such applications. To begin with, a brief introduction of oligonucleotides, AuNPs, and Oligo-AuNPs is given as follows.

1.1. Oligonucleotide

An oligonucleotide is a short nucleic acid, or polymer of nucleotides, with typically fewer than 100 bases. Although it is possible to cleave longer DNA or RNA strands to obtain them, oligonucleotides often refer to "synthetic" ones synthesized by polymerizing individual nucleotide precursors utilizing phosphoramidite chemistry. Because their chemical identity is the same as that of longer DNA or RNA except in terms of length, oligonucleotides have most of the same chemical and physical properties; they display a UV absorbance at 260 nm, form duplexes via base pairing, have binding interactions with proteins, and form monomolecular structures, such as G-quadruplexes. In addition, a variety of modified bases and nucleotide linkages (e.g., 2′-O-methylated RNA bases, phosphorothioate-modified phosphate backbones, and locked nucleic acids (LNA)) have been developed to control their properties, especially to enhance their stability.

One of the most important properties of oligonucleotides is reversible duplex formation. Based upon the pairing of the four bases (adenine (A)-thymine (T) and guanine (G)-cytosine (C)) via hydrogen bonds, an oligonucleotide sequence can reversibly bind or hybridize to the complementary sequence. During the hybridization process, its absorbance at 260 nm decreases due to hypochromism. When heated, however, the duplexed oligonucleotides dehybridize into single strands with an increase in absorbance at 260 nm. If this dehybridization, or "melting" process, is monitored as a function of temperature using UV–vis spectroscopy (at 260 nm), a broad curve (full width at half maximum (FWHM) ~10°C) is observed, whose midpoint is considered to be the melting temperature (T_m) of the duplex system. The melting temperature, which is the representative value of the stability of the duplex pair, can be controlled by changing the salt concentration of the solution, the length of the sequence, or the number of mismatched bases.

Synthetic oligonucleotides, terminally modified with functional chemical moieties such as thiols, amines, carboxyls, fluorophores, quenchers, or biotin groups, are getting more and more widely utilized in nanotechnology applications, including microarrays (9), DNA origami (10), nanomaterials assembly (7, 8, 11), and so on. In this chapter, we cover how oligonucleotides that are conjugated with nanomaterials, specifically gold nanoparticles, play a significant role as a "smart" material based upon the aforementioned properties.

1.2. Gold Nanoparticles: Chemical and Physical Properties

Colloidal gold nanoparticles (AuNPs) have been investigated by many researchers because of their unusual properties that include relatively high stability, low toxicity, catalytic activity, surface plasmon resonance (SPR), enhanced Raman signal, and facile chemical tailorability. Also, these structures can support multiple functionalities on their surface. AuNPs with diameters between 2 and 250 nm can be prepared in aqueous media in relatively monodisperse forms using a variety of synthetic methods (12, 13). The unique and intense colors associated with these AuNPs are due to their SPR. This SPR is observed in the visible region of the spectrum and is due to the collective oscillations of the free electrons in the conduction band interacting with the electromagnetic field of the incoming light. In solution, monodisperse AuNPs (~15 nm in diameter) exhibit a red color indicative of a surface plasmon absorption centered at 520 nm. Since AuNPs, especially those stabilized with citrate, are charged particles, they are exceedingly sensitive to changes in solution dielectric. For example, the addition of sodium chloride (NaCl) shields the surface charge of these AuNPs and leads to a concomitant decrease in interparticle distance and eventual particle aggregation. A solution containing coalesced AuNPs appears purple or blue in color, corresponding to a characteristic dampening and red-shifting of the SPR of these particles, to around 600 nm. This distance-dependent optical property forms the basis of certain colorimetric detection schemes that employ nanoparticle–biomolecule conjugates as probes.

1.3. General Properties of Oligo-AuNPs

An oligonucleotide–gold nanoparticle conjugate (Oligo-AuNP) is a two-component system comprised of thiolated-oligonucleotide strands and their nanoparticle scaffold. The properties of these inorganic nanoparticle core/organic biomolecular ligand hybrids stem from not only each component of the hybrids but also synergistically from the combination of those two materials. The deep red color exhibited by Oligo-AuNPs in aqueous media is from the SPR of the gold nanoparticles, which shows a narrow absorption in the visible range around 520 nm. However, the chemical recognition abilities of these structures are derived from the oligonucleotide shell. When two complementary Oligo-AuNPs are combined, they form three-dimensional networks through DNA duplex interconnects and an effective shift of the SPR (50 nm or more) takes place with a concomitant red-to-purple color change (14). Due to the reversible hybridization properties of oligonucleotides, however, the assembled Oligo-AuNPs disassemble at an elevated temperature (or reduced salt concentration) returning back to red. Importantly, melting analyses of the hybridized particle aggregates show sharper melting transitions (FWHM ~2°C) than free DNA duplexes (15). The sharpness of this melting transition is explained by a cooperative mechanism that originates from the presence of multiple DNA interconnects between the Oligo-AuNPs and the melting cascade that results as counter ions are rapidly released from the aggregate (15, 16). Further, Oligo-AuNPs exhibit particle size-dependent melting properties and enhanced binding properties, which are the representative synergetic results of cooperative interactions between the AuNP and surrounding oligonucleotides (17, 18).

In addition, Oligo-AuNPs are chemically, physically, and biologically stable, and nontoxic. While the inertness of AuNPs is well known, the biological in vivo or in vitro stability of Oligo-AuNPs had remained a question until recently. Unlike free oligonucleotide strands, those on AuNPs are resistant to enzymatic degradation by DNase (19). This high biological stability is due to the densely packed oligonucleotides on the AuNP surface, which cause steric inhibition of the enzymatic cleavage reaction. This increased resistance to nuclease degradation plays a significant role in the therapeutic applications of these structures.

1.4. Biobarcoded Oligonucleotide–Gold Nanoparticle Conjugate Probes

Biobarcode assays are characterized by their ultrahigh sensitivity and multiplexing capability for the detection of oligonucleotides or proteins. In case of oligonucleotide targets, Oligo-AuNP probes are hybridized to oligonucleotide-functionalized magnetic microparticle (MMP) probes using the target sequence as a linker; these complexes are then separated magnetically for subsequent release of the oligonucleotides from the Oligo-AuNP probes (see Fig. 1.). These released oligonucleotides or "biobarcodes" are quantitatively analyzed by the scanometric assay. The high sensitivity of the biobarcode assays is believed to originate from four reasons: (1) Because the biobarcode assay can be considered as a pseudo-homogeneous assay, the binding equilibrium of the target and the particle probes can be increased by increasing the probe concentration. (2) The dense loading of oligonucleotides (biobarcodes) on the AuNPs was found to increase the equilibrium binding constant of the Oligo-AuNP up to two orders of magnitude higher than that of unmodified free DNA (18). (3) Due to the dense loading of oligonucleotides (biobarcodes) on AuNPs, one binding event of the target sequence to the MMPs

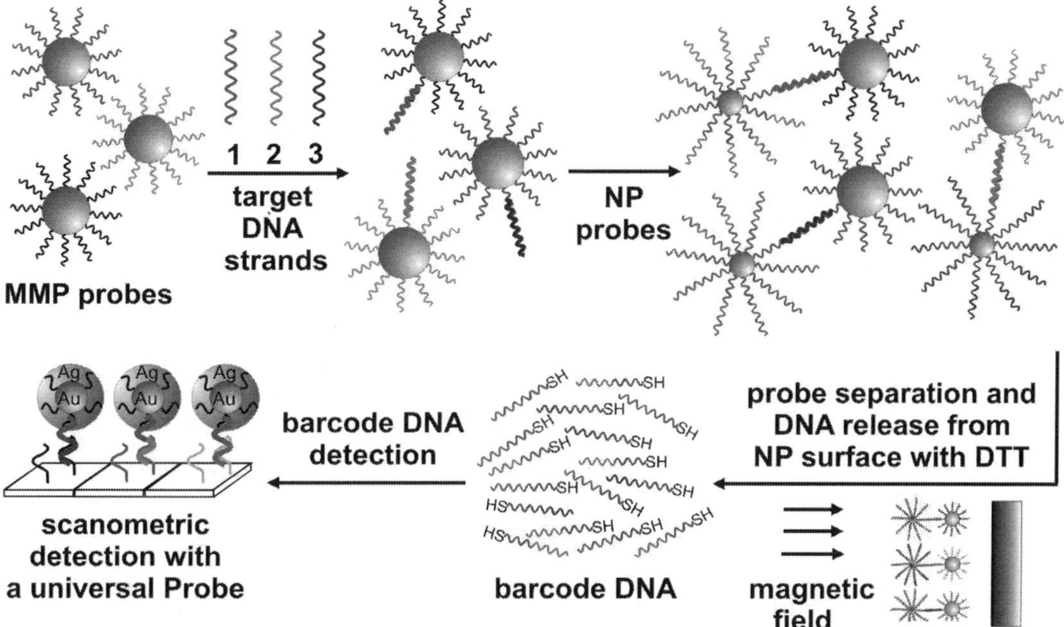

Fig. 1. Multiplexed biobarcode assay for oligonucleotide target detection. Reproduced from ref. 23 with permission from John Wiley & Sons, Inc.

and the AuNPs generates hundreds of biobarcode strands, leading to literally "hundreds of times" the target amplification. (4) Finally, these biobarcodes, as surrogate targets, are further amplified by silver enhancement during the scanometric assay. Except the second reason (2), biobarcode assays for protein targets also have the same reasons for the ultrahigh sensitivity, where antibodies on the particle probes play a role in binding the targets.

This chapter will discuss the general protocol of the biobarcode assay for oligonucleotide targets in a multiplexed form. In brief, it covers the synthesis of gold nanoparticles and their conjugates with oligonucleotides, the preparation of MMP probes, the main biobarcode assay procedure, and final scanometric assay. As mentioned above, the fundamental principles of the biobarcode assays for the detection of protein targets is very similar to what is described in this chapter for the oligonucleotide targets, except for the probe preparation that includes the conjugation of the particles with antibodies.

2. Materials

2.1. Gold Nanoparticle Synthesis

1. Gold (III) chloride trihydrate.
2. Ultrapure water (>18.2 MΩ cm).

3. Trisodium citrate.

4. Magnetic stirrer.

5. Stir bar (egg-shaped).

6. Round bottom flask.

7. Reflux condenser.

8. Heating mantle.

9. Temperature controller.

10. Concentrated hydrochloric acid (HCl, 37%).

11. Concentrated nitric acid (HNO, 70%).

12. Nylon filter (pore diameter = 0.45 μm).

2.2. Oligo-AuNP Conjugate Synthesis

1. Thiol-oligonucleotide of a given sequence (HPLC-purified) (see Note 1).

2. NAP-5 Sephadex column.

3. Dithiothreitol (DTT).

4. UV–vis spectrometer.

5. Sodium dodecyl sulfate (SDS) solution: 1 wt%

6. Phosphate buffer: 100 mM sodium phosphate $(Na_2HPO_4 + NaH_2PO_4)$, pH = 7.4.

7. Sodium chloride solution: 2.0 M NaCl.

2.3. MMP Preparation

1. Amino-functionalized MMPs (2.8 μm in diameter; Invitrogen, Carlsbad, CA).

2. Succinimidyl-4-(*p*-maleimidophenyl) butyrate (SMPB).

3. Coupling buffer: 0.2 M NaCl, 100 mM sodium phosphate, pH = 7.0.

4. Passivation buffer: 0.15 M NaCl, 150 mM sodium phosphate, pH = 8.0.

5. Separation magnets (sample volume ≥1 mL).

6. Sulfosuccinimidyl acetate (Sulfo-NHS-acetate).

7. Ammonium sulfate $((NH_4)_2SO_4)$.

8. Oligonucleotides (3′-thiol MMP-sequences).

9. Anhydrous dimethyl sulfoxide (DMSO).

2.4. Multiplexed Biobarcode Assay for DNA Targets

1. AuNPs (30 nm in diameter, Ted Pella, Inc., Redding, CA).

2. Assay buffer: 0.2 M NaCl, 0.1% Tween 20, and 10 mM sodium phosphate, pH = 7.2.

3. Separation magnets (sample volume ≥1 mL).

4. Scanometric buffer: 0.5 M NaCl, 0.01% Tween 20, and 10 mM sodium phosphate, pH = 7.2.

5. Verigene Reader System (Nanosphere, Northbrook, IL).

6. A 96-well-plate magnet.

7. Oligonucleotides (target sequences, 5′-thiol AuNP-sequences).

2.5. Scanometric Detection of DNA Targets

1. *N*-Hydroxysuccinimide (NHS) ester-activated Codelink slides (SurModics Inc., Eden Prairie, MN).

2. Printing buffer: 150 mM sodium phosphate, 0.01% SDS, pH = 8.5.

3. Capture oligonucleotide (amine-modified) and target oligonucleotide (unmodified).

4. Passivating solution: 0.2% SDS.

5. Spin dryer.

6. Two-well manual hybridization chambers (Nanosphere, Northbrook, IL).

7. Temperature-controllable incubator.

8. Scanometric buffer: 0.5 M NaCl, 0.01% Tween 20, 10 mM sodium phosphate, pH = 7.2.

9. Verigene Reader System (Nanosphere, Northbrook, IL).

10. Silver enhancing solutions (Nanosphere, Northbrook, IL).

11. Formamide.

12. Microarrayer.

13. GenePix Pro 6 software (Molecular Devices).

3. Methods

3.1. Gold Nanoparticle Synthesis

1. Clean all glassware with aqua regia (3:1 $HCl:HNO_3$, see Note 2). Rinse the glassware with 18.2 $M\Omega$ cm ultrapure water and dry in an oven prior to use.

2. Bring an aqueous solution of $HAuCl_4$ (1 mM, 500 mL) to reflux while stirring.

3. Rapidly add 50 mL 38.8 mM trisodium citrate solution to the refluxing solution.

4. Observe that the solution quickly changes color from pale yellow to deep red.

5. After waiting 15 min, allow the mixture to cool to room temperature and subsequently filter it through a 0.45 μm nylon filter.

6. Characterize the colloid using UV–vis spectroscopy (see Note 3).

3.2. Oligo-AuNP Conjugate Synthesis

1. Dissolve the lyophilized thiol-modified oligonucleotides (~40 nmol) in 0.1 M dithiothreitol (DTT) solution in phosphate buffer (0.17 M sodium phosphate, 100 µL) for 30 min (see Note 4).

2. Purify the deprotected oligonucleotides from excess DTT using a NAP-5 column. Apply less than 300 µL of the oligonucleotide–DTT mixture to a NAP-5 column, and elute about 1 mL, collecting about 4–5 drops at a time in separate microtubes.

3. Measure the amount of DNA from each aliquot by UV–vis spectroscopy and add it to AuNP solution (see Note 5).

4. Bring the solution to 0.3 M NaCl, 10 mM sodium phosphate, and 0.01% SDS, pH = 7.4 (see Note 6).

5. Typically, the system is allowed to equilibrate at least for a couple of hours.

6. Remove the excess DNA by repeated centrifugation (16,000 × g, 25 min for 13 nm AuNPs; 12,000 × g, 10 min for 30 nm AuNPs) and wash with a buffer solution (the concentration of NaCl that you will use in the subsequent assay, 10 mM phosphate, 0.005% Tween 20, pH = 7.4) (see Note 7).

7. Store the Oligo-AuNP probes at 4°C prior to use (see Note 8).

3.3. MMP Probe Preparation

1. Wash the MMPs (30 mg/mL, 1 mL) twice with anhydrous dimethyl sulfoxide (DMSO, 1 mL).

2. Prepare a solution of SMPB (50 mg) in DMSO (15 µL) prior to the reaction (see Note 9).

3. Add the SMPB/DMSO solution to the magnetic beads.

4. Allow the reaction between the primary amino group of MMPs and the N-hydroxysuccinimide (NHS) ester of SMPB to proceed for 4 h with gentle shaking at room temperature.

5. Separate the beads magnetically and wash them three times with DMSO (10 µL) and two times with coupling buffer (10 µL).

6. Reduce the disulfide bonds in all 3′-thiol MMP-sequences using DTT prior to mixing with the SMPB-activated MMPs (see Note 10).

7. Prepare solutions (5 µM) of each of the freshly cleaved oligonucleotides in coupling buffer.

8. Add 300 µL of each oligonucleotide to each of the washed SMPB-activated magnetic beads.

9. Allow the reaction between the maleimide group and the thiol-oligonucleotide to proceed at 20°C for 1 h under constant vortex.

10. Place the oligo-functionalized MMPs on a magnet. Remove the supernatant, and wash the beads twice with coupling buffer and then twice with passivation buffer.

11. Use the supernatant to determine the coupling efficiency. Measure the absorbance at 260 nm and compare it with that before the oligonucleotide-functionalization of MMPs (see Note 11).

12. Passivate the surface of the MMP probes by adding a freshly prepared solution of sulfo-NHS-acetate (100 mg) in passivation buffer (40 µL) (see Note 12). Allow the passivation process to proceed for 30 min at room temperature with mild shaking.

13. Wash the MMP probes twice with passivation buffer, twice with assay buffer, and store them at 4°C in assay buffer at a concentration of 30 mg/mL.

3.4. Multiplexed Biobarcode Assay for DNA Targets

1. Prepare Oligo-AuNPs by conjugating oligonucleotides with the 30 nm-AuNPs (see Note 13) and finally redisperse them in assay buffer ([Oligo-AuNP]~5 nM).

2. Prepare MMP multiplexing solution by combining all four MMP probes in assay buffer (final total MMP probe concentration is 1.56 mg/mL).

3. Begin the assay by mixing assay buffer (150 µL), MMP multiplexing solution (40 µL), and the appropriately mixed target solution (10 µL).

4. Heat the mixture at 45°C for 30 min and 25°C for 3 h under constant vortex to allow hybridization between the MMP probes and the target oligonucleotides (see Note 14).

5. Plate the reaction tube on a 96-well-plate magnet and wash the MMP–target complexes twice with assay buffer (200 µL).

6. Prepare Oligo-AuNP probe solution composed of all four NP probes (NP multiplexing solution) by diluting equal volumes of each NP probe (5 nM) in assay buffer containing formamide (15%) to a final total NP concentration of 200 pM (see Note 15).

7. Add the Oligo-AuNP multiplexing solution (50 µL) to the MMP–target complexes and allow hybridization to proceed at 25°C for 75 min under vortex.

8. Wash the mixture seven times with assay buffer (200 µL) using a 96-well-plate magnet to remove nonspecifically bound Oligo-AuNPs and any free barcode oligonucleotides.

9. Release the barcode oligonucleotides from the Oligo-AuNP probes by the addition of DTT (75 µL, 0.5 M) in scanometric buffer.

10. Collect the released barcode oligonucleotides from the mixture by removing MMPs using a magnet and analyze them quantitatively by the scanometric DNA detection method (see Note 16).

3.5. Scanometric Detection of DNA Targets

1. Dissolve amine-terminated capture DNA sequence (5′ H_2N-DNA sequence 3′) in printing buffer. The final concentration of the capture DNA sequence is 100 μM (see Note 17).

2. Print the DNA spots on a Codelink slide using a microarrayer. Try to keep the humidity in the arrayer as low as reasonably possible (see Note 18).

3. Hydrolyze the unreacted NHS groups of the slides or "passivate" the slides with passivating solution at 50°C for 20 min.

4. Spin-dry the slides.

5. Inject target oligonucleotides (20 μL) into hybridization chambers attached to a microarray slide prepared as above.

6. Hybridize the oligonucleotides to the capture strands on the slide in an incubator at 60°C for 15 min, at 37°C for 30 min, and at 25°C for 15 min with mild shaking (see Note 19).

7. Disassemble the chambers and rinse the slides copiously with scanometric buffer, spin them dry, and reassemble them with an unused hybridization chamber.

8. Introduce the Oligo-AuNP chip-probe solution (20 μL, [Oligo-AuNP] = 500 pM) into the hybridization chambers and allow the probes to hybridize at 37°C for 45 min (see Note 20).

9. Disassemble, wash three times using scanometric buffer, and spin-dry the hybridization chambers.

10. Perform the silver enhancement of the AuNPs by applying the silver enhancing solution (2 mL; Nanosphere, Northbrook, IL) on the dried chips for 3 min. Terminate the reaction by washing the slides with ultrapure water (18.2 MΩ cm) (see Note 21).

11. Spin-dry the slides and image them with the Verigene Reader System (Nanosphere, Northbrook, IL), which records the scattered light from the developed spots. Alternatively use a flatbed scanner to read out the scattering signal. Examples of the data obtained during readout are shown in Fig. 2.

4. Notes

1. Four exemplary sequences (Au1–Au4) for the AuNP probes used in biobarcode assays are given as an example.

    Au1: 5′ HS-TACGAGTTGAGAATC-TACCACATCATC CAT 3′

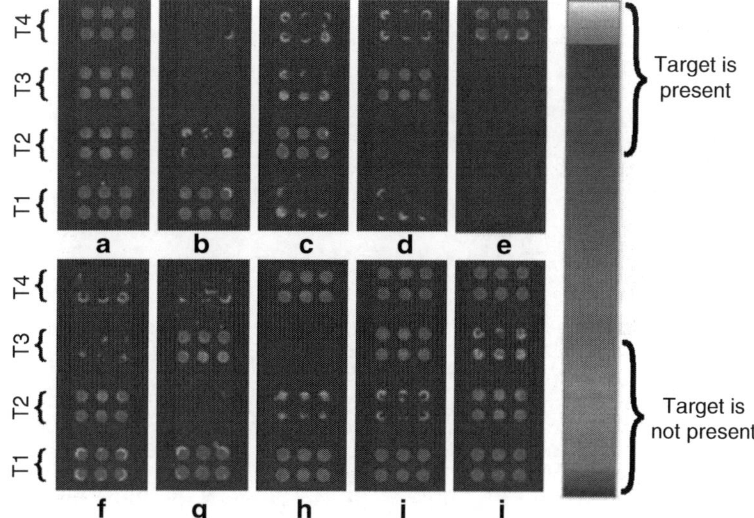

Fig. 2. Scanometric detection of the biobarcode-oligonucleotides released from the 30-nm Oligo-AuNP probes for ten different samples. (**a**) All targets are present; (**b**) T1; (**c**) T2; (**d**) T3; (**e**) T4; (**f**) T1 and T2; (**g**) T1 and T3; (**h**) T1 and T4; (**i**) T1, T3, and T4; and (**j**) T1, T2, and T4. The gray-scale images from the Verigene Reader System were converted into colored ones using GenePix Pro 6 software (Molecular Devices). The diameter of the spots on the chip is 200 μm. Reproduced from ref. 23 with permission from John Wiley & Sons, Inc.

Au2: 5′   HS-TACGAGTTGAGAATC-CTGATTACTATT GCA 3′

Au3: 5′   HS-TACGAGTTGAGAATC-TTGTTGATACTG TTC 3′

Au4: 5′   HS-TACGAGTTGAGAATC-TGCATCCAGGTC ATG 3′

2. Aqua regia is an extremely powerful oxidizing solution which is the only acidic solution that can be used to dissolve gold. Use extreme caution when working with aqua regia, because it generates harmful chlorine (Cl_2) and nitrogen oxide (NOx) gases and can cause severe tissue damage.

3. The characteristic SPR band of monodispersed particles is located around 520 nm. Particles with a more uniform size distribution have narrower absorption bands. Note that aggregated 13 nm gold nanoparticles, as discussed above, display a flattened and red-shifted absorption peak at approximately 600 nm.

4. Usually, the DNA sequences are 15- to 20-base pairs in length, complementary to a DNA target of interest, and modified with either 3′ or 5′ terminal sulfur-containing groups in the form of monothiols, cyclic disulfides (20), or trithiols (21). The terminal monothiol group of each oligonucleotide

is protected in the oxidized form (disulfide), which needs to be reduced by DTT to the monothiol form before reaction with gold nanoparticles.

5. The final oligonucleotide and AuNP concentrations are ~4 µM and ~6 nM, respectively.

6. The final concentrations of sodium phosphate (10 mM), sodium chloride (0.3 M), and SDS (0.01%) are achieved by adding the appropriate amounts of the concentrated solutions described in Subheading 2.2 to the oligonucleotide–gold nanoparticle mixture. The number of oligonucleotides per gold nanoparticle (15 nm in diameter) can be increased (~50–100 strands per particle) by increasing the NaCl concentration (from 0.1 to 1 M NaCl, respectively). Sodium ions screen the repulsions between the negatively charged oligonucleotide strands. The oligonucleotide loading also increases when the sequence has a polyethylene glycol (PEG) unit near the thiol group, or the mixture of oligonucleotides and gold nanoparticles is sonicated (22).

7. The removal of the excess DNA based upon centrifugation is repeated typically four times. More centrifugation steps can be performed when high purity is required, depending on the experimental purposes.

8. Oligo-AuNPs are stable and biochemically active for about 2 months when stored at 4°C.

9. Caution should be taken to prevent exposure of the solution to light.

10. The reduction of the disulfide bonds by DTT and the purification of the thiol-oligonucleotides are performed following the procedure described in Subheading 3.2. Four thiolated sequences (M1–M4) are given as an example; the sequence of these strands could be modified according to the experimental purposes [23). The A_{10} portion near the thiol group is a spacer and does not participate in hybridization.

M1: 5′ ATAACTGAAAGCCAA-A_{10-}SH 3′

M2: 5′ TCTTCCGTTACAACT-A_{10-}SH 3′

M3: 5′ TCCAACATTTACTCC-A_{10-}SH 3′

M4: 5′ TTATTCCAAATATCTTCT-A_{10-}SH 3′

11. Typically, there are on average 3×10^5 oligonucleotides per MMP.

12. Sulfo-NHS acetate acylates primary amino groups in basic media.

13. The conjugation of the oligonucleotide with the AuNP probe is performed following Subheading 3.2, except that the final

NaCl concentration is 0.2 M. Four thiolated sequences (Au1–Au4) are given as an example; these sequences could be modified according to the experimental purposes (23). The "TACGAGTTGAGAATC" portion is common to all the sequences and is used for hybridization with the Oligo-AuNP chip-probes in the scanometric assay.

Au1: 5′ HS-TACGAGTTGAGAATC-TACCACATCATCCAT 3′

Au2: 5′ HS-TACGAGTTGAGAATC-CTGATTACTATTGCA 3′

Au3: 5′ HS-TACGAGTTGAGAATC-TTGTTGATACTGTTC 3′

Au4: 5′ HS-TACGAGTTGAGAATC-TGCATCCAGGTCATG 3′

14. Four unmodified target sequences (T1–T4) are given as an example; the sequence of these strands could be modified according to the experimental purposes. The initial reaction temperature (45°C) was determined to completely dehybridize the target sequences from the oligonucleotides on MMPs, leading to better duplex formation at 25°C in the next step.

T1: 5′ TTGGCTTTCAGTTAT-ATGGATGATGTGGTA 3′

T2: 5′ AGTTGTAACGGAAGA-TGCAATAGTAATCAG 3′

T3: 5′ GGAGTAAATGTTGGA-GAACAGTATCAACAA 3′

T4: 5′ AGAAGATATTTGGAATAA-CATGACCTGGAT GCA 3′

15. Formamide increases the stringency at which hybridization occurs since it competes with the complementary DNA strand for hydrogen bonding interactions. The addition of 15% formamide to the reaction mixture resulted in optimal selectivity of the system.

16. The scanometric detection of DNA is performed following the procedure in Subheading 3.5.

17. Only use HPLC-purified oligonucleotides. Amine contaminants can reduce coupling.

18. Codelink Slides must be stored desiccated, because the NHS leaving group on the Codelink slide also reacts with H_2O, resulting in a decrease in the coupling efficiency of the amine-modified oligonucleotides. Before and after each spotting run, always wash the pins 3–4 times to be sure that there is no salt build-up on them.

19. The incubation temperatures were controlled to completely dehybridize the biobarcode sequences from the capture sequences on the chip at 60°C and to enhance the hybridization by gradually decreasing the temperatures (37→25°C). This process is called "annealing."

20. The hybridization temperature is determined empirically not to dehybridize the probes at too high temperature, nor to hybridize the probes nonspecifically at too low temperature.

21. The staining time needs to be optimized depending on the number of hybridized AuNP probes. Longer staining time than enough saturates the signal intensity, while too short staining time cannot enhance the low signal intensity enough to be visualized.

References

1. Vlahou, A. and Fountoulakis, A. (2005) Proteomic approaches in the search for disease biomarkers. *J. Chromatogr. B* **814**, 11–19.

2. Rigaut, G., Shevchenko, A., Rutz, B., Wilm, M., Mann, M., and Seraphin, B. (1999) A generic protein purification method for protein complex characterization and proteome exploration. *Nat. Biotechnol.* **17**, 1030–1032.

3. Seubert, P., Vigopelfrey, C., Esch, F., Lee, M., Dovey, H., Davis, D., et al. (1992) Isolation and quantification of soluble Alzheimers beta-peptide from biological-fluids. *Nature* **359**, 325–327.

4. DeRisi, J. L., Iyer, V. R., and Brown, P. O. (1997) Exploring the metabolic and genetic control of gene expression on a genomic scale. *Science* **278**, 680–686.

5. Rosi, N. L. and Mirkin, C. A. (2005) Nanostructures in biodiagnostics. *Chem. Rev.* **105**, 1547–1562.

6. Wang, H., Yang, R., Yang, L., and Tan, W. (2009) Nucleic acid conjugated nanomaterials for enhanced molecular recognition. *ACS Nano* **3**, 2451–2460.

7. Niemeyer, C. M. and Simon, U. (2005) DNA-based assembly of metal nanoparticles. *Eur. J. Inorg. Chem.* **18**, 3641–3655.

8. Mirkin, C. A., Letsinger, R. L., Mucic, R. C., and Storhoff, J. J. (1996) A DNA-based method for rationally assembling nanoparticles into macroscopic materials. *Nature* **382**, 607–609.

9. Taton, T. A., Mirkin, C. A., and Letsinger, R. L. (2000) Scanometric DNA array detection with nanoparticle probes. *Science* **289**, 1757–1760.

10. Anderson, E. S., Dong, M., Morten, M. N., Kasper, J., Subramani, R., Mamdouh, W., et al. (2009) Self-assembly of a nanoscale DNA box with a controllable lid. *Nature* **459**, 73–76.

11. Alivisatos, A. P., Johnsson, K. P., Peng, X., Wilson, T. E., Loweth, C. J., Bruchez, M. P., et al. (1996) Organization of "nanocrystal molecules" using DNA. *Nature* **382**, 609–611.

12. Grabar, K. C., Freeman, R. G., Hommer, M. B., and Natan, M. J. (1995) Preparation and characterization of Au colloid monolayers. *Anal. Chem.* **67**, 735–743.

13. Frens, G. (1973) Controlled nucleation for regulation of particle-size in monodisperse gold suspensions. *Nat. Phys. Sci.* **241**, 20–22.

14. Daniel, M. C. and Astruc, D. (2004) Gold nanoparticles: assembly, supramolecular chemistry, quantum-size-related properties, and applications toward biology, catalysis, and nanotechnology. *Chem. Rev.* **104**, 293–346.

15. Jin, R., Wu, G., Li, Z., Mirkin, C. A., and Schatz, G. C. (2003) What controls the melting properties of DNA-linked gold nanoparticle assemblies? *J. Am. Chem. Soc.* **125**, 1643–1654.

16. Storhoff, J. J., Lazarides, A. A., Mucic, R. C., Mirkin, C. A., Letsinger, R. L., and Schatz, G. C. (2000) What controls the optical properties of DNA-linked gold nanoparticle assemblies? *J. Am. Chem. Soc.* **122**, 4640–4650.

17. Lee, J.-S., Stoeva, S. I., and Mirkin, C. A. (2006) DNA-induced size-selective separation of mixtures of gold nanoaparticles. *J. Am. Chem. Soc.* **128**, 8899–8903.

18. Lytton-Jean, A. K. R. and Mirkin, C. A. (2005) A thermodynamic investigation into the binding properties of DNA functionalized gold nanoparticle probes and molecular fluorophore probes. *J. Am. Chem. Soc.* **127**, 12754–12755.

19. Rosi, N. L., Giljohann, D. A., Thaxton, C. S., Lytton-Jean, A. K. R., Han, M. S., and Mirkin, C. A. (2006) Oligonucleotide-modified gold nanoparticles for intracellular gene regulation. *Science* **312**, 1027–1030.

20. Letsinger, R. L., Elghanian, R., Viswanadham, G., and Mirkin, C. A. (2000) Use of a steroid cyclic disulfide anchor in constructing gold nanoparticle-oligonucleotide conjugates. *Bioconjug. Chem.* **11**, 289–291.

21. Li, Z., Jin, R. C., Mirkin, C. A., and Letsinger, R. L. (2002) Multiple thiol-anchor capped DNA-gold nanoparticle conjugates. *Nucleic Acids Res.* **30**, 1558–1562.

22. Hurst, S. J., Lytton-Jean, A. K. R., and Mirkin, C. A. (2006) Maximizing DNA loading on a range of gold nanoparticle sizes. *Anal. Chem.* **78**, 8313–8318.

23. Stoeva, S. I., Lee, J.-S., Thaxton, C. S., and Mirkin, C. A. (2006) Multiplexed DNA detection with bio-barcoded nanoparticle probes. *Angew. Chem. Int. Ed.* **45**, 3303–3306.

Chapter 3

Molecular Detection of Biomarkers and Cells Using Magnetic Nanoparticles and Diagnostic Magnetic Resonance

Jered B. Haun, Tae-Jong Yoon, Hakho Lee, and Ralph Weissleder

Abstract

The rapid and sensitive detection of molecular targets such as proteins, cells, and pathogens in biological specimens is a major focus of ongoing medical research, as it could promote early disease diagnoses and the development of tailored therapeutic strategies. Magnetic nanoparticles (MNP) are attractive candidates for molecular biosensing applications because most biological samples exhibit negligible magnetic susceptibility, and thus the background against which measurements are made is extremely low. Numerous magnetic detection methods exist, but sensing based on magnetic resonance effects has successfully been developed into a general detection platform termed diagnostic magnetic resonance (DMR). DMR technology encompasses numerous assay configurations and sensing principles, and to date magnetic nanoparticle biosensors have been designed to detect a wide range of targets including DNA/mRNA, proteins, enzymes, drugs, pathogens, and tumor cells with exquisite sensitivity. The core principle behind DMR is the use of MNP as proximity sensors that modulate the transverse relaxation time of neighboring water molecules. This signal can be quantified using MR imagers or NMR relaxometers, including miniaturized NMR detector chips that are capable of performing highly sensitive measurements on microliter sample volumes and in a multiplexed format. The speed, sensitivity, and simplicity of the DMR principle, coupled with further advances in NMR biosensor technology should provide a high-throughput, low-cost, and portable platform for large-scale parallel sensing in clinical and point-of-care settings.

Key words: Magnetic nanoparticles, Molecular biosensing, Diagnostic magnetic resonance, Magnetic relaxation switches

1. Introduction

Magnetic nanoparticles (MNP) offer unique advantages for molecular detection of biological targets compared to other platforms. For example, MNP are inexpensive to produce, physically and chemically stable, biocompatible, and environmentally safe.

Sarah J. Hurst (ed.), *Biomedical Nanotechnology: Methods and Protocols*, Methods in Molecular Biology, vol. 726,
DOI 10.1007/978-1-61779-052-2_3, © Springer Science+Business Media, LLC 2011

In addition, biological samples exhibit virtually no magnetic background, and thus measurements can be performed in turbid or otherwise visually obscured samples. To date, numerous methods have been developed to sense biomolecules using magnetic labels (1), but one that has achieved considerable success in biomedicine is based on magnetic resonance. In this case, the MNP serve as proximity sensors that accelerate the relaxation rate of neighboring water molecules following exposure to a magnetic field. This phenomenon has been exploited for magnetic resonance imaging to obtain detailed anatomical information.

When magnetic resonance is used for molecular detection of biomolecules and cells outside of the body, it is referred to as diagnostic magnetic resonance (DMR) (2, 3). DMR assays employ affinity molecule-conjugated MNP to bind molecular targets and effect a change in proton relaxation rate by one of two methods (see Fig. 1). The first makes use of the phenomenon of magnetic relaxation switching (MRSw), in which molecular targets are used to self-assemble MNP into clusters and thereby cause a corresponding change in the bulk relaxation rate (4, 5). The second mode involves tagging large structures such as whole cells (2, 3, 6). In both cases, the binding interactions are performed homogeneously in solution and make use of the built-in amplification that magnetic resonance offers through its effect on billions of neighboring water molecules. These factors make DMR faster than techniques that require solid-phase immobilization, diffusion of nanoparticles to sensing elements, or discrete amplification steps. The magnetic resonance signal can be quantified by NMR or MRI as a decrease in longitudinal (spin–lattice, T_1) or transverse (spin–spin, T_2) relaxation times. Typically T_2 is used because the transverse relaxivity (r_2) is significantly greater than the longitudinal relaxivity (r_1) for most MNP.

To date, DMR has successfully been used to detect a wide variety of biomolecular targets with high sensitivity and specificity, including DNA, mRNA, proteins, enzyme activity, metabolites, drugs, pathogens, and tumor cells (see Table 1). Notably, detection thresholds in the femtomolar biomarker concentration and near single-cell range have been documented in complex media such as cell lysates, sputum, and whole blood (3, 4, 6). The magnetic resonance signal can be read-out using MRI scanners or benchtop NMR systems. Recently, however, a miniaturized, chip-based NMR (μNMR) detector system was developed that houses all components for DMR detection in a handheld, portable device (2, 3, 6). The μNMR chip device is capable of performing measurements on microliter sample volumes and contains multiple detection coils to enable parallel sensing of numerous biomarkers. The μNMR chip thus represents a critical advance in DMR technology that could facilitate detection of disease markers in sample-limited clinical specimen.

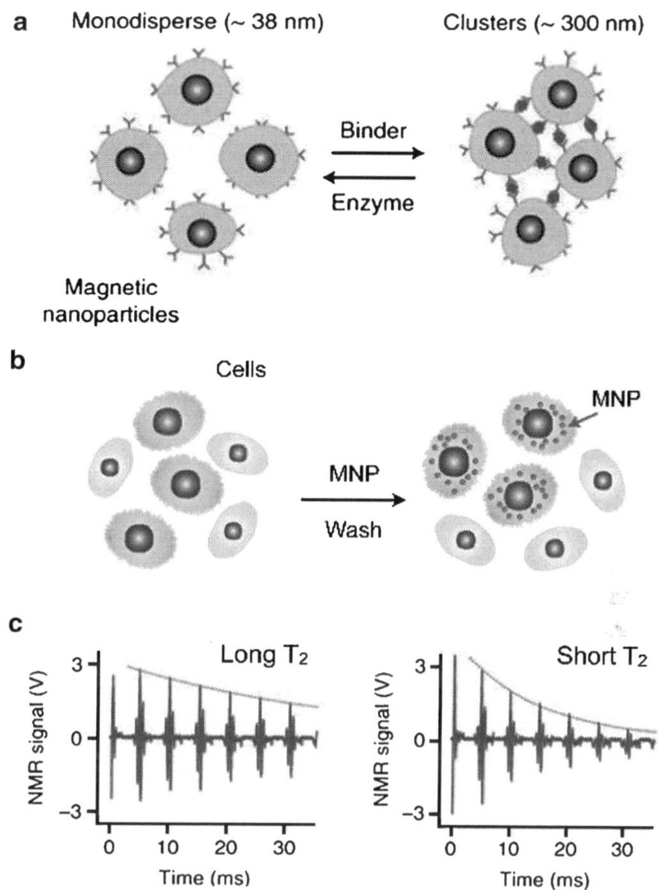

Fig. 1. DMR sensing principles. (**a**) Magnetic relaxation switching (MRSw) involves the assembly of MNP into clusters or disassembly of preformed clusters by the action of a target biomolecule. Clustered MNP dephase the nuclear spins of neighboring water molecules more efficiently than evenly dispersed MNP, shortening the bulk transverse relaxation time (T_2). (**b**) Tagging cells with MNP imparts a magnetic moment that is proportional to the number of nanoparticles bound. Following washing procedures to remove unbound MNP, the magnetic moment can be measured as a decrease in T_2 relaxation time. (**c**) Representative NMR output depicting the shortening of T_2 relaxation time that accompanies MNP clustering (MRSw) or cellular tagging. Modified with permission from ref. 2, copyright (2008) Nature Publishing Group.

2. Materials

2.1. Preparation for Bioconjugation of Affinity Molecules to MNP

1. Superparamagnetic nanoparticles: Synthesized (7, 8) or purchased commercially (Miltenyi Biotec, Auburn, CA; Ocean NanoTech, Springdale, AK) with primary amine functional groups or conjugated with proteins such as avidin or protein A.

2. Affinity molecule: Lyophilized or dissolved in buffer. This sample should be free of contaminants that would interfere with coupling interactions (see Note 1).

Table 1
DMR biosensors/applications to date

Class	Target	Mode	Magnetic nanoparticle sensor	References
DNA	Telomeres	MRSw (A)	CLIO-(CCCTAA)$_3$	(24)
RNA	GFP	MRSw (A)	CLIO-ATTTGCCGGTGT and CLIO-TCAAGTCGCACA	(4)
Protein	GFP	MRSw (A)	CLIO-antibody (anti-GFP)	(4)
	Avidin	MRSw (A)	CLIO-biotin	(4)
	β-HCG	MRSw (A)	CLIO-antibody (anti-β-HCG)	(25)
	Telomerase	MRSw (A)	CLIO-antibody (anti-telomerase)	(26)
	CA-125	MRSw (A)	CLIO-antibody (anti-CA-125)	(2)
	VEGF	MRSw (A)	CLIO-antibody (anti-VEGF)	(2)
	α-Fetoprotein	MRSw (A)	CLIO-antibody (anti-α-fetoprotein)	(2)
Enzyme activity	Caspase 3	MRSw (D)	CLIO-GDEVDG-biotin; CLIO-avidin	(4)
	*Bam*H1	MRSw (D)	CLIO-TTA-CGC-CTAGG-ATC-CTC and CLIO-AAT-GCG-GGATCC-TAC-GAG	(27)
	DNA methylase	MRSw (D)	Methylated *Bam*H1 sensor	(27)
	Renin	MRSw (D)	Biotin-IHPFHLVIHTK-biotin; CLIO-avidin	(28)
	Trypsin	MRSw (D)	Biotin-(G)$_4$RRRR(G)$_3$K-biotin or biotin-GPARLAIK-biotin; CLIO-avidin	(28)
	MMP-2	MRSw (D)	Biotin-GGPLGVRGK-biotin; CLIO-avidin	(28)
	Telomerase	MRSw (A)	CLIO-AATCCCAATCCC and CLIO-AATCCCAATCCC	(26)
	Peroxidases	MRSw (D)	CLIO-phenol or CLIO-tyrosine	(29)
Small molecule	Drug	MRSw (D)	CLIO-antibody (anti-D-phenylalanine)	(30)
	Folate	MRSw (D)	CLIO-antibody (anti-folate)	(31)
	Glucose	MRSw (D)	CLIO-Concavalin	(31)
	HA peptide	MRSw (D)	CLIO-antibody (anti-HA)	(31)
	Calcium	MRSw (A)	CLIO-Calmodulin and CLIO-M13 or CLIO-chelator	(32, 33)
Organism	Herpes virus	MRSw (A)	CLIO-antibody (anti-HSV1)	(34)
	Adenovirus-5	MRSw (A)	CLIO-antibody (anti-adenovirus-5)	(34)
	MAP	MRSw (A)	CLIO-antibody (anti-MAP)	(35)
	Staphylococcus aureus	MRSw (A)	CLIO-Vancomycin	(2)
	MTB/BCG	Tagging	CLIO-antibody or CB-antibody (anti-BCG)	(6)
Human cell	Tumor cell lines	Tagging	CLIO-antibody (anti-Her2/*neu*, EGFR, EpCAM)	(2)
	Murine tumor biopsies	Tagging	Mn-MNP-antibody (anti-Her2/*neu*, EGFR, EpCAM)	(3)

MRSw (A): MRSw assay using assembly of MNP into clusters. MRSw (D): MRSw assay based on disassembly of pre-clustered MNP

3. Phosphate-buffered saline (PBS): 137 mM NaCl, 2.7 mM KCl, 10 mM Na_2HPO_4, and 1.8 mM KH_2PO_4 in de-ionized water, adjust pH to 7.4 with NaOH. PBS can also be purchased, but be sure that divalent cations are not present (see Note 2).

4. Ethylenediamine tetraacetic acid (EDTA) buffer: 10× Solution with 100 mM EDTA in the PBS buffer listed in item 3, Subheading 2.1. EDTA is necessary for thiol-based chemistries to prevent formation of disulfide bonds between free thiol groups.

5. Bicarbonate buffer: 0.1 M $NaHCO_3$, adjust pH to 10 with NaOH.

6. Dimethylformamide (DMF) [if desired, dimethylsulfoxide (DMSO) can be used interchangeably].

7. Thiol chemistry bioconjugations will require one or more of the following reagents:

    (a) Sulfosuccinimidyl-4-[N-maleimidomethyl]cyclohexane-1-carboxylate (sulfo-SMCC, available in convenient no-weigh format from Thermo Fisher, Waltham, MA).

    (b) N-Succinimidyl-S-acetylthioacetate (SATA, Thermo Fisher).

    (c) β-Mercaptoethylamine (MEA, Sigma Aldrich, St. Louis, MO).

8. Deacetylation buffer: 0.5 M Hydroxylamine·HCl, 25 mM EDTA in the PBS buffer listed in item 3, Subheading 2.1, adjust pH to 7.4 with NaOH. This reagent is required for point (b) of item 7, Subheading 2.1 only.

9. Click chemistry bioconjugations can be performed using commercially available reagents including succinimidyl ester or iodoacetamide conjugated azide and alkyne moieties (Invitrogen, Carlsbad, CA).

10. Amicon Ultra centrifugal filtration device, 15 mL volume, 100,000 MWCO (Millipore, Billerica, MA).

11. Sephadex G-50 (GE Healthcare, Piscataway, NJ) or similar gel filtration media.

12. PD-10 (GE Healthcare) or Zeba (5 mL, Thermo Fisher) desalting columns.

13. Coomassie protein stain (Thermo Fisher).

2.2. Coupling, Processing, and Characterization

1. Copper catalyst buffer: 10× Solution with 50 mM sodium ascorbate, 10 mM $Cu^{(II)}SO_4$, 50 mM of a polytriazole ligand (see Note 3).

2. Sephadex G-100 Superfine (GE Healthcare) or similar gel filtration media.

3. Fe digestion buffer: 6 M HCl, 0.3% H_2O_2 in de-ionized water.

4. Micro BCA™ Protein Assay Kit (Thermo Fisher).

5. 96-Well plate (Becton Dickinson, Franklin Lakes, NJ).

2.3. MRSw Assay for Soluble Analyte Sensing

1. Analyte samples can be obtained from numerous sources, including unpurified biological samples such as cell culture supernatants, cell lysates, milk, sputum, and whole blood. For disassembly assays, the MNP are typically pre-clustered using purified cross-linkers such as oligonucleotides or proteins prior to addition of analyte.

2.4. Magnetic Tagging of Cells

1. PBS with bovine serum albumin (BSA) (PBS+): 1 g/L BSA added to the PBS buffer listed in item 3, Subheading 2.1, sterile filter.

2. Cell samples can be obtained from any source of interest, such as in vitro cultures or in vivo specimen, but should be suspended as single cells. This can be accomplished using various enzymes (trypsin, collagenase, etc.) and/or EDTA (see Note 4). See Note 5 for total cell requirements.

2.5. Detection Using NMR

1. 5 or 10 mm NMR tubes (Thermo Fisher).

2. Benchtop NMR relaxometer (i.e., Minispec, Bruker Optics, Billerica, MA). Alternatively, custom devices such as miniaturized NMR (μNMR) detectors have been used (2, 3, 6). The indicated options both operate at approximately 20 MHz and 0.5 T.

2.6. Detection Using MRI

1. 384-Well plate (Becton Dickinson).

2. Clinical or experimental MRI scanner operating at 1–11 T [i.e., ClinScan (7 T) or PharmScan (4.5 T), Bruker BioSpin, Billerica, MA].

3. Software to analyze T_2 images (i.e., OsiriX Viewer, available at http://www.osirix-viewer.com).

3. Methods

The MNP used for magnetic resonance detection must be super-paramagnetic, meaning that they become magnetized when placed in an external magnetic field but lose their magnetic moment when the field is removed (no magnetic remanence). The bulk of DMR assays have been performed using iron oxide nanoparticles, in particular cross-linked iron oxide (CLIO) (7, 9). CLIO has a magnetic core composed of a monocrystalline iron oxide nanoparticle (MION), and the MION core is caged in cross-linked dextran. Such iron oxide nanoparticles can be obtained in various forms from commercial sources. Aside from standard iron oxide options, it has been demonstrated that doping iron oxide with metals such as manganese, cobalt, and nickel can increase magnetization, and thus detection sensitivity (10). Likewise, a core–shell nanoparticle composed of an elemental iron core and protective iron oxide shell has also been shown to possess greater magnetic susceptibility than iron oxide (11). Manganese-doped iron oxide nanoparticles (Mn-MNP) and elemental Fe core nanoparticles (cannonballs, CB) have been validated for DMR applications and significantly improve detection sensitivity (3, 6). Micron-sized particles composed of many iron oxide cores embedded in a polymer matrix have been used for DMR assays as well (12, 13), but their extremely high magnetization is offset by rapid settling out of solution, limiting utility.

Regardless of the type of superparamagnetic core used, a polymer coating is required to render the core water soluble, prevent aggregation, and provide chemical functional groups for bioconjugation. Most often the functional groups are primary amines, but carboxylic acids or thiols are also used. The protocols listed below detail the attachment of affinity molecules to amine-terminated MNP; treatments of MNP with other surface-based functional groups can be found elsewhere (14). Subheading 3.1 or 3.2 can also be followed to first attach avidin, proteins A or G, secondary antibodies, or fluorescent molecules (see Note 6) to the MNP or alternatively MNP can be purchased with these moieties already conjugated to their surface. In these cases, skip to Subheading 3.3 for attachment of affinity molecules modified with biotin or antibody Fc domains. The choice of affinity molecule is dependent on the target and application of interest. To date, strategies have included the attachment of oligonucleotides, antibodies, peptides, small molecules, metal chelators, and natural biological binding partners, as listed in Table 1.

3.1. Bioconjugation Using Thiol Chemistry

1. These instructions are intended for the attachment of affinity molecules that contain a free thiol group, such as peptides with a terminal cysteine residue and proteins that are appropriately modified to bear a thiol group (mild reduction of disulfide bonds or conversion of a primary amine to a thiol group).

2. Dilute amino-MNP with PBS to approximately 0.5 mg Fe/mL concentration and adjust the pH to 8.0 using the bicarbonate buffer. Lower Fe concentrations can be used but a different purification process is required (see Note 7).

3. Add at least 0.1 mg sulfo-SMCC per mg Fe (approximately 30-fold molar excess for CLIO containing 30 amines/MNP, see Note 8). The addition of more sulfo-SMCC will not

negatively affect the reaction. Incubate the reaction mixture while shaking at room temperature for 1–2 h (see Note 9).

4. Concentrate the MNP by adding to a centrifugal filtration device and centrifuging for 10 min at $2,000 \times g$. The final volume is usually about 0.2 mL.

5. Remove excess sulfo-SMCC reagent using a desalting column packed with Sephadex G-50. The total column volume should be at least 50 times greater than the sample volume to ensure effective separation. The resulting maleimide-MNP should be used as soon as possible, but can be stored at 4°C for up to 24 h if necessary.

6. Prepare the affinity molecule by dissolving the sample in or diluting it with the 10× EDTA buffer, and additional PBS if necessary, to obtain a final sample volume of 1 mL containing EDTA at 1× concentration. If the affinity molecule contains a free thiol functional group, it can be used directly (skip to Subheading 3.3).

7. For proteins that contain a disulfide bond that can be reduced without affecting binding function (such as most antibodies), add 1 mg of MEA (26 mM final concentration) and incubate for 2 h at 37°C. Alternatively, the primary amine groups of certain proteins can be converted to protected sulfhydryls using SATA; these groups can then be deprotected to yield free sulfhydryls. In this case, dissolve the SATA in DMF and add it to the protein sample at two- to fivefold molar excess (see Note 10). Incubate the mixture for 1–2 h at room temperature (see Note 9), then add 0.1 mL deacetylation buffer, and incubate for another 1.5 h.

8. Purify these protein affinity molecules by desalting with a gravity (PD-10) or centrifugal column (Zeba). First, equilibrate the column with PBS containing 1× EDTA buffer before applying the sample. For PD-10 columns, collect ~1 mL fractions and check for protein by mixing 50 μL sample with 50 μL Coomassie reagent (a color change will be observed if protein is present). Pool positive fractions. For the Zeba columns, use the flow-through directly (skip to Subheading 3.3).

3.2. Bioconjugation Using "Click" Chemistry

1. These instructions are intended as a general method to attach affinity molecules to amine-modified MNP using bioorthogonal "click" chemistries (15). The most widely used "click" reaction is the copper-catalyzed azide-alkyne cycloaddition (16, 17), which has been used to conjugate small molecules and antibodies to MNP (18, 19), along with many other applications. Recently, a new catalyst-free cycloaddition between a 1,2,4,5-tetrazine and *trans*-cyclooctene dienophile

(20) was successfully used in the same capacity (21). Regardless of the reaction pair employed, they can be used interchangeably on the affinity molecule of choice and MNP, and therefore will generically be referred to as Click Reactants 1 and 2. They can be attached using amine- or thiol-reactive moieties, such as a succinimidyl ester or iodoacetamide/maleimide, respectively.

2. Dilute amino-MNP with PBS containing 5% DMF to a concentration of 0.5 mg Fe/mL. Adjust the pH to 8.0 using the bicarbonate buffer.

3. Dissolve amine-reactive Click Reactant 1 with DMF and add a tenfold molar excess relative to the total amine content per MNP. For example, 2 mg of CLIO with 30 amines/particle would require a minimum 1.4 μmol amine-reactive Click Reagent 1 (see Note 8). Incubate the reaction mixture while shaking at room temperature for 2 h.

4. Concentrate and desalt as indicated in steps 4 and 5, Subheading 3.1.

5. Prepare the affinity molecule by dissolving the sample in or diluting it with PBS to 0.8 mL. Add 0.1 mL sodium bicarbonate buffer. Dissolve Click Reactant 2 (amine- or thiol-reactive) in DMF at 10 mg/mL and add to the reaction solution at two- to fivefold molar excess (see Note 10). Include additional DMF to achieve a total concentration of 10% if necessary. Incubate at room temperature for 1–2 h (see Note 9).

6. Purify the affinity molecule as described in step 8, Subheading 3.1, but omit EDTA from the wash solution because it is only necessary if free thiol groups are present in the sample. For small molecules or peptides, HPLC purification may be required.

3.3. Coupling, Processing, and Characterization

1. Combine the appropriate MNP (conjugated with maleimide, Click Reactant 1, avidin, protein A/G, or secondary antibody) with the appropriate affinity molecule (conjugated with a free thiol group, Click Reactant 2, biotin, or antibody Fc domain) samples. If necessary, adjust the reaction volume with PBS such that the final MNP concentration is less than 0.5 mg Fe/mL to prevent cross-linking of nanoparticles (see Note 10). The affinity molecule should be added in excess to increase MNP valency, and thus binding activity, and prevent cross-linking of MNP for cases in which the affinity molecule contains multiple coupling domains. In general, a fivefold molar excess is sufficient, but a tenfold excess is preferred. For the copper-catalyzed azide-alkyne cycloaddition, the copper catalyst buffer must also be added to the reaction mixture at 1× concentration.

2. Incubate for 4 h at room temperature or overnight at 4°C.

3. For thiol conjugations, add ammonium chloride or cysteine at 1 µM final concentration to cap unreacted maleimide groups and prevent nonspecific binding to biological samples (see Note 11). Incubate 30 min at room temperature.

4. Concentrate by adding to a centrifugal filtration device and centrifuging for at least 10 min at $2,000 \times g$.

5. Remove excess affinity molecules using a size-exclusion column packed with Sephadex G-100. Again, the total column volume should be at least 50 times greater than the sample volume to ensure effective separation.

6. Quantify the MNP concentration based on Fe content by digesting with hydrochloric acid to yield $FeCl_3$ and determining the Fe absorbance at 410 nm. Dilute the MNP solution with at least nine volumes of Fe digestion buffer and incubate at 50–60°C for 1 h. Determine the absorbance at 410 nm and convert to Fe concentration using an extinction coefficient of $1,370$ M^{-1} cm^{-1} and the appropriate dilution factor. Alternatively, a quick estimate of the MNP concentration can be obtained by measuring the absorbance at 410 nm and calibrating with a known standard (i.e., the original MNP stock solution).

7. For protein and some peptide affinity molecules, successful conjugation can readily be determined using the Micro BCA™ Protein Assay. Combine equal volumes (minimum 50 µL) of the MNP sample solution and the Micro BCA working reagent in a 96-well plate. Incubate for 2 h at 37°C, cool to room temperature, and measure the absorbance of the reacted BCA reagent at 562 nm. The absorbance can be converted to a concentration using a calibration curve prepared from a stock sample of the affinity molecule with known concentration, or BSA protein as an estimate. The number of molecules per MNP can then be calculated using the molecular weight, the MNP concentration from step 6, Subheading 3.3, and the molecular weight of the MNP (see Note 8).

8. Use dynamic light scattering, if desired, to ensure that the MNP did not aggregate during conjugation procedures.

3.4. MRSw Assay for Soluble Analyte Sensing

1. This section describes the detection of soluble molecules based on the principle of MRSw, in which the molecular target induces assembly or disassembly of MNP into clusters and causes a change in the bulk T_2 relaxation time (see Fig. 2). Disassembly MRSw assays are the preferred method for the detection of enzymes (cleavage of specific sites in the cross-bridges) and small molecules (competitive binding). In this case, MNP must first be clustered, which has been accomplished using several different strategies (see Table 1).

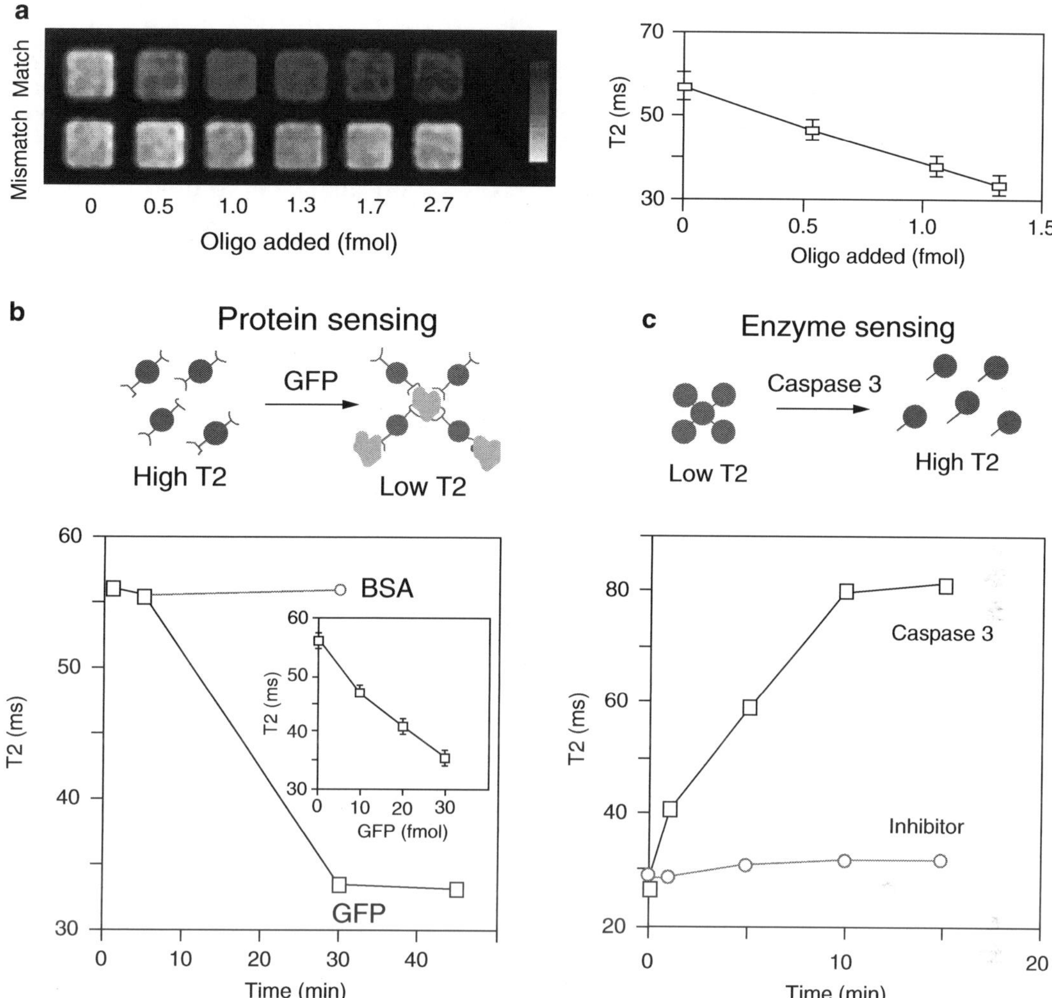

Fig. 2. DMR detection of biomolecules using magnetic relaxation switching (MRSw). (**a**) Detection of an oligonucleotide target using MNP conjugated with complementary oligonucleotide sequences. The left panel displays a T_2-weighted MR image of a 384-well plate containing varying amounts of target or mismatched oligonucleotide and constant levels of MNP. Hybridization to the target causes the MNP to cluster, resulting in a corresponding decrease in T_2 relaxation time. (**b**) Detection of GFP protein using MNP conjugated with a polyclonal antibody specific for GFP. T_2 relaxation time, which was determined using a benchtop NMR relaxometer, decreased linearly with GFP concentration, but was not affected by the concentration of a control protein (BSA). (**c**) Detection of caspase 3 enzymatic activity using MNP that were clustered using a linker containing the peptide sequence DEVD. Introduction of caspase 3 resulted in rapid cleavage of the peptide sequence and an increase in T_2 relaxation time, which was abrogated by the addition of a caspase 3 inhibitor. Reproduced with permission from ref. 4, copyright (2002) Nature Publishing Group.

2. Dilute the affinity molecule-MNP sample with PBS to 10–20 µg/mL total Fe concentration (see Note 12).

3. For preclustering MNP prior to a disassembly assay, add the cross-linking agent at the optimal concentration to induce aggregation as determined by experimentation (see Note 13). Incubate for at least 1 h at room temperature. Longer incubations

may produce more pronounced aggregation depending on the rate of cluster formation. Confirmation of MNP aggregation can be obtained by dynamic light scattering at this point. Then, combine equal volumes of the aggregated MNP solution and analyte sample.

4. For aggregation assays, add an equal volume of analyte sample to the MNP solution.

5. Incubate for at least 1 h at room temperature and proceed to analysis (Subheading 3.6 or 3.7).

3.5. Magnetic Tagging of Cells

1. This section describes the detection of biomarkers on the surface of intact, suspended cells by tagging with MNP to impart magnetic susceptibility (see Fig. 3). Similar methods can be used to label adherent cells, followed by disruption and suspension. In addition, intracellular markers can be tagged if the cells are first permeabilized (see Note 14).

2. Wash the cell sample by centrifuging at $300 \times g$ for 5 min, aspirating the supernatant, and resuspending in 0.5 mL PBS+.

3. Add the affinity molecule-MNP sample to the cell suspension. When using antibodies as the affinity molecule, a final MNP concentration of 100 nM (~45 µg/mL total Fe concentration for CLIO, see Note 8) should be sufficient (see Note 15).

4. Incubate for 30 min at room temperature on an orbital shaker.

5. Remove unbound MNP using two rounds of centrifugation as described in step 2, Subheading 3.5, but use 1 mL of ice-cold PBS+ each time.

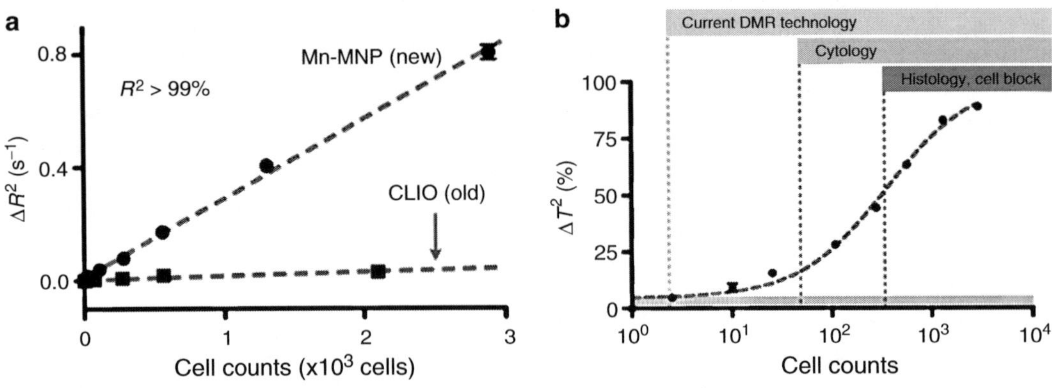

Fig. 3. DMR detection of tumor cells using the tagging method. (**a**) Her2/*neu* was detected on breast cancer cells (BT474) using an anti-Her2/*neu* antibody that was conjugated to CLIO and Mn-MNP nanoparticles. Transverse relaxation rate ($r_2 = 1/T_2$) was measured using a miniaturized NMR (µNMR) detector, and varied proportionally with cell number and the magnetic susceptibility of the nanoparticle employed. (**b**) As few as two cells could be detected using the Mn-MNP nanoparticle, well above the detection threshold of established clinical methods (cytology and histology). Reproduced with permission from ref. 3, copyright (2009) National Academy of Sciences, USA.

6. Resuspend in the minimum volume PBS+ necessary for the detection method of choice (see Subheadings 3.6 and 3.7).

3.6. Detection Using NMR

1. Magnetic resonance signal from MRSw or tagged cell samples can be detected using NMR relaxometers operating at low frequency and magnetic field strength. Benchtop systems such as the Bruker Minispec (20 MHz, <1 T) measure T_1 and T_2 relaxation times of samples in NMR tubes (see Fig. 2b, c). Alternatively, μNMR devices can be used to detect T_1 or T_2 within microfluidic channels (see Fig. 3).

2. For benchtop NMR systems, load sample into 5 (>0.3 mL) or 10 mm (>0.5 mL) NMR tubes. For μNMR devices, sample volumes as low as 1 μL have been achieved using microfluidic elements (3).

3. Measure T_2 relaxation time (and T_1 if desired). For both benchtop and miniaturized systems, T_1 relaxation time is measured using inversion recovery pulse sequences. For T_2 measurement, Carr–Purcell–Meiboom–Gill (CPMG) spin-echo pulse sequences are employed to compensate for the spatial inhomogeneity of the external magnetic field. The typical echo time is 2–5 ms and the repetition time is $\approx 5 \times T_1$ to ensure full recovery of nuclear spins (2).

3.7. Detection Using MRI

1. Clinical or experimental MRI scanners can be used to detect magnetic resonance signals from samples loaded into multi-well plates (see Fig. 2a). MRI scanners employ strong magnetic fields (1–11 T) generated by superconducting magnets, and use sophisticated data acquisition schemes.

2. Load 50 μL sample into a 384-well plate.

3. Obtain a T_2 map using T_2 spin-echo sequences with variable echo times. Typically, echo time is varied between 25 and 1,000 ms, and the repetition time ranges between 2,000 and 3,000 ms.

4. Analyze the T_2 image using the appropriate software.

4. Notes

1. If the MNP or affinity molecule storage buffer contains contaminants that will interfere with downstream processing (i.e., Tris or carrier protein for amine reactions, sodium azide for click chemistry), they must first be removed by desalting or chromatography.

2. It is important to avoid divalent cations when working with nanoparticles because they can crosslink the nanoparticles, leading to aggregation. Care should also be taken that chemical modifications to the nanoparticles, including conjugation

of proteins or peptides, do not alter the surface chemistry such that aggregation results.

3. A polytriazole ligand such as bathophenanthroline disulfonic acid should be added to the copper catalyst buffer to stabilize the reduced form of copper [$Cu^{(I)}$] that is required for catalytic activity (22).

4. When using an enzyme to disrupt tissues, caution should be taken that the biomarker of interest is not affected by the activity of the enzyme.

5. The total cell requirement for tagging assays is dependent on the target expression level and the detection sensitivity of the magnetic resonance sensor platform. For NMR measurements, detection thresholds using CLIO are in the range of 10,000 cells using a benchtop relaxometer (Minispec) and 1,000 cells using the miniaturized NMR (μNMR) for a high expression level marker (millions of copies per cell, see Fig. 3). Use of higher magnetization MNP can decrease the detection threshold to near single-cell for the μNMR.

6. Fluorophores can aid significantly in optimizing the binding of the MNP conjugates to cells using fluorescence microscopy or flow cytometry and in tracking the MNP concentration during processing steps using a fluorometer. Fluorophores should be attached prior to affinity molecule conjugation. Since excess reagents are not required for this conjugation, it is easier to control the degree of modification and MNP cross-linking is not an issue. Care should be taken that functional groups remain for affinity molecule conjugation, however. After dye conjugation, the free dye can be removed by desalting and the amount of dye attached can be determined by absorbance (known extinction coefficient) or fluorescence (compared to a standard) measurement. Following dye conjugation, the nanoparticles are typically referred to as magneto-fluorescent nanoparticles (MFNP).

7. Iron oxide MNP are brown and visible to the naked eye at concentrations above 100 μg Fe/mL. Therefore, when working with samples above this concentration, column purifications can be performed manually. For lower concentration samples, purifications should be performed using an automated system (i.e., AKTA fPLC from GE Healthcare) so that absorbance (410 nm) can be monitored.

8. The molecular weight of the MNP can be calculated based on the number of Fe atoms per core and the molecular mass of Fe (55.85 g/mol). For example, CLIO have approximately 8,000 Fe atoms/core (23), thus the estimated average molecular weight is 447,000 g/mol.

9. For modifications of MNP and affinity molecules using amine-reactive N-hydroxy-succinimidyl (NHS) esters, reaction times

of 1–2 h are common since the NHS groups hydrolyze in water. However, longer incubations may increase the reaction yield somewhat and should not adversely affect the final product provided that stability is not an issue.

10. MNP aggregation can result if the affinity molecule contains multiple coupling sites (e.g., such as multiple thiol, click reagent, or biotin molecules on a single macromolecular protein). For this reason, the affinity molecule modifications call for a minimal excess of amine-reactive reagent (i.e., two- to fivefold). These recommended values may need to be adjusted to yield approximately 1–2 coupling moieties per affinity molecule. Also of note, excess thiols can self-react to form disulfide bonds, and therefore should be capped with a reagent such as iodoacetamide.

11. Unreacted maleimide and thiol groups remaining after thiol couplings can potentially react with biomolecules or cells, increasing background adhesion and thus decreasing detection specificity. In some cases, capping of free thiol groups using iodoacetamide can help improve binding specificity. However, this should be performed following the purification steps described in steps 4 and 5, Subheading 3.3. Capping is not required when using a bioorthogonal click coupling chemistry.

12. For CLIO, the 10–20 μg/mL Fe recommended for MRSw assays equates to approximately 20–40 nM and typically registers a T_2 relaxation time of 50–100 ms. Due to the extra dilution factor associated with disassembly MRSw assays, the initial MNP concentration used is usually 20 μg/mL.

13. MNP clustering is governed by the equivalence principle, which dictates that clustering is greatest when a multivalent binder (i.e., MNP) and cross-linking agent are present at equimolar concentrations (13). In this case, the pertinent molar concentration is the number of affinity molecules per MNP, not the number of MNP.

14. Intracellular markers can be detected by adding 0.1% saponin to the PBS+ buffer. In this case, longer incubations are recommended (1 h) to increase MNP penetration into the cell, as well as extended washes at room temperature to allow unbound MNP to diffuse out of the cells.

15. The concentration of affinity molecule-MNP used for cell tagging can be lowered below 100 nM if the binding kinetics of the affinity molecule is sufficiently high. For example, a concentration of 10 nM would be sufficient if the equilibrium dissociation constant (K_D) of the interaction is subnanomolar. Likewise, the concentration should be increased if the binding kinetics is poor (i.e., micromolar K_D). The ideal concentration should be determined experimentally for each affinity molecule-MNP of interest.

Acknowledgments

The authors thank Nikolay Sergeyev for technical guidance regarding MNP synthesis and characterization and Dr. Neal K. Devaraj for assistance with click chemistry protocols.

References

1. Tamanaha, C. R., Mulvaney, S. P., Rife, J. C., and Whitman, L. J. (2008) Magnetic labeling, detection, and system integration. *Biosens. Bioelectron.* **24**, 1–13.

2. Lee, H., Sun, E., Ham, D., and Weissleder, R. (2008) Chip-NMR biosensor for detection and molecular analysis of cells. *Nat. Med.* **14**, 869–874.

3. Lee, H., Yoon, T. J., Figueiredo, J. L., Swirski, F. K., and Weissleder, R. (2009) Rapid detection and profiling of cancer cells in fine-needle aspirates. *Proc. Natl Acad. Sci. USA* **106**, 12459–12464.

4. Perez, J. M., Josephson, L., O'Loughlin, T., Hogemann, D., and Weissleder, R. (2002) Magnetic relaxation switches capable of sensing molecular interactions. *Nat. Biotechnol.* **20**, 816–820.

5. Josephson, L., Perez, J. M., and Weissleder, R. (2001) Magnetic nanosensors for the detection of oligonucleotide sequences. *Angew. Chem. Int. Ed. Engl.* **40**, 3204–3206.

6. Lee, H., Yoon, T. J., and Weissleder, R. (2009) Ultrasensitive detection of bacteria using core-shell nanoparticles and an NMR-filter system. *Angew. Chem. Int. Ed. Engl.* **48**, 5657–5660.

7. Josephson, L., Tung, C. H., Moore, A., and Weissleder, R. (1999) High-efficiency intracellular magnetic labeling with novel superparamagnetic-Tat peptide conjugates. *Bioconjug. Chem.* **10**, 186–191.

8. Lu, A. H., Salabas, E. L., and Schuth, F. (2007) Magnetic nanoparticles: synthesis, protection, functionalization, and application. *Angew. Chem. Int. Ed. Engl.* **46**, 1222–1244.

9. Harisinghani, M. G., Barentsz, J., Hahn, P. F., Deserno, W. M., Tabatabaei, S., van de Kaa, C. H., et al. (2003) Noninvasive detection of clinically occult lymph-node metastases in prostate cancer. *N. Engl. J. Med.* **348**, 2491–2499.

10. Lee, J. H., Huh, Y. M., Jun, Y. W., Seo, J. W., Jang, J. T., Song, H. T., et al. (2007) Artificially engineered magnetic nanoparticles for ultra-sensitive molecular imaging. *Nat. Med.* **13**, 95–99.

11. Peng, S., Wang, C., Xie, J., and Sun, S. (2006) Synthesis and stabilization of monodisperse Fe nanoparticles. *J. Am. Chem. Soc.* **128**, 10676–10677.

12. Koh, I., Hong, R., Weissleder, R., and Josephson, L. (2008) Sensitive NMR sensors detect antibodies to influenza. *Angew. Chem. Int. Ed. Engl.* **47**, 4119–4121.

13. Koh, I., Hong, R., Weissleder, R., and Josephson, L. (2009) Nanoparticle-target interactions parallel antibody-protein interactions. *Anal. Chem.* **81**, 3618–3622.

14. Sun, E. Y., Josephson, L., Kelly, K. A., and Weissleder, R. (2006) Development of nanoparticle libraries for biosensing. *Bioconjug. Chem.* **17**, 109–113.

15. Kolb, H. C., Finn, M. G., and Sharpless, K. B. (2001) Click chemistry: diverse chemical function from a few good reactions. *Angew. Chem. Int. Ed. Engl.* **40**, 2004–2021.

16. Rostovtsev, V. V., Green, L. G., Fokin, V. V., and Sharpless, K. B. (2002) A stepwise huisgen cycloaddition process: copper(I)-catalyzed regioselective "ligation" of azides and terminal alkynes. *Angew. Chem. Int. Ed. Engl.* **41**, 2596–2599.

17. Tornoe, C. W., Christensen, C., and Meldal, M. (2002) Peptidotriazoles on solid phase: [1,2,3]-triazoles by regiospecific copper(I)-catalyzed 1,3-dipolar cycloadditions of terminal alkynes to azides. *J. Org. Chem.* **67**, 3057–3064.

18. Sun, E. Y., Josephson, L., and Weissleder, R. (2006) "Clickable" nanoparticles for targeted imaging. *Mol. Imaging* **5**, 122–128.

19. Thorek, D. L., Elias, D. R., and Tsourkas, A. (2009) Comparative analysis of nanoparticle-antibody conjugations: carbodiimide versus click chemistry. *Mol. Imaging* **8**, 221–229.

20. Devaraj, N. K., Upadhyay, R., Haun, J. B., Hilderbrand, S. A., and Weissleder, R. (2009) Fast and sensitive pretargeted labeling of cancer cells through a tetrazine/trans-cyclooctene cycloaddition. *Angew. Chem. Int. Ed. Engl.* **48**, 7013–7016.

21. Haun, J. B., Devaraj, N. K., Hilderbrand, S. A., Lee, H., and Weissleder, R. (2010) Bioorthogonal chemistry amplifies nanoparticle binding and enhances the sensitivity of cell detection. *Nat. Nanotechnol.* **5**, 660–665.

22. Chan, T. R., Hilgraf, R., Sharpless, K. B., and Fokin, V. V. (2004) Polytriazoles as copper(I)-stabilizing ligands in catalysis. *Org. Lett.* **6**, 2853–2855.

23. Reynolds, F., O'Loughlin, T., Weissleder, R., and Josephson, L. (2005) Method of determining nanoparticle core weight. *Anal. Chem.* **77**, 814–817.

24. Grimm, J., Perez, J. M., Josephson, L., and Weissleder, R. (2004) Novel nanosensors for rapid analysis of telomerase activity. *Cancer Res.* **64**, 639–643.

25. Kim, G. Y., Josephson, L., Langer, R., and Cima, M. J. (2007) Magnetic relaxation switch detection of human chorionic gonadotrophin. *Bioconjug. Chem.* **18**, 2024–2028.

26. Perez, J. M., Grimm, J., Josephson, L., and Weissleder, R. (2008) Integrated nanosensors to determine levels and functional activity of human telomerase. *Neoplasia* **10**, 1066–1072.

27. Perez, J. M., O'Loughin, T., Simeone, F. J., Weissleder, R., and Josephson, L. (2002) DNA-based magnetic nanoparticle assembly acts as a magnetic relaxation nanoswitch allowing screening of DNA-cleaving agents. *J. Am. Chem. Soc.* **124**, 2856–2857.

28. Zhao, M., Josephson, L., Tang, Y., and Weissleder, R. (2003) Magnetic sensors for protease assays. *Angew. Chem. Int. Ed. Engl.* **42**, 1375–1378.

29. Perez, J. M., Simeone, F. J., Tsourkas, A., Josephson, L., and Weissleder, R. (2004) Peroxidase substrate nanosensors for MR imaging. *Nano Lett.* **4**, 119–122.

30. Tsourkas, A., Hofstetter, O., Hofstetter, H., Weissleder, R., and Josephson, L. (2004) Magnetic relaxation switch immunosensors detect enantiomeric impurities. *Angew. Chem. Int. Ed. Engl.* **43**, 2395–2399.

31. Sun, E. Y., Weissleder, R., and Josephson, L. (2006) Continuous analyte sensing with magnetic nanoswitches. *Small* **2**, 1144–1147.

32. Atanasijevic, T., Shusteff, M., Fam, P., and Jasanoff, A. (2006) Calcium-sensitive MRI contrast agents based on superparamagnetic iron oxide nanoparticles and calmodulin. *Proc. Natl Acad. Sci. USA* **103**, 14707–14712.

33. Taktak, S., Weissleder, R., and Josephson, L. (2008) Electrode chemistry yields a nanoparticle-based NMR sensor for calcium. *Langmuir* **24**, 7596–7598.

34. Perez, J. M., Simeone, F. J., Saeki, Y., Josephson, L., and Weissleder, R. (2003) Viral-induced self-assembly of magnetic nanoparticles allows the detection of viral particles in biological media. *J. Am. Chem. Soc.* **125**, 10192–10193.

35. Kaittanis, C., Naser, S. A., and Perez, J. M. (2007) One-step, nanoparticle-mediated bacterial detection with magnetic relaxation. *Nano Lett.* **7**, 380–383.

Chapter 4

Real-Time Quantum Dot Tracking of Single Proteins

Jerry C. Chang and Sandra J. Rosenthal

Abstract

We describe a single quantum dot tracking method that can be used to monitor individual proteins in the membrane of living cells. Unlike conventional fluorescent dyes, quantum dots (fluorescent semiconductor nanocrystals) have high quantum yields, narrow emission wavelengths, and excellent photostability, making them ideal probes in single-molecule detection. This technique has been applied to study the dynamics of various membrane proteins including glycine receptors, nerve growth factors, kinesin motors, and γ-aminobutyric acid receptors. In this chapter, a basic introduction and experimental setup for single quantum dot labeling of a target protein is given. In addition, data acquisition and analysis of time-lapse single quantum dot imaging with sample protocols are provided.

Key words: Quantum dot, Biological labeling, Biophysics, Protein trafficking, Single-particle tracking, Fluorescence microscopy

1. Introduction

Over the past decade, research utilizing semiconductor quantum dots (qdots) has continuously shown huge potential for in vitro and in vivo biological imaging (1–6). The great interest among researchers is due to the unique photophysical properties inherent to qdots, such as narrow emission wavelengths, excellent photostability, and exceptionally high quantum yields (7). These unique properties allow qdots to serve as ideal probes in biological detection. Another interesting property which makes qdots a near-perfect candidate for single-molecule studies is the fluorescent intermittency, or blinking, phenomenon (8). Taking advantage of this unique property, conventional biochemical studies such as Western blot analysis (9) and antibody affinity assays (10) are being successfully pushed forward toward the molecular scale.

Sarah J. Hurst (ed.), *Biomedical Nanotechnology: Methods and Protocols*, Methods in Molecular Biology, vol. 726,
DOI 10.1007/978-1-61779-052-2_4, © Springer Science+Business Media, LLC 2011

In 2003, the Dahan group published the first study using single qdots for the detection of glycine receptors in living cells (11). Since that time, a few groups have successfully expanded single qdot tracking for several different targets including nerve growth factors (12, 13), kinesin motors (14), and γ-aminobutyric acid (GABA) receptors (15). More recently, single qdot detection for an in vivo mouse tumor model has been demonstrated (16).

This chapter describes the principles, methodologies, and basic experimental protocols for the generation of single quantum dot tracking in living cells. The first part describes the protocol for imaging system calibration using spin-coated single qdots. The second part describes a protocol for targeted labeling of single qdots in living cells. However, site-specific labeling of target protein is a difficult task and it is not possible to provide a universal protocol for this purpose. Therefore, only the protocol derived in our laboratory is provided. The third part describes the protocols for the image processing and data analysis of single quantum dot tracking (see Fig. 1). In general, time-lapse images of live cells labeled with single qdots are first acquired using an optical microscope [e.g., epifluorescence, confocal, or total internal reflection fluorescence (TIRF) microscope]. The spot positions of single qdots are subsequently derived from the time-lapse imaging data.

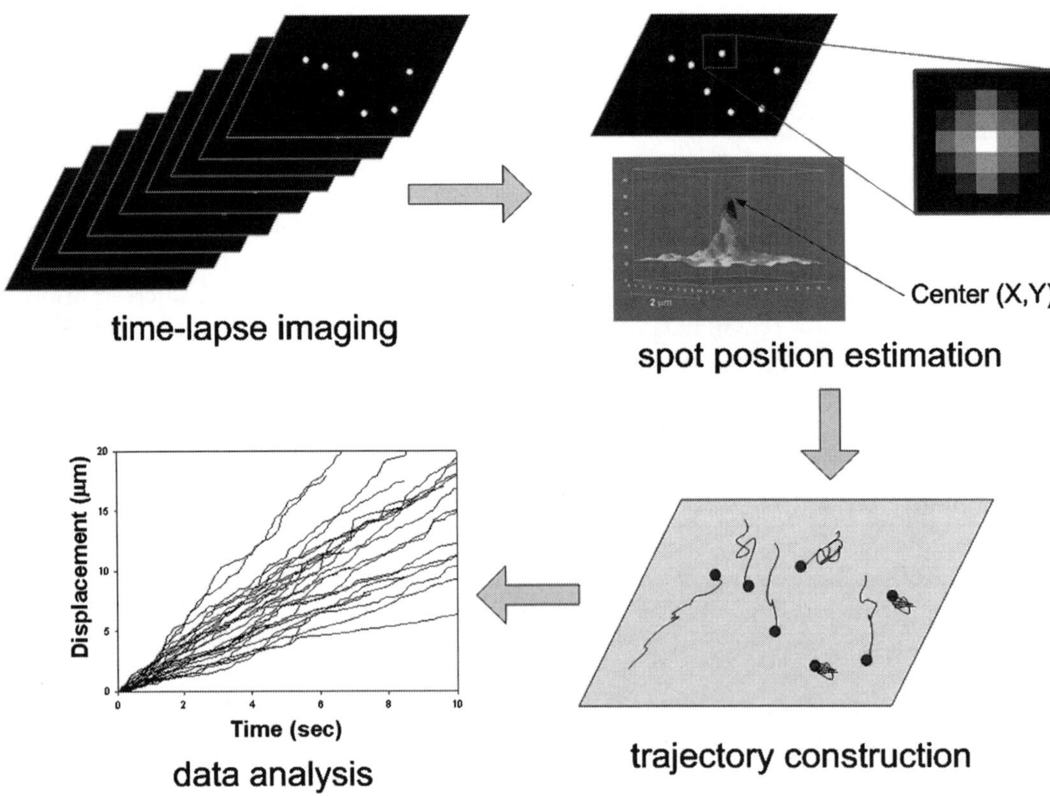

Fig. 1. Approach to single quantum dot tracking.

The latter step allows for the generation of single-molecule trajectories of the target proteins. After trajectory data analysis, diffusion dynamic behaviors of target proteins can be obtained.

2. Materials

2.1. Reagents

1. Biotinylated small molecule or antibody against an extracellular epitope.
2. Qdot streptavidin conjugate (1 µM, Invitrogen Corporation, Carlsbad, CA).
3. 35-mm cell culture dishes with coverglass bottom (MatTek Corporation, Ashland, MA).
4. HeLa Cells.
5. Dulbecco's Modified Eagle's Medium (DMEM) (GIBCO, Invitrogen Corporation, Carlsbad, CA).
6. Phenol red-free DMEM (GIBCO, Invitrogen Corporation, Carlsbad, CA).
7. Penicillin–streptomycin antibiotic mixture (100×) (GIBCO, Invitrogen Corporation, Carlsbad, CA).
8. Fetal bovine serum (FBS) (GIBCO, Invitrogen Corporation, Carlsbad, CA).
9. L-Glutamine (GIBCO, Invitrogen Corporation, Carlsbad, CA).

2.2. Equipment, Software, and Accessories

1. Fluorescent microscope (see Subheading 3.1).
2. Microscope mounted heating chamber.
3. Microscope image acquisition and analysis software (MetaMorph 7.6, Molecular Devices, Sunnyvale, CA).
4. Technical computing software for numerical analysis (Matlab R2008b, MathWorks, Natick, MA).
5. Spin coater.

3. Methods

The methods described below outline (1) the preparation of the nanoconjugate probes, (2) time-lapse imaging for single quantum dot tracking, and (3) trajectory data construction and analysis. The general principle provided in this protocol can be applied to different optical imaging systems including epifluorescence, confocal, and TIRF microscopes. Readers seeking a more detailed comparison of epifluorescence, confocal, and TIRF microscopy for single-molecule experiments are advised to consult Lang et al. (17).

3.1. Imaging System Calibration Using Spin-Coated Single Qdots

A general strategy to identify whether the optical system is able to detect individual qdots is to carry out time-lapse imaging of a very dilute qdot solution to avoid interparticle reactions. Individual qdots are characterized by their unique blinking properties (8, 18). As an example, the emission intensity of a single quantum dot is shown in Fig. 2, in which a single quantum dot blinks completely on and off during a time-lapse sequence of 60 s at a 20-Hz frame rate. When the fluorescence is produced by an aggregate structure consisting of several qdots, such blinking effects are completely canceled out. The protocol described below is based on a custom-built Zeiss Axiovert 200M inverted fluorescence microscope together with a charge-coupled device (CCD) camera (Cool-Snap$_{HQ2}$, Roper Scientific, Trenton, NJ). To track single qdots on the fly, the acquisition rate should be set at 10 Hz or higher. However, the imaging rate is usually limited by the frame readout time of the camera. This particular CCD is chosen due to its decent 60% quantum efficiency (QE) throughout the entire visible spectrum (450–650 nm) with a frame rate >20 at 512×512 pixels (see Photometrics CCD specification document at http://www.photomet.com). A more advanced back-illuminated electron multiplying CCD (EMCCD) with sub-millisecond temporal resolution will be a much better choice since the EMCCD can achieve single-photon sensitivity with exceptionally high frame rates at 10 MHz (19). Imaging should be performed with a high-resolution (63× or 100×) oil-immersion objective lens with a numerical aperture of 1.30 or greater.

1. Prepare a clean microscope glass slide coverslip (or 35-mm culture dish with coverslip in the bottom).

2. Add one drop (20 µL) of 1 nM Qdot® Streptavidin conjugate solution onto the coverslip.

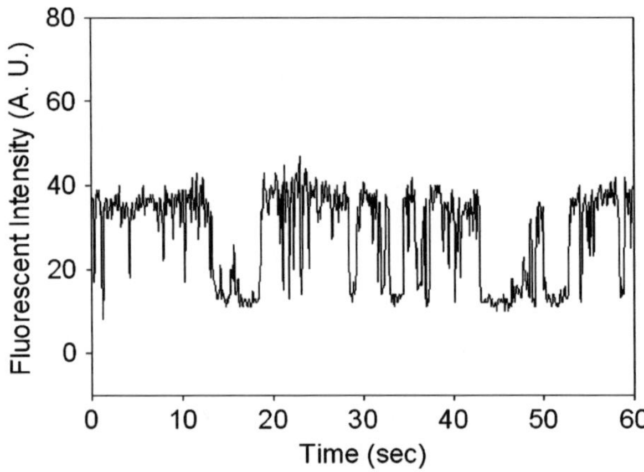

Fig. 2. Intensity of fluorescence as a function of time measured from a single quantum dot (data taken with a Zeiss LSM 5 Live Confocal Microscope at a 20-Hz frame rate). As displayed in the figure, the intensity trajectory of a single qdot displays two dominant states: an "on" state and an "off" state, termed blinking.

3. Spin cast the qdot solution on the coverslip for 30 s at 2,000 rpm or less (see Note 1).

4. Mount the coverslip on the microscope stage.

5. Acquire time-lapse images (10 Hz, 60 s).

In order to obtain a well-dispersed nanoparticle sample and minimize the number of nanoparticle aggregates, the volume and concentration of Qdot® Streptavidin conjugate (strep-qdot) solution in step 2 will need to be adjusted depending on the adhesion property between the coverslip and the qdots. In our experiments using Qdot® 655 Streptavidin conjugate solution with MatTek dishes, less than 2% of the detected fluorescent spots observed were aggregates.

3.2. Cell Culture

HeLa cells are cultured in DMEM supplemented with 10% FBS, 2 mM L-glutamine, 100 units/mL penicillin, and 100 mg/mL streptomycin and maintained at 37°C with 5% CO_2. For single quantum dot labeling studies, cells are plated at a density of 1×10^5 cells/mL in 35-mm coverslip-buttoned culture dish as viable at the single-cell level.

3.3. Single Quantum Dot Labeling and Real-Time Imaging in Live Cells

Single quantum dot labeling can be prepared through either a direct labeling (one step) procedure or an indirect (two step) protocol (see Note 2). In the direct labeling procedure, the target-specific biotinylated probe (small-molecule ligand or antibody) is premixed with the strep-qdots to make ligand-qdot nanoconjugates. Therefore, the cellular labeling strategy could be performed in one step in which the live cell sample is treated with a target-specific nanoconjugate prior to fluorescent imaging. However, this procedure might lead to multivalent quantum dot–protein interactions (see Note 3). Other covalent conjugation strategies are also available for the immobilization of different surface modified functional groups, such as carboxylic acids, on the surface of water-soluble quantum dots. For alternative conjugation methods, refer to Note 4.

In the two-step procedure, the cell sample is first incubated with biotinylated ligand or antibody. This incubation should yield the desired specific binding between the target protein and biotinylated ligand or antibody. After appropriate washing steps, strep-qdot is added as the fluorescent tag for the single-molecule imaging. For the labeling of a transfected cell sample, the standard protocol given below should be followed:

1. Prepare a 35-mm coverslip-buttoned culture dish with cells that have reached about 50% confluence (see Subheading 3.2).

2. Wash the cells gently three times with phenol red-free culture medium by repeatedly pipetting out.

3. Incubate cells with a biotinylated small molecule (0.5–2 μM) or antibody (1–10 μg/mL) in red-free DMEM for 20 min at 37°C.

4. Wash cells gently three times with phenol red-free culture medium by repeatedly pipetting out.

5. Incubate the cells with Qdot Streptavidin conjugate (0.5–1 nM) in phenol red-free culture medium for 5 min at 37°C (see Note 5).

6. Wash the cells at least three times with phenol red-free culture medium.

7. Place the culture dish on the microscope stage with mounted heating chamber.

8. Acquire time-lapse images at room temperature or 37°C (see Note 6).

3.4. Single Quantum Dot Localization and Trajectory Construction

After gathering a series of time-lapse images of live cells labeled with single qdots (see Fig. 3), trajectories of individual target proteins can be extracted by postimage processing and analysis. Data analysis can by complicated by the fact that many laboratories develop custom-made routines for single-molecule data analysis. The following methods describe not only basic principles but also rather available software routines which we believe are suitable for single qdot tracking.

3.4.1. Estimation of 2D Position with Subpixel Accuracy

In optical microscopy, the observed intensity distribution from a single point source is described by the point spread function (PSF). The theoretical PSF can be calculated:

$$\mathrm{PSF}(r) = C \times \left[\frac{2J_1\left((2\pi/\lambda)NAr\right)}{(2\pi/\lambda)NAr} \right]^2, \tag{1}$$

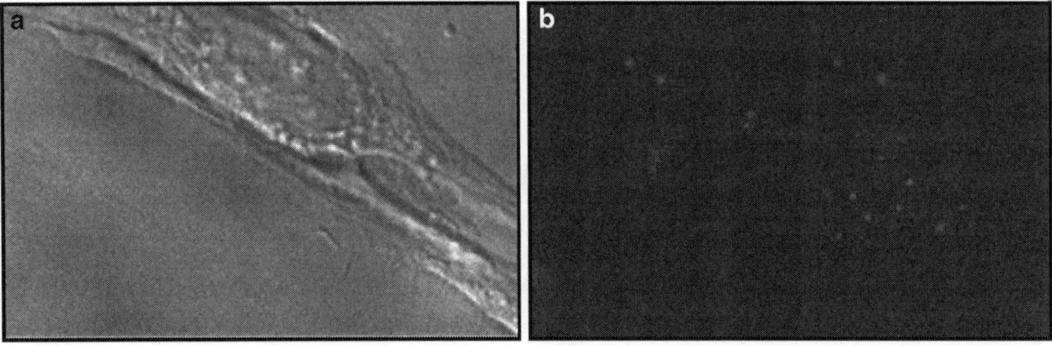

Fig. 3. Example images of membrane proteins labeled with single qdots (**a**: bright field image, **b**: fluorescence image). Note that the blinking phenomenon in the time-lapse images should be used as a signature to confirm if the individual spots represent single molecules.

where r is the distance from the origin, NA is the numerical aperture, C defines the intensity value at $r=0$, J_1 is the first-order Bessel function, and λ is the wavelength of light. As a subwavelength nanostructure, fluorescent images of single qdots are known to fulfill the PSF and have been fit well with 2D Gaussians (20). In order to calculate a subpixel estimate of single qdot position, the general method is to consider the intensity distribution of a single qdot as an elliptical Gaussian intensity function and calculate the local maximum intensity with Gaussian interpolation:

$$x_0 = \frac{\ln(z_{x-1}) - \ln(z_{x+1})}{2[\ln(z_{x+1}) - 2\ln(z_x) + \ln(z_{x-1})]}, \qquad (2)$$

where z_x is the local maximum intensity, and z_{x-1} and z_{x+1} are the two neighbors. The maximum of the interpolated curve is located at x_0 with respect to the index X of the maximum sample (see Fig. 4).

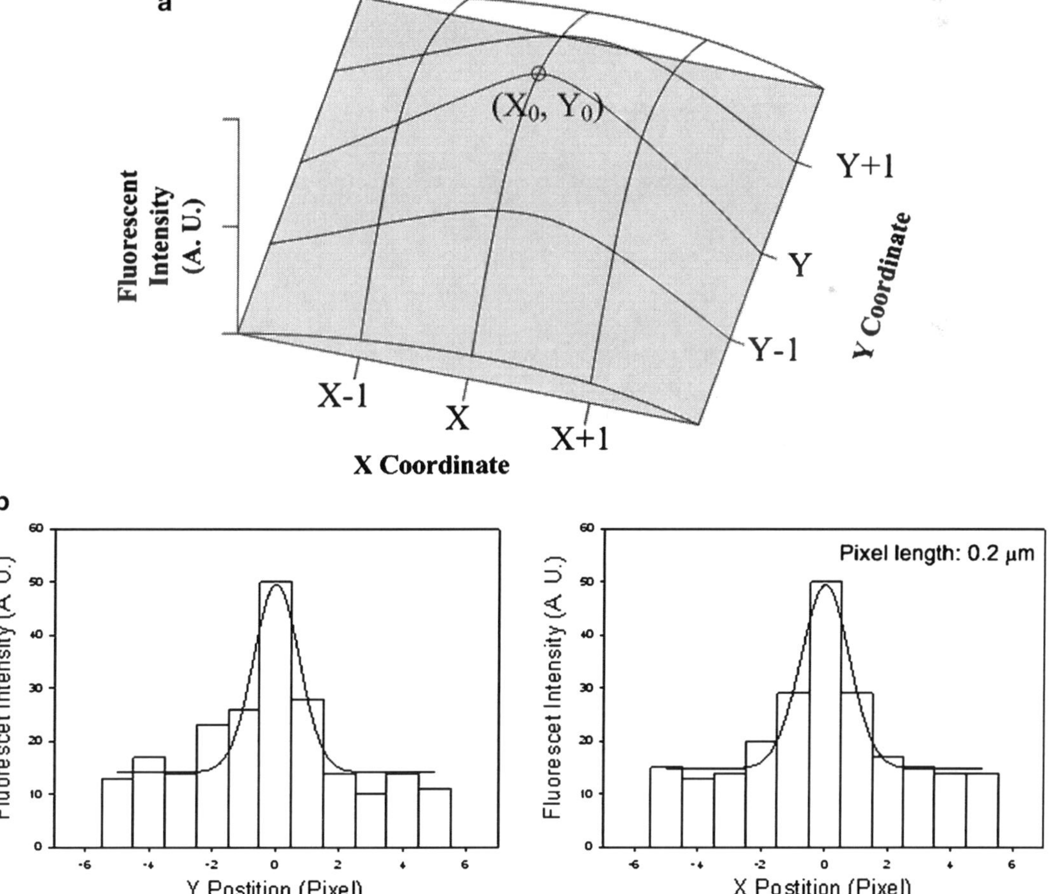

Fig. 4. (a) Schematic of 2D Gaussian regression of a single quantum dot fluorescent image. (b) Example of a direct Gaussian fit along the x- and y-axis to the intensity distribution of a single quantum dot fluorescent image. Note that a much more accurate localization in the center can be obtained by matching the Gaussian function to fit the experimental intensity data.

A tracking routine based on the least-square 2D Gaussian regression for subpixel position estimation in single-particle tracking (SPT) was initially developed by Crocker and Grie (21). The source code and an excellent introduction can be found at http://www.physics.emory.edu/~weeks/idl/. Matlab versions of these routines have been made by Blair and Dufresne and are available at http://physics.georgetown.edu/matlab/ as detailed, step-by-step tutorials. These routines have been successfully used to calculate the diffusion dynamics of single qdots at silica surfaces in static and flow conditions (22).

3.4.2. Trajectory Construction and 2D Displacement Generation

After x and y coordinates of the selected single qdots were determined in each frame, step displacements of each quantum dot, which represent the time course of the corresponding moment are determined by the following formula (see Fig. 5):

$$\Delta d_n = \left\{ [x(n+1) - x(n)]^2 + [y(n+1) - y(n)]^2 \right\}^{(1/2)}, \qquad (3)$$

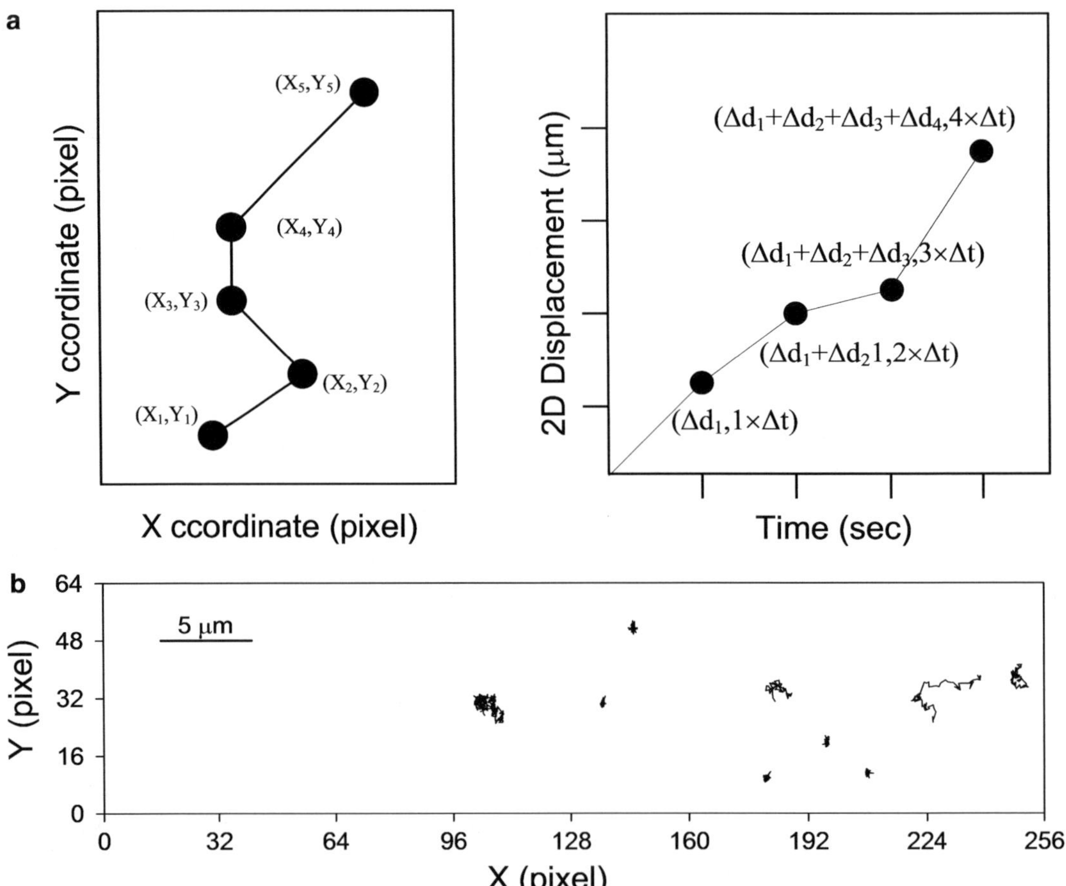

Fig. 5. (**a**) Schematic of XY trajectory and 2D step displacement of single quantum dot tracking. (**b**) An example of the XY trajectories of individual transmembrane proteins labeled with single qdots in living cells.

where $x(n)$ and $y(n)$ denote the position in frame (n), Δd_n indicates that a single step takes place during a single lag time Δt from time point $\Delta t \times (n)$ to $\Delta t \times (n+1)$.

3.5. Diffusion Behavior Analysis

In addition to the standard XY trajectory and 2D displacement plots, mean-square displacement (MSD) for individual trajectories should also be used for probing the single protein diffusion dynamics in a plasma membrane. MSD can be obtained according to the formula given below:

$$MSD(n\delta t) = (N-1-n)^{-1} \sum_{j=1}^{N-1-n} \left\{ \begin{array}{l} [x(j\Delta t + n\Delta t) - x(j\Delta t)]^2 \\ -[y(j\Delta t + n\Delta t) - y(j\Delta t)]^2 \end{array} \right\}, \quad (4)$$

where $x(t)$ and $y(t)$ are the position of particle at time t. $x(j\Delta t + n\Delta t)$ and $y(j\Delta t + n\Delta t)$ are the position following a time interval $n\Delta t$, N is total number of frames, Δt is time-resolution, n is the number of time intervals, and j is a positive integer.

Once the MSD values for individual trajectories were obtained, the diffusion coefficient could be calculated through the best fit of the MSD with the theoretical model listed below (see Fig. 6). For an excellent overview with details of the theoretical models of membrane diffusion dynamics as well as the principle of SPT, see Saxton and Jacobson (23).

$$MSD(t) = 4D\,\Delta t \quad \text{Normal diffusion,} \quad (5)$$

$$MSD(t) = 4D\,\Delta t^{\alpha} \quad \text{Anomalous diffusion,} \quad (6)$$

$$MSD(t) = 4D\,\Delta t + (v\Delta t)^2 \quad \text{Directed motion with diffusion,} \quad (7)$$

$$MSD(t) = \langle r_c^2 \rangle \left[1 - A_1 \exp\left(-\frac{4A_2 D\Delta t}{\langle r_c^2 \rangle} \right) \right] \quad \text{Corralled motion,} \quad (8)$$

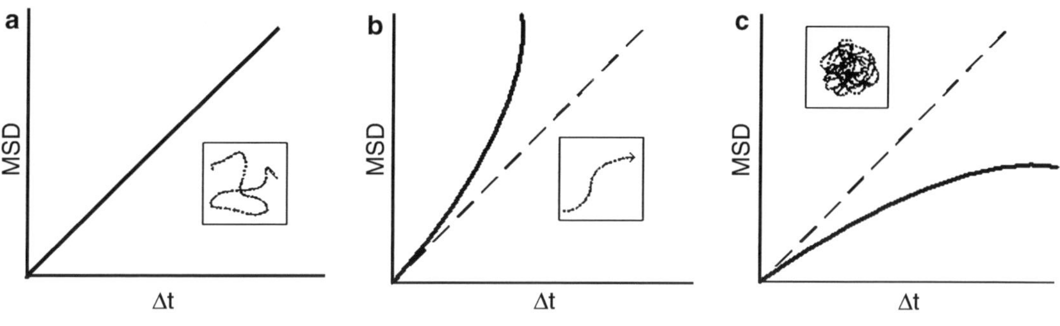

Fig. 6. Different types of diffusion behavior and their corresponding mean-square displacement (MSD) curve for (**a**) normal diffusion, (**b**) direct motion, and (**c**) corralled motion. The *insets* to the plots schematically show expected 2D trajectories. Note that the diffusion behavior and MSD curve of anomalous diffusion are expected between normal diffusion and corralled motion.

where D denotes the diffusion coefficient, Δt is time-resolution, α is the anomalous diffusion exponent, v is velocity, $\langle r_c^2 \rangle$ indicates the corral size, and A_1 and A_2 are constants determined by the corral geometry.

Due to its simplicity and ability to produce rapid results, the random walk model (normal diffusion, Eq. 5) is commonly used to estimate the diffusion coefficient in single quantum dot tracking experiments (11). However, in biological interpretation, anomalous diffusion (Eq. 6) was reported as the dominant model of diffusion dynamics in the plasma membrane (24, 25). The standard method to obtain the diffusion coefficient from the anomalous diffusion model is through a linear fit, as indicated below, since this fit converges more consistently than the variable power-law fit (25):

$$\log[MSD(t)] = \alpha \log(\Delta t) + \log(4D) \qquad (9)$$

4. Notes

1. The spinning force is not critical in this step. In most cases, a rotational speed as low as 500 rpm is sufficient to achieve a uniformly spread.

2. It must be emphasized that the key parameter for preparation of the functionalized qdot probe for single-molecule tracking is to use a target-specific small molecule or antibody with low nonspecific binding. All labeling protocols should employ three negative control experiments to demonstrate binding specificity prior to the single qdot tracking experiments: (1) labeling of a parental cell lacking target protein expression, (2) labeling of a target protein-expressing cell preblocked with an inhibitor or antibody specific to the intended target, and (3) incubation of a target protein-expressing cell with quantum dots only (without a target-specific small molecule or antibody). These negative controls will provide an independent assessment of whether the observed single qdot labeling was the result of nonspecific interactions arising at the cell membrane–qdot interface, or the result of specific interactions with the intended target.

3. For one-step labeling, it is important to remember that each strep-qdot is typically conjugated with 5–10 streptavidins (Qdot® Streptavidin Conjugates User Manual, Invitrogen, CA). Using a nanoconjugate probe prepared from the one-step procedure may result in more than one protein binding to a single qdot, which leads to multivalent quantum dot–protein interactions where the movement of single quantum dot would not be able to reflect individual membrane protein

trafficking. In this case, the concentration of the ligand in the reaction mixture is important in determining the final ligand/ quantum dot ratio of the coupling. Incubation of biotinylated ligand with strep-qdot at a 1:1 molar ratio was typically used. Recently, Howarth et al. published a modified protocol which provide a much better solution in the case where monovalent instead of multivalent streptavidin-conjugated qdots were prepared (26).

4. In contrast to qdot probes prepared by using Qdot® Streptavidin Conjugate, Qdot® Antibody Conjugation Kits, and ITK™ Carboxyl Quantum Dots can also be used for direct conjugation. The experimental procedures used to perform direct conjugation and purification are described in detail elsewhere (27, 28). Other general protocols are also available on Invitrogen's web site (http://www.invitrogen.com).

5. An average of 10–20 qdot-labeled proteins/cell is expected to be presented.

6. The experiment is carried out at room temperature to settle lower imaging rate with decreasing particle velocity, however, imaging at 37°C is closer to the physiological condition.

Acknowledgments

The authors wish to thank their colleagues in the group, especially Dr. James McBride, Dr. Michael Schreuder, Albert Dukes, and Oleg Kovtun, for all the helpful discussions and suggestions. We thank Dr. David Piston for helpful advice with single quantum dot tracking experimental setup. This work was supported by grants from National Institutes of Health (R01EB003778).

References

1. Bruchez, M., Moronne, M., Gin, P., Weiss, S., and Alivisatos, A. P. (1998) Semiconductor nanocrystals as fluorescent biological labels. *Science* **281**, 2013–2016.

2. Chan, W. C. and Nie, S. (1998) Quantum dot bioconjugates for ultrasensitive nonisotopic detection. *Science* **281**, 2016–2018.

3. Alivisatos, P. (2004) The use of nanocrystals in biological detection. *Nat. Biotechnol.* **22**, 47–52.

4. Kim, S., Lim, Y. T., Soltesz, E. G., De Grand, A. M., Lee, J., Nakayama, A., et al. (2004) Near-infrared fluorescent type II quantum dots for sentinel lymph node mapping. *Nat. Biotechnol.* **22**, 93–97.

5. Rosenthal, S. J., Tomlinson, I., Adkins, E. M., Schroeter, S., Adams, S., Swafford, L., et al. (2002) Targeting cell surface receptors with ligand-conjugated nanocrystals. *J. Am. Chem. Soc.* **124**, 4586–4594.

6. Liu, W., Choi, H. S., Zimmer, J. P., Tanaka, E., Frangioni, J. V., and Bawendi, M. (2007) Compact cysteine-coated CdSe(ZnCdS) quantum dots for *in vivo* applications. *J. Am. Chem. Soc.* **129**, 14530–14531.

7. McBride, J., Treadway, J., Feldman, L. C., Pennycook, S. J., and Rosenthal, S. J. (2006) Structural basis for near unity quantum yield core/shell nanostructures. *Nano Lett.* **6**, 1496–1501.

8. Nirmal, M., Dabbousi, B. O., Bawendi, M. G., Macklin, J. J., Trautman, J. K., Harris, T. D., et al. (1996) Fluorescence intermittency in single cadmium selenide nanocrystals. *Nature* **383**, 802–804.

9. Scholl, B., Liu, H. Y., Long, B. R., McCarty, O. J., O'Hare, T., Druker, B. J., et al. (2009) Single particle quantum dot imaging achieves ultrasensitive detection capabilities for Western immunoblot analysis. *ACS Nano* **3**, 1318–1328.

10. Temirov, J. P., Bradbury, A. R., and Werner, J. H. (2008) Measuring an antibody affinity distribution molecule by molecule. *Anal. Chem.* **80**, 8642–8648.

11. Dahan, M., Levi, S., Luccardini, C., Rostaing, P., Riveau, B., and Triller, A. (2003) Diffusion dynamics of glycine receptors revealed by single-quantum dot tracking. *Science* **302**, 442–445.

12. Cui, B., Wu, C., Chen, L., Ramirez, A., Bearer, E. L., Li, W.-P., et al. (2007) One at a time, live tracking of NGF axonal transport using quantum dots. *Proc. Natl Acad. Sci. USA.* **104**, 13666–13671.

13. Rajan, S. S., Liu, H. Y., and Vu, T. Q. (2008) Ligand-bound quantum dot probes for studying the molecular scale dynamics of receptor endocytic trafficking in live cells. *ACS Nano* **2**, 1153–1166.

14. Courty, S., Luccardini, C., Bellaiche, Y., Cappello, G., and Dahan, M. (2006) Tracking individual kinesin motors in living cells using single quantum-dot imaging. *Nano Lett.* **6**, 1491–1495.

15. Bouzigues, C., Morel, M., Triller, A., and Dahan, M. (2007) Asymmetric redistribution of GABA receptors during GABA gradient sensing by nerve growth cones analyzed by single quantum dot imaging. *Proc. Natl Acad. Sci. USA.* **104**, 11251–11256.

16. Tada, H., Higuchi, H., Wanatabe, T. M., and Ohuchi, N. (2007) *In vivo* real-time tracking of single quantum dots conjugated with monoclonal anti-HER2 antibody in tumors of mice. *Cancer Res.* **67**, 1138–1144.

17. Lang, E., Baier, J., and Köhler, J. (2006) Epifluorescence, confocal and total internal reflection microscopy for single-molecule experiments: a quantitative comparison. *J. Microsc.* **222**, 118–123.

18. Yao, J., Larson, D. R., Vishwasrao, H. D., Zipfel, W. R., and Webb, W. W. (2005) Blinking and nonradiant dark fraction of water-soluble quantum dots in aqueous solution. *Proc. Natl Acad. Sci. U.S.A.* **102**, 14284–14289.

19. Zhang, L., Neves, L., Lundeen, J. S., and Walmsley, I. A. (2009) A characterization of the single-photon sensitivity of an electron multiplying charge-coupled device. *J. Phys. B At. Mol. Opt. Phys.* **42**, 114011.

20. Michalet, X., Fabien, P., Thilo, D. L., Maxime, D., Marcel, P. B., Alivisatos, A. P., et al. (2001) Properties of fluorescent semiconductor nanocrystals and their application to biological labeling. *Single Mol.* **2**, 261–276.

21. Crocker, J. and Grier, D. (1996) Methods of digital video microscopy for colloidal studies. *J. Colloid Interface Sci.* **179**, 298–310.

22. Rife, J. C., Long, J. P., Wilkinson, J., and Whitman, L. J. (2009) Particle tracking single protein-functionalized quantum dot diffusion and binding at silica surfaces. *Langmuir* **25**, 3509–3518.

23. Saxton, M. J. and Jacobson, K. (1997) Single-particle tracking: applications to membrane dynamics. *Annu. Rev. Biophys. Biomol. Struct.* **26**, 373–399.

24. Saxton, M. J. (2008) A biological interpretation of transient anomalous subdiffusion. II. Reaction kinetics. *Biophys. J.* **94**, 760–771.

25. Martin, D. S., Forstner, M. B., and Kas, J. A. (2002) Apparent subdiffusion inherent to single particle tracking. *Biophys. J.* **83**, 2109–2117.

26. Liu, W., Howarth, M., Greytak, A. B., Zheng, Y., Nocera, D. G., Ting, A. Y., et al. (2008) Compact biocompatible quantum dots functionalized for cellular imaging. *J. Am. Chem. Soc.* **130**, 1274–1284.

27. Warnement, M. R., Faley, S. L., Wikswo, J. P., and Rosenthal, S. J. (2006) Quantum dot probes for monitoring dynamic cellular response: reporters of T cell activation. *IEEE Trans. Nanobioscience* **5**, 268–272.

28. Warnement, M. R., Tomlinson, I. D., Chang, J. C., Schreuder, M. A., Luckabaugh, C. M., and Rosenthal, S. J. (2008) Controlling the reactivity of ampiphilic quantum dots in biological assays through hydrophobic assembly of custom PEG derivatives. *Bioconjug. Chem.* **19**, 1404–1413.

Chapter 5

Titanium Dioxide Nanoparticles in Advanced Imaging and Nanotherapeutics

Tijana Rajh, Nada M. Dimitrijevic, and Elena A. Rozhkova

Abstract

Semiconductor photocatalysis using nanoparticulate TiO_2 has proven to be a promising technology for use in catalytic reactions, in the cleanup of water contaminated with hazardous industrial by-products, and in nanocrystalline solar cells as a photoactive material. Metal oxide semiconductor colloids are of considerable interest because of their photocatalytic properties. The coordination sphere of the surface metal atoms is incomplete and thus traps light-induced charges, but also exhibits high affinity for oxygen-containing ligands and gives the opportunity for chemical modification. We use enediol linkers, such as dopamine and its analogs, to bridge the semiconductors to biomolecules such as DNA or proteins. Nanobio hybrids that combine the physical robustness and chemical reactivity of nanoscale metal oxides with the molecular recognition and selectivity of biomolecules were developed. Control of chemical processes within living cells was achieved using TiO_2 nanocomposites in order to develop new tools for advanced nanotherapeutics. Here, we describe general experimental approaches for synthesis and characterization of high crystallinity, water soluble 5 nm TiO_2 particles and their nanobio composites, methods of cellular sample preparation for advanced Synchrotron-based imaging of nanoparticles in single cell X-ray fluorescence, and a detailed experimental setup for application of the high-performance TiO_2-based nanobio photocatalyst for targeted lysis of cancerous or other disordered cells.

Key words: TiO_2, Nanoparticles, Surface reconstruction, Photocatalysis, Charge transfer complex, Hybrid composites, DNA, Antibody, Targeted cancer therapy, Synchrotron X-ray fluorescence

1. Introduction

The impact of nanoscience and nanotechnology on cell manipulation and actuation is critically dependent on the creation of new classes of functionally and physically integrated hybrid materials that incorporate nanoparticles and biologically active molecules. These hybrid bioinorganic composites integrate both inorganic materials and biological entities via multivalent lock-and-key

Sarah J. Hurst (ed.), *Biomedical Nanotechnology: Methods and Protocols*, Methods in Molecular Biology, vol. 726,
DOI 10.1007/978-1-61779-052-2_5, © Springer Science+Business Media, LLC 2011

interactions, offering opportunities to impact diverse applications ranging from quantum computation, energy transduction and site-selective catalysis to advanced nanomedicine. Hybrid materials that combine collective properties of crystalline materials and localized properties of biomolecules are of special interest because they present basic functional units capable of carrying out site-selective redox chemistry within intracellular machinery.

In order to develop new tools for biomedicine and biotechnology that carry out redox chemistry, we focus on these special classes of functional nanomaterials that mimic the exquisite control over energy and electron transfer that occurs in natural energy transducing processes (1). Nanobio hybrids that combine the physical robustness and chemical reactivity of nanoscale metal oxides with the molecular recognition and selectivity of biomolecules were developed. DNA and monoclonal antibodies were utilized to direct TiO_2 nanoparticles to the specific cells with target molecules. Photo-induced charge separation was then employed to control and manipulate processes within the cells and to alter their functioning. These hybrid TiO_2–DNA nanocomposites form a basis for the development of novel artificial restriction enzymes and single nucleotide polymorphism (SNP) genotyping with high specificity in vitro or in vivo (2, 3).

Recently, we developed polychromatic visible-light inducible nanobio hybrid systems based on 5 nm TiO_2 nanocrystals covalently tethered to a biological vehicle (4) capable of selective recognition of cancer-associated antigens (see Figs. 1–3). Similar to "classic" photodynamic therapy (PDT), our approach includes three main components: light, oxygen, and a photoreactive material. The hybrid semiconductor particles absorb energy from light, which is then transferred to molecular oxygen, producing cytotoxic reactive oxygen species (ROS) (4, 5). The advantages of nanoscale photosensitization compared to "classical" PDT are the result of a synergistic combination of the advanced physical properties of inorganic materials and the targeting abilities of biomolecules and the multiple functions of drugs and imaging payloads in one ideal therapeutic system. Furthermore, nanoparticles may overcome biological barriers.

Approaches for tethering biomolecules to the surface of TiO_2 particles utilize the ability of oxygen-containing functional groups, such as carboxy-, hydroxyl-, and phosphate, to bind to the surface of nanoparticles. Our strategy to construct bio-TiO_2 hybrids is based on dihydroxybenzenes, for example, dopamine (DA) or its natural metabolite 3,4-dihydroxyphenylacetic acid (DOPAC), as linkers (see Fig. 1) (6–8). As a result of the presence of two OH– groups in the *ortho* position, the catecholate group forms a strong bidentate complex with the coordinatively unsaturated Ti atoms at the surface of nanoparticles (7). Furthermore, it has been

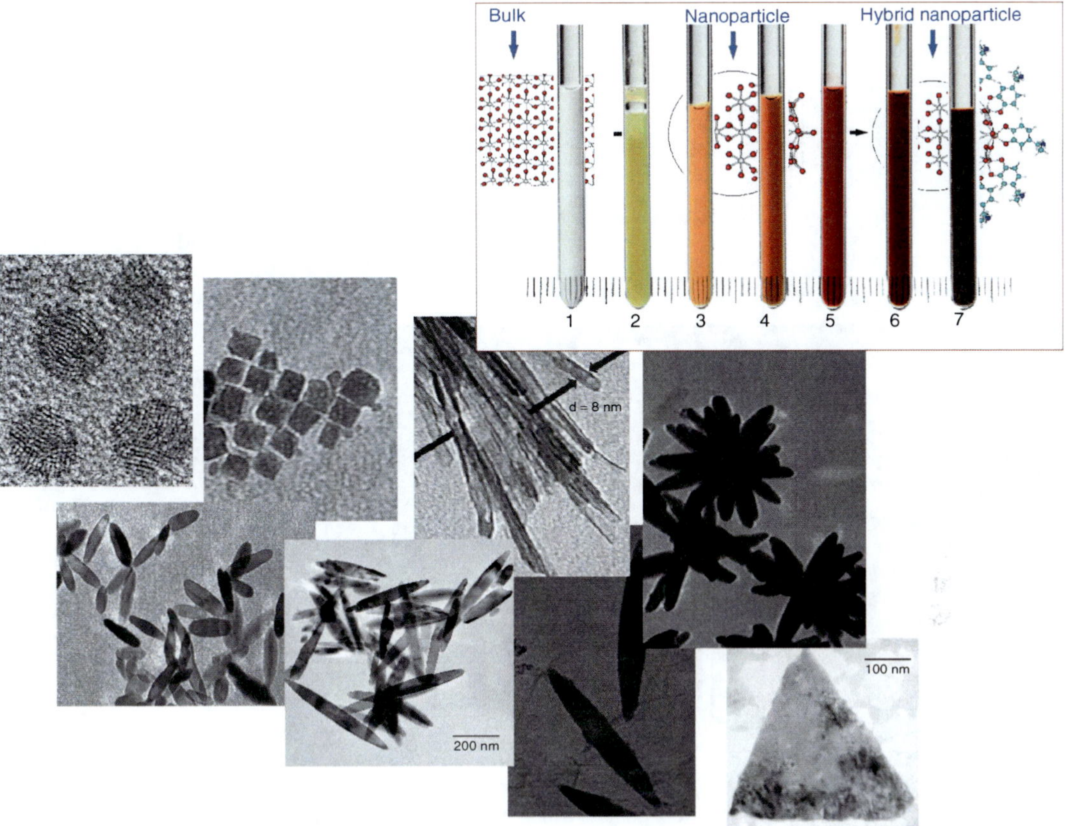

Fig. 1. TiO$_2$ nanoscale materials. *Top*: Surface-modified 45-Å TiO$_2$ nanoparticles with different bidentate ligands: (1) bare TiO$_2$, (2) salicylic acid, (3) dihydroxycy-clobutenedione, (4) vitamin C, (5) alizarin, (6) dopamine, and (7) *tert*-butyl catechol. Reprinted with permission from ref. 6. *Bottom*: TEM images of a library of TiO$_2$ nanoscale materials. Modified with permission from Dimitrijevic, N. M., Saponjic, Z. V., Rabatic, B. M., Poluektov, O. G., and Rajh, T. (2007) Effect of size and shape of nanocrystalline TiO$_2$ on photogenerated charges. An EPR study. *J. Phys. Chem. C.* **111**, 14597–14601.

shown that when DNA or proteins are covalently bound to DA, DA acts as a conductive bridge between the biomolecules and TiO$_2$ nanocrystals, allowing transport of photogenerated holes to the biomolecules (9, 10). Chemisorption of DA or DOPAC serves two important purposes: first, it "heals" the semiconductor surface, enabling absorption in a *visible* part of solar spectrum (see Fig. 1), and, second, DA or DOPAC modify the particle surface with amino- or carboxylic functional groups, which are useful for further covalent tethering to a biomolecule (see Fig. 2). Readily available water-soluble reagents for carbodiimide coupling allow tethering of biological molecules to the particle surface with retained photo-physical and biological recognition properties in the resulting hybrid system.

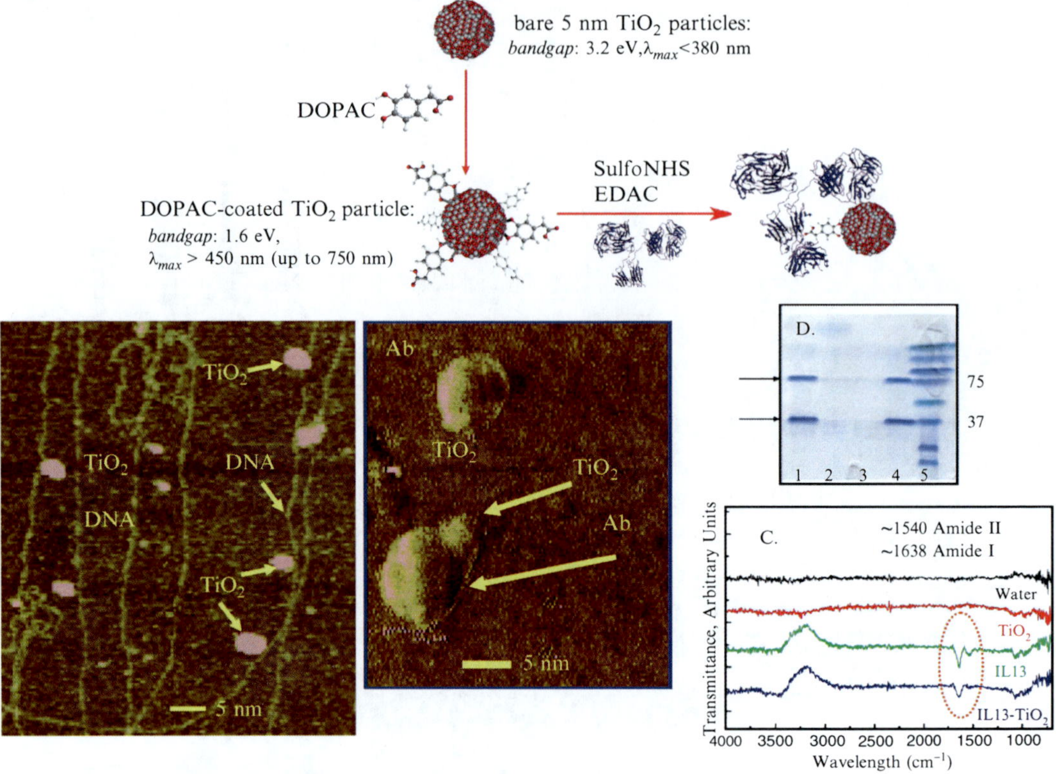

Fig. 2. Synthesis of TiO₂ nanobio hybrids. *Top*: General scheme. *Bottom left and middle*: Examples of TiO₂-biomolecule hybrids (AFM images). *Bottom right*: Complete conversion of a free antibody in the course of the carbodiimide coupling to the TiO₂–DOPAC is confirmed by SDS-PAGE (denaturing conditions) analysis of the final conjugate (4), reaction mixture supernatant (2) and all washing solutions (3). (1) and (5) are free antibody and Precision Plus Protein Standards (BioRad), respectively. FTIR spectrum of the conjugate (*blue line*) contains amide bands typical of immunoglobulins. Modified with permission from ref. 4.

The ability to manipulate molecules and materials with nanometer resolution is crucial in the field of nanotechnology. In order to understand and control complex behavior of nanoscale materials within biological system, such as cellular machinery, advanced imaging techniques are required. Thus, interactions of 5 nm TiO₂ nanoparticles functionalized with DNA or an antibody with cellular compartments (nucleus and mitochondria) or membrane proteins were studied by X-ray fluorescence microscopy (XFM) using the Advanced Photon Source (4, 11). Third-generation synchrotrons with spatially coherent high-brilliance X-rays allow elemental mapping of biological specimens in near-native environments with submicrometer to dozens of nanometers spatial resolution, which provides valuable complementary information to light microscopy. An exposure of a sample to hard X-rays results in the emission of characteristic "secondary" (or fluorescent) X-rays or distinctive "signatures" of each element comprising the sample.

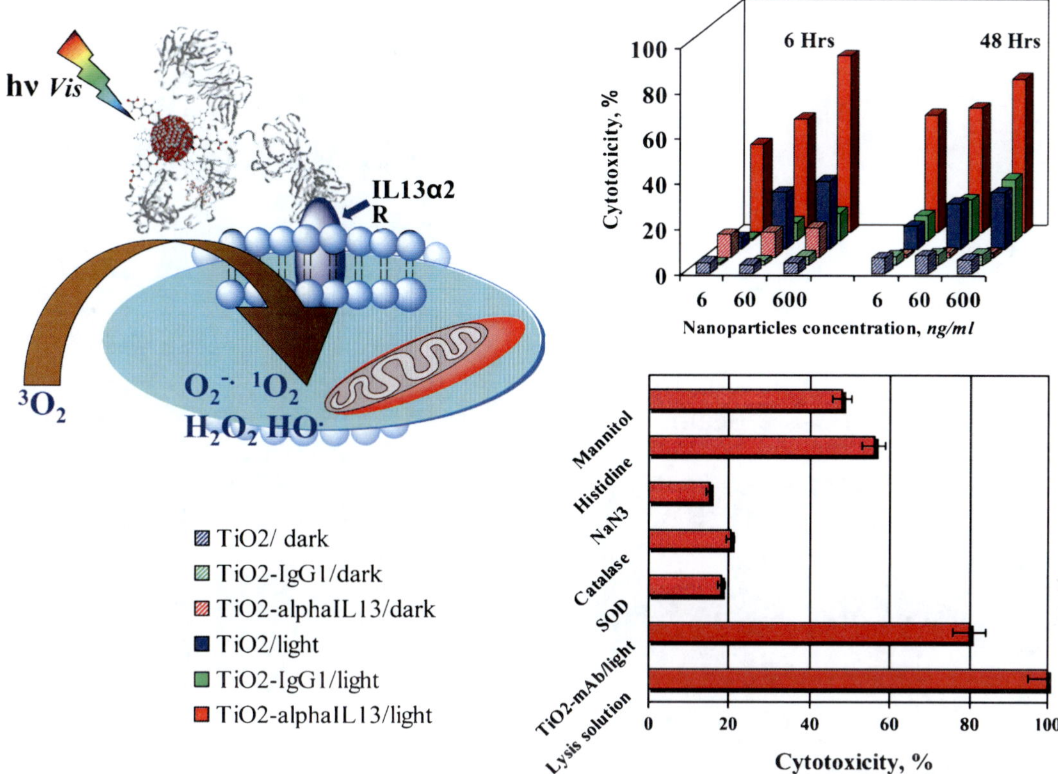

Fig. 3. General concept of the TiO_2-based photocatalytic cell lysis. *Left*: Nanobiocomposites consisted of 5 nm TiO_2 and IL13R-recognizing antibody linked via DOPAC linker to recognize and bind exclusively to surface IL13R. Visible light photo-excitation of the nanobio hybrid in an aqueous solution results in the formation of various ROS. ROS, mainly superoxide, cause cell membrane damage, permeability changes, and cell death. *Right*: Phototoxicity of the TiO_2–mAb toward A172 glioblastoma cells (*top*) and same cytotoxicity in the presence of various ROS quenchers (*bottom*). Isotype-matched negative control antibody immunoglobulin IgG1, either conjugated or unconjugated, did not recognize isolated or cellular IL13a2R and did not show photo-induced toxicity. Modified with permission from ref. 4.

XFM allows imaging of tiny nanoparticles composed of nonbiogenic elements including titanium in single cell and, therefore, these can be used to label and map cellular structures of interest (see Fig. 4). Furthermore, this method makes possible imaging of the redistribution of cellular endogenous trace elements (e.g., iron, manganese, calcium) as response to external stimulus, for example, intracellular redox elements relocalization during photo-induced apoptosis.

Semiconductor TiO_2 is well known as a photocatalyst in the degradation of organic substrates and the deactivation of microorganisms and even viruses ((4) and references therein). Under ultraviolet light (UV) excitation, TiO_2 nanoparticles of various sizes, morphologies, and solubility have been reported to exhibit cytotoxicity toward some tumor cells. Owing to the significantly

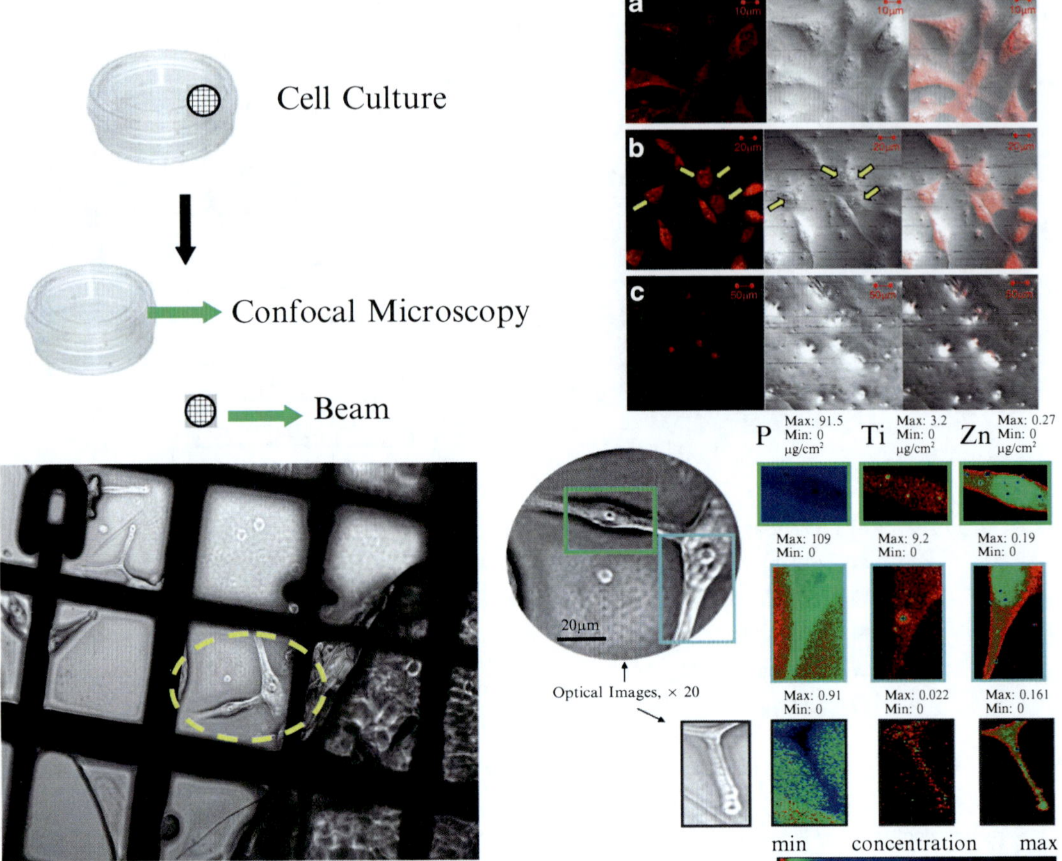

Fig. 4. Advanced X-ray imaging of the TiO_2 nanobio hybrid within a single cell. *Top left*: Cells are grown in Petri dishes with attached carbon/formvar-coated gold-finder grids. *Bottom left*: Optical micrograph of cells attached to the grid. *Top right*: Laser confocal microscopy images of cell undergoing various stages of photo-induced apoptosis. Control cells with TiO_2–mAb, but no light applied (**a**), and after 30 min (**b**) or 90 min (**c**) following light exposure. *Right bottom*: X-ray fluorescence imaging of the TiO_2–mAb binding to the single glioblastoma cell (representative images of the high antigen overexpressing A172 line). Elemental distribution of biogenic phosphorus and zinc are used to sketch cells and nucleus. The intensity of the elemental images was displayed using a prism color table in logarithm scale, which is shown in the *bottom right*. The maximum and minimum threshold values in micrograms per square centimeter are given above each frame. Scans were obtained by using 10.0-keV incident energy with dwell times of 1 s per pixel and 1-μm steps through the sample. Modified with permission from ref. 4.

improved photoreactivity the catecholate-modified TiO_2 nanoparticles represent promising materials for nanotherapeutics as they can be induced by the *visible part of the solar spectrum* closely approaching the optimal spectral window for biological tissue penetration of 800 nm (9, 12). We recently demonstrated successful applicability of TiO_2–DA (DOPAC)-based nanobio composites in targeted photo-induced lysis of brain cancer or pathogenic T-cells associated with psoriasis (4, 13).

Here, we describe general experimental approaches for synthesis and characterization of high crystallinity, water soluble

5 nm TiO$_2$ (14) particles and their nanobio composites for functional integration with biomolecules (4, 14). Methods of cellular sample preparation for advanced Synchrotron-based imaging of nanoparticles in single cell X-ray fluorescence are also described (4, 15, 16). Finally, a detailed experimental setup for application of the high-performance TiO$_2$-based nanobio photocatalyst for targeted lysis of cancerous or other disordered cells is presented (4, 13).

2. Materials

All chemicals of the highest grade available were obtained from Sigma-Aldrich and used without further purification. Milli-Q water was used in all experiments (see Note 1).

2.1. Cell Culture

The human malignant glioma cell lines A172MG and U87MG (American Type Culture Collection, Manassas, VA) and normal human astrocytes (Cambrex-Clonetics, East Rutherford, NJ) are routinely grown in Dulbecco's Modified Eagle's Medium (DMEM) with 4.5 g/L glucose and L-glutamine, supplemented with 10% fetal bovine serum (FBS; Mediatech, Herndon, VA) in a humidified atmosphere with 5% CO$_2$ at 37°C.

2.2. Synthesis of High Crystallinity "Bare" 5 nm TiO$_2$ Particles

1. Milli-Q water, out-gas with argon or nitrogen to remove oxygen.
2. Titanium tetrachloride.
3. 2,000 MW Cutoff dialysis cassettes (Pierce, Rockford, IL).
4. 0.2 M LiOH.
5. Isopropyl glycidyl ether (also called 1,2-epoxy-3-isopropoxypropane).

2.3. TiO$_2$-Biomolecule Synthesis and Characterization

1. 10 mM Sodium phosphate buffer (PBS), pH = 6.2.
2. Monoclonal anti-human-IL13Ralpha2 or Mouse IgG1 Isotype Control (R&D Systems, Minneapolis, MN). Antibodies (500 µg) are reconstituted in 50 µL sterile PBS pH = 7.4 right before conjugation to the pre-activated nanoparticles to final concentration 10 mg/mL.
3. 11 mM DOPAC in 10 mM PBS, pH = 6.2.
4. 44 mM N-Hydroxysulfosuccinimide (Sulfo-NHS) (Pierce, Rockford, IL) in 10 mM PBS, pH = 6.2.
5. 50 mM 1-Ethyl-3-[3-dimethylaminopropyl]carbodiimide hydrochloride (EDAC) (Sigma-Aldrich, St. Louis, MO) in 10 mM PBS, pH = 6.2.
6. 2-Mercaptoethanol.

7. 2,000 MW Cutoff dialysis cassettes (Pierce, Rockford, IL).

8. 10 mM PBS, pH = 7.4.

9. 25 mM Glycine.

10. 4–20% Tris–HCl precast gel and prestained molecular weight markers "Kaleidoscope markers" (Bio-Rad, Hercules, CA).

2.4. Photo-Induced Cell Lysis

1. 10 mM PBS, pH = 7.4.

2. Petri dishes (12-well cell culture cluster, tissue culture treated, Costar®).

3. Cermax® PE300BFM 300 W Xenon Lamp, UV filter (Perkin Elmer Optoelectronics). An IR filter was custom-made in the Argonne glass blowing shop, but any cylinder of sufficient size filled with water can be used.

4. Standard LDH test: the CytoTox-ONE™ Homogeneous Membrane Integrity Assay (Promega, Madison, WI).

5. Stock solutions of radical scavengers: 10,000 U/mL catalase, 1,000 U/mL superoxide dismutase (SOD), 5 M mannitol, 200 mM NaN_3, and 1 M histidine (His).

2.5. Cell Sample Preparation for X-Ray Fluorescence

1. TEM 100-mesh, carbon/formvar-coated gold-finder grids (Electron Microscopy Sciences, PA).

2. Glass bottom 35 × 10 mm clear wall cell culture dishes (*PELCO®*).

3. 4% Paraformaldehyde solution in 10 mM phosphate-buffered saline, pH = 7.2 (Affymetrix/USB). The pH point is not well buffered, and tricky to reach. If necessary, adjust solution pH with concentrated HCl or with 1 M NaOH. To get 1% paraformaldehyde, dilute 1 mL of the 4% paraformaldehyde to 4 mL with 10 mM PBS.

4. 0.5% Dodecyltrimethylammonium chloride (DOTMAC)/1% paraformaldehyde. Dilute 4 mL 4% paraformaldehyde and 0.5 mL 10% DOTMAC in 5.5 mL 10 mM PBS, pH = 7.4.

5. 1% BSA in 10 mM PBS, pH = 7.4.

6. 10 mM PBS, pH = 7.4.

7. 20 mM Piperazine-N,N'-bis(2-ethanesulfonic acid) (PIPES)/200 mM sucrose buffer. Prepare about 50 mL of this buffer by dissolving PIPES and sucrose in Milli-Q-water. Adjust pH to 7 only by adding acetic acid. Do not use any other base or acid.

2.6. Laser Confocal Imaging

1. Glass bottom 35 × 10-mm clear wall cell culture dishes (*PELCO®*).

2. MitoTracker Red CMXRos (Molecular Probes, Invitrogen).

3. Zeiss LSM 510 Meta Confocal Microscope.

3. Methods
(See Note 2)

3.1. Synthesis of High Crystallinity Bare 5 nm TiO₂ Particles

1. Add 5.0 mL of prechilled (~4°C) titanium tetrachloride dropwise to 200 mL ice water under vigorous stirring until the white fog disappears. *Attention! This step requires chemical fume hood.*

2. Dialyze this solution against 2 L Milli-Q water using 2,000 MW cutoff dialysis cassettes at 4°C for 3 days, changing water daily, allowing slow growth of the particles, until the pH of the colloid reaches ~3.5 (see Note 3).

3. Dilute the colloid ~20 times (final colloid concentration ~0.015 M) and add 100 µL isopropyl glycidyl ether.

4. Inject 1 mL LiOH and, mix vigorously to adjust pH to ~9–12.

5. Dialyze this solution against 2 L against a buffer with desired pH values (e.g., for carboxylic group activation an optimal buffer with pH = 6.3).

6. Keep colloidal solutions of particles at 4°C for at least 6 months before use for perfecting crystallinity by aging.

3.2. TiO₂-Biomolecules Synthesis and Characterization

1. Mix 320 µL 11 µM TiO₂ particles with 16 µL 11 mM DOPAC (final concentration 550 µM) in 10 mM phosphate buffer pH = 6.2, to reach a final particle/DOPAC ratio of 1/100. The slightly yellow TiO₂–DOPAC complex immediately formed was monitored by observing the characteristic ligand-to-particle charge-transfer (LPCT) band at 420 nm (shoulder) as a result of chemisorption of DOPAC molecules on the TiO₂ surface. Optical absorption spectra were recorded using "Nanodrop ND-1000" or Perkin Elmer Lambda 950 UV/Vis and LS55 spectrometers.

2. Mix 6 µL 44 mM Sulfo-NHS and 4 µL 50 mM EDAC in the same buffer with the TiO₂–DOPAC. Incubate the reaction mixture for 1 h at room temperature with continuous gentle shaking in the dark, and then add 1 µL 2-mercaptoethanol to quench excess EDAC.

3. Place the reaction mixture into 2,000 MW cutoff dialysis bags (each 1 mL in size) and dialyze against 1 L 10 mM phosphate buffer, pH = 7.4 for 1 h.

4. Mix 500 µg of an antibody (mAb) (see Note 4) in 50 µL 10 mM phosphate buffer, pH = 7.4 with the pre-activated TiO₂–DOPAC particles (TiO₂–DOPAC–N-hydroxysulfosuccinimide ester) and incubate for 4–6 h at room temperature with continuous gentle mixing in the dark.

5. Add to the reaction mixture 25 µL 25 mM glycine and incubate for 15 min to quench the remaining active sites on the

particle surface. Remaining unquenched active sites were reported to reduce an antibody's biorecognition activity.

6. Dialyze the reaction mixture for 6 h using the same dialysis system.

7. Spin-wash the final TiO_2–mAb conjugate four times with 100 μL 10 mM phosphate buffer, pH = 7.4, to remove any unbound protein. Resuspend the final slightly yellowish pellet in 300 μL of the same buffer. The resulting colloidal suspension is stable at 4°C for up to 1 month.

8. Verify complete conversion of the free mAb by comparative SDS/2-mercaptoethanol-denaturation polyacrylamide gel electrophoresis of the final TiO_2–mAb conjugate and all spin/washing solutions (see Fig. 2). AFM TiO_2–mAb imaging revealed an average TiO_2 particle/mAb ratio as ~1.5/1 (see Fig. 2). The FTIR spectrum of the TiO_2–mAb conjugate contained new bands at 1638 (amide I) and 1540 (amide II) cm^{-1} typical for immunoglobulins (see Fig. 2).

3.3. Photo-Induced Cell Lysis and Cell Viability Examination

1. Grow cells in 12-well plates to reach 10^5 cells per well and then wash them three times with PBS (see Note 5).

2. Add aliquots of 6–600 ng/mL TiO_2–mAb or the antibody-free TiO_2 particles to cells and incubate for 1 h at the same culture condition (see Note 6). After 1 h, wash the cells thoroughly (six times) with PBS to eliminate any unbound nanomaterial.

3. Illuminate the plates for 5 min by focused polychromatic visible light with Cermax® PE300BFM 300 W Xenon Lamp. A UV filter (Orion Lighting Systems) should be used to cut wavelengths below 380 nm off. The intensity of incident light focused on the Petri dish ($D = 3.5$ cm) reached 60 mW/cm² as measured by a power energy meter (Scientech 372, Boulder, CO). A water filter ($D\,3.5 \times H\,20$ cm) is used to remove infrared radiation and exclude hyperthermia cytotoxic effects (see Note 7).

4. After light treatment, leave the cells to recover in a humidified atmosphere with 5% CO_2 at 37°C. Assess cell viability in 6, 24, and 48 h by standard LDH testing.

5. Perform control experiments to verify the origin of photo-induced cytotoxicity in cultures exposed to the TiO_2–mAb. Control experiments include cell cultures exposed to (1) TiO_2–mAb but not illuminated (negative control), (2) free TiO_2 particles with illumination (negative control), (3) TiO_2–IgG1 isotype antibody conjugates with illumination (negative control to demonstrate specific binding of the TiO_2–mAb nanoparticles to cancer cells), and (4) focused light without nanoparticles (a background control to estimate nanoparticles-driven phototoxicity). ROS-scavengers of H_2O_2 (final concentration: 100 U/mL catalase), SOD

(10 U/mL), hydroxyl radical (50 mM mannitol), and singlet oxygen (2 mM NaN_3 or 10 mM His) are used in control experiments for the identification of ROS involved in the cytotoxicity. An example of the results produced is shown in Fig. 3.

<table>
<tr><td>

3.4. Cell Samples Preparation for X-Ray Fluorescence Elemental Analysis

</td><td>

1. Culture cells at the standard conditions in growth medium in glass bottom, 35×10 mm, clear wall cell culture dishes (*PELCO®*) with 100-mesh, carbon/formvar-coated gold-finder grids (Electron Microscopy Sciences, PA) attached to their bottom until a sufficient amount of cells are attained on the grid (~10 K per grid) (see Note 8).

2. Simultaneously fix and permeabilize cells by incubation in 0.5% DOTMAC/1% paraformaldehyde in 10 mM PBS for 5 min (see Note 9).

3. Add 1% paraformaldehyde in 10 mM PBS, incubate cells for 20 min.

4. Block the cells for 90 min in 1% BSA to reduce nonspecific interaction with the antibody.

5. Add 600 ng/mL TiO_2–mAb conjugate and incubate for 1 h.

6. Add dyes at this stage to stain cellular compartments, if desired.

7. Wash specimens six times with sterile PBS followed by the removal of residual PBS by several washes in 20-mM pipes, pH 7.2–200 mM sucrose (see Note 10). Gently remove excess liquid using Kimwipes, air-dry.

8. Capture optical images of the cells using optical microscope (e.g., Zeiss LSM 510 Meta Confocal Microscope) with 40× or 20× objectives before X-ray analysis.

9. Keep samples dry in a desiccator for use in up to 3–4 months. An example of sample preparation and the results produced are shown in Fig. 4.

</td></tr>
</table>

4. Notes

1. All solutions should be prepared in water that has a resistivity of 18.2 MΩ cm and total organic content of less than five parts per billion. This standard is referred to as "water" in this text.

2. Since nanoscale materials can exhibit properties different from bulk materials with the same chemical composition and not all their properties and health effects are yet known, nanoscale materials must be considered of unknown toxicity. Therefore, as all new compounds, or those of unknown toxicity, nanoscale materials should be considered as materials both

acutely and chronically toxic. Habitual safe practices should be directed to minimizing of exposure by avoiding skin contact and inhalation through proper clothing and ventilation (17).

3. If allowed to dry out, the solution gradually turns into a transparent soft gel, which continues to change with time, from translucent to slight white, shrinking to split in the end. In step 2, the pH of the colloid solution should not reach higher than 3.5–4 to avoid the TiO_2 isoelectric point at pH = 5.0. Rapid injection of LiOH then allows the adjustment of the pH to 10–12, avoiding isoelectric point and TiO_2 precipitation.

4. This protocol can be adapted for other biomolecules, including DNA and PNA functionalized with an amino-group. For tethering biomolecules containing carboxylic groups TiO_2–DA should be used instead of TiO_2–DOPAC.

5. This protocol can be adapted for many other cell culture lines.

6. Cells can also be incubated with nanoparticles or nanobio conjugates at lower temperatures of 4°C to avoid possible internalization of particles.

7. In all experiments, the cell culture solution temperature could be remotely monitored with an infrared camera (e.g., model: ICI 7320 with 0.038 K temperature sensitivity, Infrared Cameras Inc. or similar). When water filter is applied, the temperature variations should not be more than 1–2°C during the light exposure. This excludes hyperthermia as a possible mechanism of cell damage.

8. The carbon/formvar-coated gold-finder grids are gently attached to a bottom of a dish using scotch-tape. Blunt-ended tweezers are very convenient for handling very fragile grids. Glass window area should be avoided as it will be used for laser confocal imaging.

9. This method of cell fixation and preparation can be applied for any advanced microcopy, including X-ray fluorescence. This technique minimizes disruption of plasma membrane microstructures and cellular compartments as well as metal ion topology and it does not significantly alter typical cellular trace element content relative to fixation by plunge freezing (4, 15, 16).

10. For a final washing any buffers and solutions containing sodium, potassium, chloride, etc. must be avoided.

Acknowledgments

Work at the Center for Nanoscale Materials was supported by the US Department of Energy, Office of Science, Office of Basic Energy Sciences, under Contract No. DE-AC02-06CH11357.

We are thankful to our colleagues Drs. B. Lai, L. Finney, S. Vogt and J. Maser from Advanced Photon Source, Argonne National Laboratory and collaborators Drs. M. S. Lesniak and I. V. Ulasov from the University of Chicago, Brain Tumor Center.

References

1. Thurnauer, M. C., Dimitrijevic, N. M., Poluektov, O. G., and Rajh, T. (2004) Photoinitiated charge separation: from photosynthesis to nanoparticles. *Spectrum* **17**, 10–15.

2. Liu, J., de la Garza, L., Zhang, L., Dimitrijevic, N. M., Zuo, X. B., Tiede, D. M., et al. (2007) Photocatalytic probing of DNA sequence by using TiO_2/dopamine-DNA triads. *Chem. Phys.* **339**, 154–163.

3. Liu, J., Saponjic, Z. V., Dimitrijevic, N. M., Luo, S., Preuss, D., and Rajh, T. (2006) Hybrid TiO_2 nanoparticles: an approach for developing artificial restriction enzymes. in: *Colloidal Quantum Dots for Biomedical Applications* (Osinski, M., Yamamoto, K., and Jovin, T. M. eds.). Bellingham, WA. *SPIE Proc Series* **6096**, 60960F.

4. Rozhkova, E. A., Ulasov, I. V., Lai, B., Dimitrijevic, N. M., Lesniak, M. S., and Rajh, T. (2009) A high-performance nanobio photocatalyst for targeted brain cancer therapy. *Nano Lett.* **9**, 3337–3342.

5. Dimitrijevic, N. M., Rozhkova, E. A., and Rajh, T. (2009) Dynamics of localized charges in dopamine-modified TiO_2 and their effect on the formation of reactive oxygen species. *J. Am. Chem. Soc.* **131**, 2893–2899.

6. Rajh, T., Chen, L. X., Lukas, K., Liu, T., Thurnauer, M. C., and Tiede, D. M. (2002) Surface restructuring of nanoparticles: an efficient route for ligand-metal oxide crosstalk. *J. Phys. Chem. B* **106**, 10543–10552.

7. Redfern, P. C., Zapol, P., Curtiss, L. A., Rajh, T., and Thurnauer, M. C. (2003) Computational studies of catechol and water interactions with titanium oxide nanoparticles. *J. Phys. Chem. B* **107**, 11419–11427.

8. Saponjic, Z. V., Dimitrijevic, N. M., Tiede, D. M., Goshe, A. J., Zuo, X., Chen, L. X., et al. (2005) Shaping nanometer-scale architecture through surface chemistry. *Adv. Mater.* **17**, 965–971.

9. Rajh, T., Saponjic, Z., Liu, J., Dimitrijevic, N. M., Scherer, N. F., Vega-Arroyo, M., et al. (2004) Charge transfer across the nanocrystalline-DNA interface: probing DNA recognition. *Nano Lett.* **4**, 1017–1023.

10. Dimitrijevic, N. M., Saponjic, Z. V., Rabatic, B. M., and Rajh, T. (2005) Assembly and charge transfer in hybrid TiO2 architectures using biotin-avidin as a connector. *J. Am. Chem. Soc.* **127**, 1344–1345.

11. Paunesku, T., Rajh, T., Wiederrecht, G., Maser, J., Vogt, S., Stojicevic, N., et al. (2003) Biology of TiO_2-oligonucleotide nanocomposites. *Nat. Mater.* **2**, 343–346.

12. de la Garza, L., Saponjic, Z. V., Dimitrijevic, N. M., Thurnauer, M. C., and Rajh, T. (2006) Surface states of titanium dioxide nanoparticles modified with enediol ligands. *J. Phys. Chem. B* **110**, 680–686.

13. Rajh, T., Dimitrijevic, N. M., Elhofy, A., and Rozhkova, E. A. (2010) Hybrid TiO_2 based nanocomposites-new tools for biotechnology and biomedicine. in: *Handbook of Nanophysics: Functional Nanomaterials* (Sattler, K. D. ed.). CRC, Boca Raton, FL.

14. Rajh, T., Ostafin, A. E., Micic, O. I., Tiede, D. M., and Thurnauer, M. C. (1996) Surface modification of small particle TiO_2 colloids with cysteine for enhanced photochemical reduction: an EPR study. *J. Phys. Chem.* **100**, 4538–4545.

15. Nakamura, F. (2001) Biochemical, electron microscopic and immunohistological observations of cationic detergent-extracted cells: detection and improved preservation of microextensions and ultramicroextensions. *BMC Cell Biol.* **2**, 10.

16. Finney, L., Mandava, S., Ursos, L., Zhang, W., Rodi, D., Vogt, S., et al. (2007) X-ray fluorescence microscopy reveals large-scale relocalization and extracellular translocation of cellular copper during angiogenesis. *Proc. Natl. Acad. Sci. USA* **104**, 2247–2252.

17. Committee on Prudent Practices for Handling, Storage, and Disposal of Chemicals in Laboratories, National Research Council. (1995) *Prudent Practices in the Laboratory Handling and Disposal of Chemicals.* National Academy Press, Washington, DC.

Chapter 6

Surface Modification and Biomolecule Immobilization on Polymer Spheres for Biosensing Applications

Chris R. Taitt, Lisa C. Shriver-Lake, George P. Anderson, and Frances S. Ligler

Abstract

Microspheres and nanospheres are being used in many of today's biosensing applications for automated sample processing, flow cytometry, signal amplification in microarrays, and labeling in multiplexed analyses. The surfaces of the spheres/particles need to be modified with proteins and other biomolecules to be used in these sensing applications. This chapter contains protocols to modify carboxyl- and amine-coated polymer spheres with proteins and peptides.

Key words: Microspheres, Nanoparticles, Immobilization, Latex spheres

1. Introduction

While encapsulation of enzymes in hydrophilic microspheres has long been the standard procedure for industrial bioprocessing (e.g., food, pharmaceuticals, and cosmetics) and a variety of commercial products are available (1), microspheres and nanospheres with recognition molecules on the surface have only become widely utilized over the last decade. These structures are being used in analytical and biosensing, drug delivery, and affinity purification applications. As integral components of analytical methods, recognition molecules are currently being immobilized on magnetic spheres for target separation (2–6), on fluorescent microspheres for use in flow cytometric analysis (7–11), on latex spheres for dynamic light scatter-based assays (12–14), and on gold colloids for the generation of colored signals (15, 16). Fluorescent or metallic spheres have been used for signal amplification in optical and electrochemical assays (17–20).

Sarah J. Hurst (ed.), *Biomedical Nanotechnology: Methods and Protocols*, Methods in Molecular Biology, vol. 726, DOI 10.1007/978-1-61779-052-2_6, © Springer Science+Business Media, LLC 2011

The methods used for immobilization vary, depending on the composition of the recognition molecule and the chemistry of the bead surface (bead will be used interchangeably with microsphere and/or nanoparticle throughout this chapter). In general, there are three basic approaches for immobilization of molecules on spheres, depending on the chemistry of the bead surface. If a metal is exposed on the surface, organic molecules containing a binding moiety are preferred (i.e., a thiol group to bind to a gold surface or a histidine oligomer to bind to exposed nickel (16, 21, 22)). If the surface is glass or silicon, organosilanes are bound to activated silanol groups on the surface and serve as anchors for cross-linkers and/or biomolecules (23, 24). Finally, if the spheres are composed of a synthetic polymer such as polystyrene or polymethylmethacrylate (PMMA) [7, 25–28), liquid crystals (29), or natural biopolymers such as chitosan or collagen (30, 31), chemical reactions compatible with both the complex surface and the molecule being immobilized must be used.

Thus, a variety of approaches have been developed to attach biomolecules to polymer spheres of nanometer (nm) and micron diameter. In general, the goals are (1) to provide a bond that is stable, (2) to prevent secondary adsorption of the biomolecule to the bead with attendant denaturation, (3) to expose the active site of the biomolecule to the environment away from the bead surface, and (4) to immobilize the biomolecule at high density.

The primary choices for achieving a stable bond are to covalently attach the biomolecule to the bead directly or through an intermediate layer or to attach it through a high affinity interaction such as avidin–biotin linkage or oligonucleotide hybridization. The trapping of biomolecules onto microspheres/nanospheres using electrostatic layers such as polyethyleneimine and polystyrene has also been documented (32–34), but this method may inhibit interactions between some recognition molecules and their target ligands, especially if the targets are large.

Secondary adsorption is not a problem if the polymer spheres are composed of a hydrophilic polymer, as is the case with natural biopolymers, but can be a problem if they are composed of hydrophobic polymers, such as PMMA, nylon, and polystyrene (latex). In the latter case, hydrophilic molecules such as protein, dextran, or polyethylene glycol (PEG) can be linked to the surface of the bead, and the recognition molecule can be linked to this intermediate layer. In addition to preventing the denaturation of the attached recognition molecule, these hydrophilic layers also help prevent nonspecific adsorption of nontarget molecules from complex samples (35–37). When the recognition molecules are relatively small (e.g., antimicrobial peptides, haptens, or oligosaccharides), a spacer can be placed between them and the bead surface (or the intermediate layer) to improve the access of the active site to its target in the surrounding fluid. In many cases, an intermediate layer that prevents nonspecific adsorption can also serve as the spacer.

Finally, in most assay procedures, it is helpful to immobilize the recognition molecules at relatively high density in order to maximize target binding (38). If direct binding or binding through a spacer or intermediate layer does not provide sufficient density, molecular brushes or dendrimers can be used to increase the number and density of attachment sites on the surface of the spheres (39–41). In this chapter, we use the term "scaffold" to refer to the intermediate layer between the bead and the recognition molecules that either prevents nonspecific adsorption or increases the density of reactive groups at the bead surface.

Several companies (Luminex Corporation (42), Spherotech (43), Bangs Lab (44), Invitrogen (45), and Sigma-Aldrich (46)) sell microspheres and nanospheres with reactive groups on the surface as well as fluorescent dyes or ferromagnetic material in the bead core. The established protocols provided by the manufacturers are usually for direct immobilization of large molecules exhibiting amine groups (see Fig. 1a, analogous to the manufacturers' protocols). We have found that this is the simplest and most direct method to bind a protein to a bead functionalized with a carboxyl group. For large proteins, there are sufficient lysines with terminal amino groups that the majority of molecules will be oriented so that sufficient active sites are available to bind the targets. This protocol can also be used for the attachment of avidin and subsequent noncovalent attachment of a biotinylated recognition molecule (29). Biotinylation may provide better control over the orientation of the molecule than direct immobilization, particularly for small molecules. Moreover, a long-chain biotin (e.g., EZ-Link NHS-LC-LC-biotin, EZ-Link maleimide-PEG11-biotin, or EZ-Link biotin-PEO-LC-amine) (47) can add a short spacer to move the recognition molecule further from the avidin binding pocket and into the surrounding fluid. Low molecular weight haptens have been attached to large proteins or dextran and then subsequently attached to the carboxyl groups on the bead surface through the protein using this scheme (48, 49).

A convenient variation on this same basic chemistry can be performed in which the carboxyl groups on the microsphere/nanoparticle surface are modified by the attachment of diamine scaffolds to create a surface layer of amino groups (see Fig. 1b). Recognition molecules can then be attached through the carboxyl groups rather than through terminal amino groups. In some cases, this reverses the molecular orientation at the surface to provide better exposure for the active sites. For recognition molecules of less than 1,000 molecular weight, attaching the appropriate end to the bead is frequently critical to its activity (50, 51).

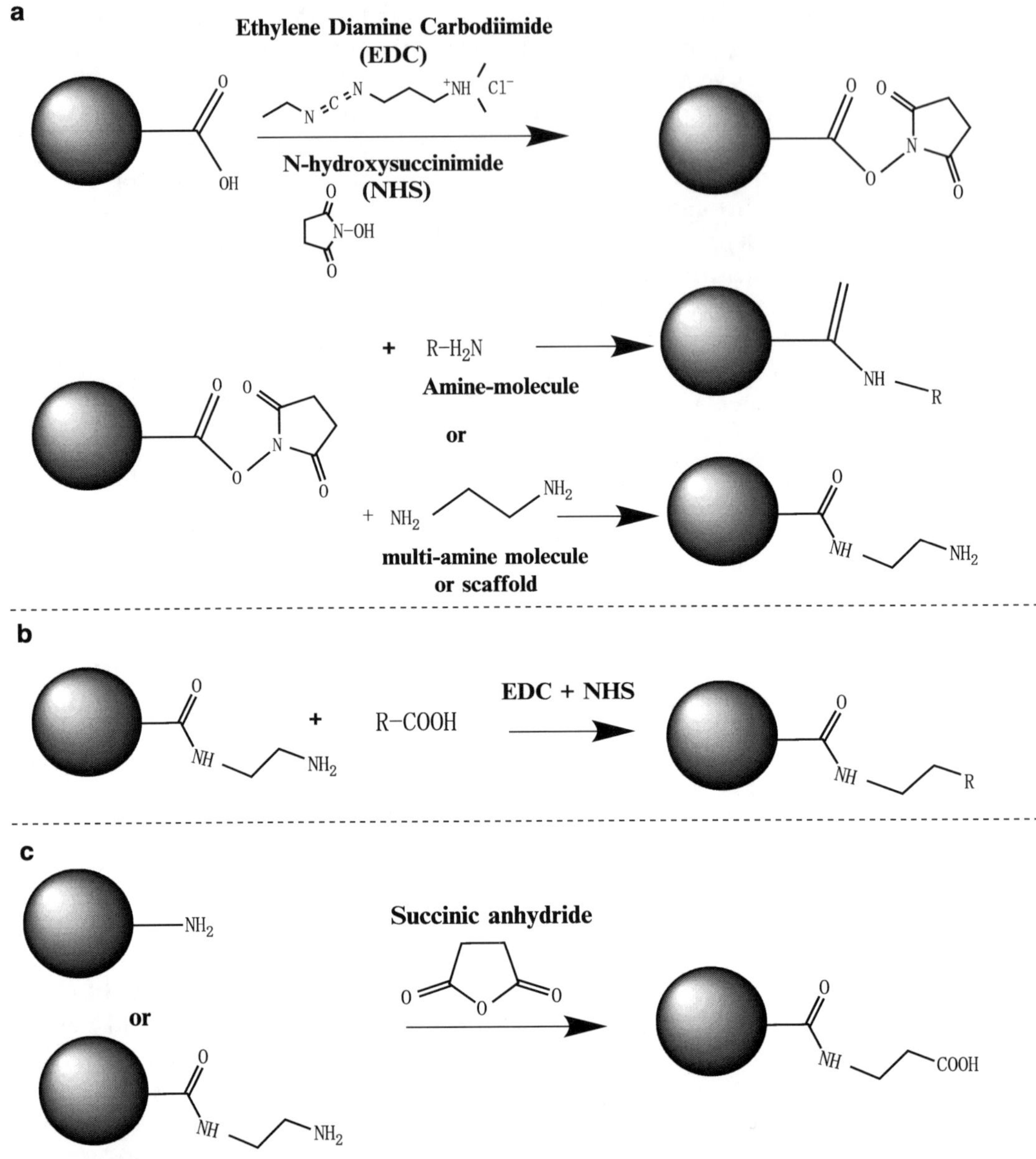

Fig. 1. Three schemes (**a–c**) for generating biomolecule-coated microspheres/nanoparticles.

We also show that an attachment protocol for spheres that have amine groups on the surface rather than carboxyl groups (see Fig. 1c). This protocol can also be employed after carboxylated spheres have been modified with a protein [such as bovine serum albumin (BSA)], dextran, or a dendrimer (52) to create a scaffold of amino groups on the surface. Once reactive carboxyl groups are attached to the amino groups, the same procedures are employed as in Fig. 1a.

These three protocols, described in detail below, are specified for spheres with micron diameters (microspheres) and a centrifugation step is used to separate the spheres from the reagents/solutions. These protocols work equally well with nm-diameter spheres. However, for smaller spheres (nanospheres), bead separation can be accomplished with dialysis, desalting gel-filtration chromatography, centrifugation in Centricon columns, or simple quenching of the reaction. For example, the activity of sulfo-N-hydroxysuccinimide can be quenched by the addition of 2-mercaptoethanol or terminated by desalting (53).

2. Materials

2.1. Attachment of Amine-Containing Molecules to COOH-Microspheres (See Fig. 1a)

1. Microspheres: Luminex's xMAP-compatible or other microspheres can be purchased from the following vendors: MiraiBio, Spherotech, Invitrogen, and Sigma-Aldrich.

2. Activation buffer: 0.1 M sodium phosphate buffer, pH = 6.

3. EDC solution: 50 mg/mL 1-ethyl-3-[3-dimethylaminopropyl]carbodiimide hydrochloride (EDC), in dimethyl sulfoxide (DMSO). This solution must be used within 2 h of preparation (see Note 1).

4. Sulfo-NHS Solution: 50 mg/mL N-hydroxysulfosuccinimide (sulfo-NHS) in deionized H_2O. Prepare the same quantity of solution as used for EDC solution, above.

5. Biomolecule solution (e.g., protein, peptide, ethylenediamine, and aminodextran): 10 μg/mL to 1 mg/mL (depending on the MW and availability of the amine-containing biomolecule) in phosphate-buffered saline (PBS), pH = 7.4 (see Note 2). Generally, a large excess of the biomolecule relative to the bead assures the attachment of a higher density of molecules, but lower amounts may also be effective.

6. PBS, pH = 7.2–7.4.

7. PBST: PBS supplemented with 0.05% Tween-20.

8. PBSTB: PBS supplemented with 0.05% Tween-20 and 1 mg/mL BSA.

2.2. Attaching COOH-Containing Molecules to NH₂-Microspheres ("Upside-Down EDC Reaction") (See Fig. 1b)

1. Dialysis tubing: Dialysis tubing with an appropriate molecular weight cutoff can be obtained from many manufacturers and should be fully rehydrated before use per manufacturer's instructions. Use of dialysis clips (versus tying the tubing) greatly simplifies the protocol and allows the user to perform EDC/NHS activation of proteins within the same dialysis bag.

2. Biomolecule solution: 10 μg/mL to 1 mg/mL carboxyl-containing protein/biomolecule in PBS, pH = 7.4. Higher or

lower concentrations of biomolecule can be used as needed. Total protein solution volume should be 1 mL or less.

3. Activation buffer: 0.1 M phosphate buffer, pH = 6.

4. EDC solution: 50 mg/mL EDC in DMSO. This solution must be used within 2 h of preparation (see Note 1).

5. Sulfo-NHS solution: 50 mg/mL sulfo-NHS in deionized H_2O. Prepare the same quantity as for the EDC solution.

6. 100× 2-Mercaptoethanol: 35 µL 2-mercaptoethanol in 965 µL deionized water (0.5 M 2-mercaptoethanol) (see Note 3).

7. Amine-decorated microspheres: Microspheres treated with an appropriate di- or multivalent amine-containing molecule (e.g., ethylenediamine and aminodextran) as described in Subheading 3.1 or ones commercially available (43).

8. NHS coupling buffer: 100 mM sodium borate, pH = 8.0. Other basic or neutral buffers may work as well or better depending on the nature of the ligand.

9. PBS, pH = 7.2.

10. PBST: PBS supplemented with 0.05% Tween-20.

11. PBSTB: PBS supplemented with 0.05% Tween-20 and 1 mg/mL BSA.

12. Small molecule possessing carboxyl moiety, e.g., fumonisin, aflatoxin, or peptide.

2.3. Converting NH₂-Microspheres into COOH-Microspheres (See Fig. 1c)

1. Amine-decorated microspheres: From step 10 of Subheading 3.1.3, coated using aminodextran or ethylenediamine.

2. Conversion buffer: 100 mM sodium borate, pH = 9.0.

3. Succinic anhydride.

4. PBS, pH = 7.2.

---

3. Methods

3.1. Attachment of Amine-Containing Molecules (Including Scaffolds) to COOH-Microspheres (See Fig. 1a)

Attachment of amine-containing molecules to carboxyl-decorated microspheres is simple and straightforward when using an EDC/NHS linkage. The presence of NHS stabilizes the amine-reactive intermediate (O-acylisourea) created by the reaction of EDC with carboxyls by converting the intermediate to an amine-reactive sulfo-NHS ester. We have adapted the protocol used by Luminex and other commercial sources for microspheres (42–46) to simplify the physical manipulations and to minimize the loss of microspheres during preparation. Further, this protocol can be used to convert carboxyl-decorated microspheres into amine-decorated microspheres (to be used in further coupling reactions) if the biomolecule being attached possesses multiple amine moieties, including ethylene diamine ($H_2N-CH_2-CH_2-NH_2$), polyoxyethylene bis(amine) ($H_2N-[-CH_2-CH_2-O-]n-CH_2-CH_2-NH_2$), or aminodextran. Likewise surface carboxyls can be converted to thiols using cystamine ($HN_2-CH_2-CH_2-S-S-CH_2-CH_2-NH_2$) with subsequent reduction of the disulfide by a reducing agent such as tris(2-carboxyethyl) phosphine (TCEP).

3.1.1. Microsphere Preparation: Washing

In this step, microspheres are washed and exchanged into an appropriate buffer for EDC chemistry.

1. Resuspend microspheres in stock bottle by vortexing and brief bath sonication.

2. Pipet 100 µL of beads into Eppendorf tube ($\sim 1.5 \times 10^6$ microspheres; scale as desired).

3. Centrifuge for 4 min at $18,400 \times g$ or 14,000 rpm in an Eppendorf centrifuge.

4. Remove the supernatant and put it into a secondary tube (referred to as the "chase tube"). The use of this chase tube minimizes the loss of microspheres during handling.

5. Resuspend microspheres in 100 µL Activation buffer in the primary tube.

6. Repeat centrifugation of both the primary and chase tubes.

7. Discard supernatant from the chase tube, move the supernatant from the primary tube to the chase tube.

8. Resuspend microspheres in 100 µL Activation buffer in the primary tube.

9. Repeat centrifugation of the primary and chase tubes.

10. Discard supernatant from the chase tube, move the supernatant from the primary tube to the chase tube.

11. Repeat centrifugation of the chase tube.

12. Discard the supernatant from the chase tube.

13. Resuspend microspheres in the chase tube in 40 µL Activation buffer.

14. Resuspend microspheres in the primary tube in 80 µL Activation buffer.

15. Combine the solution in the chase tube with that in the primary tube.

3.1.2. Microsphere Activation with EDC/ Sulfo-NHS

In this section, the protocol describes the activation of the carboxyl groups on the microspheres with EDC/sulfo-NHS to form an NHS-ester. This intermediate is more stable than the O-acylisourea intermediate that is formed in the absence of NHS.

1. Add 15 µL EDC solution and 15 µL sulfo-NHS solution to the microspheres in the primary tube. The final concentrations of each will be ~5 mg/mL (25 mM).

2. Mix well by briefly vortexing and bath sonicating. Keep in the dark for 20 min.

3. Centrifuge for 4 min at $18{,}400 \times g$.

4. Remove the supernatant from the primary tube and pipette into the chase tube.

5. Resuspend microspheres in the primary tube in 100 µL Activation buffer.

6. Repeat centrifugation of the primary and chase tubes.

7. Discard supernatant from the chase tube and transfer supernatant from the primary tube to the chase tube.

8. Resuspend the microspheres in the primary tube in 100 µL Activation buffer. Repeat centrifugation of the primary and chase tubes.

9. Discard supernatant from the chase tube and transfer supernatant from the primary tube to chase tube.

10. Resuspend the microspheres in the primary tube in 100 µL PBS.

11. Repeat centrifugation of the primary and chase tubes.

12. Discard the supernatant from the chase tube.

13. Resuspend microspheres in the chase tube with PBS (~50 µL) and vortex/sonicate.

14. Transfer the contents of the chase tube to the primary tube and vortex/sonicate to mix.

3.1.3. Cross-linking of Activated Microspheres to Amine-Containing Biomolecules/Linkers or Scaffolds

In this step, the NHS-ester active group on the bead binds to amine-containing biomolecules in solution. The protocol described is appropriate for creating a single type of "decorated" microsphere. Alternatively, the resuspended microspheres can be aliquoted and each aliquot is incubated with a different biomolecule to create batches of microspheres functionalized with different biomolecules. For immobilization of proteins, we typically use 100 µg (~1 mg/mL) protein for 1.5×10^6 microspheres. Less can be used, but using an excess of the biomolecule at a high concentration helps to ensure good coupling efficiency. To calculate the minimum amount of protein to use, determine the surface area of the microspheres to be coated and the footprint occupied by each protein; for most couplings, a tenfold excess of protein or more should be used. At the end of the reaction, the biomolecule-decorated microspheres are resuspended and washed in an appropriate buffer. The composition of this final buffer will depend on whether or not additional coupling reactions are desired.

1. Add 100 μL biomolecule solution to the resuspended microspheres (see Subheading 3.1.2, step 15).

2. Vortex and bath-sonicate the microspheres to ensure they are well suspended.

3. Incubate for at least 2 h in the dark. Rotate the tube to keep the microspheres suspended or alternatively, mix frequently (every 10–15 min). Alternatively, an overnight incubation can be performed.

4. Centrifuge for 4 min at 18,400 × g.

5. Remove supernatant and discard.

6. Resuspend microspheres in 100 μL (a) PBST or PBSTB if no additional couplings are desired, (b) Activation buffer if additional EDC-mediated coupling is desired, and (c) other type of coupling buffers as required for additional coupling reactions.

7. Centrifuge for 4 min at 18,400 × g.

8. Remove supernatant and discard.

9. Wash microspheres once by resuspending in 100 μL appropriate buffer (see step 6) and centrifuging for 4 min at 18,400 × g.

10. Resuspend microspheres in 100 μL buffer and proceed with additional coupling reactions. Alternatively, store microspheres at 4°C in the dark for several days to several years or more, depending on the stability of the immobilized biomolecule. Larger storage volumes may be desirable depending on the application.

3.2. Attaching COOH-Containing Molecules to NH₂-Microspheres ("Upside-Down" EDC Reaction) (See Fig. 1b)

Two protocols are described for the attachment of protein (see Subheading 3.2.1) or small molecules (see Subheading 3.2.2) to microspheres possessing pendant amine moieties. These amine-decorated microspheres can be prepared by treating carboxyl-decorated microspheres (as supplied by the manufacturer) with a di- or multivalent amine-containing "scaffold." The protocols below are essentially the inverse reaction from that detailed in Subheading 3.1 – essentially an "upside-down" EDC reaction.

3.2.1. Attachment of Proteins to NH₂-Decorated Microspheres

Direct attachment to activated microspheres requires that the protein solution not contain any other amine/carboxyl reactive groups (e.g., glycine, Tris, and BSA). For protein solutions supplied in an amine/carboxyl-containing buffer, the original buffer must be removed prior to coupling. For this reason, a dialysis step is included to exchange the buffer and remove any interfering components. If the protein is supplied or can be satisfactorily diluted into the activation buffer it may be possible to proceed to step 3.

1. Place protein solution (~1 mg/mL) into a dialysis membrane having a suitable MW cutoff for the protein.

2. Dialyze against >100 times the sample volume of 10% Activation buffer for ~1 h. If the protein has previously been in an amine-containing buffer, it will be necessary to change the buffer at least three times (after 1 h, after 3 h, and after overnight incubation) (see Note 2).

3. Add 1/50 volume each of EDC and sulfo-NHS solutions to the protein solution in the dialysis bag. The final concentration of each will be 1 mg/mL (5 mM).

4. Dialyze for 1 h against fresh 10% Activation buffer.

5. After 1 h, add 1/100 volume 100× 2-mercaptoethanol solution to the protein solution; the final concentration of mercaptoethanol in the sample will be 5 mM (see Note 4) to stop the reaction.

6. During protein dialysis, prepare the microspheres either purchased as NH_2-microspheres or prepared using step 10 of Subheading 3.1.3. Wash the amine-decorated microspheres twice by resuspending in 100 µL NHS coupling buffer and centrifuging for 4 min at 18,400×g.

7. After quenching of the EDC reaction (see step 5) or removing the unreacted EDC (see Note 4), add an equal volume of NH_2-decorated microspheres to the target protein.

8. Incubate the microspheres with the protein for at least 2 h.

9. Centrifuge the mixture for 4 min at 18,400×g.

10. Discard the supernatant and wash the protein-coated microspheres twice by resuspension in 100 µL PBST or PBSTB and centrifugation for 4 min at 18,400×g.

11. Store the protein-coated microspheres at 4°C in the dark.

3.2.2. Homogeneous Protocol for Small Carboxyl-Containing Molecules

Many small molecules such as peptides, phycotoxins, and mycotoxins possess carboxyl moieties. If these molecules do not possess disulfide bridges and are not in the presence of amine-based buffers, they can be attached to amine-coated microspheres without dialysis.

1. Dissolve or dilute the molecule of interest into an appropriate volume of Activation buffer (typically, 0.1–1 mg/mL in ≤1 mL).

2. Add 1/50th volume each of EDC and sulfo-NHS solutions. The final concentration of each will be 1 mg/mL (5 mM).

3. Incubate for 1 h at room temperature.

4. During EDC activation, prepare the amine-decorated microspheres. Wash the amine-decorated microspheres twice by

resuspension in 100 μL NHS coupling buffer and centrifugation for 4 min at 18,400×*g*. Resuspend the microspheres in NHS coupling buffer.

5. Add 1/100th volume of 100× 2-mercaptoethanol solution to the EDC/sulfo-NHS reaction (see Note 3). The final concentration of mercaptoethanol will be 5 mM.

6. Mix the resuspended microspheres with the EDC-activated small molecule.

7. Incubate the microspheres with the activated small molecule for at least 2 h at room temperature.

8. Centrifuge the mixture for 4 min at 18,400×*g*.

9. Discard the supernatant and wash the coated microspheres twice by resuspension in 100 μL PBST or PBSTB and centrifugation for 4 min at 18,400×*g* each time.

10. Store the small molecule-coated microspheres at 4°C in the dark.

3.3. Making Protein–Small Molecule Conjugates

In some cases, the user may desire to create a multivalent conjugate comprising a protein scaffold and many attached small molecules before its attachment to microspheres. These conjugates may be useful because the conjugate orients the small molecules correctly, provides a higher avidity surface for subsequent immunoassays, or presents the antigen in a highly immunogenic form for antibody generation. There are a number of methods for conjugating small molecules with proteins, particularly with regard to creating immunogens (49). In our laboratory, we have coupled mycotoxins and explosives to proteins, such as BSA or keyhole limpet hemocyanin (KLH). We have also employed the EDC chemistry described above with small molecules, using the protein in place of the microspheres. Once these small molecule–protein conjugates have been formed, the protein can be attached to the microspheres through an amine as described in the protocols above. It is often also convenient to couple a protein, such as ovalbumin, BSA, or hen egg lysozyme, to the microsphere as the intermediate layer and then link the small molecule to the immobilized protein using EDC chemistry (see Subheading 3.1). In that case, the use of chase tubes is not typically necessary.

3.4. Converting NH₂-Microspheres into COOH-Microspheres (See Fig. 1c)

Sometimes a user desires a higher density of carboxyl moieties than is provided on commercially available microspheres, or alternatively, the attached moiety may be unstable if directly immobilized to the microsphere surface using EDC-based, zero-length attachment chemistry. In these cases, the user may attach an intervening layer of ethylenediamine (to add a short linker) or a multivalent scaffold such as aminodextran (to increase the density of available sites for immobilization) using EDC chemistry (see Subheading 3.1). The resulting amine-decorated surfaces can then be converted back into carboxylated surfaces using the protocol below. Once converted, the standard EDC/sulfo-NHS protocol described in Subheading 3.1 can be used to attach an amine-containing biomolecule to the diamine or multivalent scaffold. This can easily be accomplished by the use of succinic anhydride.

1. Remove approximately 50–100 µL amine-decorated microspheres from step 10, Subheading 3.1.3.

2. Centrifuge microspheres for 4 min at $18,400 \times g$ and discard supernatant.

3. Resuspend microspheres in 0.5 mL Conversion buffer and centrifuge again.

4. Resuspend microspheres in 0.5 mL Conversion buffer.

5. Add ~1 mg solid succinic anhydride directly to the resuspended microspheres.

6. Sonicate and vortex the sample until all the succinic anhydride is dissolved.

7. Incubate for 30 min with intermittent mixing.

8. Centrifuge the microspheres for 4 min at $18,400 \times g$ and discard supernatant.

9. Wash once by resuspending in 0.5 mL Conversion buffer and centrifuging.

10. Repeat steps 4–8 twice more for a total of three treatments with succinic anhydride.

11. Wash once by resuspending in 0.5 mL PBS or desired buffer and centrifuging.

12. Finally, resuspend the microspheres in the desired buffer: PBS if storing or Activation buffer if further coupling chemistry is going to be performed (see Subheading 2.2). The carboxy-coated microspheres can then be exposed to proteins/biomolecules using the protocols described above (see Subheading 3.1).

3.5. Experimental Results

Many immunoassays have been developed for the detection of small antigens including mycotoxins, phycotoxins, and explosives that possess either a single epitopic site or several overlapping epitopes. For example, while ochratoxin possesses a single carboxyl moiety and must be immobilized using that group, fumonisin possesses both amine and carboxyl groups and can be immobilized in multiple orientations (see Fig. 2). In these assays, appropriate presentation of the small molecule is critical for optimal assay performance. Furthermore, when a competitive format is utilized, the density of the immobilized species is also important. If the density of immobilized species is too high, assay sensitivity

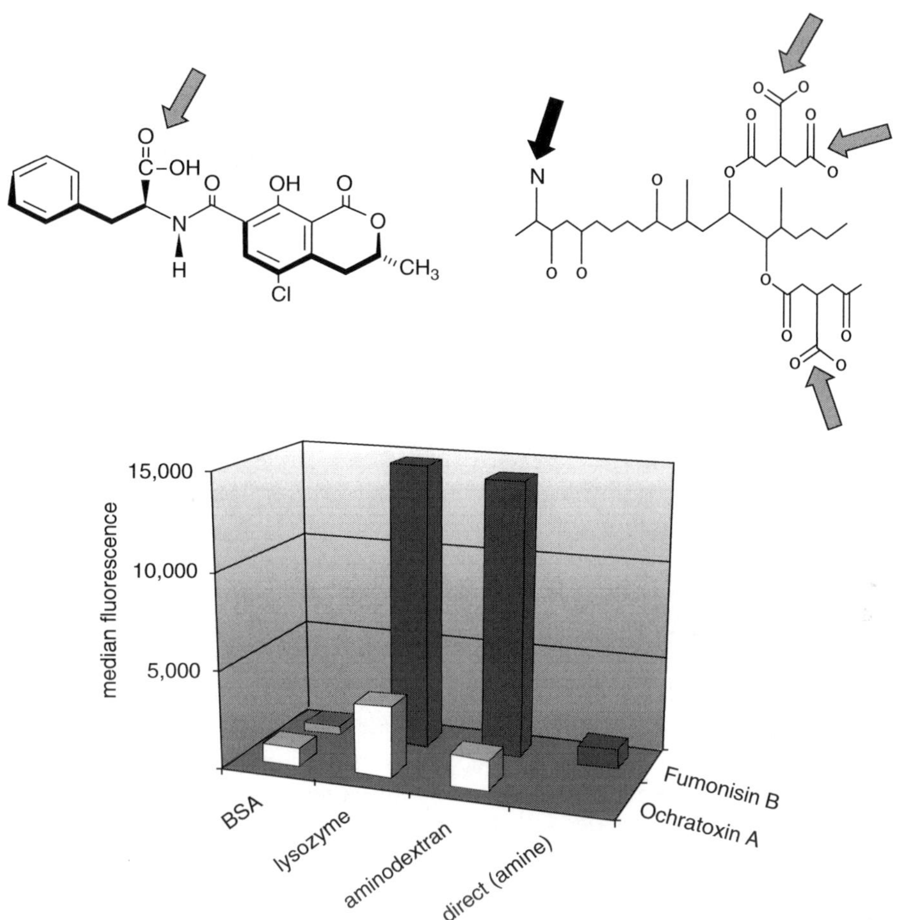

Fig. 2. *Top*: Chemical structures of ochratoxin A (*left*) and fumonisin B1 (*right*). *Gray arrows* point to carboxyl moieties that can be linked to amine groups through EDC chemistry. The *black arrow* points to the unique primary amine on fumonisin B1; this amine can be linked to carboxyl moieties using EDC chemistry. *Lower panel*: Median fluorescence of labeled anti-fumonisin monoclonal antibodies (10 µg/mL, *gray bars*) and anti-ochratoxin antibodies (10 µg/mL, *white bars*) bound to microspheres coated with fumonisin or ochratoxin, respectively, that were immobilized through amine-rich scaffolds (BSA, lysozyme, and aminodextran) or directly.

will be poor; if too low, the signals generated will be too low to produce a robust assay.

In order to develop sensitive and robust assays utilizing immobilized small molecules, microspheres were decorated with immunogens or capture reagents using different methods, and their performance in immunoassays was used to compare the presentation and functionality of the immobilized species. Each mycotoxin (ochratoxin and fumonisin) was immobilized onto samples of microspheres decorated with three different types of amine-rich scaffolds (i.e., BSA, lysozyme, and aminodextran). First, the method outlined in Subheading 3.1 was used to attach the amine-rich scaffolds to carboxy-functionalized microspheres, then the method outlined in Subheading 3.2 was used to attach the carboxyl groups of the mycotoxins to these scaffolds. Also, fumonisin B1 was coupled directly to the carboxy-functionalized microspheres via its single amine group. Anti-fumonisin antibodies bound poorly to microspheres on which fumonisin had been immobilized by its single primary amine, indicating that that amine may be in a critical part of the epitopic region (see Fig. 2). Interestingly, however, anti-fumonisin and anti-ochratoxin antibodies bound very well to microspheres on which the respective mycotoxins were immobilized through lysozyme or aminodextran scaffolds, although poor binding was observed when they were immobilized through BSA. Since the molar proportion of lysines (through which the mycotoxins were attached) is roughly equivalent with BSA and lysozyme, and only half as high in the aminodextran used here, the relative affinities of the antibodies for aminodextran- or lysozyme-immobilized mycotoxins is not a simple density issue.

Small molecules have also been used as affinity capture reagents in assays designed to complement immunoassays; such capture biomolecules include aptamers (54–57), carbohydrates and glycolipids (58–64), and antimicrobial peptides (50, 65–67). Here too, both the presentation and density of immobilized capture molecules are critical for assay performance. We immobilized several antimicrobial peptides (polymyxin B, melittin, cecropin A, and magainin II) as biological recognition elements for *Escherichia coli*; these molecules have previously been used to detect *E. coli* and other Gram-positive and Gram-negative bacteria in biosensor assays (see Fig. 3). Each peptide was immobilized onto microspheres coated with various scaffolds or linkers. Scaffolds offering the possibility of higher density linkages included BSA, lysozyme, aminodextran, and poly-D-lysine. After immobilization onto the microspheres, pendant amines on these scaffolds were converted to carboxyl moieties using succinic anhydride (see Subheading 3.4), and subsequently linked to the peptides using standard EDC chemistry (see Subheading 3.1.2). Additional linkers including ethylenediamine and PEG diamine were also immobilized using standard EDC chemistry, converted to carboxyl-terminated linkers, and finally linked to the peptides' amines via standard EDC chemistry. The final linker tested was a glycine tetrapeptide; immobilization of antimicrobial peptides using this $(gly)_4$ linker was accomplished through two rounds of direct EDC coupling.

In general, *E. coli* exhibited the highest binding to melittin-coated microspheres *independent* of the scaffold or linker used with most effective immobilization on poly-D-lysine scaffolds (see Fig. 3). Cecropin A also exhibited optimal binding when immobilized through poly-D-lysine. However, we observed that *E. coli* bound best to polymyxin B when aminodextran was used as scaffold. Overall, no clear patterns of preferential immobilization were observed among the different peptides, indicating the importance of optimizing the attachment chemistry for each system.

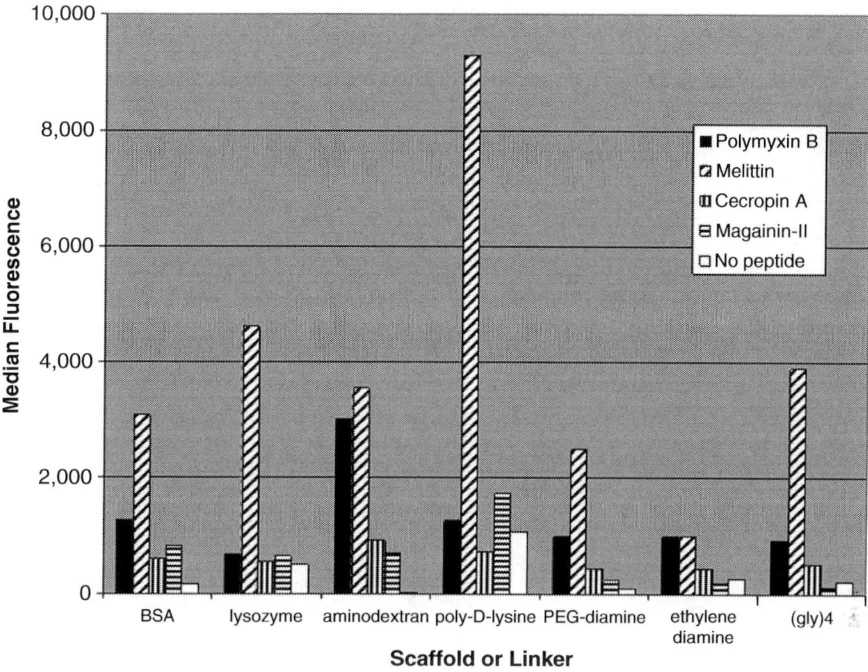

Fig. 3. *Escherichia coli* assays using antimicrobial peptides as capture reagents. Biotinylated *E. coli* was incubated for 30 min with beads coated with various antimicrobial peptides (polymyxin B, melittin, cecropin A, and magainin II), washed, and then interrogated with streptavidin–phycoerythrin conjugate. Shown are median fluorescence values for each bead set. The antimicrobial peptides were immobilized onto beads indirectly using amine-rich scaffolds (BSA, lysozyme, aminodextran, and poly-ᴅ-lysine), amine-terminated linkers (PEG diamine and ethylenediamine), or a tetrapeptide possessing both a single carboxyl and single amine [(gly)$_4$]. Nonspecific binding of *E. coli* to the scaffold/tether material is shown as *white bars* ("no peptide").

4. Notes

1. EDC is highly unstable in the presence of water. It should be stored in a dessicator at –20°C. The dessicator and vial inside should be allowed to warm to room temperature before opening to avoid water condensation.

2. This solution must have no amine-containing components other than the biomolecule itself; amine-containing buffers such as Tris and glycine should not be used. If Tris or another unwanted amine is present, it will be necessary to dialyze or remove by gel filtration prior to reacting with the microspheres or poor coupling would result.

3. 2-Mercaptoethanol is toxic. Further, it has a distinctive unpleasant odor and causes irritation to nasal passageways and respiratory tract upon inhalation. For these reasons, it should be used in a chemical fume hood and unused/waste solutions should be disposed of as hazardous waste.

4. 2-Mercaptoethanol forms a stable complex with unreacted EDC and quenches the activation reaction. However, mercaptoethanol may also inactivate certain proteins. If inactivation of the protein is problematic, activated protein can be purified away from unreacted EDC using a Centricon (Millipore) or gel filtration, provided that these methods are performed as rapidly as possible.

Acknowledgments

This work was supported by the Defense Threat Reduction Agency and ONR/NRL 6.2 work unit 6336. The views are those of the authors and do not represent opinion or policy of the Naval Research Laboratory, US Navy, or Department of Defense.

References

1. Brady, D. and Jordaan, J. (2009) Advances in enzyme immobilisation. *Biotechnol. Lett.* **31**, 1639–1650.

2. Bi, S., Zhou, H., and Zhang, S. S. (2009) Bio-bar-code functionalized magnetic nanoparticle label for ultrasensitive flow injection chemiluminescence detection of DNA hybridization. *Chem. Commun.* 5567–5569.

3. Liu, C. X., Stakenborg, T., Peeters, S., and Lagae, L. (2009) Cell manipulation with magnetic particles toward microfluidic cytometry. *J. Appl. Phys.* **105**, 102014-1–102014-11.

4. Ugelstad, J., Stenstad, P., Kilaas, L., Prestvik, W. S., Herje, R., Berge, A., et al. (1993) Monodisperse magnetic polymer particles – New biochemical and biomedical applications. *Blood Purif.* **11**, 349–369.

5. Yang, H., Li, H. P., and Jiang, X. P. (2008) Detection of foodborne pathogens using bioconjugated nanomaterials. *Microfluid. Nanofluid.* **5**, 571–583.

6. Safarik, I. and Safarikova, M. (2009) Magnetic nano- and microparticles in biotechnology. *Chem. Pap.* **63**, 497–505.

7. Dunbar, S. A. (2006) Applications of Luminex (R) xMAP (TM) technology for rapid, high-throughput multiplexed nucleic acid detection. *Clin. Chim. Acta* **363**, 71–82.

8. Lehmann, S., Dupuy, A., Beaudeux, J. L., and Lizard, G. (2009) Multiplexed analysis for identification and evaluation of novel biomarkers in biological fluids, tissue and cell extracts. *Ann. Biol. Clin. (Paris)* **67**, 381–393.

9. Muldrew, K. L. (2009) Molecular diagnostics of infectious diseases. *Curr. Opin. Pediatr.* **21**, 102–111.

10. Tait, B. D., Hudson, F., Cantwell, L., Brewin, G., Holdsworth, R., Bennett, G., et al. (2009) Review article: Luminex technology for HLA antibody detection in organ transplantation. *Nephrology* **14**, 247–254.

11. Vignali, D. A. A. (2000) Multiplexed particle-based flow cytometric assays. *J. Immunol. Methods* **243**, 243–255.

12. Du, B. A., Li, Z. P., and Cheng, Y. Q. (2008) Homogeneous immunoassay based on aggregation of antibody-functionalized gold nanoparticles coupled with light scattering detection. *Talanta* **75**, 959–964.

13. Liu, X., Dai, Q., Austin, L., Coutts, J., Knowles, G., Zou, J. H., et al. (2008) A one-step homogeneous immunoassay for cancer biomarker detection using gold nanoparticle probes coupled with dynamic light scattering. *J. Am. Chem. Soc.* **130**, 2780–2782.

14. Sanchez-Martinez, M. L., Aguilar-Caballos, M. P., and Gomez-Hens, A. (2009) Homogeneous immunoassay for soy protein determination in food samples using gold nanoparticles as labels and light scattering detection. *Anal. Chim. Acta* **636**, 58–62.

15. Thaxton, C. S., Rosi, N. L., and Mirkin, C. A. (2005) Optically and chemically encoded nanoparticle materials for DNA and protein detection. *MRS Bull.* **30**, 376–380.

16. Wilson, R. (2008) The use of gold nanoparticles in diagnostics and detection. *Chem. Soc. Rev.* **37**, 2028–2045.

17. Cha, B. H., Lee, S. M., Park, J. C., Hwang, K. S., Kim, S. K., Lee, Y. S., et al. (2009) Detection of hepatitis B virus (HBV) DNA at femtomolar concentrations using a silica nanoparticle-enhanced microcantilever sensor. *Biosens. Bioelectron.* **25**, 130–135.

18. Seydack, M. (2005) Nanoparticle labels in immunosensing using optical detection methods. *Biosens. Bioelectron.* **20**, 2454–2469.

19. Wang, J. L., Munir, A., Li, Z. H., and Zhou, H. S. (2009) Aptamer-Au NP conjugates-enhanced SPR sensing for the ultrasensitive sandwich immunoassay. *Biosens. Bioelectron.* **25**, 124–129.

20. Wu, Y. F., Liu, S. Q., and He, L. (2009) Electrochemical biosensing using amplification-by-polymerization. *Anal. Chem.* **81**, 7015–7021.

21. Kuramitz, H. (2009) Magnetic microbead-based electrochemical immunoassays. *Anal. Bioanal. Chem.* **394**, 61–69.

22. Nagaoka, T., Shiigi, H., and Tokonami, S. (2007) Highly sensitive and selective chemical sensing techniques using gold nanoparticle assemblies and superstructures. *Bunseki Kagaku* **56**, 201–211.

23. Weetall, H. H. (1993) Preparation of immobilized proteins covalently coupled through silane coupling agents to inorganic supports. *Appl. Biochem. Biotechnol.* **41**, 157–188.

24. Marquette, C. A., Cretich, M., Blum, L. J., and Chiari, M. (2007) Protein microarrays enhanced performance using nanobeads arraying and polymer coating. *Talanta* **71**, 1312–1318.

25. Chen, Z. J., Peng, K., and Mi, Y. L. (2007) Preparation and properties of magnetic polystyrene microspheres. *J. Appl. Polym. Sci.* **103**, 3660–3666.

26. Yalcin, G., Elmas, B., Tuncel, M., and Tuncel, A. (2006) A low, particle-sized, nonporous support for enzyme immobilization: Uniform poly(glycidyl methacrylate) latex particles. *J. Appl. Polym. Sci.* **101**, 818–824.

27. Bake, K. D. and Walt, D. R. (2008) Multiplexed spectroscopic detections. *Annu. Rev. Anal. Chem.* **1**, 515–547.

28. Kang, K., Kan, C., Yeung, A., and Liu, D. (2006) The immobilization of trypsin on soap-free P(MMA-EA-AA) latex particles. *Mater. Sci. Eng. C* **26**, 664–669.

29. Spillmann, C. M., Naciri, J., Anderson, G. P., Chen, M. S., and Ratna, B. R. (2009) Spectral tuning of organic nanocolloids by controlled molecular interactions. *ACS Nano* **3**, 3214–3220.

30. Kuo, S. M., Chang, S. J., Yao, C. H., and Manousakas, I. (2009) A perspective view on the preparation of micro- and nanoparticulates of biomaterials from electrostatic and ultrasonic methods. *Biomed. Eng. Appl. Basis Commun.* **21**, 343–353.

31. Nobs, L., Buchegger, F., Gurny, R., and Allemann, E. (2004) Current methods for attaching targeting ligands to liposomes and nanoparticles. *J. Pharm. Sci.* **93**, 1980–1992.

32. Al-Saadi, A., Yu, C. H., Khurtoryanskiy, V. V., Shih, S. J., Crossley, A., and Tsang, S. C. (2009) Layer-by-layer electrostatic entrapment of protein molecules on superparamagnetic nanoparticle: A new strategy to enhance adsorption capacity and maintain biological activity. *J. Phys. Chem. C* **113**, 15260–15265.

33. Kato, N. and Caruso, F. (2005) Homogeneous, competitive fluorescence quenching immunoassay based on gold nanoparticle/polyelectrolyte coated latex particles. *J. Phys. Chem. B* **109**, 19604–19612.

34. Subramani, C., Ofir, Y., Patra, D., Jordan, B. J., Moran, I. W., Park, M. H., et al. (2009) Nanoimprinted polyethyleneimine: A multimodal template for nanoparticle assembly and immobilization. *Adv. Funct. Mater.* **19**, 2937–2942.

35. Choi, S. W., Kim, W. S., and Kim, J. H. (2003) Surface modification of functional nanoparticles for controlled drug delivery. *J. Dispersion Sci. Technol.* **24**, 475–487.

36. Rossi, S., Lorenzo-Ferreira, C., Battistoni, J., Elaissari, A., Pichot, C., and Delair, T. (2004) Polymer mediated peptide immobilization onto amino-containing N-isopropylacrylamide-styrene core-shell particles. *Colloid Polym. Sci.* **282**, 215–222.

37. Sheng, Y., Liu, C. S., Yuan, Y., Tao, X. Y., Yang, F., Shan, X. Q., et al. (2009) Long-circulating polymeric nanoparticles bearing a combinatorial coating of PEG and water-soluble chitosan. *Biomaterials* **30**, 2340–2348.

38. Iannone, M. A. and Consler, T. G. (2006) Effect of microsphere binding site density on the apparent affinity of an interaction partner. *Cytometry A* **69A**, 374–383.

39. Mouaziz, H., Braconnot, S., Ginot, F., and Elaissari, A. (2009) Elaboration of hydrophilic aminodextran containing submicron magnetic latex particles. *Colloid Polym. Sci.* **287**, 287–297.

40. Shen, Y., Kuang, M., Shen, Z., Nieberle, J., Duan, H. W., and Frey, H. (2008) Gold nanoparticles coated with a thermosensitive hyperbranched polyelectrolyte: Towards smart temperature and pH nanosensors. *Angew. Chem. Int. Ed. Engl.* **47**, 2227–2230.

41. Shi, X. Y., Thomas, T. P., Myc, L. A., Kotlyar, A., and Baker, J. R. (2007) Synthesis,

characterization, and intracellular uptake of carboxyl-terminated poly(amidoamine) dendrimer-stabilized iron oxide nanoparticles. *Phys. Chem. Chem. Phys.* **9**, 5712–5720.

42. Luminex Corporation. (2009) Microspheres/ Reagents.

43. Spherotech Inc. (2009) Microspheres.

44. Bangs Laboratories Inc. (2009) Polymer Microspheres.

45. Invitrogen Corporation. (2009) Beads and Microspheres.

46. Sigma-Aldrich Company. (2009) Latex Beads.

47. ThermoScientific – Pierce. (2009) Biotin Labeling of Antibodies and Other Proteins.

48. Anderson, G. P., Lamar, J. D., and Charles, P. T. (2007) Development of a Luminex based competitive immunoassay for 2,4,6-trinitrotoluene (TNT). *Environ. Sci. Technol.* **41**, 2888–2893.

49. Anderson, G. P. and Goldman, E. R. (2008) TNT detection using llama antibodies and a two-step competitive fluid array immunoassay. *J. Immunol. Methods* **339**, 47–54.

50. Kulagina, N. V., Shaffer, K. M., Ligler, F. S., and Taitt, C. R. (2007) Antimicrobial peptides as new recognition molecules for screening challenging species. *Sens. Actuators B Chem.* **121**, 150–157.

51. Ngundi, M. M., Kulagina, N. V., Anderson, G. P., and Taitt, C. R. (2006) Nonantibody-based recognition: Alternative molecules for detection of pathogens. *Expert Rev. Proteomics* **3**, 511–524.

52. Jain, K. K. (2006) Nanoparticles as targeting ligands. *Trends Biotechnol.* **24**, 143–145.

53. Hermanson, G. T. (2009) *Bioconjugate Techniques* (2nd ed.). Elsevier Science and Technology Books, Boulder, CO.

54. Porschewski, P., Grattinger, M. A. M., Klenzke, K., Erpenbach, A., Blind, M. R., and Schafer, F. (2006) Using aptamers as capture reagents in bead-based assay systems for diagnostics and hit identification. *J. Biomol. Screen.* **11**, 773–781.

55. Cheng, A. K. H., Ge, B., and Yu, H. Z. (2007) Aptamer-based biosensors for label-free voltammetric detection of lysozyme. *Anal. Chem.* **79**, 5158–5164.

56. Minunni, M., Tombelli, S., Gullotto, A., Luzi, E., and Mascini, M. (2004) Development of biosensors with aptamers as bio-recognition element: The case of HIV-1 Tat protein. *Biosens. Bioelectron.* **20**, 1149–1156.

57. Cheng, A. K. H., Sen, D., and Yu, H. Z. (2009) Design and testing of aptamer-based electrochemical biosensors for proteins and small molecules. *Bioelectrochemistry* **77**, 1–12.

58. Spangler, B. D. and Tyler, B. J. (1999) Capture agents for a quartz crystal microbalance-continuous flow biosensor: Functionalized self-assembled monolayers on gold. *Anal. Chim. Acta* **399**, 51–62.

59. Ngundi, M. M., Taitt, C. R., and Ligler, F. S. (2007) Crosslinkers modify affinity of immobilized carbohydrates for cholera toxin. *Sensor Lett.* **5**, 621–624.

60. Ngundi, M. M., Taitt, C. R., McMurry, S. A., Kahne, D., and Ligler, F. S. (2006) Detection of bacterial toxins with monosaccharide arrays. *Biosens. Bioelectron.* **21**, 1195–1201.

61. Rowe-Taitt, C. A., Cras, J. J., Patterson, C. H., Golden, J. P., and Ligler, F. S. (2000) A ganglioside-based assay for cholera toxin using an array biosensor. *Anal. Biochem.* **281**, 123–133.

62. Song, X. D., Shi, J., and Swanson, B. (2000) Flow cytometry-based biosensor for detection of multivalent proteins. *Anal. Biochem.* **284**, 35–41.

63. Moran-Mirabal, J. M., Throckmorton, D., Singh, A. K., and Craighead, H. G. (2004) Micropatterning of functional lipid domains for toxin detection. *Mater. Res. Soc. Symp. Proc.* **EXS-1**, H3.20.21–H23.20.23.

64. Chen, H., Zheng, Y., Jiang, J.-H., Wu, H.-L., Shen, G.-L., and Yu, R.-Q. (2008) An ultrasensitive chemiluminescence biosensor for cholera toxin based on ganglioside-functionalized supported lipid membrane and liposome. *Biosens. Bioelectron.* **24**, 684–689.

65. Kulagina, N. V., Anderson, G. P., Ligler, F. S., Shaffer, K. M., and Taitt, C. R. (2007) Antimicrobial peptides: New recognition molecules for detecting botulinum toxins. *Sensors* **7**, 2808–2824.

66. Kulagina, N. V., Lassman, M. E., Ligler, F. S., and Taitt, C. R. (2005) Antimicrobial peptides for detection of bacteria in biosensor assays. *Anal. Chem.* **77**, 6504–6508.

67. Taitt, C. R., North, S. H., and Kulagina, N. H. (2009) Antimicrobial peptide arrays for detection of inactivated biothreat agents. *Methods Mol. Biol.* **570**, 233–255.

Chapter 7

Multivalent Conjugation of Peptides, Proteins, and DNA to Semiconductor Quantum Dots

Duane E. Prasuhn, Kimihiro Susumu, and Igor L. Medintz

Abstract

Semiconductor nanocrystals or quantum dots (QDs) have become well-established as a unique nanoparticle scaffold for bioapplications due to their robust luminescent properties. In order to continue their development and expand this technology, improved methodologies are required for the controllable functionalization and display of biomolecules on QDs. In particular, efficient routes that allow control over ligand loading and spatial orientation, while minimizing or eliminating cross-linking and aggregation are needed. Two conjugation approaches are presented that address these needs: (1) polyhistidine-based metal-affinity self-assembly to QD surfaces and (2) carbodiimide-based amide bond formation to carboxy-functionalized polyethylene glycol or PEGylated QDs. These approaches can be successfully employed in the construction of a variety of QD-biomolecule constructs utilizing synthetic peptides, recombinant proteins, peptides, and even modified DNA oligomers.

Key words: Quantum dots, Semiconductor nanocrystal, Bioconjugation, Nanoparticle, Self-assembly, EDC coupling, Peptide, Protein, Polyhistidine, Metal affinity, Fluorescence, Biosensing

1. Introduction

1.1. Biomolecule-Functionalized, Luminescent Semiconductor Quantum Dots

One of the most prominent research areas in nanotechnology continues to be the development of nanoparticle systems for use in biomedical and bioinorganic nanoscale engineering applications. These systems offer several advantages over those that are conventionally used; for example, nanoparticles can be imbued with multiple molecular functionalities, allowing them to perform both a therapeutic and a diagnostic function simultaneously (1). Luminescent semiconductor nanocrystals, or quantum dots (QDs), have now become well-established fluorophores and are one of the most popular types of nanomaterials being explored for use in in vitro and in vivo bioapplications (2–5). This utility derives from their unique

Sarah J. Hurst (ed.), *Biomedical Nanotechnology: Methods and Protocols*, Methods in Molecular Biology, vol. 726, DOI 10.1007/978-1-61779-052-2_7, © Springer Science+Business Media, LLC 2011

95

photophysical properties which include broad absorption spectra with large molar extinction coefficients ($\varepsilon \sim 10^5$–10^6 $M^{-1}cm^{-1}$); narrow, symmetrical photoluminescence (PL) spectra (35–45 nm full width at half-maximum), which can be tuned from the UV to near-IR as a function of QD core size and constituent material (the PL peaks red-shift as the radius of the nanoparticle is increased); high quantum yields of up to 40–60%, effective Stokes shifts of 300–400 nm; large multiphoton action cross-sections with typical values of 8,000 to >20,000 Goeppert-Mayer (GM) units at 800 nm (6); access to facile multiplexing configurations; and unparalleled resistance to photo- and chemical degradation as compared to conventional organic fluorophores and fluorescent proteins (3, 7–9). Further, these properties allow QDs to function as excellent fluorescence resonance energy transfer (FRET) donors in a variety of biosensing configurations (7).

Core/shell QD materials are typically used in biological applications, where a wider band-gap semiconductor shell functions to passivate the core material, prevent leaching, and improve fluorescence quantum yield (3, 8–11). These nanomaterials are commonly synthesized from binary combinations of semiconductor materials, including ZnS, CdS, CdSe, InP, CdTe, PbS, PbTe, and PbSe and can have a range of emissions from the UV to near-IR depending upon the constituents used (3, 8–11). The reactions used to synthesize these structures are usually carried out in organic solution at high temperatures using pyrophoric precursors and the resulting particles are stabilized with hydrophobic organic ligands that lack intrinsic aqueous solubility (10, 11). This presents an issue for the use of such structures in biological applications, since water-soluble antibodies, proteins, peptides, or DNA must be attached to the QD surface to achieve a conjugate structure capable of targeting and biorecognition in an aqueous environment. Therefore, the QD surfaces require further modification with bifunctional molecular ligands that imbue them with aqueous solubility (3, 8) and that sometimes display pertinent groups (such as carboxyls or amines) for subsequent chemical coupling to biomolecules of interest. The two major strategies for accomplishing this are either the use of bifunctional amphiphilic polymers that interdigitate with the native organic surface ligands or hydrophilic molecules that replace the native layer via "cap" exchange to provide a hydrophilic particle surface (3, 8, 12).

Unfortunately, most strategies for functionalizing the surface of QDs with biomolecules provide little control over their loading and/or spatial orientation. For instance, although biotin-Avidin chemistry is commonly used, it is always constrained by the need to label both participants appropriately, their susceptibility to cross-linking, and the presence of large multivalent Avidin protein (64 kDa) intermediaries in the conjugate (13). Also, with very few exceptions, adsorption or noncovalent association (through electrostatic or hydrophobic interactions) of natural/engineered proteins or nucleic acids generally yields poor control over loading and results in heterogeneity in spatial orientation and mixed functional avidity (3, 4, 8). Functional groups commonly targeted for biomodification, such as carboxyls and amines, are ubiquitous in biological molecules and the resulting functionalization with QDs, or other nanoparticles, can be heterogeneous unless appropriate issues are considered. Reviews expanding on many of these points are available for the interested reader (3, 4, 8–10).

Thus, alternative QD conjugation chemistries that are efficient, targeted, utilize functionalities orthogonal to common biological moieties (i.e., do not cross-react), minimize heterogeneity in attachment, and allow for control over conjugate loadings can greatly increase the utility of these nanomaterial fluorophores. Two such chemistries for functionalizing QDs are presented here: (1) polyhistidine-based self-assembly which is applicable to proteins, peptides, and modified DNA and (2) carbodiimide-based covalent coupling, which is more suited to peptides and DNA. When appropriately implemented, these chemistries can provide high-affinity attachments with control over the valence and orientation of biomolecules attached to a QD particle.

1.2. Self-Assembly of Polyhistidine-Appended Biomolecules to Quantum Dots

Many proteins are recombinantly engineered to express N- or C-terminal hexahistidine $(His)_6$ sequences so that they can be purified over Ni^{2+}-nitrilotriacetic acid (NTA) chelate media (14, 15). This technology is based on the high-affinity interactions of the histidine side-chain imidazolium groups with chelated divalent cations, including those of Ni, Cu, Fe, Cr, Co, Zn, and Mn. Over the last few years, the same type of approach has been used to synthesize and characterize biomolecule–semiconductor QD conjugates where the polyhistidine-appended biomolecule is attached to the surface of the semiconductor QD by taking advantage of the metal-affinity interaction between, for example, the Zn-rich surface of a CdSe/ZnS core/shell QD and the histidine side-chain of a biomolecule of interest (see Fig. 1a) (17). The self-assembly that occurs rapidly (on the order of seconds to minutes after mixing), is equally efficient when using QDs capped with small, negatively charged dihydrolipoic acid (DHLA) or neutral polyethylene glycol (PEG) ligands, has a high-affinity equilibrium constant in solution (K_a ~1 nM^{-1}), and allows for control over the number of substrate molecules arrayed on a single QD through modulation of the molar ratios of the participants used (i.e., number of nanoparticles to number of biomolecules) (17). In some cases, such self-assembly can also provide control over the spatial orientation of the biomolecule on the QD surface (18). This metal-affinity driven strategy has already been demonstrated for assembling proteins, peptides, and appropriately modified peptide–DNA conjugates on QDs to create a variety of sensor systems that

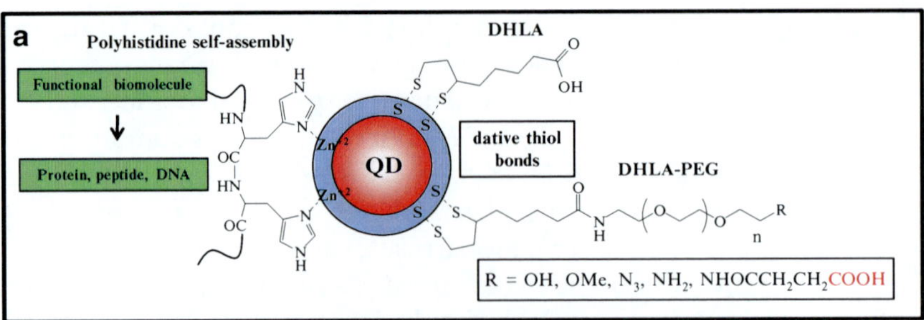

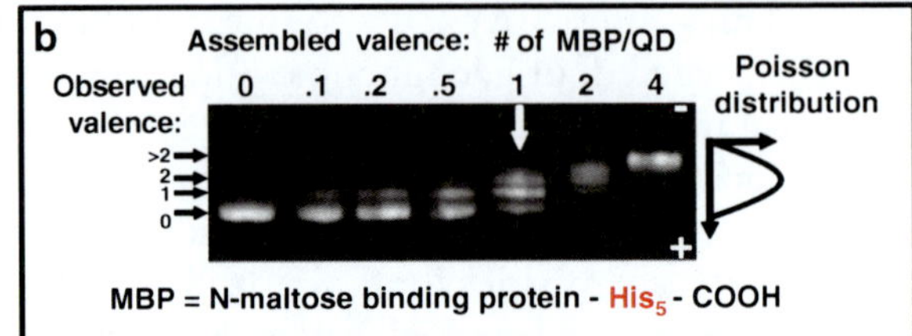

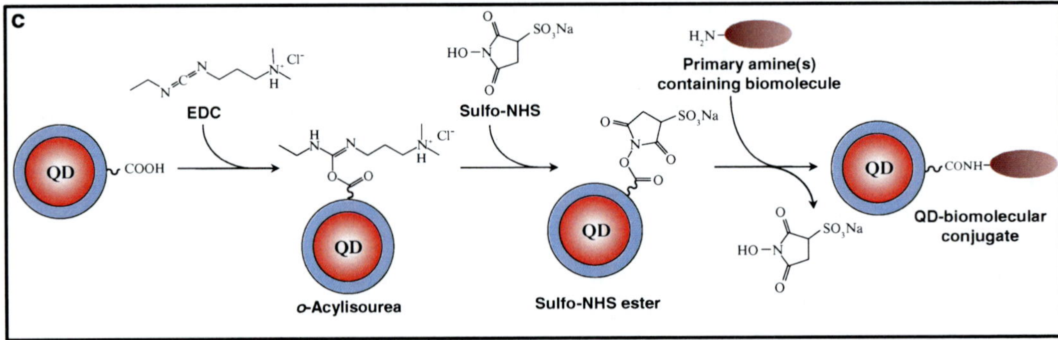

Fig. 1. (**a**) General representation of a decorated QD nanocrystal (CdSe core in *red* and ZnS overcoat in *blue*), not to scale. The QDs are made more hydrophilic by cap-exchanging with dihydrolipoic acid (DHLA) or a PEG derivative that can display a variety of terminal functionalities (DHLA-PEG, NHCO is an amide bond). A polyhistidine-appended functional biomolecule (only two histidine residues shown) coordinated to the QD surface by metal-affinity driven self-assembly is also displayed. It should be noted that although two adjacent histidines are depicted as involved in coordination to two differing Zn^{2+} ions, the exact nature of the complexation probably involves multiple, varying coordination configurations. (**b**) An agarose gel showing the separation of self-assembled QD-protein bioconjugates with different numbers of proteins per conjugate. In this example, maltose binding protein (MBP, MW ~ 44 kDa) with a C-terminal pentahistidine sequence was self-assembled to DHLA-QDs. At small ratios, samples show several mobility shift bands due to the Poisson distribution as expected. The *white arrow* indicates a sample valence of 1 which can be expected to manifest three types of ratios: ~33% with 0 molecules/QD, ~33% with 1 molecule/QD, and ~33% with >1 molecule/QD (mostly 2/QD). These conjugates merge into a single band indicative of a homogeneous distribution of conjugate sizes as the average protein to-QD ratio increases. Figure adapted from ref. 16 with permission from the American Chemical Society. (**c**) General schematic for the conjugation of aminated biomolecules to PEGylated-QDs (PEG molecules terminate in carboxyls) through amide bond formation using an EDC-mediated reaction with a sulfo-NHS ester intermediary step (13).

target nutrients, explosives, and other small molecules (3, 7, 19–22) and is being steadily adopted by many research groups (23, 24).

Several factors should be considered before undertaking this type of bioconjugation with QDs. First, the relevant biomolecules must display clearly available polyhistidine sequences (i.e., not internal or sterically hindered). As mentioned, proteins are commonly expressed with recombinant $(His)_6$-sequences and nascent peptides can be synthesized to include them. DNA needs to be chemically linked to the $(His)_6$-sequences and several strategies for achieving this have been demonstrated with terminally-modified oligonucleotides (17, 19–21, 25). Similar to protein purification, four to six consecutive His residues are sufficient for high-affinity attachment to QDs (17). Due to the small ligand size, proteins, peptides, and DNA can be assembled to dihydrolipoic acid (DHLA)-functionalized QDs, while the larger size of the PEGylated ligands may sterically preclude assembly of larger globular proteins (see Fig. 1a). When these types of strategies were demonstrated with commercial QD preparations, the addition of small amounts of Ni^{2+} ions that presumably interacted with surface functional groups (principally carboxyls), was necessary to form a functional Ni^{2+} chelate structure (26). Further, QDs have been covalently modified with NTA groups on their surface ligands to facilitate similar attachments (27).

In this approach, the number of (sometimes fluorescently labeled) biomolecules attached to the QDs can usually be determined through two methods. The first is through physical separation of the conjugates using an agarose gel (16). Under appropriate conditions, even mono-labeled QD species can be easily resolved (see Fig. 1b) (28). In the second method, changes in FRET between the central QD donor and increasing numbers of acceptor dye-labeled biomolecules attached to the QD are monitored and the number of self-assembled biomolecules is then determined by comparison to a known or previously established "calibration curve" (see Fig. 2) (19). Poisson distribution kinetics determines the average number of biomolecules attached per QD since self-assembly is a random process (29). This may only be an issue when working at low valences of <4 biomolecules per QD as unlabeled QDs can be assumed to be present. At valences of >4, the actual distributions will more closely match predictions. For example, when assembling a nominal ratio of 1 biomolecule per QD, the resulting bioconjugates can be expected to manifest three types of ratios: ~ 33% with 0 molecules/QD, ~ 33% with 1 molecule/QD, and ~33% with >1 molecule/QD (mostly 2/QD) (see Fig. 1b).

1.3. Carbodiimide Coupling to Polyethylene Glycol Functionalized Quantum Dots

Amide bond formation using 1-ethyl-3-(3-dimethylaminopropyl) carbodiimide (EDC) chemistry is one of the most common bioconjugation chemistries employed for modifying biomolecules including peptides, proteins, and DNA (see Fig. 1c) (13). The popularity of this chemistry arises from the fact that the two targeted groups (amines and carboxyls) are ubiquitous to proteins and peptides and can be chemically added to DNA; the same functional groups can be displayed on most nanoparticle surfaces in a relatively facile manner. Also, the reagents for performing this chemistry can be purchased in large quantities at relatively cheap

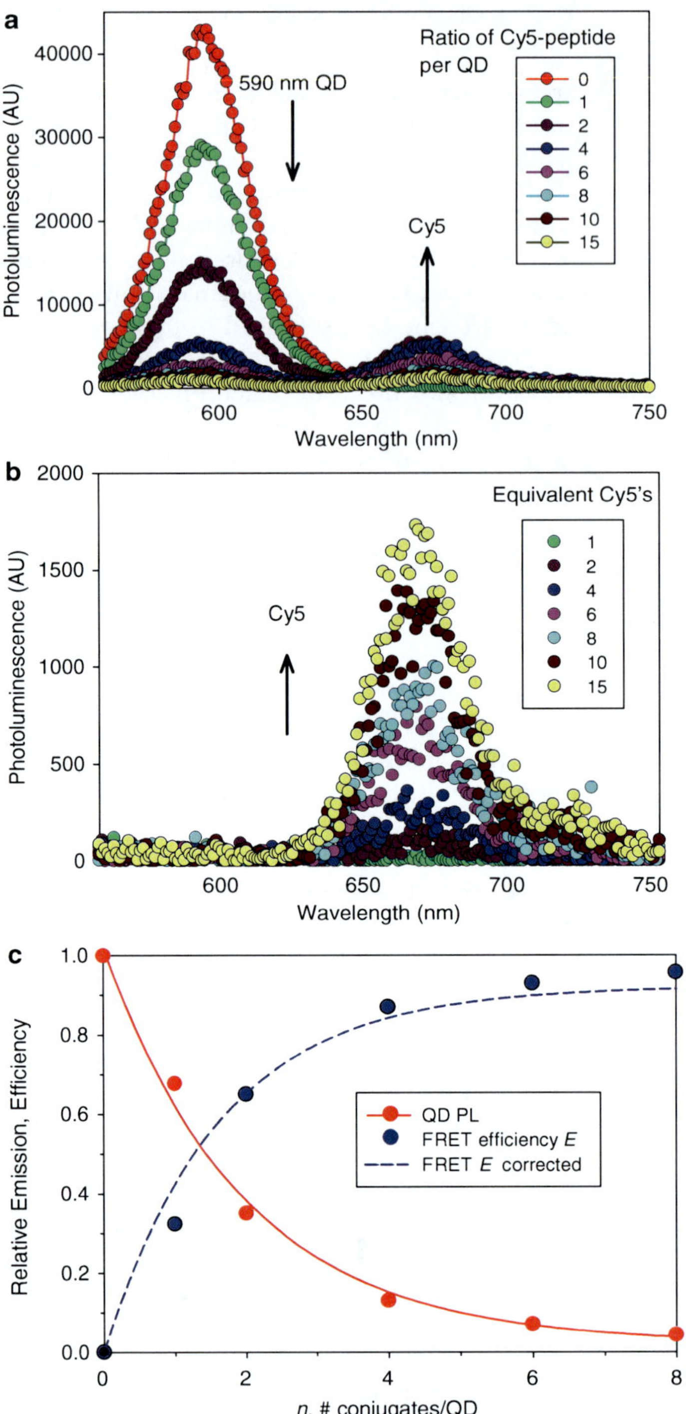

Fig. 2. (**a**) Representative FRET spectral results for a QD–Cy5-labeled acceptor system (QD donor emission maxima at 590 nm and Cy5 acceptor emission maxima at 670 nm) for an increasing number of Cy5-labeled His_6-peptides self-assembled per QD. The QD donor PL shows the characteristic losses in magnitude as the Cy5-acceptor emission rises. (**b**) The direct-acceptor excitation spectrum for equivalent amounts of Cy5-labeled His_6-peptides alone. Note the significant increase in FRET-sensitized Cy5 emission at ~670 nm when coordinated to the QDs. All spectra were taken using 350 nm excitation. (**c**) Normalized QD PL loss vs. n (acceptor valence), corresponding FRET efficiency, and Poisson-corrected FRET efficiency for the QD–Cy5-labeled His_6-peptide system. The QD-dye separation distance or r value determined for this system using Eq. 7.3 is ~43Å and the corrected value using Eq. 7.4 is 45Å. This type of data would, for example, be useful to quantitatively monitor enzymatic cleavage of the peptide while attached to the QD in a protease sensor configuration (19, 22).

prices. Indeed, the QDs used in some of the seminal biomedical demonstrations where functionalized with proteins for cellular uptake by reacting to the thiol-alkyl-COOH ligands utilizing EDC chemistry (30). Subsequent studies, however, showed that QDs capped with DHLA or other similar short-chain thiol-alkyl-COOH ligands often encounter a loss of solubility and macroscopic aggregation at the neutral to acidic pHs required for this chemistry. This aggregation results from the protonation of the terminal carboxyl groups responsible for maintaining colloidal stability (31). The more recent development of pH stable, PEG-based QD solubilizing ligands displaying a variety of terminal functionalities (including carboxylic acids) circumvent many of the previous issues (12).

Again, several factors are important for consideration when undertaking this type of bioconjugation with QDs. The chemistry *a priori* should work equally well with QDs displaying carboxyls and targeting amines on the biomolecules as well as in the converse configuration. The best results to date, however, have been consistently achieved utilizing QDs displaying carboxyls and targeting either peptides or DNA site-specifically modified with a unique, terminal amine. This type of set-up provides for control over biomolecule orientation on the QD since the attachment can only be in one configuration. EDC chemistry for attaching proteins to QDs will usually result in mixed orientation/avidity and cross-linking due to the multiple possible molecular conformations. These issues can be addressed by testing several different ratios of proteins to QDs to optimize the outcome. Although EDC can be used directly in the reaction with both participants, better results are achieved in combination with sulfo-NHS to first activate the carboxylic acids (13). EDC has a very short usable half-life in aqueous solution and it has been found that the reactions can be "recharged" with additional aliquots of fresh EDC without deleterious effects to provide higher conjugation yields. The number of moieties attached per QD can be controlled empirically through the molar excess of biomolecule used and usually requires several iterative attempts if a specific ratio is required.

Successful bioconjugation can be confirmed by monitoring changes in QD mobility on agarose or polyacrylamide gels, by using Förster resonance energy transfer (FRET) interactions, or by UV absorbance if the biomolecule is dye-labeled (see Figs. 2 and 3) (12) although there are intrinsic limitations to each. In fact, we commonly utilize side-by-side reactions where a dye-labeled peptide or DNA is attached as a positive control. The difference in appearance of the QD-protein conjugate profiles relative to control QD samples in gel analysis are used as inferential evidence to verify or confirm conjugation (see Fig. 3). This data does not, however, tell us the actual ratio of proteins conjugated/QD nor if they are still functional. Further, as QD absorbance

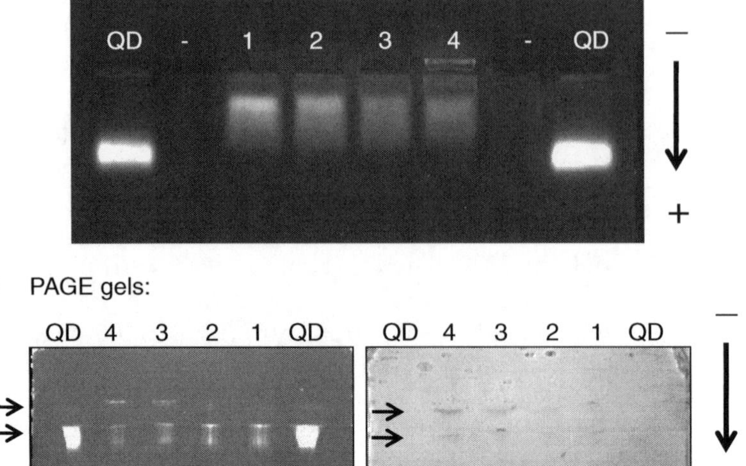

Fig. 3. Representative agarose (**a**) and polyacrylamide gels (**b**) for PEGylated QDs derivatized with a ~25 kDa protein via EDC coupling chemistry. QD designates the nanoparticles alone. QDs were reacted with a 3×, 8×, 16×, and 32× excess molar ratio of proteins as designated by samples 1–4, respectively. In the agarose gel, the differences in migration and intensity correlate with the formation of a much larger molecular weight species. In the PAGE gels, the staining of proteins by Coomassie, along with the QD fluorescence, correlate to the newly formed species, especially the slower moving, higher molecular weight QD-protein composites.

dramatically increases as a continuum towards the UV, this will most often mask the commonly used DNA/protein absorptions at 260 and 280 nm, respectively, severely limiting this portion of the spectrum for analysis of component absorptions.

2. Materials

2.1. Self-Assembly of Polyhistidine-Appended Biomolecules to Quantum Dots

1. Dissolve approximately 1 mg of peptide (here, a peptide sequence $CSTRIDEANQRATKLP_9SH_6$ that terminates with a C-terminal His_6 sequence was utilized as an example) in 100 µL of 1× phosphate buffered saline (PBS; 137 mM NaCl, 10 mM phosphate, 2.7 mM KCl; Fisher), pH = 7.4 (see Note 1). This example is similar to that described in ref. 22.

2. Two vials of Cy5-maleimide monoreactive dye (GE Healthcare; Piscataway, NJ) dissolved in 100 µL dimethyl sulfoxide (DMSO) total volume (see Note 2).

3. Three mini-columns containing Ni-NTA Agarose resin (Qiagen; Valencia, CA) (see Note 3).

4. 300 mM imidazole dissolved in 1× PBS, pH = 7.4.

5. Oligonucleotide purification cartridge (OPC; Invitrogen; Carlsbad, CA) equilibrated with 3 mL acetonitrile and 3 mL 2 M triethylammonium acetate (TEAA, Invitrogen; Carlsbad, CA).

6. 70% acetonitrile/water (v/v).

7. 100 mM TEAA.

8. 1 μM aqueous solution of QDs surface functionalized with dihydrolipoic acid (DHLA) or DHLA-PEG$_{600}$ (see Note 4).

9. 96-Well microtiter plates (polystyrene with nonbinding surface, Corning; Corning, NY).

2.2. EDC Coupling to PEGylated QDs for Bioconjugation

1. 100 mM Lissamine rhodamine B ethylenediamine dye (Invitrogen; Carlsbad, CA) (for this example) in 1× PBS, pH = 7.4.

2. 100 mM *N*-hydroxysulfosuccinimide (sulfo-NHS; Pierce Biotechnology, Rockford, IL) in 1× PBS, pH = 7.4.

3. 500 mM 1-ethyl-3-(3-dimethylaminopropyl)carbodiimide (EDC; Pierce Biotechnology) in 1× PBS, pH = 7.4.

4. 1 μM aqueous solution of QDs (emission at ~590 nm) capped with DHLA-PEG$_{750}$-OMe (methoxy)/DHLA-PEG$_{600}$-COOH (19:1 ratio of Methoxy:COOH terminated ligand – equivalent to QDs functionalized with 5% surface carboxyls) (see Note 5).

5. Disposable PD-10 desalting gel columns (GE Healthcare, Piscataway, NJ) equilibrated with 1× PBS, pH = 7.4.

3. Methods

3.1. Self-Assembly of Polyhistidine-Appended Biomolecules to Quantum Dots and Förster Resonance Energy Transfer Characterization of the Resulting Conjugates

1. Mix peptide solution with reactive dye solution in a 1.5-mL Eppendorf tube and incubate overnight (16–18 h) in the dark at room temperature or alternatively at 4°C with constant agitation using a benchtop vortexer.

2. Load the peptide–dye mixture over three consecutive columns of Ni-NTA agarose using 1-mL syringes (see Note 6).

3. Wash each column with 10 mL 1× PBS to remove excess dye.

4. Elute the Cy5-peptide with ~2 mL 300 mM imidazole per column.

5. Load the dye-modified peptide eluent (in imidazole) into an OPC containing reverse phase media (see Note 7).

6. Rinse the cartridge with 50 mL of 100 mM TEAA and 50 mL deionized water (see Note 8).

7. Elute the substrate from the OPC with a 70% acetonitrile/water solution (1–2 mL).

8. Concentrate the sample as necessary. In these experiments, a DNA120 speed-vacuum centrifuge (GMI; Ramsey, MN) was used.

9. Measure the absorption spectra of the Cy5-peptide and quantify the amount of labeled peptide (see Note 9).

10. Aliquot the dye–peptide into 1.5-mL Eppendorf tubes, then dry and store the samples as a pellet at –20°C in a dessicator (can be stored >6 months).

11. When ready to use, resuspend the dye–peptide in 10% DMSO/water (v/v) yielding a final concentration of 10 µM.

12. To multiple 1.5-mL Eppendorf tubes, add 20 µL 1 µM QD-DHLA-PEG$_{600}$ particles (equivalent to 20 pmols of QD per sample in 100 µL or 0.2 µM).

13. To each tube, add the necessary volume of dye–peptide solution to yield the desired ratio of dye–peptide to QD (i.e., 2.0 µL for 1 peptide/QD, 4.0 µL for 2 peptides/QD, etc.) in a final volume of 100 µL of 1× PBS buffer. Mix/vortex and briefly centrifuge to collect the samples in the bottom of the tubes. Table 1 shows an example of the volumes of reagents and concentrations necessary to achieve desired peptide loadings.

14. Incubate the samples at room temperature for at least 15 min.

15. Transfer 100 µL of each reaction into a microtiter 96-well plate and measure the emissions on a plate reader with an excitation at 350 nm. Here, readings were taken on a Tecan Safire Dual Monochromator Multifunction Microtiter Plate Reader (Tecan; Durham, NC). Representative emission

Table 1
Reagent volumes utilized for QD-peptide self-assembly

Reagent (µL)	Desired ratio of peptide/QD							
	0	1	2	4	6	8	10	15
QD (1 µM)	20	20	20	20	20	20	20	20
Peptide (10 µM)	–	2	4	8	12	16	20	30
10× PBS	10	10	10	10	10	10	10	10
H$_2$O	70	68	66	62	58	54	50	40
Total	100	100	100	100	100	100	100	100

results are presented in Fig. 2 (see Note 10). Most laboratory benchtop fluorometers can also be used.

16. Analyze the resulting FRET interactions (Fig. 2) as follows:

First, determine the Förster separation distance R_0 (defined as the spacial separation distance between donor and acceptor corresponding to 50% energy transfer efficiency) for the particular QD-donor dye-acceptor pair that is being utilized using (32):

$$R_0 = \left(\frac{[9,000 \times (\ln 10)]\kappa_p^2}{128\pi^5 n_D^4 N_A} Q_D I \right)^{1/6}, \tag{1}$$

where, n_D is the refractive index of the medium, Q_D is the QD PL quantum yield (determined through comparison of QD sample emission with the emission of a standard compound of known quantum yield; typical values for the reported QDs are 20–30%), I is the integral of the spectral overlap function, κ_p^2 is the dipole orientation factor ($\kappa_p^2 = 2/3$ is used for self-assembled donor–acceptor pairs with random dipole orientations) and N_A is Avogadro's number (18, 19, 32–34).

Then, extract the average energy transfer efficiency E from the fluorescence data for the biomolecule-QD conjugates using the expression:

$$E = \frac{(F_D - F_{DA})}{F_D}, \tag{2}$$

where, F_D and F_{DA} are, respectively, the fluorescence intensities of the QD donor alone and donor in the presence of the acceptor(s) (i.e., the labeled biomolecules) (32) (Fig. 2).

Then, if analyzed within the Förster dipole–dipole formalism, the energy transfer efficiency data can be fit to the expression (34):

$$E = \frac{nR_0^6}{nR_0^6 + r^6}, \tag{3}$$

where, R_0 was determined using Eq. 1, n is the average number of fluorescently labeled biomolecules per QD and r is the QD-donor dye-acceptor center-to-center separation distance. The metal-His driven self-assembly of biomolecules to these nanocrystals provides conjugates with a centrosymmetric distribution of acceptors around a QD. For QDs self-assembled with dye-labeled proteins and peptides, this distribution is usually characterized by a constant average center-to-center separation distance, r (19, 34). For conjugates having small numbers of acceptors ($n < 5$), heterogeneity in the conjugate valence can be accounted for by using a Poisson distribution function, $p(N,n)$, when fitting the efficiency data (29):

$$E = \sum_n p(N,n)E(n) \qquad p(N,n) = N^n \frac{e^{-N}}{n!}, \qquad (4)$$

where, n designates the exact numbers of acceptors (valence) for conjugates with a nominal average valence of N (see Note 11).

3.2. EDC Coupling to PEGylated QDs for Bioconjugation

1. Here, an amine-containing dye is conjugated to a QD. To a 1.5-mL Eppendorf tube, add 118.8 µL of a QD-DHLA-PEG$_{750}$-OMe/DHLA-PEG$_{600}$-COOH (19:1) stock solution. This is similar to the example highlighted in ref. 12.

2. Add EDC (77 µL), sulfo-NHS (46.2 µL), and 1× PBS (81 µL).

3. Incubate the reaction at room temperature in the dark with orbital shaking for ~10 min.

4. Lissamine rhodamine B ethylenediamine (77 µL) is then added to yield a final reaction volume of 400 µL with final concentrations of 1.93 µM, 19.3 mM, 57.8 mM, and 19.3 mM for QD, EDC, sulfo-NHS, and the dye, respectively (see Note 12).

5. Allow the mixture to react at room temperature in the dark with orbital shaking for ~2 h (see Note 13). The EDC and sulfo-NHS in the reaction can be recharged through the addition of further aliquots as needed.

6. Load the mixture onto a PD-10 desalting gel column and elute the QD conjugate with 1× PBS (see Note 14).

7. Concentrate the sample as needed and run on a UV-visible spectrophotometer to determine conjugate loading per QD if the biomolecule is dye-labeled. If a protein is used, agarose or polyacrylamide gels can provide inferential evidence that the QDs have been conjugated (see Fig. 3), but not exact loading numbers or verification of activity.

4. Notes

1. The presented work utilizes a peptide as an example substrate for conjugation to QDs. These peptides are commonly prepared by standard solid-phase synthesis on Rink amide resin and can be subsequently modified with a dye molecule using

cysteine–maleimide chemistry. The methodology, however, works in conjunction with a variety of bioconjugation chemistries and can be extended to other polyhistidine-appended biomolecules, such as recombinantly expressed proteins and peptides, and DNA oligomers (17, 21, 25).

2. The labeling kit, as purchased, contains enough dye to label 1 mg of protein. Thus, to ensure complete labeling, use excess dye (i.e., two or more vials).

3. Mini-columns with syringe attachments at both ends were prepared by loading ~1 mL of Ni-NTA Agarose resin into empty oligonucleotide purification cartridges (OPC, Applied Biosystems).

4. CdSe/ZnS core/shell QDs capped with hydrophobic, organic ligands (generally a mixture of trioctylphosphine/trioctylphosphine oxide (TOP/TOPO) and hexadecylamine) were synthesized using organometallic procedures as previously reported (11, 31, 35). The nanocrystals were made water-soluble by exchanging the TOP/TOPO ligands with polyethylene glycol-appended dihydrolipoic acid (DHLA) or DHLA-PEG$_{600}$ (PEG with M_{W}~600 or M_{W}~750) through standard methods (12, 31, 36, 37). The above procedures can be extended to cap particles of various sizes (diameters of ~4 to >10 nm) with DHLA and recent publications have also utilized commercially available QDs (23, 38, 39).

5. Mixed surface CdSe/ZnS core/shell QDs were prepared as mentioned in Note 4. The nanocrystals were made water-soluble by exchanging the TOP/TOPO ligands with a 19:1 mixture of methoxy- and carboxy-terminated polyethylene glycol-appended dihydrolipoic acid (DHLA-PEG$_{750}$-OMe/DHLA-PEG$_{600}$-COOH, respectively) through standard methods (12). Using QDs prepared with higher percentages of carboxylated ligands on their surface will result in correspondingly higher surface functionalization efficiencies.

6. To facilitate purification, the reaction mixture is diluted with 200–400 µL buffer. Then, the entire reaction mixture is passed through a single column 10–15 times to ensure maximum binding. The remaining filtrate is sequentially passed through a second column 10–15 times and then a third column 10–15 times. If necessary, more columns could be used to ensure maximum product isolation, but generally all substrate binding has occurred within the initial passes through the two columns since ~1 mL of NTA resin can bind ~1 mg of labeled protein.

7. Generally, a single loading through an OPC is sufficient. If, however, there is still color in the filtrate (or one is working with a colorless substrate), a second/third loading onto an

equilibrated OPC may be desired/necessary for maximum product isolation.

8. The function of this step is twofold: it removes the imidazole (which can quench QD photoluminescence), and it desalts/lyophilizes the peptide prior to long-term storage.

9. This reactive dye has an extinction coefficient of $250,000 \ M^{-1} \ cm^{-1}$ at 649 nm.

10. Some biomolecules, such as large proteins, may have difficulty accessing the QD surface for self-assembly due to steric interactions with the PEG ligands. Recombinant addition of a flexible, linker sequence between the polyhistidine residues and the main protein structure can lead to enhanced coordination. QDs cap-exchanged with far smaller DHLA do not experience such steric issues.

11. Due to the nature and scope of this publication, only a cursory discussion is presented on the FRET analyses of these QD systems. A more thorough explanation and discussion of this topic can be obtained in refs. 7, 19, 34.

12. The 10 min incubation prior to dye addition is expected to allow some pre-formation of the amine-reactive NHS-ester intermediate. These are example reaction concentrations; optimized reagent concentrations can vary. The reader is referred to refs. 12, 13, 37 for a discussion of considerations regarding reaction optimization.

13. At times, a nonfluorescent precipitate may form after reaction. The exact nature of this material is unknown, but it is believed to be some sort of salt by-product. Also, longer reaction times are possible, but gradual photoluminescence loss can occur if extended for much longer than a few hours (i.e., overnight).

14. As reported, this reaction is done at a QD concentration and volume that allows monitoring of the emission using a UV lamp during column elution. If less material is required, fractions (1 mL or smaller) can be collected and the fraction containing the desired eluent can be identified using a UV-visible spectrophotometer. Samples can be concentrated using a speed vacuum as well.

Acknowledgments

I. L. M. and K. S. acknowledge DTRA/ARO, ONR, NRL, and the NRL-NSI for financial support. D. E. P. acknowledges an ASEE fellowship through NRL.

References

1. Niemeyer, C. M. (2001) Nanoparticles, proteins, and nucleic acids: Biotechnology meets materials science. *Angew. Chem. Int. Ed.* **40**, 4128–4158.

2. Alivisatos, P. (2004) The use of nanocrystals in biological detection. *Nat. Biotechnol.* **22**, 47–52.

3. Medintz, I., Uyeda, H., Goldman, E., and Mattoussi, H. (2005) Quantum dot bioconjugates for imaging, labeling and sensing. *Nat. Mater.* **4**, 435–446.

4. Michalet, X., Pinaud, F. F., Bentolila, L. A., Tsay, J. M., Doose, S., Li, J. J., et al. (2005) Quantum dots for live cells, *in vivo* imaging, and diagnostics. *Science* **307**, 538–544.

5. Parak, W. J., Pellegrino, T., and Plank, C. (2005) Labelling of cells with quantum dots. *Nanotechnology* **16**, R9–R25.

6. Clapp, A. R., Pons, T., Medintz, I. L., Delehanty, J. B., Melinger, J. S., Tiefenbrunn, T., et al. (2007) Two-photon excitation of quantum-dot-based fluorescence resonance energy transfer and its applications. *Adv. Mater.* **19**, 1921–1926.

7. Medintz, I. L. and Mattoussi, H. (2009) Quantum dot-based resonance energy transfer and its growing application in biology. *Phys. Chem. Chem. Phys.* **11**, 17–45.

8. Klostranec, J. M. and Chan, W. C. W. (2006) Quantum dots in biological and biomedical research: Recent progress and present challenges. *Adv. Mater.* **18**, 1953–1964.

9. Alivisatos, A. P., Gu, W., and Larabell, C. A. (2005) Quantum dots as cellular probes. *Annu. Rev. Biomed. Eng.* **7**, 55–76.

10. Murray, C. B., Kagan, C. R., and Bawendi, M. G. (2000) Synthesis and characterization of monodisperse nanocrystals and close-packed nanocrystal assemblies. *Annu. Rev. Mater. Sci.* **30**, 545–610.

11. Dabbousi, B. O., Rodriguez-Viejo, J., Mikulec, F. V., Heine, J. R., Mattoussi, H., Ober, R., et al. (1997) (CdSe)ZnS core-shell quantum dots: synthesis and optical and structural characterization of a size series of highly luminescent materials. *J. Phys. Chem. B* **101**, 9463–9475.

12. Susumu, K., Uyeda, H. T., Medintz, I. L., Pons, T., Delehanty, J. B., and Mattoussi, H. (2007) Enhancing the stability and biological functionalities of quantum dots via compact multifunctional ligands. *J. Am. Chem. Soc.* **129**, 13987–13996.

13. Hermanson, G. T. (2008) *Bioconjugate Techniques (2nd Ed.)* Academic Press, San Diego.

14. Hochuli, E., Bannwarth, W., Dobeli, H., Gentz, R., and Stuber, D. (1988) Genetic approach to facilitate purification of recombinant proteins with a novel metal chelate adsorbent. *Biotechnology* **6**, 1321–1325.

15. Schmitt, J., Hess, H., and Stunnenberg, H. G. (1993) Affinity purification of histidine-tagged proteins. *Mol. Biol. Rep.* **18**, 223–230.

16. Pons, T., Uyeda, H. T., Medintz, I. L., and Mattoussi, H. (2006) Hydrodynamic dimensions, electrophoretic mobility and stability of hydrophilic quantum dots. *J. Phys. Chem. B* **110**, 20308–20316.

17. Sapsford, K. E., Pons, T., Medintz, I. L., Higashiya, S., Brunel, F. M., Dawson, P. E., et al. (2007) Kinetics of metal-affinity driven self-assembly between proteins or peptides and CdSe-ZnS quantum dots. *J. Phys. Chem. C* **111**, 11528–11538.

18. Medintz, I. L., Konnert, J. H., Clapp, A. R., Stanish, I., Twigg, M. E., Mattoussi, H., et al. (2004) A fluorescence resonance energy transfer derived structure of a quantum dot-protein bioconjugate nanoassembly. *Proc. Nat. Acad. Sci. USA* **101**, 9612–9617.

19. Medintz, I. L., Clapp, A. R., Brunel, F. M., Tiefenbrunn, T., Uyeda, H. T., Chang, E. L., et al. (2006) Proteolytic activity monitored by fluorescence resonance energy transfer through quantum-dot-peptide conjugates. *Nat. Mater.* **5**, 581–589.

20. Berti, L., D'Agostino, P. S., Boeneman, K., and Medintz, I. L. (2009) Improved peptidyl linkers for self-assembling semiconductor quantum dot bioconjugates. *Nano Res.* **2**, 121–129.

21. Medintz, I. L., Berti, L., Pons, T., Grimes, A. F., English, D. S., Alessandrini, A., et al. (2007) A reactive peptidic linker for self-assembling hybrid quantum dot-DNA bioconjugates. *Nano Lett.* **7**, 1741–1748.

22. Sapsford, K. E., Pons, T., Medintz, I. L., and Mattoussi, H. (2006) Biosensing with luminescent semiconductor quantum dots. *Sensors* **6**, 925–953.

23. Dennis, A. M. and Bao, G. (2008) Quantum dot-fluorescent protein pairs as novel fluorescence resonance energy transfer probes. *Nano Lett.* **8**, 1439–1445.

24. Ipe, B. I., Lehnig, M., and Niemeyer, C. M. (2005) On the generation of free radical species from quantum dots. *Small* **1**, 706–709.

25. Pons, T., Medintz, I. L., Sapsford, K. E., Higashiya, S., Grimes, A. F., English, D. S., et al. (2007) On the quenching of semiconductor quantum dot photoluminescence by proximal gold nanoparticles. *Nano Lett.* **7**, 3157–3164.

26. Yao, H., Zhang, Y., Xiao, F., Xia, Z., and Rao, J. (2007) Quantum dot/bioluminescence resonance energy transfer based highly sensitive detection of proteases. *Angew. Chem. Int. Ed.* **46**, 4346–4349.

27. Kim, J., Park, H. -Y., Kim, J., Ryu, J., Kwon, D. Y., Grailhe, R., et al. (2008) Ni-nitrilotriacetic acid-modified quantum dots as a site-specific labeling agent of histidine-tagged proteins in live cells. *Chem. Commun.* (16), 1910–1912.

28. Sperling, R. A., Pellegrino, T., Li, J. K., Chang, W. H., and Parak, W. J. (2006) Electrophoretic separation of nanoparticles with a discrete number of functional groups. *Adv. Funct. Mater.* **16**, 943–948.

29. Pons, T., Medintz, I. L., Wang, X., English, D. S., and Mattoussi, H. (2006) Solution-phase single quantum dot fluorescence resonant energy transfer sensing. *J. Am. Chem. Soc.* **128**, 15324–15331.

30. Chan, W. C. W. and Nie, S. (1998) Quantum dot bioconjugates for ultrasensitive nonisotopic detection. *Science* **281**, 2016–2018.

31. Mattoussi, H., Mauro, J. M., Goldman, E. R., Anderson, G. P., Sundar, V. C., Mikulec, F. V., et al. (2000) Self-assembly of CdSe-ZnS quantum dot bioconjugates using an engineered recombinant protein. *J. Am. Chem. Soc.* **122**, 12142–12150.

32. Lakowicz, J. R. (2006) *Principles of Fluorescence Spectroscopy (3rd Ed.)* Springer, New York.

33. Medintz, I. L., Clapp, A. R., Mattoussi, H., Goldman, E. R., Fisher, B., and Mauro, J. M. (2003) Self-assembled nanoscale biosensors based on quantum dot FRET donors. *Nat. Mater.* **2**, 630–638.

34. Clapp, A. R., Medintz, I. L., Mauro, J. M., Fisher, B. R., Bawendi, M. G., and Mattoussi, H. (2004) Fluorescence resonance energy transfer between quantum dot donors and dye-labeled protein acceptors. *J. Am. Chem. Soc.* **126**, 301–310.

35. Peng, Z. A. and Peng, X. (2001) Formation of high-quality CdTe, CdSe, and CdS nanocrystals using CdO as precursor. *J. Am. Chem. Soc.* **123**, 183–184.

36. Mei, B. C., Susumu, K., Medintz, I. L., Delehanty, J. B., Mountziaris, T. J., and Mattoussi, H. (2008) Modular poly(ethylene glycol) ligands for biocompatible semiconductor and gold nanocrystals with extended pH and ionic stability. *J. Mater. Chem.* **18**, 4949–4958.

37. Mei, B. C., Susumu, K., Medintz, I. L., and Mattoussi, H. (2009) Polyethylene glycol-based bidentate ligands to enhance quantum dot and gold nanoparticle stability in biological media. *Nat. Protoc.* **4**, 412–423.

38. Dif, A., Henry, E., Artzner, F., Baudy-Floc'h, M., Schmutz, M., Dahan, M., et al. (2008) Interaction between water-soluble peptidic CdSe/ZnS nanocrystals and membranes: Formation of hybrid vesicles and condensed lamellar phases. *J. Am. Chem. Soc.* **130**, 8289–8296.

39. Liu, W., Howarth, M., Greytak, A. B., Zheng, Y., Nocera, D. G., Ting, A. Y., et al. (2008) Compact biocompatible quantum dots functionalized for cellular imaging. *J. Am. Chem. Soc.* **130**, 1274–1284.

Chapter 8

A Single SnO$_2$ Nanowire-Based Microelectrode

Jun Zhou, Yaguang Wei, Qin Kuang, and Zhong Lin Wang

Abstract

SnO$_2$ nanowires are synthesized via the chemical-vapor-deposition process using gold as the catalyst and characterized by X-ray powder diffraction, field-emission scanning electron microscopy, high-resolution transmission electron microscopy, and cathodoluminescence. Finally, a new type of microelectrode based on a single SnO$_2$ nanowire is fabricated. This microelectrode is expected to have promising applications in various chemical and biomedical nanosensors.

Key words: SnO$_2$, Nanowire, Microelectrode, Chemical sensor, Biosensor

1. Introduction

One-dimensional (1D) nanowires are ideal building blocks for constructing nanosized devices due to their high surface to volume ratio and their special physical and chemical properties (1). One of the most active research areas in nanoscience is the design and fabrication of chemical and biosensing devices based off of these wires due to their diverse practical and potential applications (2–5). To improve the sensing characteristics, a general route is to make chemical and biomedical sensors at the nanoscale, taking advantage of the large surface areas of nanoscale structures. Chemical nanosensors based on carbon nanotubes (6, 7), silicon nanowires (1), and ceramic (8) nanostructures are of particular interest.

SnO$_2$ is an n-type semiconductor and nanostructures made from this material possess many unique optical and electrical properties. SnO$_2$ has a wide band gap of 3.6 eV at 300 K, and possesses remarkable receptivity variation in gaseous environments, high-optical transparency in the visible range (up to 97%),

Sarah J. Hurst (ed.), *Biomedical Nanotechnology: Methods and Protocols*, Methods in Molecular Biology, vol. 726,
DOI 10.1007/978-1-61779-052-2_8, © Springer Science+Business Media, LLC 2011

111

low resistivity, and excellent chemical stability. These properties make SnO_2 nanowires well suited for chemical sensors and transparent conducting electrodes.

In this chapter, we focus on the synthesis of SnO_2 nanowires and the fabrication of a single SnO_2 nanowire-based microelectrode (9). The SnO_2 nanowires are synthesized via the chemical vapor deposition (CVD) process using gold as the catalyst and characterized by X-ray powder diffraction (XRD), scanning electron microscopy (SEM), high-resolution transmission electron microscopy (HRTEM), and cathodoluminescence (CL). Then, we describe the fabrication of a single SnO_2 nanowire-based microelectrode.

2. Materials

2.1. Synthesis of SnO_2 Nanowires

1. Growth system: a homemade CVD experimental apparatus including a horizontal quartz reaction chamber (Φ 50 mm $\times$ 400 mm), a DC controller, a gas supply control system, and a rotary pump system. The details of such an apparatus can be found in (10).

2. Temperature detector: infrared radiation thermometric indicator (Thermalert TX, Raytek).

3. Precursor gas: SnH_4 mixed with N_2 at a fixed ratio of 8% and stored in an air pocket (SY-42, Sanhe Medical Instrument Co. Ltd).

4. Substrates: n-type Si (100) wafers (20 mm $\times$ 5 mm $\times$ 0.3 mm).

5. Catalyst: 5-nm Au film was deposited on the Si substrates by means of ion sputtering using Au as the target.

6. Ultrasonic system (Branson 2510).

7. Cleaning solution: acetone (Analytical Reagent from Alfa Aesar) and ethanol.

8. Deionized (DI) water.

2.2. Fabrication of Microelectrodes

1. Equipment for patterning of Au electrodes: hotplate, Solitec Spin Coater, Karl Suss MJB-3 Mask Aligner, CVC E-beam evaporator.

2. Materials for patterning of Au electrodes: Au (99.99% purity), AZ5214 photoresist, strip 315 developer, acetone, ethanol, deionized water.

2.3. Single Nanowire Microelectrode Fabrication

1. Probe station including two tungsten probes, X–Y stage, and microscopy (Cascade Microtech, Inc., Beaverton, OR).

2. Function generator (Stanford research DS345, Sunnyvale, CA).

3. 50-mL glass bottle.

4. Scalpel.

5. Wire bonder (West Bond 7476D079, West Bond Inc., Anaheim, CA).

6. Gold wire (Φ 25 μm).

7. Ethanol.

8. Platinum.

9. Pt deposition: FEI Nova Nanolab 200 FIB/SEM.

10. Support chip.

---

3. Methods

3.1. Synthesis of SnO$_2$ Nanowires

1. Clean n-type Si (100) wafers (20 mm × 5 mm × 0.3 mm) (see Note 1) with acetone, ethanol, and DI water sequentially for 5 min each in an ultrasonic bath.

2. Deposit 5 nm Au on the top surface by thermal evaporation process. The gold was used as catalyst for the growth of SnO$_2$ nanowires.

3. Prepare the precursor SnH$_4$ by mixing with N$_2$ at a fixed ratio of 8% and store in an air pocket.

4. Place the Si substrates at the center of the horizontal quartz reaction chamber.

5. Pump down the reaction chamber to 1–2 Torr using a rotary pump.

6. Heat the Si substrates from room temperature to 770°C within several seconds by adjusting the DC through the Si substrates. The temperature of the Si substrates was measured using an infrared radiation thermometric indicator (see Note 2).

7. Introduce the SnH$_4$/N$_2$ mixture to the reaction chamber with the flow rate of 20 sccm (see Note 3). The deposition usually takes 5–10 min.

8. Cool the reaction chamber to room temperature naturally.

3.2. Characterization of SnO$_2$ Nanowires

1. Characterize the structure of the SnO$_2$ nanowires by XRD (Panalytical X-pert) using Cu Kα radiation (see Fig. 1).

2. Characterize the morphology of the SnO$_2$ nanowires using field-emission scanning electron microscopy (FESEM) and HRTEM equipped with energy dispersion spectroscopy (EDS) (see Fig. 2 and Note 4).

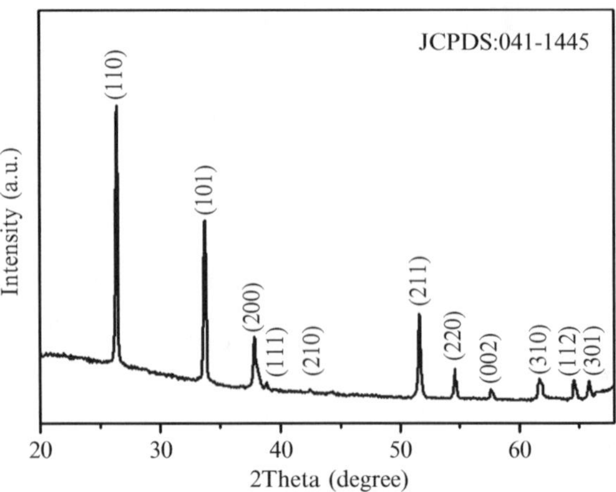

Fig. 1. XRD pattern of the as-synthesized SnO_2 nanowires. The diffraction peaks agree well with that of bulk SnO_2 of rutile structure (JCPDS: 041-1445). Reproduced from ref. 9 with permission from the American Chemical Society.

Fig. 2. (**a**) Typical SEM image of SnO_2 nanowires. (*Top right*) TEM image and (*bottom right*) corresponding SAED pattern of a single SnO_2 nanowire. Reproduced from ref. 9 with permission from the American Chemical Society.

3. Characterize the crystalline quality and presence of the defect structure of the SnO_2 nanowires using a cathodoluminescence (CL) system at room temperature (see Note 5).

3.3. Fabrication of Microelectrodes

1. Ultrasonically clean the SiO_2/Si wafer using acetone, ethanol, and DI water for 5 min each in sequence then blow dry the wafer with pure nitrogen gas (see Note 6).

2. Spin coat the photoresist onto the wafer at f376×*g* (4,000 rpm) for 40 s then bake the film at 80°C for 2 min.

3. Pattern the photoresist coated wafer using a Karl Suss MJB-3 Mask Aligner (UV patterning).

4. Develop the patterned photoresist using strip 315 developer for 2 min.

5. Deposit 50 nm of Au on the patterned substrate using a CVC E-beam evaporator.

6. Remove the photoresist by immersing the substrate into acetone for 10 min.

3.4. Fabrication of a Single SnO$_2$ Nanowire Microelectrode

1. Pour 10 mL ethanol into the glass bottle. Scratch the as-synthesized SnO$_2$ nanowires from the substrate into the bottle using a scalpel. Then, ultrasonicate the sample for 15 min to disperse the nanowire bundles into individual nanowires (see Note 7). Finally, the SnO$_2$ nanowire suspension is formed.

2. Place the Au microelectrode on the plate of the probe station. By adjusting the X–Y stage, contact the two tungsten probes to the two 500 μm × 500 μm Au pads, which are 10 μm apart.

3. Add a 1-μL droplet of SnO$_2$ nanowire suspension (nanowire concentration ~1 mg/mL) between the two electrodes.

4. Turn on the function generator and apply a 5-V and 1-MHz AC signal between the two tungsten probes (see Note 8).

5. Allow the solution to dry, then deposit two Pt (4 μm × 1 μm × 0.2 μm) pads by FIB on the two ends of the SnO$_2$ nanowire on the Au microelectrodes to improve the electrical contact (see Note 9 and Fig. 3a).

6. Place the device onto a supporting chip and connect the two electrodes to the supporting chip by Au wire bonding (see Fig. 3b).

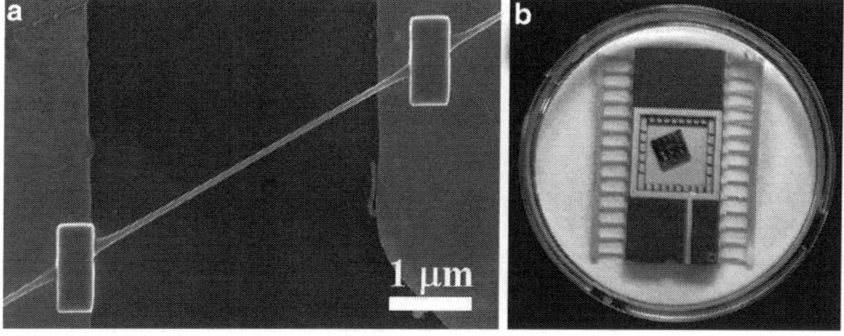

Fig. 3. (**a**) SEM image of a single SnO$_2$ nanowire which was fixed on the Au microelectrodes by Pt pads that were deposited using FIB. (**b**) Optical image of a real device connected on a support chip. Reproduced from ref. 9 with permission from the American Chemical Society.

4. Notes

1. These wafers can be used as substrates to grow SnO_2 nanowires since their melting points are higher than 770°C.

2. For this application, 770°C is an ideal temperature for the growth of high-quality SnO_2 nanowires. When the deposition temperature is less than 700°C, the grown SnO_2 nanowires are too large in diameter (around 300–500 nm). When the deposition temperature is over 850°C, the grown SnO_2 nanowires are too long and easily twisted with each other.

3. The chemical reaction of this process is as follows: $SnH_4 + 2O_2 \rightarrow SnO_2 + 2H_2O$.

4. Typically, the diameter and length of the SnO_2 nanowires range from 50 to 300 nm and up to tens of micrometers, respectively (Fig. 2a). The TEM image (Fig. 2b) shows that there is an Au nanoparticle on the tip of each SnO_2 nanowire, revealing the vapor–liquid–solid (VLS) growth mechanism. The selective area electron diffraction (SAED) pattern (Fig. 2c) indicates that the SnO_2 nanowires are single crystalline with a [001] growth direction.

5. The near band edge emission, which was expected around 320 nm, was not detected. But, a broad blue luminescent peak centered at approximately 470 nm was detected, indicating that the as-synthesized SnO_2 NWs possess a large number of oxygen vacancies in the crystal.

6. It is critical to keep the wafer wet when moving it from one solution to another, otherwise residue will attach to the wafer, which will be difficult to remove.

7. The ultrasonication time should not be longer than 15 min. Otherwise, the SnO_2 nanowires will break into short pieces and then be unable to bridge the gap between the Au electrodes.

8. This signal generates an alternating electrostatic force on the SnO_2 nanowires in the solution. Under the electrical polarization force, the SnO_2 nanowires are placed in contact with the two electrodes. This method is called the "dielectrophoresis technique." By precisely controlling the concentration of the SnO_2 nanowires in the solution, a circuit can be made where only a single SnO_2 nanowire bridges the two electrodes. The 5-V and 1-MHz AC signal is optimized for fabricating single nanowire devices by the dielectrophoresis method.

9. The deposition current should be very low (~10 pA). Otherwise the whole SnO_2 nanowire will be contaminated by the Ga vapor and cause the device to short circuit.

Acknowledgments

The work was partially sponsored by the NSF, DARPA, and the NIH.

References

1. Cui, Y., Wei, Q. Q., Park, H. K., and Lieber, C. M. (2001) Nanowire nanosensors for highly sensitive and selective detection of biological and chemical species. *Science* **293**, 1289–1292.
2. Hahm, J. and Lieber, C. M. (2004) Direct ultrasensitive electric detection of DNA and DNA sequence variations using nanowire nanosensors. *Nano Lett.* **4**, 51–54.
3. Patolsky, F., Zheng, G. F., Hayden, O., Lakadamyali, M., Zhuang, X. W., and Lieber, C. M. (2004) Electrical detection of single viruses. *Proc. Natl. Acad. Sci. U.S.A.* **101**, 14017–14022.
4. Zheng, G. F., Patolsky, F., Cui, Y., Wang, W. U., and Lieber, C. M. (2005) Multiplexed electrical detection of cancer markers with nanowire sensor arrays. *Nat. Biotechnol.* **23**, 1294–1301.
5. Patolsky, F., Timko, B. P., Yu, G. H., Fang, Y., Greytak, A. B., Zheng, G. F., et al. (2006) Detection, stimulation, and inhibition of neuronal signals with high-density nanowire transistor arrays. *Science* **313**, 1100–1104.
6. Kong, J., Franklin, N. R., Zhou, C. W., Chapline, M. G., Peng, S., Cho, K. J., et al. (2000) Nanotube molecular wires as chemical sensors. *Science* **287**, 622–625.
7. Qi, P. F., Vermesh, O., Grecu, M., Javey, A., Wang, Q., and Dai, H. J. (2003) Toward large arrays of multiplex functionalized carbon nanotube sensors for highly sensitive and selective molecular detection. *Nano Lett.* **3**, 347–351.
8. Kim, I., Rothschild, A., Lee, B. H., Kim, D. Y., Jo, S. M., and Tuller, H. L. (2006) Ultrasensitive chemiresistors based on electrospun TiO$_2$ nanofibers. *Nano Lett.* **6**, 2009–2013.
9. Kuang, Q., Lao, C. S., Wang, Z. L., Xie, Z. X., and Zheng, L. S. (2007) High-sensitivity humidity sensor based on a single SnO$_2$ nanowire. *J. Am. Chem. Soc.* **129**, 6070–6071.
10. Kuang, Q., Jiang, Z. Y., Xie, Z. X., Lin, S. C., Lin, Z. W., Xie, S. Y., et al. (2005) Tailoring the optical property by a three-dimensional epitaxial heterostructure: a case of ZnO/SnO$_2$. *J. Am. Chem. Soc.* **127**, 11777–11784.

Chapter 9

Biosensing Using Nanoelectromechanical Systems

Ashish Yeri and Di Gao

Abstract

Nanoelectromechanical systems (NEMS) correlate analyte-binding events with the mechanical motions of devices in nanometer scales, which in turn are converted into detectable electrical or optical signals. Biosensors based on NEMS have the potential to achieve ultimate sensitivity down to the single-molecule level, provide rapid and real-time detection signals, be operated with extremely low power consumption, and be mass produced with low cost and high reproducibility. This chapter reviews fundamental concepts in NEMS fabrication, actuation and detection, and device characterization, with examples of using NEMS for sensing DNA, proteins, viruses, and bacteria.

Key words: Biosensors, Microelectromechanical systems, Micromachining, Microfabrication, Nanofabrication, MEMS

1. Introduction: Evolution of MEMS to NEMS for Biosensing

The development of the atomic force microscope (AFM) by Binnig, Quate, and Gerber in 1986 gave rise to the field of force spectroscopy whereby intermolecular forces such as van der Waals forces, Casimir's forces, dissolution forces in liquid, and even single-molecule rupture forces on the order of piconewtons could be measured by a tiny mechanical cantilever. When applied to the study of biomolecules, force spectroscopy can be used to measure the attractive forces between biomolecules and the discrete steps in the rupture of these bonds. Inspired by AFM, researchers started to use microelectromechanical systems (MEMS), where the displacement of a mechanical component is driven and sensed by electronic signals, for biosensing. Using MEMS devices, the presence of a biological analyte can be detected either based on the deflection of a mechanical component such as a cantilever, where analyte binding to a "functionalized" cantilever produces a

Sarah J. Hurst (ed.), *Biomedical Nanotechnology: Methods and Protocols*, Methods in Molecular Biology, vol. 726,
DOI 10.1007/978-1-61779-052-2_9, © Springer Science+Business Media, LLC 2011

deflection of the cantilever beam, or from the change of the resonant frequency of a mechanical component in the resonant mode, where analyte binding causes a decrease in the resonant frequency of vibration of the cantilever beam (1).

Although MEMS biosensors have been well studied for over a decade, they are plagued by a number of problems. For example, cantilever deflection-based MEMS devices require a large amount of molecules to be bound by the device to transduce the change into measurable quantities. Also, these devices are unable to respond to the rapidly changing forces of biomolecule interactions that have time scales on the order of microseconds; this could be critical when trying to distinguish between specific and nonspecific interactions. These and other factors including the potential for increased sensitivity, reduced power consumption, and reduced cost (2) have motivated the reduction in size of existing MEMS devices. Miniaturization also might allow these devices to be implanted in the human body for medical diagnostic applications.

2. NEMS Fabrication

2.1. Top-Down Approach for Fabrication of NEMS

The fabrication processes of NEMS biosensors can be broadly classified into two categories: top-down and bottom-up approaches. The top-down approach is directly borrowed from the microfabrication technologies. The only difference between top-down fabrication processes on micrometer and nanometer scales is the lithography process. For fabrication of MEMS, photolithography using either visible or ultraviolet light is typically employed to pattern micrometer-scale device features; for fabrication of NEMS, more advanced lithography techniques, using electron beams, focused ion beams, atomic force microscope tips, or pre-synthesized nanostructures, for example, are employed to define the device features.

The two most common top-down microfabrication techniques are bulk micromachining and surface micromachining. In bulk micromachining, fabrication of mechanical elements is achieved by etching the substrate. Figure 1a presents a schematic of a typical bulk micromachining process. The structural material such as Si_3N_4, SiC, or GaAs is deposited via chemical vapor deposition (CVD), for instance, onto the substrate and then patterned by lithography. The mechanical elements then are released from the substrate by removing the underlying substrate material. In surface micromachining, a sacrificial material is deposited and patterned between the structural material and the substrate (see Fig. 1b). Instead of etching away the substrate material, the sacrificial layer is then removed by etching to release the structure.

Advanced and innovative lithography techniques are required to transfer the top-down approaches used in microfabrication to

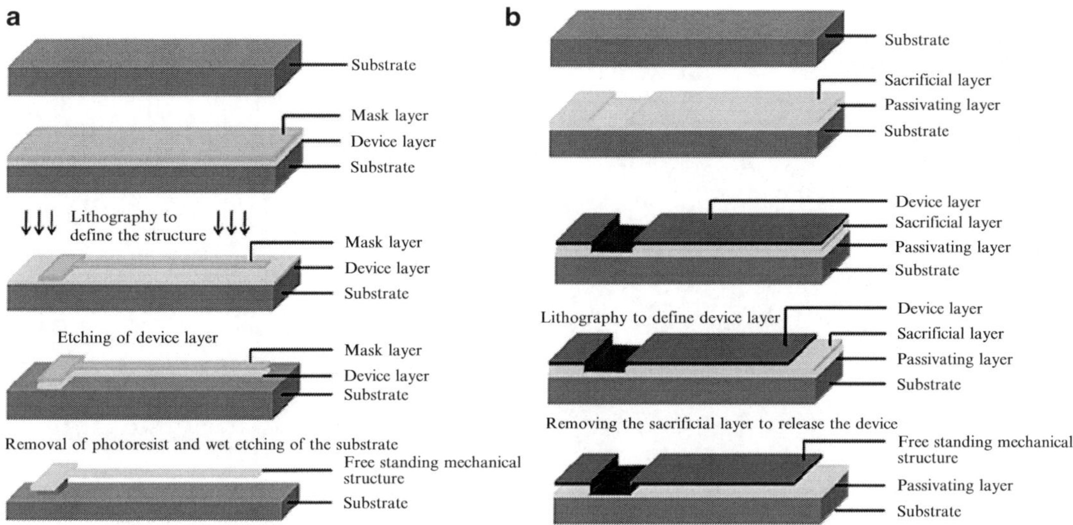

Fig. 1. Basic depiction of (**a**) bulk micromachining and (**b**) surface micromachining processing steps leading to suspended devices.

nanofabrication strategies. Electron beam (e-beam) lithography, which is capable of defining features of almost any geometry down to tens of nanometers in size, is the most commonly employed technique to fabricate NEMS using the top-down approach. Figure 2 shows an example where e-beam lithography is used to fabricate NEMS devices consisting of a Fabry–Perot interferometer and a doubly clamped beam array on a silicon-on-insulator (SOI) substrate (3). After coating the silicon with poly(methyl methacrylate) (PMMA), e-beam lithography was used for patterning. Aluminum was evaporated (35 nm) on the resist as a mask for CF_4 reactive ion etching (RIE), and then the silicon oxide was etched using buffered hydrofluoric acid. The thickness of the mechanical element and the distance between the mechanical element and the substrate are defined by the thickness of the silicon layer and the thickness of the insulator layer of the SOI substrates, respectively. The lateral dimension of the mechanical element is defined by e-beam lithography. In another example, the critical dimension of the mechanical element in NEMS is defined by presynthesized nanostructures such as nanofibers. Figure 3 shows such an example, where a silicon nitride nanowire less than 200 nm wide is fabricated using a PMMA nanofiber synthesized by electrospinning (4). After patterning the electrodes, the nanofiber is deposited between them, serving as a mask to define the width of the silicon nitride nanowire during the etching process.

2.2. Bottom-Up Approach for Fabrication of NEMS

In the bottom-up approach, the mechanical element of the NEMS (e.g., nanowires, nanotubes) is synthesized through chemical reactions, and the dimension of these nanostructures is defined during the synthesis process instead of by lithography techniques.

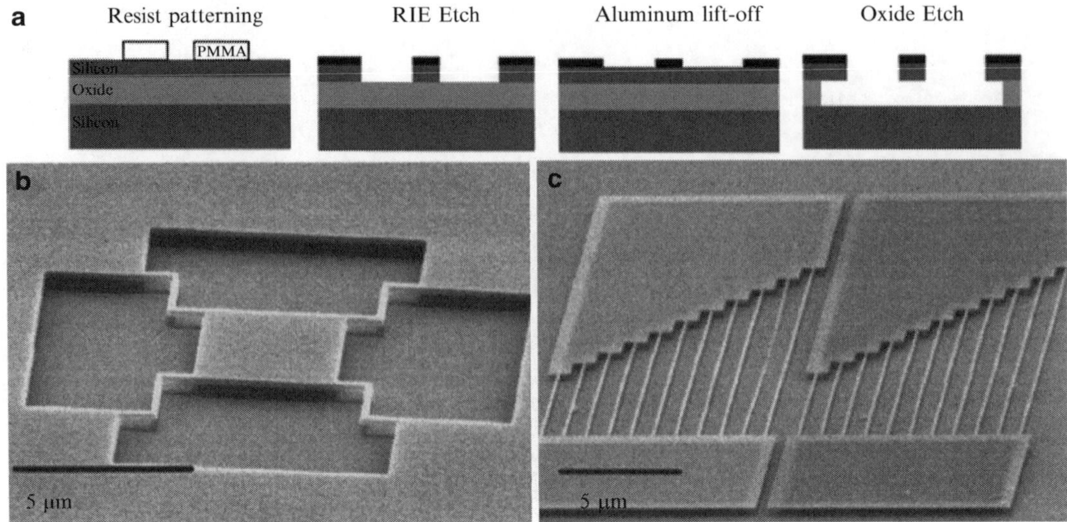

Fig. 2. Example of NEMS device fabrication using e-beam lithography through the top-down approach. (**a**) Schematic of fabrication steps. (**b**) A 200 nm thick square Si pad suspended with silicon nanowires 100 nm wide. The distance between the pad and the silicon substrate is 400 nm. (**c**) 50 nm thick and 200 nm (*left*) and 120 nm (*right*) wide suspended Si nanowires of lengths varying from 7 to 16 μm. Reprinted with permission from ref. 3.

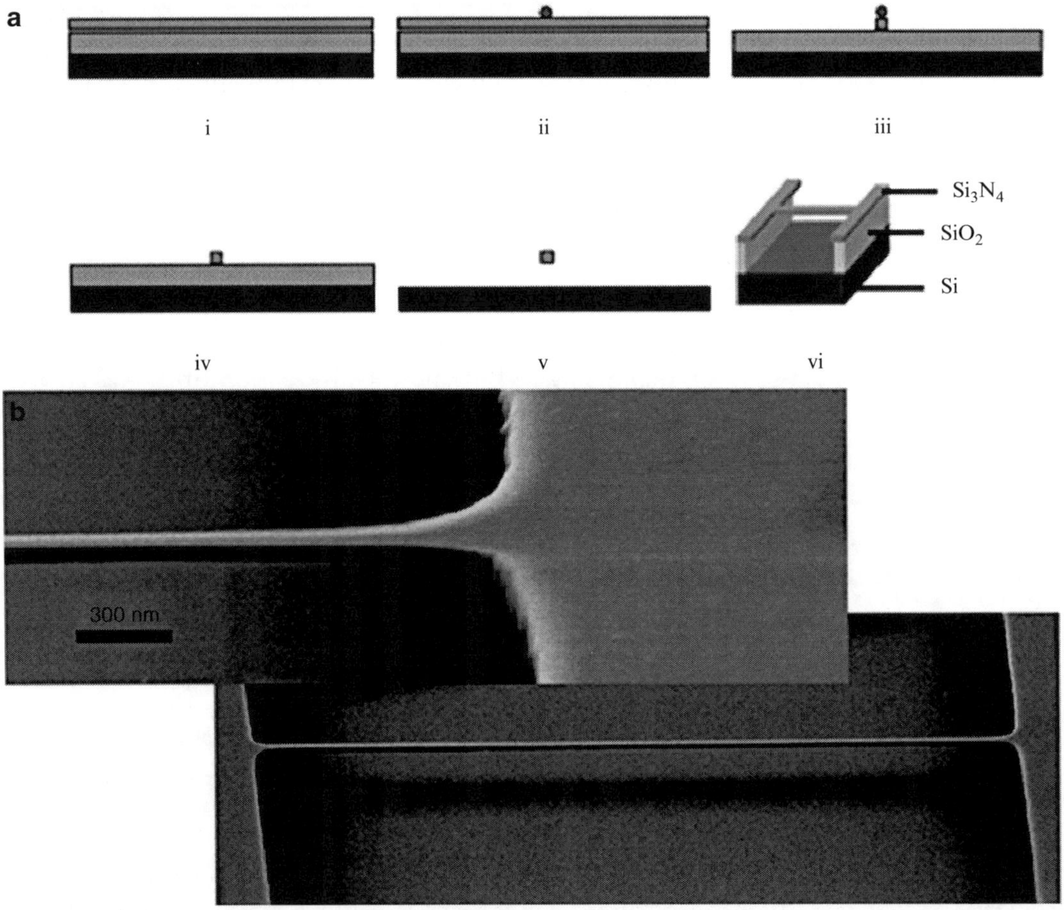

Fig. 3. Example of NEMS device fabrication using presynthesized nanostructures. A resonator based on a silicon nitride nanowire is fabricated using a polymer nanofiber synthesized by electrospinning. (**a**) Schematic of the processing steps. (**b**) SEM images of a 15 μm long, less than 200 nm wide, doubly clamped silicon nitride nanowire. Reprinted with permission from refs. 6 and 4, respectively.

In order for the NEMS device to function, these nanostructures must be positioned between electrodes which are used to drive and sense their mechanical motion. These electrodes are typically on the micrometer scale and fabricated by micromachining processes. It is a challenging task to position nanostructures between micrometer-sized electrodes; it is typically accomplished using one of the following two routes.

The first route involves placing presynthesized nanostructures on prefabricated electrodes, either by random dispersion or by using micromanipulators, and then welding the nanostructure to the electrode by e-beam lithography or focused ion beam deposition. This route has been widely used for fabricating electronic sensors based on 1D nanomaterials, where there is no mechanical motion of the nanostructures. However, because the anchor, by which the nanostructure is attached to the electrode, is not rigid enough for high-frequency resonators, this technique is rarely used for fabrication of NEMS devices. A variant method of this approach is to calcinate the presynthesized nanostructure after it is placed on the electrode, during which the chemical composition of the nanostructure may change. Figure 4 shows such an example, where a silica nanofiber with a resonant frequency of ~ 10 MHz is made by first placing a polymer nanofiber, presynthesized by electrospinning, on top of the electrodes and then calcinating the fiber at 850°C (4–6).

The second route involves direct chemical synthesis of nanostructures at a specific location on the device, preferably in a predetermined direction and architecture (7, 8). The advantages of this route are that the construction of the NEMS devices is simplified, and the arduous task of postsynthesis assembly and the

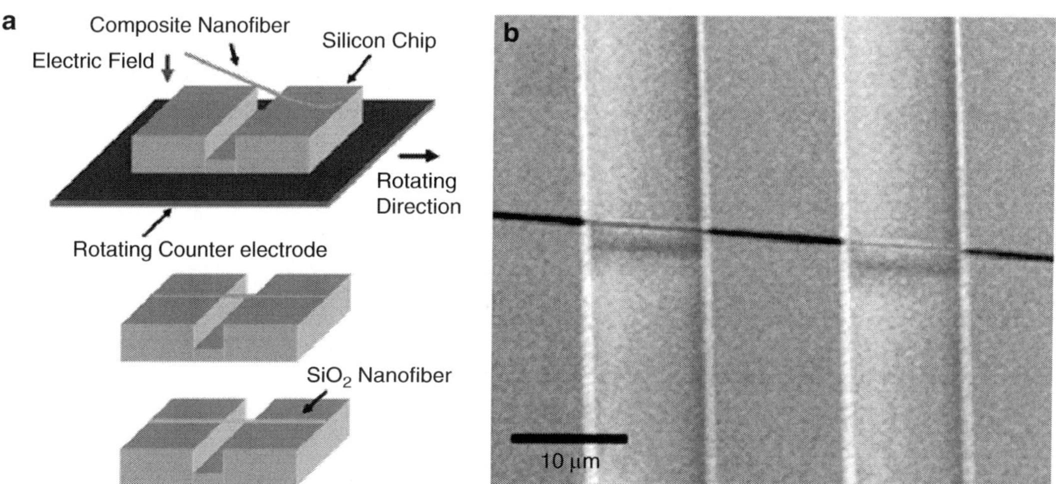

Fig. 4. Example of NEMS device fabrication by placing presynthesized nanostructures onto prefabricated electrodes through the bottom-up approach. (a) Diagram showing steps involved in fabricating a silica nanofiber with a resonant frequency of ~10 MHz by first depositing a polymer nanofiber synthesized by electrospinning on top of the electrodes and then calcinating the fiber at 850°C. (b) SEM image of a silica nanofiber anchored to the electrodes. Reprinted with permission from ref. 5.

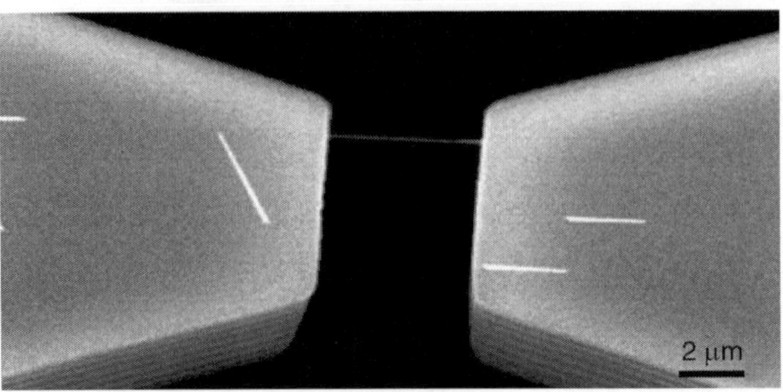

Fig. 5. Scanning electron microscopy (SEM) image of a Si nanowire resonator grown by chemical vapor deposition. The doubly clamped device is suspended over a trench fabricated on a silicon-on-insulator (SOI) substrate. Reprinted with permission from ref. 9.

potential deterioration of the nanostructure during the process are eliminated (7). Figure 5 shows an example of NEMS fabricated through this route, in which a doubly clamped single crystalline Si nanowire is suspended between two microelectrodes (9). The microelectrodes are fabricated by top-down microfabrication techniques, while the Si nanowire is directly synthesized between the electrodes by a metal-catalyzed CVD process, which is a bottom-up approach. The location and the diameter of the nanowire are controlled by the location and size of the Au catalytic nanoparticle, which is deposited onto the sidewall of one electrode prior to the growth of the wire.

3. NEMS Actuation and Detection

NEMS devices convert electrical inputs to the sensor into mechanical motion and vice versa. Similar to MEMS devices, a typical NEMS device consists of an input transducer and an output transducer; the input transducer converts the electrical inputs into mechanical motion and the output transducer senses the motion or displacement of the sensor and converts it into electrical or optical signals. Although the configuration of NEMS is similar to that of MEMS, actuation and detection methods used for MEMS devices do not always work well with NEMS devices. In some cases, optical techniques are not compatible with the small length scale of NEMS devices; in others, parasitic impedances make electrical detection challenging. This section summarizes some of the actuation and detection methods that have been reported in the literature.

3.1. Actuation Methods

3.1.1. Thermal Actuation

When a current is passed through a NEMS resonator, it induces surface heating and thermal expansion of the structural material. The NEMS device may be actuated by this process utilizing customized structural designs (e.g., by using a bilayer structure consisting of two layers of materials with different thermal expansion coefficients). Superimposing an AC voltage onto a constant DC voltage leads to sinusoidal heating of the mechanical structure. Because of the very small dimension of the NEMS device, the time scale of the heating and cooling of the mechanical structure could be small enough (on the order of ns to ps) for actuation of high-frequency NEMS resonators. When it is not convenient to use the resonator itself as the heating source, thermal actuation may also be carried out by fabricating on-chip Joule-heating resistors near the resonators (8).

3.1.2. Magnetomotive Actuation

When an alternating current $i(\omega)$ is passed through the NEMS cantilever beam in the presence of a strong magnetic field B, the beam experiences a Lorentz force of $f(\omega) = lBi(\omega)$, where l is the length of the beam. The direction of the force is perpendicular to both the AC and the magnetic field. Cleland et al. (10) have demonstrated magnetomotive actuation of a doubly clamped beam; AC is passed through the beam and a magnetic field perpendicular to the direction of the AC induces out-of-plane beam vibrations.

3.1.3. Piezoelectric Actuation

Piezoelectric actuation can be carried out by either exciting the NEMS device using an external (off-chip) piezoelectric device or embedding piezoelectric materials into the NEMS device. Off-chip piezoelectric actuators have been used to detect thiolated SAMS (11) and viruses (12), but such actuators typically have a low Q-factor (see Subheading 4.2.2), and the requirement of an external piezoelectric device is undesirable for miniaturized devices. Direct piezoelectric actuation can be accomplished by embedding the piezoelectric material inside the mechanical structure. For example, Lee et al. have actuated Si_3N_4 cantilever beams at frequencies on the order of 10^4 Hz with very high mass sensitivities (13, 14) by depositing lead zirconate titanate (PZT, a piezoelectric material) on them.

3.1.4. Electrostatic Actuation

When a voltage is applied across two plates of a capacitor, an attractive force develops between them. This attractive force increases nonlinearly as the separation between the plates decreases. Since the separation between the plates in MEMS and NEMS devices is very small, the electrostatic force is sufficiently large to actuate the mechanical elements of the devices. For example, a doubly clamped beam can be actuated by applying a potential between the beam and a nearby gate electrode (15); an out-of-plane resonance of a single crystalline Si mesh suspended by nanobeams has been induced by an electric force perpendicular to the plane (16).

3.1.5. Other Actuation Methods

Actuation can also be carried out by thermal noise and the fluctuations in ambient air. Ilic et al. has reported detection of *Escherichia coli* in both air and vacuum (17, 18) using the transverse vibrations resulting from thermal noise. This technique is simple, requiring no external oscillators or complex fabrication techniques. A cantilever beam also has been excited at its resonance frequency by heating from a laser source (19). Here, a 670-nm diode laser was employed to actuate a cantilever beam, and optical interferometry was used to detect its deflection.

3.2. Detection Methods

3.2.1. Optical Detection

Focused laser beams are commonly used to detect the deflection of cantilevers in MEMS sensors. When the analyte of interest binds to the cantilever and produces a deflection, a laser beam focused on the cantilever detects this deflection. This simple detection scheme does not scale down well to NEMS devices due to the diffraction of the laser at these smaller length scales. Another common optical detection technique is based on interferometry, where interference of light is used to sense very minute changes in deflection of the mechanical element in the NEMS device. For example, Michelson interferometry-based methods use the interference of the laser beam reflected from the structure with a stable reference laser beam; Fabry–Perot interferometry-based methods use the refractive index difference between the structural material of the NEMS device and the substrate (3, 20, 21). Although interferometry is able to detect subwavelength movement of mechanical structures, the strong diffraction effects associated with the very small dimensions of NEMS devices are still prohibitive in making optical interferometry applicable for detecting motions of NEMS devices. In addition, micrometer-scale optic fiber cables are generally used for detection with MEMS devices; these would be too large for use with NEMS devices. Thus, when faced with the practical problems associated with focusing lasers onto subwavelength structures, the high power losses associated with scattering by surface defects (22), and the troublesome positioning of the optic fiber, optical detection systems are limited for use with NEMS devices.

3.2.2. Electrostatic or Capacitive Detection

The deflection of a charged resonator, acting as one of the plates of a capacitor, causes a change in the capacitance, which is inversely proportional to the distance from the substrate. This change of capacitance can be detected by custom-designed circuitry. This method offers very high detection sensitivity for both MEMS and NEMS devices. One problem associated with capacitive detection though is the parasitic capacitance, which is in parallel to the actuating capacitance; this reduces the efficiency of detection at high frequencies. Balanced bridge techniques (15), for example, can be used to overcome the effects of parasitic capacitance. Capacitive detection schemes are very sensitive to the dielectric constant of the medium, and changes in this parameter can have unfavorable effects.

3.2.3. Piezoresistive Detection

Piezoresistive materials, such as doped Si, undergo changes in electrical resistance when a strain is applied on them or when mechanical motion imposes a strain. Piezoresistive detection is usually carried out with a Wheatstone network (e.g., with two of four resistors placed on a cantilever). Piezoresistive materials have great potential to be used in NEMS devices because such devices can be operated without the aid of external equipment, can be made in reduced size, and have the ability to operate in opaque liquids such as blood. However, care must be taken to prevent overheating of the device due to Joule heating, which would cause undesirable mechanical deformations.

4. NEMS Based on Cantilever and Doubly Clamped Beams

Beams with at least one dimension in the nanometer scale, including both cantilevers (with one end anchored) and doubly clamped (with both ends anchored) beams, have simple geometries and are relatively easy to fabricate and model. Therefore, they are widely used as mechanical elements in NEMS sensors. This section introduces some fundamental concepts for cantilever- and doubly clamped beam-based NEMS.

4.1. Cantilever Beams Operated in Deflection Mode

The deflection of cantilever beams upon deposition of metals was first studied by G.G. Stoney (23) who gave the relationship (known as Stoney's formula) between the radius of curvature of the beam (R) and the surface stress (σ) as

$$R = \frac{Et^2}{6\sigma(1-v)},\tag{1}$$

where E is the Young's modulus, t is the thickness of the substrate, and v is the Poisson's ratio of the substrate. The same principles are applied to the deflection of cantilever beams used as sensors. Analyte binding to the surface of the cantilever beam causes a surface stress which decreases the radius of curvature of the beam (see Fig. 6a). Typically, only one side of the cantilever beam is functionalized to trap the analyte, which causes the beam to bend in one direction. Although, it is debated if the surface stress σ from analyte binding is the sole cause of the bending of the cantilever. Also, the location of analyte binding could in fact increase the stiffness of the material, causing the radius of curvature to increase. Cantilever beams operated in the static mode or deflection mode have the advantage that they can be operated in liquid media without a significant loss of sensitivity (see Note 1).

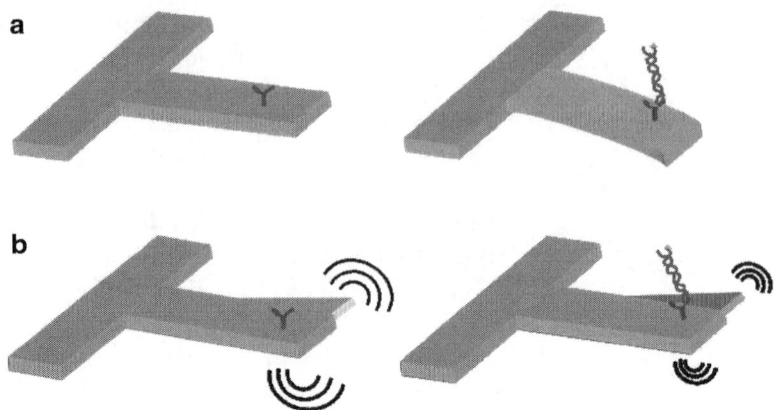

Fig. 6. Schematic showing the two different modes of operation for cantilever-based biosensors. The cantilever is functionalized on only the top side with biomolecules that capture the target molecule. (**a**) Deflection-based sensing: the cantilever on the left is the reference cantilever and analyte binding is shown on the right; the deflection with respect to the reference is due to the surface stress induced by the analyte binding. (**b**) Resonance-based sensing: the shift in the resonance frequency of the cantilever on the right with respect to the reference on the left is related to the analyte binding.

4.2. Cantilever and Doubly Clamped Beams Operated in Resonant Mode

4.2.1. Frequency

NEMS resonators typically can achieve frequencies on the order of MHz with some in the GHz range. By treating a NEMS resonator as a harmonic oscillator, its resonant frequency (f_0), in the absence of external forces and damping, is given by:

$$f_0 = \frac{1}{2\pi\sqrt{k/m_{eff}}}, \tag{2}$$

where k is the spring constant and m_{eff} is the effective mass of the cantilever. The spring constant is proportional to the flexural rigidity, D, of the resonator, which is given by the product of the Young's modulus, E, and the moment, I. Accordingly, the fundamental out-of-plane resonance frequency (f_0) of a cantilever or a doubly clamped beam has the following relationship to the geometry and the material properties of the beam:

$$f_0 \propto \sqrt{\frac{E}{\rho}}\frac{t}{L^2}, \tag{3}$$

where t is the thickness and L is the length of the beam and ρ is the mass density. Based on Eq. 3, materials with a high ratio of E to ρ are preferred for NEMS devices to obtain high frequencies. To improve the electrical conductance of the beam, a metallic layer is often added onto the semiconducting material. In such cases, the above relationship becomes:

$$f_0 \propto \frac{1}{L^2}\sqrt{\frac{E_1 I_1 + E_2 I_2}{\rho_1 A_1 + \rho_2 A_2}}, \tag{4}$$

where the subscripts 1 and 2 refer to the semiconducting and metallic layers, respectively.

Upon mass addition (Δm) to the resonator, f_0 changes by Δf, which based on Eq. 2 can be written by a first order approximation as:

$$\Delta f = \frac{-\Delta m}{2m_{eff}} f_0 \, . \qquad (5)$$

assuming that Δf is attributed only to Δm, k remains unchanged, and the added mass is uniformly distributed over the beams. According to Eq. 5, by measuring the resonant frequency before and after loading, the mass of the analyte added to the resonator may be calculated (see Fig. 6b).

Because their effective mass is very small, NEMS devices have the potential to be very sensitive mass sensors. However, the interpretation of experimental data could be very complicated due to the fact that both the added mass and the binding of analyte to the resonator may induce surface stress. Further, the distribution of mass and the induced surface stress may not be uniform across the resonator. The frequency shift associated with the surface stress induced by analyte binding has been analyzed by treating it as a one dimensional axial force; however, this treatment has been the source of much debate. When nonuniform binding occurs (e.g., the analyte binds only to the tips of a cantilever), the relationship between the frequency shift and the position of the bound analyte becomes very complex. Depending upon the position of the adsorbed mass, the NEMS device could experience a negative frequency shift (at the tip of the cantilever beam) or a positive frequency shift (near the clamped end). This result has been shown experimentally with adsorbed bacteria (see Fig. 7) (24, 25) and proteins (26). A theoretical model was discussed by Tamayo et al. in ref. 27. When the analyte concentration in solution is very low, the resonant frequency shift may be enhanced by the addition of mass labels in certain biosensor applications.

4.2.2. Q-Factor

The quality factor (Q-factor) is defined as the ratio of the resonance frequency of the oscillator to the bandwidth. It is inversely proportional to the energy dissipation factor D, which is defined in terms of energy dissipated versus energy stored in the oscillator as:

$$D = \frac{1}{Q} = \frac{\text{Energy dissipated per oscillation}}{2\pi \; \text{Energy stored in the oscillator}} \, . \qquad (6)$$

The Q-factor can be used to characterize the performance of NEMS devices. When the NEMS devices are used as sensors, the sensitivity of the sensor is directly determined by the Q-factor;

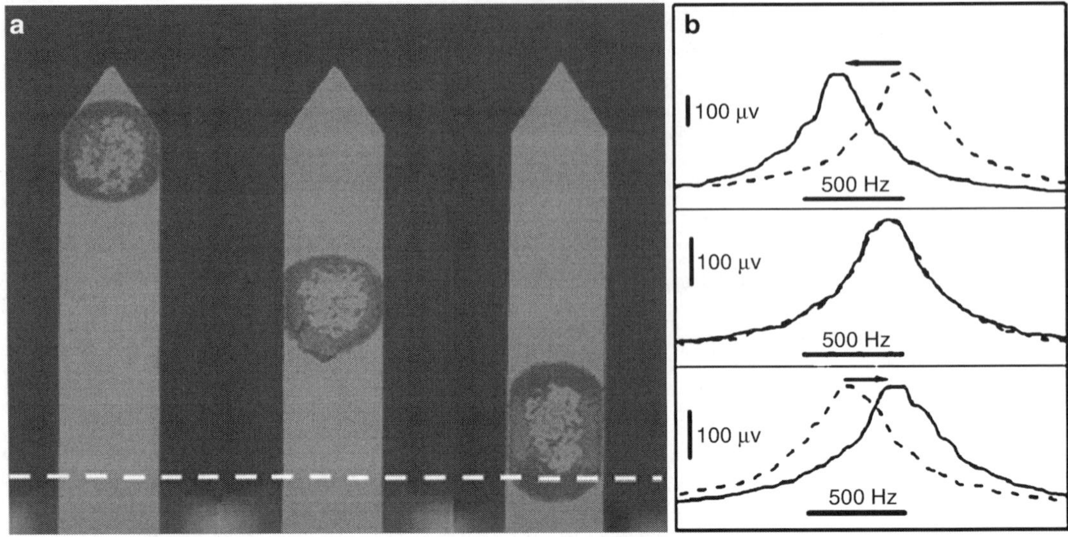

Fig. 7. Dependency of frequency shift on the location of the mass addition on cantilever beams. (**a**) Optical micrographs of bacteria deposited at three different locations on the cantilever (500 µm long, 100 µm wide, and 1 µm thick) which are 390, 200, and 73 µm, respectively, away from the position of clamping (shown by *dotted line*). (**b**) Frequency response of the cantilevers where the location of bacteria is as shown in (**a**). The dotted line is before the deposition of bacteria and the continuous line is after bacteria adsorption. The measurement was performed in air and the resonant frequencies are in the range of 7–8.5 kHz. The resonant frequency decreased by 4%, remained the same, and increased by 3.6% after the deposition of bacteria near the end, near the center, and near the clamping, respectively. Reprinted with permission from ref. 24.

the least mass (Δm_{min}) that could be detected with resonant sensors is given by (28, 29):

$$\Delta m_{min} = 2m_{eff}\sqrt{\frac{B}{\omega Q^{-DR/20}}},\tag{7}$$

where m_{eff} is the effective mass of the resonator, B is the measurement bandwidth, ω is the fundamental resonant frequency, and DR is a dynamic range term.

Q-factors of NEMS devices strongly depend on the design of the devices and process conditions. Typically, the Q-factors of NEMS resonating devices are in the range of 10^3–10^5 in vacuum (30). NEMS devices fabricated from silicon (mono or polycrystalline), GaAs, and SiC have all shown very high Q-factors. As the dimensions of the resonator decrease, the Q-factor decreases because the higher surface-to-volume ratio will cause more surface losses (see Note 2). This enables the NEMS resonator to operate at a very low power level with very high force sensitivity. Efforts to increase the Q-factor of submicron resonators include high-temperature annealing (31–33) and chemical treatments to alleviate surface loss mechanisms due to the oxidation of the substrate. Also, Q-factors may be improved by increasing the crystallinity of the materials and generating high internal residual stress. For example, devices fabricated from Si_3N_4 having an internal stress of about 1,200 MPa show a Q-factor on the order of a million

(4, 28); a comparison between the doubly clamped nanowires fabricated from the high-stress Si_3N_4 and low-stress Si_3N_4 shows that the high stress may improve Q-factors.

4.2.3. Power Level

A rough estimate of operating power levels can be given by the ratio of the thermal energy to the characteristic time scale of energy exchange between the mode and the surroundings at frequency, ω_o. The thermal energy is given by $k_B T$, where k_B is the Boltzmann's constant and T is temperature. The time scale τ is given by Q/ω_o. Therefore, the minimum operating power level of NEMS devices can be estimated by $k_B T \omega_o / Q$, which is on the order of aW to fw (10^{-18} to 10^{-15} W) for devices with Q-factors of about 1,000 to 10,000 and resonance frequencies in the MHz range.

5. Biosensor Examples Enabled by NEMS (See Note 3)

Many biodetection strategies (e.g., for DNA, proteins, and cells) rely heavily on fluorescent labels. Label-free detection techniques, enabled by technologies such as NEMS, surface plasmon resonance (SPR), and quartz crystal microbalances (QCM), have been very useful not only as sensors but also as tools for the study of the binding processes of biomolecules. Compared to SPR and QCM, NEMS devices have additional advantages that they are extremely small and have great potential to be mass-produced with high reproducibility. In this section, we look into some of the outstanding applications of NEMS devices for biosensing.

5.1. Biosensing by Cantilever-Based NEMS in the Deflection Mode

Deflection-based cantilever beams have been utilized for the detection of cells, proteins, and DNA.

5.1.1. Detection of Proteins

Microcantilever-based detection of prostate specific antigen (PSA), a tumor marker for prostate cancer, was demonstrated by Mazumdar et al. (34, 35). Using gold-coated Si_3N_4 cantilever beams ($200 \times 20 \times 0.5$ µm, $L \times W \times T$) and an optical detection system which measured the deflection of the cantilever tip, they successfully detected two forms of PSA at concentrations ranging from 0.2 ng/mL to 60 µg/mL in the presence of human serum albumin (HSA) and human plasminogen (HP), each at a concentration of 1 mg/mL. Their work also showed that surface stress, which is responsible for the deflection of the cantilever, is a function of the PSA concentration in the solution.

Arntz et al. (36) used an array of microcantilevers to detect two cardiac markers, creatin kinase and myoglobin, at a sensitivity of about 20 µg/mL. The antibodies to these cardiac markers were immobilized onto the cantilever surface, and antigen–antibody binding was detected, as the absolute deflection compared to a reference cantilever beam, by optical reflection. The reference beam also was used to eliminate thermal noise and turbulence which arose due to injection of the samples.

Also, Wee et al. (37) used a piezoresistive detection method to detect PSA and C-reactive proteins (CRP), cardiac markers, at a sensitivity of 10 ng/mL using deflection of cantilever beams. Fluorescently tagged PSA was used to verify that the change in cantilever resistance upon PSA binding was dependent only on the concentration of the PSA in solution.

5.1.2. Detection of DNA

Huber et al. (38) used gold-coated Si cantilever beams to study the binding of two DNA transcription factors, with a sensitivity of 80–100 nM, to double-stranded DNA functionalized onto the beam surface.

Fritz et al. (39) have been able to differentiate between the binding of a 12-mer single-stranded DNA (ssDNA) and a 12-mer ssDNA with a single base mismatch to the complementary ssDNA strand tethered to gold-coated Si cantilever beams, where the detection was made using optical reflection techniques. McKendry et al. (40) have demonstrated rapid DNA detection based on hybridization of a target DNA to strands on a cantilever beam. Here, only one of the cantilever beams out of a total of eight is functionalized with the complementary base sequence with respect to the analyte ssDNA. The resultant deflection of this beam was monitored by optical reflection, and the absolute deflection due to analyte binding was noted by subtracting the deflections of the other beams which were treated as reference beams.

The detection of a single-nucleotide polymorphism (SNP) was also demonstrated by Mukhopadhyay et al. (41) where the surface stresses resulting from hybridization of a fully complementary strand to a probe immobilized onto the cantilever surface was compared to the hybridization of a strand with a single base mismatch. In this case, the resistance change of the piezoresistive cantilever upon binding of the analyte was used to measure the deflection of the cantilever beam.

5.2. Biosensing by Cantilever-Based NEMS in the Resonant Mode

5.2.1. Detection of SAM Coating

The binding of thiolated self-assembled monolayers (SAMs) on gold-coated silicon cantilevers has been detected with a sensitivity of 5.5 fg by using photothermal actuation of the cantilever beam and optical interferometry to measure the resonant frequency (19). Ilic et al. have detected thiolated SAMS on a gold pad located near the cantilever tip and studied the effect of spatial mass loading on the decrease in its resonant frequency, enabling an accurate determination of the mass bound (11).

5.2.2. Detection of Proteins

PSA was also detected electrically with a sensitivity of 10 pg/mL (42) in air and of 1 ng/mL (13) in liquid by Kim and coworkers. Piezoelectric material (PZT) 0.5 µm thick was sandwiched between two Pt electrodes and was resonated with an AC voltage, which eliminates the use of an external oscillator. Similarly, Kwon et al. (13) detected activated cyclic adenosine monophosphate (cyclic AMP)-dependent protein kinase (PKA), with a sensitivity of 6.6 pM, with a fragment of PKI protein, PKI (5–24), immobilized onto the piezoelectric cantilever beam. Protein kinase is an enzyme which modifies other proteins, regulating their activity and coordinating central cellular processes. The measurement of C-reactive protein which provides a good indicator for the prediction of heart attack and stroke in postmenopausal women was also done by Lee et al. (43) on a similar format employing a PZT embedded cantilever.

Burg et al. (44) used a suspended microfluidic channel in place of a regular cantilever to address the problem that NEMS devices have low Q-factors when operated in fluids (see Fig. 8). The biomolecules bind on the inner walls of the suspended microchannel where the microchannel itself is resonated in air or vacuum. This way, Q-factors of the order of 15,000 are achieved with zeptogram mass sensitivity in vacuum. The thickness of the microchannel walls and the fluid layer was 800 nm and 1.2 μm, respectively. A continuous delivery of analyte solution was made possible by a polydimethylsiloxane (PDMS) microfluidic network. The device is actuated electrostatically, and the deflection of the microchannel is measured by optical reflection. Biotinylated

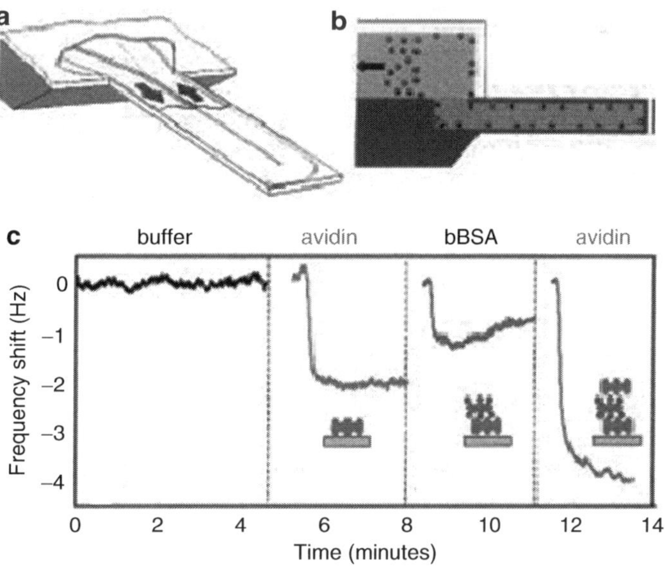

Fig. 8. Embedding microfluidic channels in a suspended resonant cantilever for biosensing. (a) Schematic showing the flow direction in the suspended microchannel. (b) Schematic of the cross-section of the microchannel showing analyte binding within the microchannel. The cantilever with the microchannel is resonated in vacuum. (c) Frequency decrease of the resonating microchannel in vacuum as a function of time due to binding of avidin, biotinylated BSA, and avidin in sequence. Reprinted with permission from ref. 44.

bovine serum albumin (bBSA) (44) was detected by functionalizing the surface of the microchannel with avidin. Then, bBSA was flowed through the channel followed by avidin again (see Fig. 8). With this system operated in air, the surface mass sensitivity was 10^{-17} g/μm^2. The detection of goat anti-mouse IgG (45) was also performed in a microchannel where biotinylated anti-goat antibodies were linked to the surface through neutravidin on a layer of poly(ethyleneglycol)-biotin grafted poly-l-lysine (PLL-PEG-biotin). Real-time binding of goat anti-mouse IgG was observed for concentrations ranging from 0.7 nM to 0.7 μM. A distinct advantage of this geometry is the ability to weigh molecules that pass through the microchannel without binding.

5.2.3. Detection of Double-Stranded DNA

Ilic et al. (46) detected a 1,587 bp dsDNA using an array of resonant NEMS cantilever beams. Si$_3$N$_4$ beams 90 nm thick and between 3.5 and 5 μm long were excited, and their resonant frequencies were detected by optical interferometry (see Fig. 9). A gold nanodot was deposited on the free end of the cantilever to which the thiolated dsDNA binds, providing a reliable calibrated response of frequency decrease with mass adsorption (since the shift in resonant frequency is dependent on the location of binding). The binding of the dsDNA was performed in solution, and the resonant frequencies were noted before and after the binding in vacuum (3×10^{-7} Torr). Q-factors in the range of 10^4 were reported, and the mass of a single dsDNA molecule 1,587 bps long was detected.

5.2.4. Detection of Viruses

Resonating silicon cantilever beams of nanoscale thickness were used to detect Vaccinia virus (average mass ~9.5 fg) by thermal and ambient noise actuation of the beams (47, 48). The detection of the resonant frequency was made by using a microscope scanning laser Doppler vibrometer.

The detection of an insect baculovirus was performed by Ilic et al. (11) using a Si$_3$N$_4$ cantilever beam 150 nm thick, 0.5 μm wide, and 6, 8, or 10 μm long with paddles of 1×1 μm. The beams were excited piezoelectrically, and signal transduction was achieved by optical interferometry. The beams were immersed in a solution of AcV1 antibody (antibody to the baculovirus) followed by immersion in the virus solution; they then were dried in nitrogen. The resonant frequencies were noted before and after the immobilization of the virus in vacuum (4×10^{-6} Torr). In calculating the mass adsorbed, it was assumed that there was no change in the flexural rigidity and that there was a linear relationship between the decrease in frequency and the mass adsorbed. Mass sensitivities of the order of 10^{-15} g/Hz were reported, and virus concentrations ranging from 10^5 to 10^7 pfu/mL were detected. Control experiments determined that only about 3.5% of the mass change resulted from nonspecific binding, showing that the assay had high specificity.

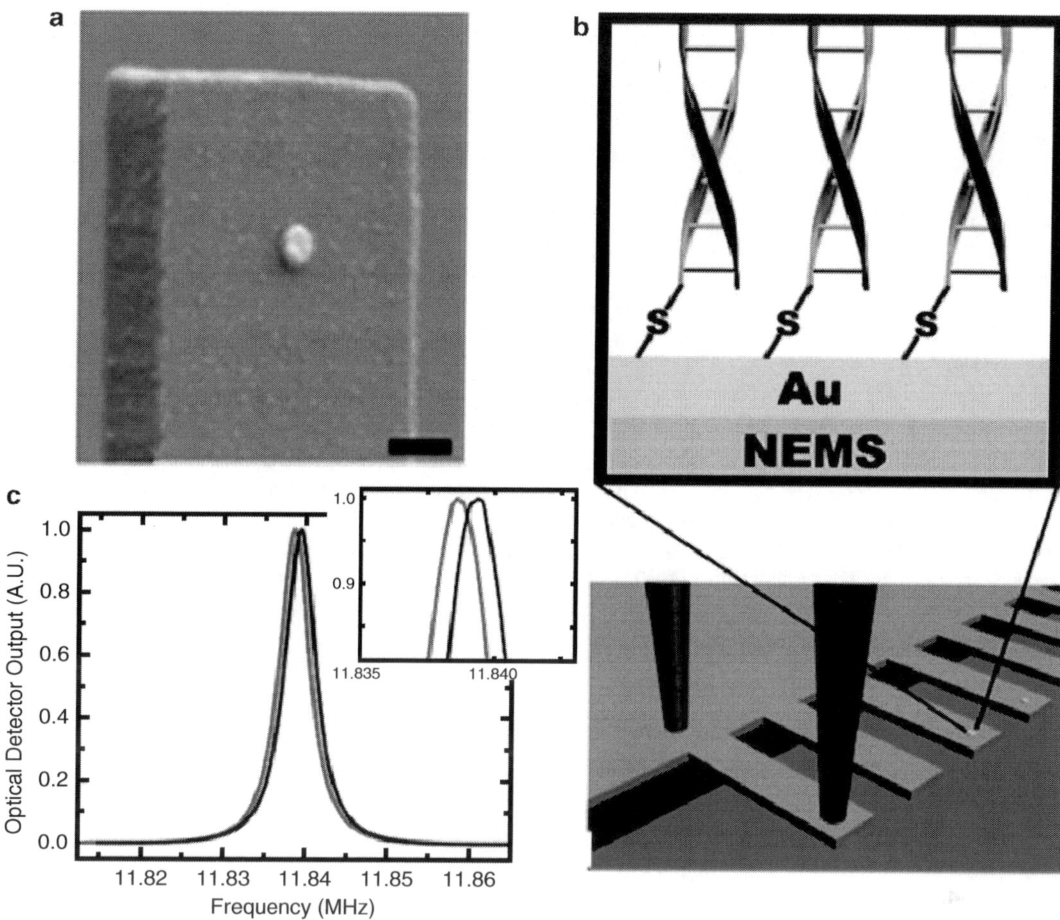

Fig. 9. Detection of dsDNA using an array of resonant cantilever beams. (a) SEM image showing the gold nanodot on the cantilever tip 300 nm away from the free end. (b) Schematic of the thiolated dsDNA binding to the Au nanodot (*top*). Schematic showing the actuating laser (415 nm) with the detecting laser (HeNe) in the foreground (*bottom*). (c) Frequency spectra in vacuum of the cantilever beam before and after binding of approximately 2 dsDNA molecules. Reprinted with permission from ref. 46.

5.2.5. Detection of Bacteria

A highly sensitive detection of *E. coli* (17, 18) was performed using Si_3N_4 cantilever beam of varying dimensions (see Fig. 10). The beams were excited to their resonance frequencies by thermal mechanical noise and ambient fluctuations in air and were detected by an optical deflection system. The antibodies to *E. coli* cells were bound to the cantilever beams by soaking them in solution and drying them. The resonance frequency was measured before and after the beam was bound by the *E. coli* cells. Again, it was assumed that there was a linear relationship between the decrease in frequency and the mass adsorbed and that there were no changes in the flexural rigidity of the beam. These assumptions are justified as the mass of *E. coli* adsorbed for varying cell concentrations is linear with the frequency change. *Escherichia coli* cells (ranging from 10^6 to 10^9 in number) were detected,

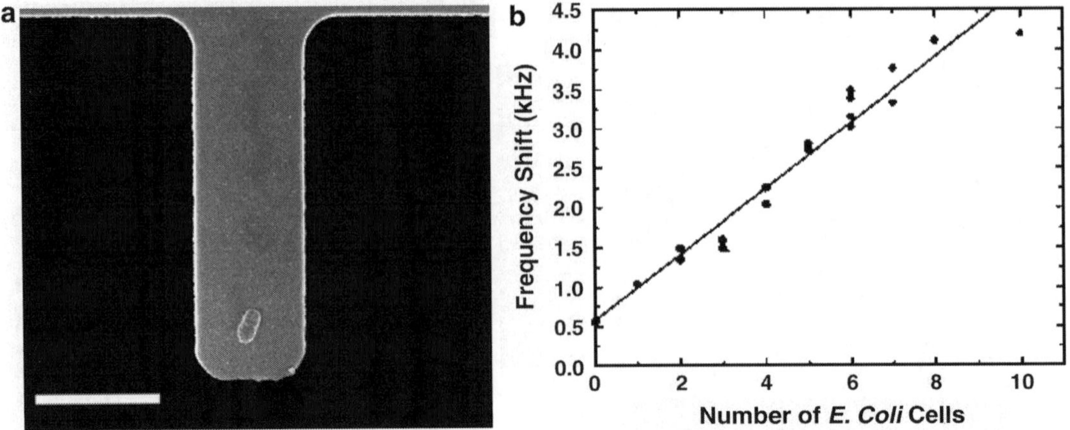

Fig. 10. Detection of *Escherichia coli* using resonating cantilever beams. (**a**) Scanning electron micrograph of a single *E. coli* 0157:H7 cell bound to the immobilized antibody layer on top of the oscillator. Length and width of the cell were estimated to be 1.43 mm and 730 nm and the thickness determined from tapping AFM to be 350 nm. (**b**) Measured frequency shift as a function of the number of *E. coli* cells. Reprinted with permission from ref. 17.

showing that the NEMS device could be operated in air without the use of external oscillator. The high dampening of oscillations in air gave rise to a very low Q-factor as compared to that for the operation of the device in vacuum, which is of the order of 10^4.

6. Notes

1. Most biosensing applications require that the sensor be operated in a liquid environment. In the resonant mode, viscous damping in liquid significantly reduces the Q-factor. When NEMS sensors are operated in liquid media, issues concerning fluid flow and mass transport also need to be taken into account.

2. Although size reduction of NEMS devices may lead to higher frequencies and improvement in mass sensitivity, as we scale down to the point where the size of the analyte becomes comparable to the size of the device, device modeling and experimental data interpretation may become considerably complex.

3. Most NEMS sensors require that the surface of the device be functionalized with a biomolecule. However, most bioreceptors are chemically and structurally complex and may lose their "biosensing" ability due to the conformation changes that they undergo when immobilized onto substrates. Also, the limited stability of the bioreceptors can adversely affect the long-term storage possibilities and reusability of the biosensor. In addition, functionalization of NEMS with biomolecules could result in losses in Q-factor, which would be detrimental

to the success of the biosensor. For example, a common way to attach biomolecules such as DNA and proteins to surfaces is by thiol-chemistry, which requires the mechanical sensing element to be coated with gold. Experimental studies (49, 50) have shown that even a very thin layer of gold, ~100 nm thick, causes a considerable decrease in the Q-factor of microcantilevers.

Acknowledgments

The authors acknowledge financial support from the National Science Foundation (CBET 0747164).

References

1. Waggoner, P. S. and Craighead, H. G. (2007) Micro- and nanomechanical sensors for environmental, chemical, and biological detection. *Lab. Chip.* **7**, 1238–1255.

2. Hierold, C. (2004) From micro- to nanosystems: mechanical sensors go nano. *J. Micromech. Microeng.* **14**, S1–S11.

3. Carr, D. W. and Craighead, H. G. (1997) Fabrication of nanoelectromechanical systems in single crystal silicon using silicon on insulator substrates and electron beam lithography. *J. Vac. Sci. Technol. B* **15**, 2760–2763.

4. Verbridge, S. S., Parpia, J. M., Reichenbach, R. B., Bellan, L. M., and Craighead, H. G. (2006) High quality factor resonance at room temperature with nanostrings under high tensile stress. *J. Appl. Phys.* **99**, 124304.

5. Kameoka, J., Verbridge, S. S., Liu, H., Czaplewski, D. A., and Craighead, H. G. (2004) Fabrication of suspended silica glass nanofibers from polymeric materials using a scanned electrospinning source. *Nano Lett.* **4**, 2105–2108.

6. Czaplewski, D. A., Verbridge, S. S., Kameoka, J., and Craighead, H. G. (2004) Nanomechanical oscillators fabricated using polymeric nanofiber templates. *Nano Lett.* **4**, 437–439.

7. Wu, Y., Cui, Y., Huynh, L., Barrelet, C. J., Bell, D. C., and Lieber, C. M. (2004) Controlled growth and structures of molecular-scale silicon nanowires. *Nano Lett.* **4**, 433–436.

8. Lange, D., Hagleitner, C., Hierlemann, A., Brand, O., and Baltes, H. (2002) Complementary metal oxide semiconductor cantilever arrays on a single chip: mass-sensitive detection of volatile organic compounds. *Anal. Chem.* **74**, 3084–3095.

9. He, R., Feng, X. L., Roukes, M. L., and Yang, P. (2008) Self-transducing silicon nanowire electromechanical systems at room temperature. *Nano Lett.* **8**, 1756–1761.

10. Cleland, A. N. and Roukes, M. L., (1996) Fabrication of high frequency nanometer scale mechanical resonators from bulk Si crystals. *Appl. Phys. Lett.* **69**, 2653–2655.

11. Ilic, B., Craighead, H. G., Krylov, S., Senaratne, W., Ober, C., and Neuzil, P. (2004) Attogram detection using nanoelectromechanical oscillators. *J. Appl. Phys.* **95**, 3694–3703.

12. Ilic, B., Yang, Y., and Craighead, H. G. (2004) Virus detection using nanoelectromechanical devices. *Appl. Phys. Lett.* **85**, 2604–2606.

13. Lee, J. H., Hwang, K. S., Park, J., Yoon, K. H., Yoon, D. S., and Kim, T. S. (2005) Immunoassay of prostate-specific antigen (psa) using resonant frequency shift of piezoelectric nanomechanical microcantilever. *Biosens. Bioelectron.* **20**, 2157–2162.

14. Kwon, H., Han, K., Hwang, K., Lee, J., Kim, T., Yoon, D., et al. (2007) Development of a peptide inhibitor-based cantilever sensor assay for cyclic adenosine monophosphate-dependent protein kinase. *Anal. Chim. Acta* **585**, 344–349.

15. Ekinci, K. L. (2005) Electromechanical transducers at the nanoscale: Actuation and sensing of motion in nanoelectromechanical systems (nems). *Small* **1**, 786–797.

16. O'Shea, S. J., Lu, P., Shen, F., Neuzil, P., and Zhang, Q. X. (2005) Out-of-plane electrostatic actuation of microcantilevers. *Nanotechnology* **16**, 602–608.

17. Ilic, B., Czaplewski, D., Zalalutdinov, M., and Craighead, H. G. (2001) Single cell detection with micromechanical oscillators. *J. Vac. Sci. Technol. B* **19**, 2825–2828.

18. Ilic, B., Czaplewski, D., Craighead, H. G., Neuzil, P., Campagnolo, C., and Batt, C. (2000) Mechanical resonant immunospecific biological detector. *Appl. Phys. Lett.* **77**, 450–452.

19. Lavrik, N. V. and Datskos, P. G. (2003) Femtogram mass detection using photothermally actuated nanomechanical resonators. *Appl. Phys. Lett.* **82**, 2697–2699.

20. Carr, D. W. and Craighead, H. G. (1998) Measurement of nanomechanical resonant structures in single-crystal silicon. *J. Vac. Sci. Technol. B* **16**, 3821–3824

21. Kouh, T., Karabacak, D., Kim, D. H., and Ekinci, K. L. (2005) Diffraction effects in optical interferometric displacement detection in nanoelectromechanical systems. *Appl. Phys. Lett.* **86**, 013106.

22. Foresi, J. S., Black, M. R., Agarwal, A. M., and Kimerling, L. C. (1996) Losses in polycrystalline silicon waveguides. *Appl. Phys. Lett.* **68**, 2052–2054.

23. Stoney, G. G. (1909) The tension of metallic films deposited by electrolysis. *Proc. R. Soc. Ser. A* **82**,172–175.

24. Ramos, D., Tamayo, J., Mertens, J., Calleja, M., and Zaballos, A. (2006) Origin of the response of nanomechanical resonators to bacteria adsorption. *J. Appl. Phys.* **100**, 106105.

25. Craighead, H. (2007) Nanomechanical systems: Measuring more than mass. *Nat. Nanotechnol.* **2**, 18–19.

26. Gupta, A. K., Nair, P. R., Akin, D., Ladisch, M. R., Broyles, S., Alam, M. A., et al. (2006) Anomalous resonance in a nanomechanical biosensor. *Proc. Natl. Acad. Sci. USA* **103**, 13362–13367.

27. Tamayo, J., Ramos, D., Mertens, J. and Calleja, M. (2006) Effect of the adsorbate stiffness on the resonance response of microcantilever sensors. *Appl. Phys. Lett.* **89**, 224104.

28. Verbridge, S. S., Craighead, H. G., and Parpia, J. M. (2008) A megahertz nanomechanical resonator with room temperature quality factor over a million. *Appl. Phys. Lett.* **92**, 013112.

29. Ekinci, K. L., Yang, Y. T., and Roukes, M. L. (2004) Ultimate limits to inertial mass sensing based upon nanoelectromechanical systems. *J. Appl. Phys.* **95**, 2682–2689.

30. Ekinci, K. L. and Roukes, M. L. (2005) Nanoelectromechanical systems. *Rev. Sci. Instrum.* **76**, 061101.

31. Yasumura, K. Y., Stowe, T. D., Chow, E. M., Pfafman, T., Kenny, T. W., Stipe, B. C., et al. (2000) Quality factors in micron- and submicron-thick cantilevers. *J. Microelectromech. Syst.* **9**, 117–125.

32. Haucke, H., Liu, X., Vignola, J. F., Houston, B. H., Marcus, M. H., and Baldwin, J. W. (2005) Effects of annealing and temperature on acoustic dissipation in a micromechanical silicon oscillator. *Appl. Phys. Lett.* **86**, 181903.

33. Yang, J., Ono, T., and Esashi, M. (2000) Surface effects and high quality factors in ultrathin single-crystal silicon cantilevers. *Appl. Phys. Lett.* **77**, 3860–3862.

34. Wu, G., Datar, R. H., Hansen, K. M., Thundat, T., Cote, R. J., and Majumdar, A. (2001) Bioassay of prostate-specific antigen (psa) using microcantilevers. *Nat. Biotechnol.* **19**, 856–860.

35. Majumdar, A. (2002) Bioassays based on molecular nanomechanics. *Dis. Markers* **18**, 167–174.

36. Arntz, Y., Seelig, J. D., Lang, H. P., Zhang, J., Hunziker, P., Ramseyer, J. P., et al. (2003) Label-free protein assay based on a nanomechanical cantilever array. *Nanotechnology* **14**, 86–90.

37. Wee, K. W., Kang, G. Y., Park, J., Kang, J. Y., Yoon, D. S., Park, J. H., et al. (2005) Novel electrical detection of label-free disease marker proteins using piezoresistive self-sensing micro-cantilevers. *Biosens. Bioelectron.* **20**, 1932–1938.

38. Huber, F., Hegner, M., Gerber, C., Guntherodt, H., and Lang, H. (2006) Label free analysis of transcription factors using microcantilever arrays. *Biosens. Bioelectron.* **21**, 1599–1605.

39. Fritz, J., Baller, M. K., Lang, H. P., Rothuizen, H., Vettiger, P., Meyer, E., et al. (2000) Translating biomolecular recognition into nanomechanics. *Science* **288**, 316–318.

40. McKendry, R., Zhang, J., Arntz, Y., Strunz, T., Hegner, M., Lang, H. P., et al. (2002) Multiple label-free biodetection and quantitative dna-binding assays on a nanomechanical cantilever array. *Proc. Natl. Acad. Sci. USA* **99**, 9783–9788.

41. Mukhopadhyay, R., Lorentzen, M., Kjems, J., and Besenbacher, F. (2005) Nanomechanical sensing of dna sequences using piezoresistive cantilevers. *Langmuir* **21**, 8400–8408.

42. Hwang, K. S., Lee, J. H., Park, J., Yoon, D. S., Park, J. H., and Kim, T. S. (2004) *In situ* quantitative analysis of a prostate-specific antigen (psa) using a nanomechanical pzt cantilever. *Lab Chip* **4**, 547–552.

43. Lee, J. (2004) Label free novel electrical detection using micromachined pzt monolithic thin film cantilever for the detection of c-reactive protein. *Biosens. Bioelectron.* **20**, 269–275.

44. Burg, T. P. and Manalis, S. R. (2003) Suspended microchannel resonators for biomolecular detection. *Appl. Phys. Lett.* **83**, 2698–2700.

45. Burg, T. P., Godin, M., Knudsen, S. M., Shen, W., Carlson, G., Foster, J. S., et al. (2007) Weighing of biomolecules, single cells and single nanoparticles in fluid. *Nature* **446**, 1066–1069.

46. Ilic, B., Yang, Y., Aubin, K., Reichenbach, R., Krylov, S., and Craighead, H. G. (2005) Enumeration of dna molecules bound to a nanomechanical oscillator. *Nano Lett.* **5**, 925–929.

47. Gupta, A., Akin, D., and Bashir, R. (2004) Single virus particle mass detection using microresonators with nanoscale thickness. *Appl. Phys. Lett.* **84**, 1976–1978.

48. Johnson, L., Gupta, A. K., Ghafoor, A., Akin, D., and Bashir, R. (2006). Characterization of vaccinia virus particles using microscale silicon cantilever resonators and atomic force microscopy. *Sens. Actuators, B* **115**, 189–197.

49. Sandberg, R., Mølhave, K., Boisen, A., and Svendsen, W. (2005) Effect of gold coating on the q-factor of a resonant cantilever. *J. Micromech. Microeng.* **15**, 2249–2253.

50. Dohn, S. R., Sandberg, R., Svendsen, W., and Boisen, A. (2005) Enhanced functionality of cantilever based mass sensors using higher modes. *Appl. Phys. Lett.* **86**, 233501.

Chapter 10

Nano "Fly Paper" Technology for the Capture of Circulating Tumor Cells

Shutao Wang, Gwen E. Owens, and Hsian-Rong Tseng

Abstract

Some efficient diagnosis and therapy systems require the isolation and quantification of circulating tumor cells (CTCs), since these species are important "biomarkers" for monitoring cancer metastasis and prognosis. Existing techniques for isolating/counting CTCs include immunomagnetic-bead-based separation and microfluidic capture. However, some of these techniques have low capture efficiency and low specificity. Through the use of a three-dimensional (3D) nanostructured substrate – specifically, a silicon-nanowire (SiNW) array coated with epithelial-cell-adhesion-molecule antibodies (anti-EpCAM) – we show that CTCs can be captured efficiently and specifically. Unlike conventional methods for isolating CTCs that depend on collision frequency and contact duration, nanoscaled local topographic interactions between the CTCs and the substrate increase their binding and markedly enhance capture efficiency.

Key words: Circulating tumor cells, Silicon-nanopillar array, Cell capture, Cancer diagnostic

1. Introduction

Most cancer-related deaths from solid tumors are caused by metastasis (1). Although the molecular mechanisms of cancer metastases remain largely unknown, tumor cells shed from a primary tumor mass, which enter the blood stream and travel to distant tissues, signal the earliest stage of malignant progression (2, 3). These cells are known as circulating tumor cells (CTCs) (4). Conventional diagnostic imaging and serum marker detection, which can enumerate and characterize CTCs within blood, provide valuable information (4–7) for evaluating early stage cancer metastases, predicting patient prognosis, and monitoring therapeutic interventions and outcomes (8). Over the past decade, several techniques for isolating and counting CTCs have been

Sarah J. Hurst (ed.), *Biomedical Nanotechnology: Methods and Protocols*, Methods in Molecular Biology, vol. 726,
DOI 10.1007/978-1-61779-052-2_10, © Springer Science+Business Media, LLC 2011

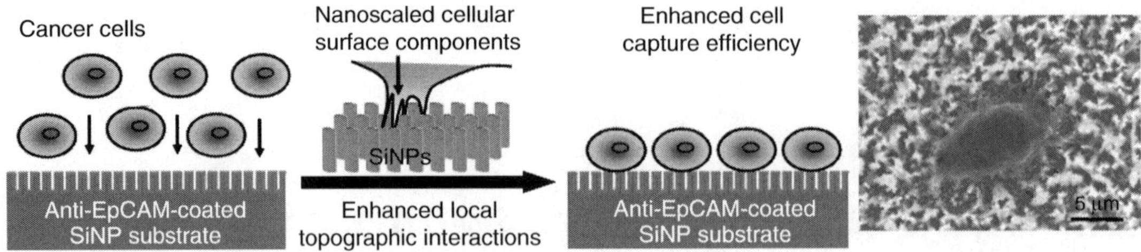

Fig. 1. A nano "fly paper" substrate for capturing CTCs: a 3D-nanostructured substrate – specifically, a silicon-nanopillar (SiNP) array coated with anti-EpCAM – enhances the local topographic interactions between nanoscaled cell surface components and the SiNP substrate, resulting in improved CTC-capture efficiency. As shown in the SEM micrograph (*right*), many interdigitated cellular protrusions (with diameters of ca. 100–200 nm) were observed on the SiNP substrates, validating the working mechanism of our SiNP-based cell-capture approach. Copyright Wiley-VCH Verlag GmbH & Co. KGaA. Reproduced from ref. 13 with permission.

developed (e.g., immunomagnetic beads (9), flow cytometry (10), microfluidics (11, 12)). These techniques allow reproducible detection of CTCs in blood when used in a clinical setting. However, the challenges that remain include improving CTC-capture efficiency, reducing the cost of measurement, and simplifying the sequential molecular analysis of captured cells.

In this chapter, we describe nano "fly paper" technology (13) (see Fig. 1) that can be employed to capture CTCs. Silicon nanopillar (SiNP)-covered substrates coated with epithelial-cell-adhesion-molecule antibodies (anti-EpCAM) exhibit outstanding efficiency for isolating viable CTCs from whole blood samples. Using a simple stationary device, CTCs are immobilized on the SiNP substrates thanks to strong topographic interactions between the substrate-grated SiNPs and the nanoscale cell surface structure.

2. Materials

2.1. Making SiNP Substrates Using Wet Chemical Etching

1. Oriented prime grade silicon wafers, p-type, resistivity of ca. 10–20 ohm cm (Silicon Quest Int'l, Santa Clara, CA). Store at room temperature.

2. Acetone (ACS reagent, SpectroGrade, 99.5%). Store at room temperature.

3. Ethanol, >99.5%. Store at room temperature.

4. Piranha solution: 4:1 (v/v) sulfuric acid/hydrogen peroxide. Store at room temperature.

5. RCA solution: 1:1:5 (v/v/v) ammonia/hydrogen peroxide/ H_2O.

6. Etching solution: 4.6 M Hydrofluoric acid (HF), 0.2 M silver nitrate in deionized water. Store at room temperature.

7. Aqua regia solution: 3:1 (v/v) hydrochloric acid/nitric acid.

2.2. Making Streptavidin-Coated SiNP Substrates

1. 4% (v/v) 3-Mercaptopropyl trimethoxysilane in ethanol (95%) (Sigma-Aldrich, St. Louis, MO). Store at room temperature.

2. 0.25 mM N-γ-maleimidobutyryloxy succinimide ester (GMBS, 98% HPLC) (Sigma-Aldrich). Store at room temperature.

3. Streptavidin (10 μg/mL) (Invitrogen, Carlsbad, CA). Store in single-use aliquots at –20°C.

4. 1× Dulbecco's phosphate-buffered saline (PBS) (Invitrogen). Store at 4°C.

2.3. Scanning Electron Microscopy of SiNP Substrates

1. 4% Glutaraldehyde (E.M. grade, Polysciences, Warrington, PA) in 0.1 M cacodylic acid sodium salt trihydrate (Sigma-Aldrich). Store at room temperature.

2. 1% Osmium tetroxide (ACS reagent, >98%, Sigma-Aldrich). Toxic! Store at room temperature.

3. 1% Tannic acid (Electron Microscopy Sciences, Hatfield, PA). Store at room temperature.

4. 0.5% Uranyl acetate (Electron Microscopy Sciences). Store at room temperature.

5. Alcohol (97%, Sigma-Aldrich).

6. Hexamethyldisilazane (HDMS) (Sigma-Aldrich). Toxic! Store at room temperature.

7. Gold plate.

2.4. Preparation of Artificial Blood Samples

1. Breast cancer cell line, MCF7 (American Type Culture Collection, Manassas, VA).

2. Reagents for cell culture: Dulbecco's Modified Eagle's Medium (DMEM, 1×) (Invitrogen), fetal bovine serum (FBS) (Fisher Scientific, Pittsburg, PA), penicillin–streptomycin (100×) (Fisher Scientific), and trypsin. Store at –20°C.

3. Citrated whole rabbit blood (Colorado Serum Company, Denver, CO).

4. Vybrant® DiD cell-labeling solution (Invitrogen). Store at 4°C.

5. Dulbecco's phosphate-buffered saline (PBS) (Invitrogen) Store at 4°C.

2.5. Capture and Immunochemistry of CTCs from Artificial Blood Samples

1. Lab-Tek chamber slides, 4-well glass, sterile (Thermo Fisher Scientific). Store at room temperature.

2. 10 μg/mL Biotinylated anti-human EpCAM/TROP1 antibody (Goat IgG, R&D Systems, Minneapolis, MN), 1% (w/v) bovine serum albumin (BSA, Sigma-Aldrich), 0.09% (w/v) sodium azide in 1× PBS. Store in single-use aliquots at –20°C.

3. 1× PBS.

4. 4% Paraformaldehyde in 1× PBS. Store at 4°C.

5. 0.2% Triton X-100 in 1× PBS. Store at 4°C.

6. 1% DAPI in 1× PBS. Store at 4°C.

2.6. Capture and Immunochemistry for CTCs Captured from Patient Samples

1. Lab-Tek chamber slides, 4-well glass, sterile (Thermo Fisher Scientific). Store at room temperature.

2. 10 µg/mL Biotinylated anti-human EpCAM/TROP1 antibody (Goat IgG, R&D Systems, Minneapolis, MN), 1% (w/v) bovine serum albumin (BSA, Sigma-Aldrich), 0.09% (w/v) sodium azide in 1× PBS. Store in single-use aliquots at –20°C.

3. 1× PBS.

4. Vacutainer tubes containing the anticoagulant ethylenediaminetetraacetic acid (ETDA).

5. 4% Paraformaldehyde in 1× PBS. Store at 4°C.

6. 0.3% Triton X-100 in 1× PBS. Store at 4°C.

7. Blocking solution: 5% normal goat serum, 0.1% Tween 20, 3% BSA in 1× PBS.

8. 1% DAPI in 1× PBS. Store at 4°C.

9. 20 µg/mL Anti-cytokeratin conjugated with phycoetythrin (CAM5.2 conjugated with PE) (BD Biosciences, San Jose, CA) in 1× PBS. Store in single-use aliquots at –20°C.

10. 20 µg/mL Anti-human CD45 conjugated with FITC (Ms IgG1, clone H130) (BD Biosciences) in 1× PBS. Store in single-use aliquots at –20°C.

---

3. Methods

Over the past decade, techniques for isolating and counting CTCs have been developed that are based on different working mechanisms (4–7). Recently, we have developed a novel 3D SiNP-based CTC-capture technology and we call this as nano "fly paper." To test the cell-capture performance of the SiNPs substrates, we prepared a cell suspension solution of an EpCAM-positive breast cancer cell line (i.e., MCF7) in cell culture medium (DMEM). After optimization of experimental parameters (e.g., CTC capture, sensitivity, and reproducibility), a clinical study with the SiNP-based CTC-capture technology was conducted using blood samples collected from patients with metastatic prostate cancer in collaboration with the Department of Urology UCLA under UCLA IRB approval (IRB #09-03-038-01). After CTC capture by the SiNP-based system, a 3-parameter immunocytochemistry protocol (for parallel staining of DAPI, FITC-labeled anti-CD45, and PE-labeled anti-cytokeratin (CAM2.5)) was applied to stain the immobilized cells. Based on the signal thresholds and size/morphology features established for model cells, the CTCs were clearly distinguishable from background immune cells. Since only 1.0 mL of blood is required for each CTC-capture study, we could perform as many as three measurements on each blood sample. Figure 2 presents an example of one successful CTC-capture study on the SiNP-based substrate using blood samples that, when tested using CellSearch™ technology, had failed to detect any CTCs.

3.1. Making SiNP Substrates Using Wet Chemical Etching

1. Cut silicon wafer into pieces 1×2 cm in dimension.

2. Ultrasonicate the cut silicon substrates in acetone for 10 min at room temperature and dry under nitrogen gas. Next, ultrasonicate the substrates in ethanol for 5 min at room temperature and dry them under nitrogen gas. These steps remove contamination (such as organic grease) from the silicon substrate.

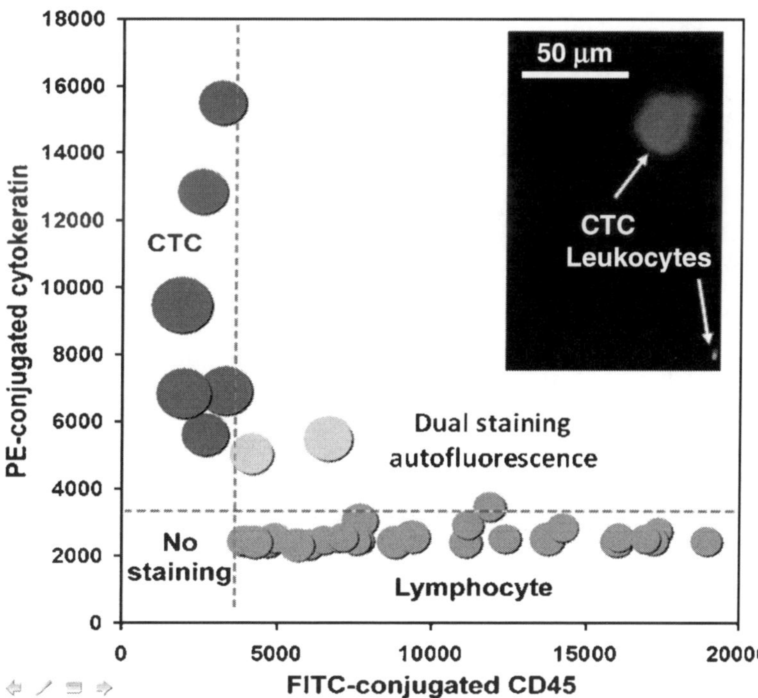

Fig. 2. A scatter plot summarizes the quantitative immunocytochemistry measured on both cytokeratin and CD45 expression levels of patient CTCs and lymphocytes captured on a 1×2 cm SiNP substrate. The size of each dot reflects the footprint of each cell. In this case, we were able to identify 6 CTCs in a 1-mL blood sample from a prostate cancer patient. The insert is a typical fluorescent image of CTCs captured on SiNP substrates.

3. To etch the silicon wafer surface, first heat the silicon substrates in boiling piranha solution for 1 h. Then, heat the substrates in boiling RCA solution for 1 h. Rinse the substrates five times with DI water. Next, place the silicon substrates in a teflon vessel and etch them with etching solution at 50°C (see Note 1).

4. Immerse the substrates in boiling aqua regia for 15 min.

5. Rinse the substrates with DI water and dry them under nitrogen gas.

3.2. Making Streptavidin-Coated SiNP Substrates

1. Place the substrates in 4% (v/v) 3-mercaptopropyl trimethylsilane in ethanol for 45 min at room temperature.

2. Treat the substrates with 0.25 mM N-γ-maleimidobutyryloxy succinimide ester (GMBS) and incubate for 30 min at room temperature.

3. Treat the substrates with streptavidin (SA) (10 μg/mL) and incubate for 30 min at room temperature.

4. Flush the substrates with 1× PBS to remove excess streptavidin.

5. Store modified substrates at 4–8°C for up to 6 months.

3.3. Scanning Electron Microscopy Observation of SiNP Substrates

1. Allow cells to incubate on the substrates for 24 h.

2. Fix cells with 4% glutaraldehyde buffered in 0.1 M sodium cacodylate and incubate cells for 1 h at 4°C. Afterward, post-fix cells using 1% osmium tetroxide for 1 h, using 1% tannic acid as a mordant.

3. Dehydrate samples through a series of alcohol concentrations (30, 50, 70, and 90%). Stain samples with 0.5% uranyl acetate and then further dehydrate the samples through a series of higher alcohol concentrations (96, 100, and 100%). Dehydrate the samples for the final time in hexamethyldisilazane (HMDS).

4. Air-dry the samples.

5. Once dry, sputter-coat the samples with gold (3–5 nm in thickness) and examine with a field emission SEM (accelerating voltage of 10 keV).

3.4. Preparation of Artificial Blood Samples

1. Stain MCF7 cells (in 1 mL DMEM without serum, 10^6 cells/mL) with DiD red fluorescent dye (5 μL, for 20 min).

2. Prepare control samples by spiking stained MCF7 cells into rabbit blood with cell densities of 1,000–1,250, 80–100, and 5–20 cells/mL.

3.5. Capture and Immunochemistry of CTCs from Artificial Blood Samples (See Fig. 3)

1. Place substrates into a size-matched 4-well Lab-Tek Chamber Slide. Drop 25 μL of biotinylated anti-EpCAM (10 μg/mL in 1× PBS with 1% (w/v) BSA and 0.09% (w/v) sodium azide) onto a 1×2 cm substrate. Incubate for 30 min. Wash with 1× PBS.

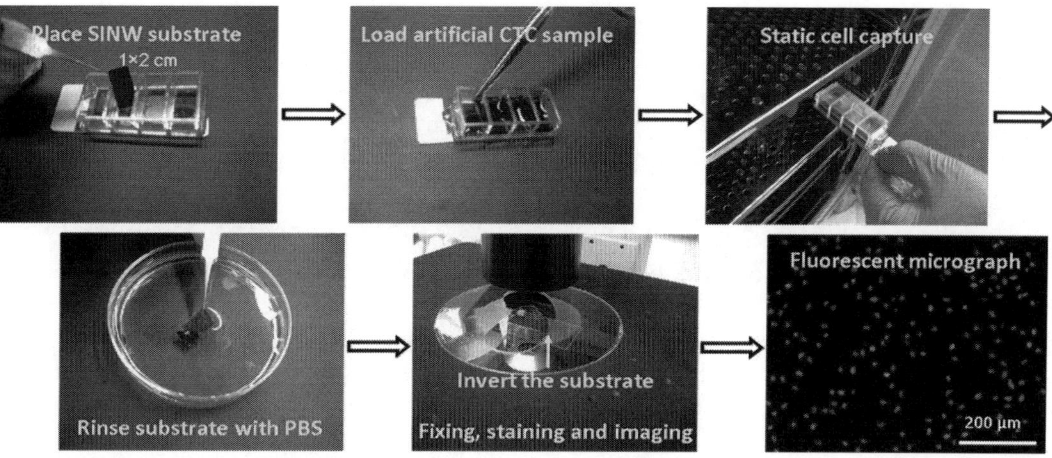

Fig. 3. A simple and convenient procedure of cell-capture experiments by using the nano "fly paper" technology. Copyright Wiley-VCH Verlag GmbH & Co. KGaA. Reproduced from ref. 13 with permission.

2. Load 1 mL control sample (see Subheading 3.4) onto each substrate.

3. Incubate for 45 min (37°C, 5% CO_2).

4. Gently wash the substrate with 1× PBS at least five times.

5. Fix cells captured on the substrates with 4% paraformaldehyde (PFA) in 1× PBS for 20 min.

6. To stain and visualize captured cells, treat the substrates with 0.2% Triton X-100 in 1× PBS and incubate for 10 min. Incubate the substrates with a DAPI solution (1% DAPI in 1× PBS) for 5 min. Wash the substrates three times with 1× PBS. Invert the substrates onto a standard cover glass.

7. Image and count cells using a Nikon TE2000 fluorescence microscope and a hemocytometer, respectively (see Note 2) (see Subheading 3.7).

3.6. Capture and Immunochemistry for CTCs Captured from Patient Samples (See Fig. 3)

1. Place substrates into a size-matched 4-well Lab-Tek Chamber Slide. Drop 25 µL of biotinylated anti-EpCAM (10 µg/mL in 1× PBS with 1% (w/v) BSA and 0.09% (w/v) sodium azide) onto a 1×2 cm substrate. Incubate for 30 min. Wash with 1× PBS.

2. Blood samples drawn from patients with advanced solid-stage tumors (as approved by IRB) are collected into vacutainer tube containing the anticoagulant ETDA. Samples should be processed immediately after collection.

3. To capture cells, load 1 mL patient sample onto each SiNP substrate. Incubate for 45 min (37°C, 5% CO_2) and then gently wash the substrates with 1× PBS at least five times.

4. Fix cells captured on the substrate by loading 200 µL of 4% paraformaldehyde (PFA) in 1× PBS onto each substrate for 20 min at room temperature. Wash each substrate three times with PBS.

5. Permeabilize cells by treating each substrate with 200 µL of 0.3% Triton X-100 in PBS for 30 min at room temperature. Subsequently, wash each substrate three times with 1× PBS.

6. Add 200 µL of blocking solution to each substrate and incubate for 1 h at room temperature. Next, add 200 µL of each fluorophore-labeled antibody solution (CAM2.5-PE and CD45-FITC) to each substrate and incubate the substrates in the dark at 4°C overnight. Wash with 200 µL of 1× PBS three times (first wash: 15 min at room temperature, second and third washes: 5 min at room temperature). Incubate substrates with 1% DAPI solution for 5 min. Wash each substrate three times with 1× PBS.

7. Gently invert the substrates, using tweezers, onto a cover glass to prepare for imaging and counting (see Note 3).

3.7. Identification of CTCs by Fluorescence Microscopy (See Note 4)

1. Select fluorescent microscope settings. Optimized exposure times: DAPI (blue) filter: 50 ms exposure time (background: ~1,300), FITC (green) filter: 300 ms exposure time (background: ~1,600), PE (red) filter: 100 ms exposure time (background: ~1,300). Set the digitizer to 1 MHz. Other settings will yield a very high background signal when using a green filter.

2. Place a sample on the microscope and focus on the edge of the substrate. Once focused, switch to the blue filter.

3. Starting in the upper right corner of the substrate, scan for nuclei that are approximately 7–20 µm in diameter at 4× or 10× magnification.

4. Increase the magnification to 10× or 20× when a putative cell has been located. Check the fluorescence intensity under blue, red, and green filters. Score a sample fluorescence intensity >2× the background fluorescence intensity as a positive result.

4. Notes

1. The nanopillar length depends on the duration of the etching step; nanopillars 10 µm in length are optimal for these experiments.

2. Cells that show dual stains (red: DiD+ and blue: DAPI+) and meet the phenotypic morphological characteristics (e.g., cell

size, shape, and nucleus size) are scored as CTCs. Cells stained by DAPI+ only are counted as nonspecific cells.

3. Cells that demonstrate dual staining (red: PE+ and blue: DAPI+) and meet standard phenotypic and morphological characteristics should be scored as CTCs. Cells that show dual staining (green: FITC+ and blue: DAPI+) should be excluded as lymphocytes/nonspecific cells (see Fig. 2). Species that demonstrate staining in all three filters (green+ and red+ and blue+) should be excluded as cellular debris.

4. Care should be taken to ensure that the substrates are never allowed to dry.

Acknowledgments

The authors appreciate the helpful discussions with Dr. Hao Wang, Dr. Jing Jiao, Dr. Ken-ichiro Kamei, Dr. Jing Sun, Kuan-Ju Chen, Gwen E. Owens, and David J. Sherman. This research was supported by NIH-NCI NanoSystems Biology Cancer Center (U54CA119347).

References

1. Steeg, P. S. (2006) Tumor metastasis: mechanistic insights and clinical challenges. *Nat. Med.* **12**, 895–904.

2. Bernards, R. and Weinberg, R. A. (2002) A progression puzzle. *Nature* **418**, 823.

3. Pantel, K. and Brakenhoff, R. H. (2004) Dissecting the metastatic cascade. *Nat. Rev. Cancer* **4**, 448–56.

4. Budd, G. T., Cristofanilli, M., Ellis, M. J., Stopeck, A., Borden, E., Miller, M. C., et al. (2006) Circulating tumor cells versus imaging--predicting overall survival in metastatic breast cancer. *Clin. Cancer Res.* **12**, 6403–6409.

5. Allard, W. J., Matera, J., Miller, M. C., Repollet, M., Connelly, M. C., Rao, C., et al. (2004) Tumor cells circulate in the peripheral blood of all major carcinomas but not in healthy subjects or patients with nonmalignant diseases. *Clin. Cancer Res.* **10**, 6897–6904.

6. Zieglschmid, V., Hollmann, C., and Böcher, O. (2005) Detection of disseminated tumor cells in peripheral blood. *Crit. Rev. Clin. Lab. Sci.* **42**, 155–196.

7. Pantel, K., Brakenhoff, R. H., and Brandt, B. (2008) Detection, clinical relevance and specific biological properties of disseminating tumour cells. Nat. Rev. Cancer **8**, 329–340.

8. Cristofanilli, M., Budd, G. T., Ellis, M. J., Stopeck, A., Matera, J., Miller, M. C., et al. (2004) Circulating tumor cells, disease progression, and survival in metastatic breast cancer. *New Engl. J. Med.* **351**, 781–791.

9. Talasaz, A. H., Powell, A. A., Huber, D. E., Berbee, J. G., Roh, K.-H, Yu, W., et al. (2009) Isolating highly enriched populations of circulating epithelial cells and other rare cells from blood using a magnetic sweeper device. *Proc. Natl. Acad. Sci. USA* **106**, 3970–3975.

10. Allan, A. L., Vantyghem, S. A., Tuck, A. B., Chambers, A. F. Chin-Yee, I. H., and Keeney, M. (2005) Detection and quantification of circulating tumor cells in mouse models of human breast cancer using immunomagnetic enrichment and multiparameter flow cytometry. *Cytometry A* **65**, 4–14.

11. Nagrath, S., Sequist, L. V., Maheswaran, S., Bell, D. W., Irimia, D., Ulkus, L., et al. (2007) Isolation of rare circulating tumour cells in cancer patients by microchip technology. *Nature* **450**, 1235–1239.

12. Adams, A. A., Okagbare, P. I., Feng, J., Hupert, M. L., Patterson, D., Gottert, J., et al.

(2008) Highly efficient circulating tumor cell isolation from whole blood and label-free enumeration using polymer-based microfluidics with an integrated conductivity sensor. *J. Am. Chem. Soc.* **130**, 8633–8641.

13. Wang, S., Wang, H., Jiao, J., Chen, K.-J., Owens, G. E., Kamei, K.-I., et al. (2009) Three-dimensional nanostructured substrates toward efficient capture of circulating tumor cells. *Angew. Chem. Int. Ed.* **48**, 8970–8973.

Chapter 11

Polymeric Nanoparticles for Photodynamic Therapy

Yong-Eun Koo Lee and Raoul Kopelman

Abstract

Photodynamic therapy is a relatively new clinical therapeutic modality that is based on three key components: photosensitizer, light, and molecular oxygen. Nanoparticles, especially targeted ones, have recently emerged as an efficient carrier of drugs or contrast agents, or multiple kinds of them, with many advantages over molecular drugs or contrast agents, especially for cancer detection and treatment. This paper describes the current status of PDT, including basic mechanisms, applications, and challenging issues in the optimization and adoption of PDT; as well as recent developments of nanoparticle-based PDT agents, their advantages, designs and examples of in vitro and in vivo applications, and demonstrations of their capability of enhancing PDT efficacy over existing molecular drug-based PDT.

Key words: Photodynamic therapy, Nanoparticle, Photosensitizer, Optical penetration depth, Polymer

1. PDT Basic Facts

Photodynamic therapy (PDT) is a minimally invasive localized treatment modality based on three key components: photosensitizer, light, and molecular oxygen. It is a relatively new therapy – tested in cells and animal models in the 1960s and 1970s, tested in clinical trials in the 1980s, and approved clinically in 1995 in the USA (1) – although its discovery dates back to the early 1900s (2, 3). Currently, PDT is clinically approved for treatment of several medical conditions, including cases of skin actinic keratosis, several forms of cancer, blindness due to age-related macular degeneration and localized bacterial infection (3–5). New medical applications of PDT continue to be discovered and clinically tried (4, 5). The PDT procedure involves the following steps: (1) the photosensitizer, after being administered systemically, or topically, preferentially accumulates in target tissues, and (2) light of

Sarah J. Hurst (ed.), *Biomedical Nanotechnology: Methods and Protocols*, Methods in Molecular Biology, vol. 726,
DOI 10.1007/978-1-61779-052-2_11, © Springer Science+Business Media, LLC 2011

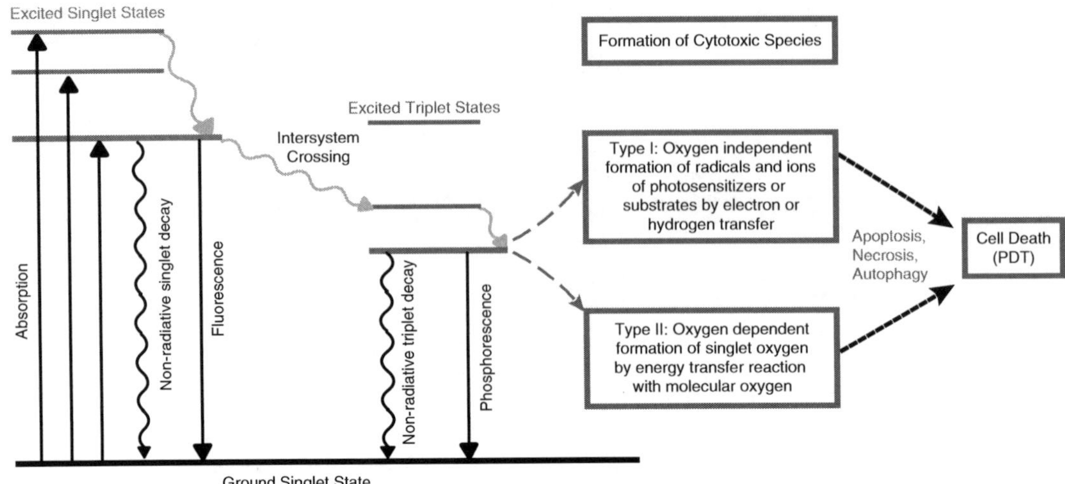

Fig. 1. Reaction kinetics of photosensitization leading to PDT (Types I and II).

the appropriate wavelength is directed onto the target tissue to induce a photodynamic reaction between an optically excited photosensitizer and surrounding oxygen (or other) molecules, so as to produce reactive oxygen species (ROS) (see Fig. 1).

Type I reactions involve a direct electron/hydrogen transfer from the photosensitizer, producing ions, or electron/hydrogen abstraction from a substrate molecule, to form free radicals. These radicals then react rapidly, usually with oxygen, resulting in the production of ROS such as superoxide anion radicals, hydrogen peroxide, and hydroxyl radicals. Type II reactions involve an energy transfer reaction between the excited triplet state photosensitizer and an oxygen molecule, resulting in the formation of "singlet oxygen," the first electronically excited, and highly reactive, state of oxygen. The reactive species thus produced, via type I or II reactions, then attack plasma membranes or subcellular organelles such as the mitochondria, plasma membrane, Golgi apparatus, lysosomes, endosomes, and endoplasmic reticulum, leading to cell death by apoptosis, necrosis, and autophagy (6). The relative contribution from type I and II reactions depends on the type of photosensitizer used, the available tissue oxygen and the type of substrate (e.g., a cell membrane or a biomolecule), as well as the binding affinity of the photosensitizer for the substrate (3). Direct and indirect evidence indicates that type II reactions, hence singlet oxygen, play a dominant role in PDT (7, 8). The reported lifetime (τ) of singlet oxygen in vivo ranges from 0.03 to 0.18 μs (8), and the estimated diffusion distance of singlet oxygen, based on a diffusion coefficient, D, for oxygen in water of 2×10^{-9} m^2/s (9) and the root-mean-square diffusion distance, $(2Dt)^{1/2}$, is only 11–27 nm. The other cytotoxic reactive species generated by the type I reactions also have short lifetimes and diffusion distances (10). This indicates that the PDT effects are confined to the illuminated area containing photosensitizers. This localization minimizes side effects related to PDT.

The PDT efficiency depends on the photosensitizer's ability to produce ROS, the oxygen availability, and the light dose (intensity and duration), as well as the photosensitizer concentration at the treated area. The PDT action on the tumor tissue level can be described by three different processes: (1) direct photodynamic tumor cell killing, (2) indirect tumor cell killing by photodynamic damage to, or shutdown of, the tumor supporting vasculature, with loss of oxygen and nutrients to the tumor, and (3) additional antitumor contributions from the inflammatory and immune responses of the host (11).

Since the photosensitizers used are of inherently low systemic toxicity and become pharmacologically active only when exposed to light, PDT can be applied selectively to the diseased tissue without significantly affecting connective tissues and underlying structures. There is less morbidity, less functional disturbance, and better cosmetic outcome than with other therapies. PDT has a low mutagenic potential and, except for sustained skin phototoxicity, few adverse effects. PDT can also be repeated at the same site without lifetime dose limitation, unlike radiotherapy. PDT can be used as a stand-alone modality or in combination with other therapeutic modalities, wherein it can be applied either before or after other cancer therapies, without compromising these treatments or being compromised itself. For example, PDT is employed for the treatment of glioblastoma multiforme (GBM) as adjuvant therapy after surgical resection of the glioblastoma mutiforme tumor, resulting in a significant survival advantage without added risk to the patient (12).

The one challenging issue for PDT is that its efficiency is inherently limited by the optical tissue penetration depth. The effective excitation light magnitude is determined by the combination of optical absorption and scattering properties of the tissue. The optical absorption properties of tissue vary significantly with wavelength (see Fig. 2). The optical scattering of tissue decreases with wavelength, and the reduced scattering coefficient (μ'_s) follows the scaling law, $\mu'_s = \lambda^{-b}$, where $b = 0.5$–2 over the visible and IR spectral region (13). For the spectral range of 450–1,750 nm, tissue scattering is, in general, more prevalent than absorption, although for the range of 450–600 nm, melanin and hemoglobin provide significant absorption, while water plays a similar role for $\lambda > 1,350$ nm (13). Therefore, the optimal optical window for PDT, as well as for optical imaging, is in the near-infrared (NIR) spectral region (600–1,300 nm), where the scattering and absorption by tissue are minimized and, therefore, the longest penetration depth can be achieved. Within this optical window, the longer the wavelength is, the deeper is the penetration depth.

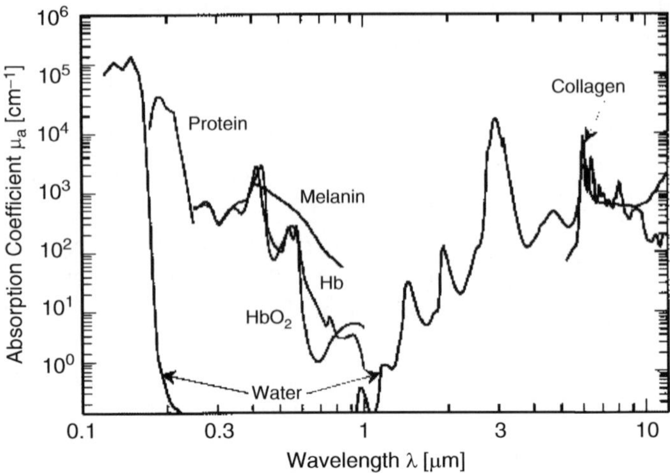

Fig. 2. Optical absorption coefficients of principal tissue chromophores in the 0.1–12 μm spectral region. Reproduced from ref. 13 with permission from the American Chemical Society.

Moreover, the optical penetration depth varies significantly with tissue type (14–16) (see Table 1). Here, the optical penetration depth into tissue is defined as the distance over which the light intensity drops to $1/e$ or 37% of the initial value. It should be noted that beyond this depth there remains light at a lower intensity, which still may be sufficient for PDT. The effective depth of PDT-induced necrosis, a more practical term for therapeutic depth of PDT activity, is approximately three times the penetration depth (17, 18). For instance, the optical penetration depth at 630 nm, the wavelength for PDT using Photofrin, is 0.2–2.9 mm, depending on organ (see Table 1), which limits PDT efficacy to less than 9 mm below the illuminated surface.

PDT has also other challenging issues regarding its clinical therapeutic efficacy, similar to challenges faced by chemotherapy and radiotherapy. Those include improvement of PDT drug efficiency, efficient delivery of the PDT drugs to the treatment area, and development of methods to maintain the therapeutic efficiency even at low oxygen levels, especially in the case of tumors.

2. Factors Controlling PDT Effectiveness

The PDT efficiency depends on multiple factors: (1) the photochemical properties of the photosensitizer such as its singlet oxygen production efficiency and tissue penetration depth of its excitation light, (2) the delivery system, (3) the biological state of the tumor including the tumor type and its microvasculature, as well as the tissue oxygenation level, (4) the physical localization

Table 1
Optical penetration depth (mm) of selected tissues

Tissue	Wavelength of light							References
	630 nm	632.8 nm	665 nm	675 nm	780 nm	835 nm	1,064 nm	
Blood		0.19		0.28	0.42	0.51		[14]
Mammary tissue		2.59		2.87	3.12	3.54		[14]
Mammary carcinoma		2.87		3.14	3.62	4.23		[14]
Mammary carcinoma in C3H/HEJ mice	2.0		2.3				3.7	[15]
Brain (postmortem)		0.92		1.38	2.17	2.52		[14]
Brain	1.6							[16]
Brain tumor (glioma)	3.1							[16]
Colon		2.48		2.73	2.91			[14]
Lung		0.81		1.09	1.86	2.47		[14]
Lung carcinoma		1.68		2.01	2.82	3.89		[14]

and the amount of photosensitizer in treated tissue, and (5) the light parameters such as light dose, light fluence rate, and the interval between drug and light administration. The success of PDT depends on improvements in the combination of all these factors, so as to circumvent the aforementioned challenging issues. The chemical, biological, and clinical progresses for those factors made so far are described below.

2.1. PDT Efficiency and Optical Tissue Penetration Depth of Photosensitizer

The hematoporphyrin derivative, Photofrin, is the first clinically approved, and currently the most widely used, PDT drug. It is approved for tumors (esophageal and endobronchial cancers) and for high-grade dysplasia in Barrett's esophagus. This so-called "first-generation" PDT drug has several drawbacks. Photofrin lacks a definite molecular structure as it is a mixture of oligomers formed by ether and ester linkages of up to eight porphyrin units (19); this makes the interpretation of its pharmacokinetic data difficult. Photofrin requires excitation light of 630 nm to achieve sufficient tissue penetration, though the absorption peak at 630 nm corresponds to the longest wavelength but weakest absorption peak of Photofrin, resulting in low PDT efficiency. Furthermore, its accumulation in skin and slow clearance cause prolonged skin photosensitivity lasting for at least 30 days but often up to 90 days or more (20). Therefore, there has been a strong need for producing photosensitizers with significantly improved properties, in comparison to Photofrin.

Ideally, a photosensitizer should (1) be a single chemical compound of definite molecular structure in a highly purified and stable form, (2) possess high photosensitizing ability, (3) possess high photostability, (4) have a relatively high absorbance at longer wavelengths, for greater tissue penetration, (5) exhibit no dark toxicity, (6) exhibit no long-term cutaneous phototoxicity, (7) accumulate selectively in the targeted tissue, and (8) undergo fast clearance from the body to avoid prolonged photosensitivity.

To date, second-generation photosensitizers that have definite structures and/or a major absorption band in the wavelength range of 650–800 nm have been developed (see Table 2).

In addition to the improvements in conventional one-photon excitation PDT as shown above, two-photon absorption induced excitation of photosensitizers has received attention lately as an alternative approach to achieve long tissue penetration depth. This approach can achieve enhanced tissue penetration depth, in comparison to the one-photon excitation approach, by combining the energy of two photons – in the range of 780–950 nm – where tissues have maximum transparency to light but where the energy of one photon is not sufficient enough to produce singlet oxygen (21). Several such dyes have been tested for use in vitro (22) and in vivo (21) but not yet for use in a clinical setting.

2.2. Localization of the Photosensitizer in Diseased Tissue or Tumor

Selective delivery of drug or imaging contrast agents to diseased target tissue or cells is one of the most important factors in determining the efficacy of the intravenously injected therapy or imaging modalities. Higher drug localization in target tissue can increase the therapeutic efficacy, even with a reduced dose. Second-generation photosensitizers have been developed to exhibit improved selectivity and body clearance rate compared to ones in the first-generation so as to reduce post-PDT photosensitivity. However, developing a way to enhance tumor selectivity is still one of the major challenges to be solved in PDT. This problem can be addressed by "third-generation" photosensitizers – photosensitizers conjugated with a targeting component that is specific toward the tumor site/cells and that can increase the selective accumulation of the photosensitizer in sites of interest as well as minimize healthy cell localization and concomitant damage (23).

2.3. Administration of PDT Drug and Light

The PDT drug can be administered to the infected area either topically or by systemic intravenous injection (see Table 2). For topical administration, there is no need for delay in the administration of light, unless the delivered PDT drug is a prodrug such as Levulan that requires 15–60 min of activation time for the treatment of actinic keratoses (24). For systemic administration, however, the location and accumulation amount of the photosensitizers depend on postinjection time. The drugs, after a short

Table 2
Second-generation photosensitizers that are clinically approved or under clinical trial

Photosensitizer	Trade name	Excitation wavelength (nm)	Method of administration	Clinical use
Temoporfin (meta-tetrahydroxy-phenylchlorin, i.e., m-THPC)	Foscan	652	Intravenous (i.v.)	Approved for the palliative treatment of patients with advanced head and neck cancer (European Union, Norway, and Iceland)
Talaporfin (mono-L-aspartyl chlorine e6)	Laserphyrin	664	i.v.	Approved for early endobronchial carcinoma (Japan)
Verteporfin (Benzoporphyrin derivative-monoacid ring)	Visudyne (liposomal formulation)	689	i.v.	Approved for the treatment of predominantly classic subfoveal choroidal neovascularization (CNV) due to age-related macular degeneration (AMD), pathologic myopia, or presumed ocular histoplasmosis (USA)
HPPH (2-(1-hexyloxyethyl)-2-devinyl pyropheophorbide-alpha)	Photochlor	665	i.v.	Phases 1 and 2 clinical studies in patients with Barrett's esophagus with high-grade dysplasia, obstructive esophageal cancer, early or late-stage lung cancer, or basal cell carcinoma
Lutetium texaphyrin (Motexafin lutetium)	Antrin Lutex Lutrin	732	i.v.	Phase I trial for locally recurrent prostate cancer; Phase I for phototherapy (PT) in patients undergoing percutaneous coronary intervention with stent deployment
ALA (5-aminolevulinic acid HCl)	Levulan	630	Topical	Prodrug for Protoporphyrin IX (PpIX). Approved for a precancerous skin condition called actinic keratosis (USA)
Methylene blue	Periowave	665	Topical	Approved for periodontal diseases, and nasal decolonization of methicillin-resistant *Staphylococcus aureus* (Canada)

postinjection time that is less than the plasma half-life of the PDT drugs, predominantly stay in the vascular compartment of the tumor. At longer postinjection times, drugs may accumulate in the extravascular compartment of the tumor due to its relatively slow leakage from the vasculature and interstitial diffusion. Therefore, the interval between the PDT drug administration and light exposure (drug-to-light interval) may play an important role in the PDT outcome. The drug-to-light interval used for current clinical PDT is quite long – for example, on the order of 40–50 and 90–110 h for Photofrin (19) and Foscan (25), respectively.

There have been efforts to enhance the PDT efficacy by changing the light or drug administration protocols. Fractionated drug-dose PDT (i.e., administration of photosensitizers at multiple time intervals before light activation) was reported to achieve a high localization of photosensitizers in both vascular and tumor cell compartments, resulting in a superior PDT effect compared to single-dose PDT (26). The light fluence rate is an important item to modulate the PDT outcome. The fluence rate during the administration of light can significantly affect photobleaching and antitumor effectiveness, by controlling tissue oxygenation during light delivery (27, 28).

3. Why Nanoparticles for PDT?

Nanoparticles have been utilized as a delivery vehicle for drugs, image contrast agents or for both, and have shown their ability to improve the efficacy of existing imaging and therapy methods, especially in cancer detection and treatment, including PDT (29). The success can be attributed to the many advantages resulting from the nanoparticles' inherent properties including their size, inert and nontoxic matrices, as well as their flexible engineerability that allows various modifications of their properties, improved targetability, and possible multifunctionality. Specifically, the advantages of using nanoparticles for PDT are described below.

First, each nanoparticle can carry a large amount of photosensitizers, offering a coherent, critical mass of destructive power for intervention. Most photosensitizers are hydrophobic and poorly water-soluble, thereby tending to form aggregates under physiological conditions. Proper loading of photosensitizers into nanoparticles of higher aqueous solubility and longer plasma residence time can make intravenous administration very easy and enhance the PDT efficiency of the system by preventing the aggregation of the photosensitizers. Note that for most photosensitizers, their dimers or aggregates have a very low singlet oxygen production efficiency (30, 31). The drug-loading procedure can be done by a variety of methods, such as encapsulation, covalent linkage, or postloading by physical adsorption through electrostatic or hydrophobic interactions. The loading degree can be controlled by the input amount of photosensitizers per nanoparticle, as well as by engineering or modifying the nanoparticles' properties, such as size, composition, charge, and hydrophobicity.

Second, the size of the nanoparticles offers certain distinct advantages for drug delivery. When drug molecules or carriers are in the blood stream, they can extravasate across the endothelium by transcapillary pinocytosis, as well as by passage through inter-endothelial cell junctions, gaps, or fenestrae (32). The molecular weight or the size of the drug molecules or carriers has been shown to be a critical factor for the endothelial permeability (32, 33). The size of fenestrae varies with the organ type – from a few angstroms for the blood–brain barrier (BBB) to 150 nm for the liver and spleen (34). However, in pathological situations, such as cancer or inflammatory tissues, the vasculature becomes leaky, with larger fenestrae, up to 780 nm in size (34). Small molecular drugs freely diffuse in and out of most normal tissues as well as tumor tissues, without any discrimination. However, nanoparticles escape the vasculature preferentially through abnormally leaky tumor blood vessels and are then subsequently retained in the tumor tissue for a long time, because tumor tissue has a very poor lymphatic drainage system, unlike normal or inflammatory tissue. Because of this penetration phenomenon, called the enhanced permeability and retention (EPR), nanoparticles can efficiently deliver therapeutic agents to tumors (35, 36). Nanoparticles between 10 and 100 nm are believed to provide the best option, because nanoparticles with a size greater than 10 nm can avoid kidney clearance, resulting in prolonged and elevated levels in the blood stream, and those with a size smaller than 100 nm can penetrate deep into tissues, without being trapped by the phagocytes (37, 38).

Third, the nanoparticles can be surface-modified to make more efficient and selective delivery to target tissues or cells. Drug localization in target tissue is known to be determined by vascular permeability and interstitial diffusion, which depend not only on the size of the drugs as described above but also on the configuration, charge, and hydrophilic or lipophilic properties of the compound, as well as the physiological properties of blood vessels (39). These properties can be relatively easily modified for the nanoparticle-based agents. The charge, hydrophilicity, or lipophilicity can be changed either by introducing polymers or monomers with characteristic functional groups (charged, hydrophilic, or lipophilic) during synthesis or by coating the nanoparticles with polymers of such characteristics. For instance, nanoparticles can be coated with polyethylene glycol (PEG) for longer plasma circulation time (40). The nanoparticle surface can also be conjugated with tumor-specific ligands, such as antibodies, aptamers, peptides, or small molecules that bind to antigens that are present on the target cells or tissues, for "active targeting." Some of the targeting moieties even help NPs to penetrate into tumor cells (41–43), or help get across physiological drug barriers such as the blood–brain barrier (BBB) (44), for the treatment of brain tumors and other central nervous system (CNS) diseases. Targeted NPs were reported to have enhanced binding affinity and specificity over targeted molecular drug and contrast agents, due to the multiple numbers of targeting ligands packed on their surface, a "multivalency effect" (45, 46).

Fourth, the NP matrix or surface coating may be able to prevent the PDT drugs from degrading in, or interacting with, the living biological environment. This is done by blocking the diffusion of large enzyme molecules into the NPs. For example, methylene blue is clinically approved for topical application to diseased tissue (see Table 2) but has not been used for systemic delivery as it gets reduced into an inactive form by plasma enzymes. By being loaded inside nanoparticles, methylene blue can be protected from enzymatic reduction (see Fig. 3) and thus becomes useful for intravenously delivered PDT applications (47).

Fifth, nanoparticles can reduce immunogenicity and side effects of the drug. The maximum tolerated dose of the drug can be increased, as the nontoxic (biocompatible) polymer reduces toxicity.

Sixth, nanoparticles can also alleviate the problem posed by the multidrug resistance (MDR) of cancer cells – which reduces the

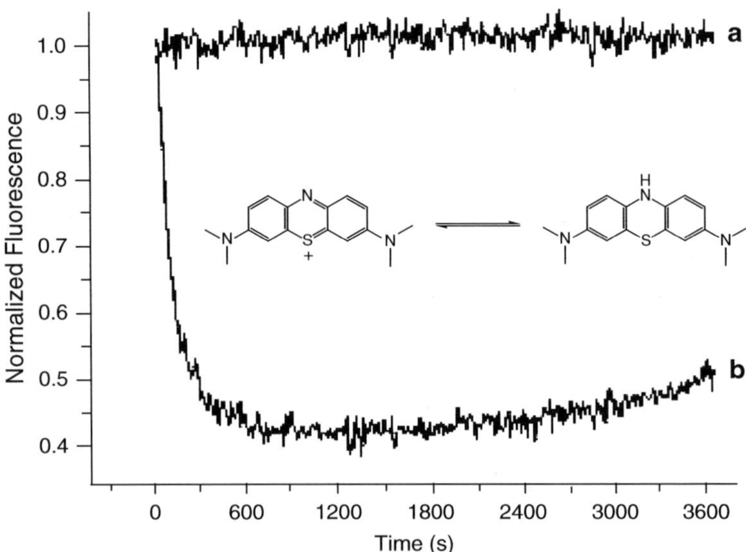

Fig. 3. Fluorescence emission at 680 nm vs. time (after subtraction of photobleaching for both curves) for: (**a**) 3 mg/mL MB-containing NPs and (**b**) 1 μg/mL MB, respectively, when mixed in PBS (pH 7.2) with 0.45 μmol NADH and 0.05 mg diaphorase. The inset shows the reduction chemistry of MB to its nonphotoactive form. Reproduced from ref. 47 with permission from Elsevier Inc.

effectivity of most drugs – by masking the drug (i.e., entrapping it within the NPs). This feature may significantly enhance the delivery of drugs that are normally excluded from tumors by MDR.

Seventh, a single nanoparticle can be made multifunctional by being loaded with multiple components such as drugs, contrast agents, targeting ligands, or other active components, enabling synergistic multimodal therapy, multimodal imaging, as well as integrated imaging and therapy that can provide an individual patient-based, image-guided therapy, by imaging the diseased tissue and then treating the tissue on-site (29, 43). Multifunctionality can be introduced to molecular agents, but this usually involves a very complicated synthetic procedure (48, 49). Sometimes such prepared molecular agents are not very useful as the dose for imaging and the dose for therapy can be quite different. Note that the nanoparticles can be loaded with multiple agents in different ratios by using standard methods that are available for each nanoplatform.

Due to the advantages listed above, nanoparticles have the potential to improve many aspects of PDT. High amounts of drug can be effectively accumulated at target tissue, thus reducing the drug dose and the possible concomitant side effects but enhancing the PDT efficiency (50, 51). Furthermore, the drug delivery time required for PDT or drug-light interval can be significantly shortened (43, 52).

4. Nanoparticle-Based PDT Agents: Designs, Materials, Preparation, and Characterization

A variety of nanoparticle matrices and photosensitizers have been utilized to develop nanoparticle-based PDT agents. The investigated photosensitizers include single photon absorption photosensitizers, such as porphyrins (e.g., Photofrin and verteporfin), chlorins, m-THPC, phthalocyanines, methylene blue and hypericin, as well as two-photon absorption PDT dyes, such as porphyrin tetra(p-toluenesulfonate). The investigated nanoparticles are of both synthetic and natural origins. Most of them are polymeric (i.e., polymer nanoparticles or polymer-coated nonpolymeric nanoparticles), although there are nanoparticles made of photosensitizers (53) or nonpolymeric nanoparticles with conjugated photosensitizers. The polymer nanoparticles include synthetic polymer nanoparticles made of polylactide–polyglycolide copolymers (PLGA) (50, 54–60), polyacrylamide (29, 43, 47, 61–66), silica (61) and organically modified silica (ormosil) (61, 67–76), and natural polymer nanoparticles made of natural proteins and polysaccharides such as albumin (77), xanthane gum (78), collagen (78), and alginate (79, 80). Nonpolymeric nanoparticles include gold nanoparticles (52, 81), iron oxide (82–86), photon upconverting nanoparticles made of $NaYF_4$ nanocrystals doped with Yb^{3+} and Er^{3+} (87, 88), quantum dots (89), and fullerenes (90) as well as nanosized biological entities such as low-density lipoproteins (22 nm) (91) or plant viruses (28 nm) (92). The polymers used for surface coating include silica/ormosil (82–84, 87), dextran (85), chitosan (86), and polyethyleneimine (PEI) (88). The nanoparticles can be prepared from monomers, by polymerization, as in the case of polyacrylamide, silica, or ormosil, or from preformed polymer, by precipitation, solvent evaporation, or covalent cross-linkage, as in the case of PLGA and natural polymer nanoparticles.

The designs of the agents are very versatile (see Fig. 4). The most common and simplest design is the nanoparticle, as an inert platform, loaded with photosensitizers. There are a couple of designs utilizing FRET for two-photon absorption PDT. In one FRET-based design, polymer-coated upconverting nanoparticles were loaded with photosensitizers (88), and in another design, photosensitizers and two-photon absorption dyes were loaded into inert nanoparticles (74). The two-photon dyes or upconverting nanoparticles absorb NIR irradiation and emit visible light to excite the photosensitizing molecules so as to avoid the tissue penetration depth limitations. There are also other designs where the nanoparticles are made completely of photosensitizers themselves (53) or where the photosensitizers are conjugated to gold particles (52, 81). Multifunctional nanoparticle agents have been also designed to perform not only PDT but also image contrast enhancement or other therapies. For instance, a PDT drug and chemotherapy drug were loaded into inert nanoplatforms for simultaneous PDT and chemotherapy (80), PDT drugs and NIR fluorescent dyes for PDT and fluorescence imaging (67), and PDT drugs and iron oxide nanocrystals or gadolinium chelate for PDT and MRI (43, 62, 65, 93). In other designs, magnetic iron oxide nanoparticles are coated with polymer shells containing photosensitizers (82–86) or conjugated with photosensitizers (94) for PDT and MRI, or PDT and hyperthermia therapy, respectively. In another design, X-ray scintillation nanoparticles are conjugated with porphyrins for simultaneous radiotherapy and PDT (95, 96). When the nanoparticle–photosensitizer conjugates are stimulated by X-rays during radiotherapy, the particles generate visible light that can activate the photosensitizers for PDT. These nanoparticle agents can be used for deep tumor treatment as X-rays can penetrate through tissue. The scintillation nanoparticles are made of several doped inorganic materials such as $LaF_3:Ce^{3+}$, $LuF_3:Ce^{3+}$ $CaF_2:Mn^{2+}$, $CaF_2:Eu^{2+}$, $BaFBr:Eu^{2+}$, $BaFBr:Mn^{2+}$, $CaPO_4:Mn^{2+}$ (95), and $LaF_3:Tb^{3+}$ (96).

Some of these nanoparticles are biodegradable, over a time scale of several months, and some are not. It should be noted that, unlike chemotherapy, the PDT efficacy does not depend on

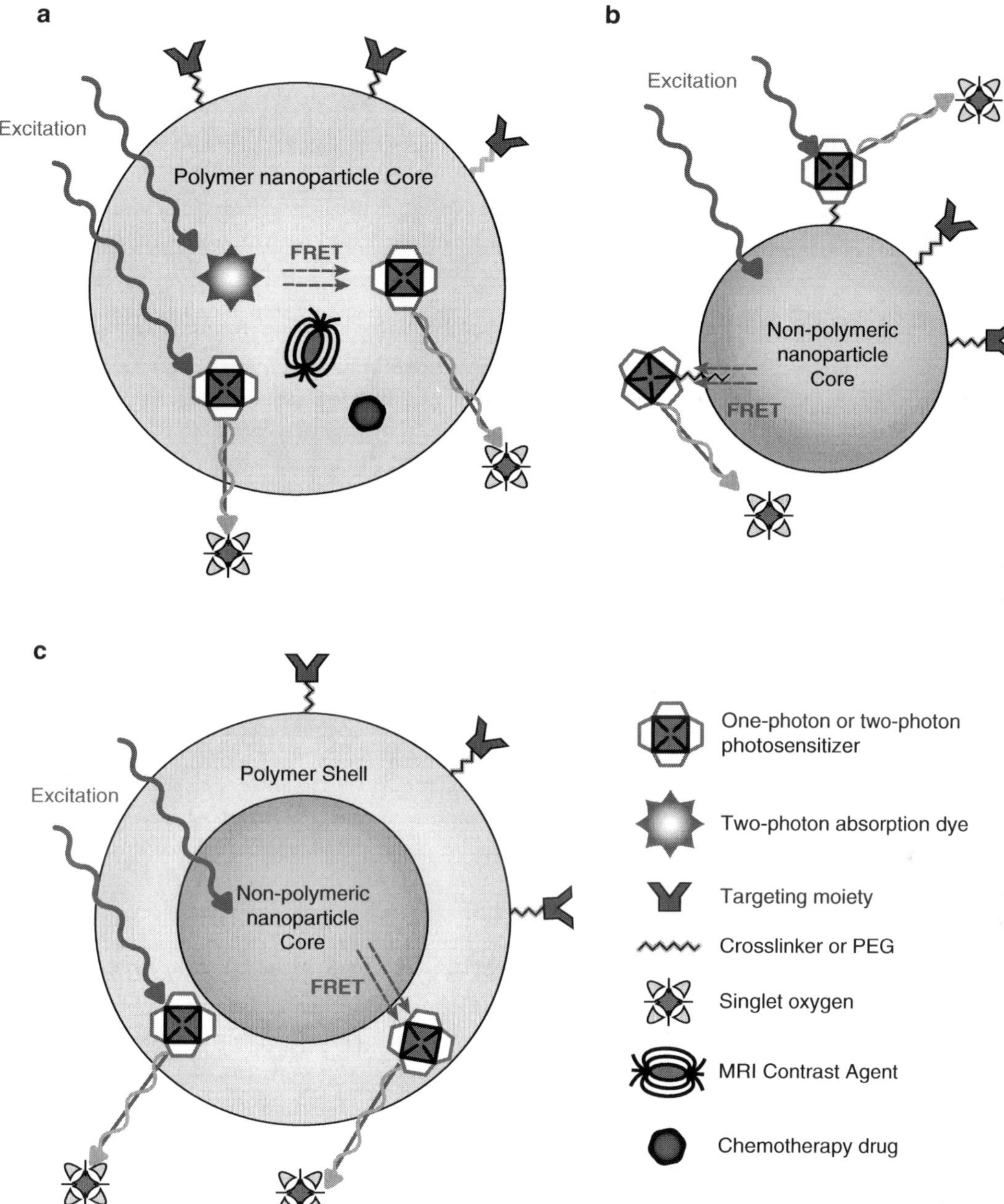

Fig. 4. Various designs of nanoparticle-based PDT agents: (**a**) polymer nanoparticles loaded with photosensitizers and other active components, (**b**) nonpolymeric nanoparticles with surface-conjugated photosensitizers, and (**c**) nonpolymeric nanoparticles with surface polymer shell loaded with photosensitizers.

the drug's release from the nanoparticles, because molecular oxygen can enter the nanoparticles and the produced singlet oxygen can diffuse out of the nanoparticle matrix to induce photodynamic reaction. In such cases, the PDT efficiency of the agents depends on the nanoparticle type (size and oxygen permeability of the matrix), as shown by a study done with three nanoparticles of different nondegradable matrices loaded with the same photosensitizer (methylene blue) (61). Nondegradable nanoparticles, however, may confront bioelimination issues for in vivo applications.

Loading of photosensitizers into nanoparticles are performed typically by one of the three methods: encapsulation, covalent linkage, or postloading. The encapsulation is a simple method wherein the photosensitizers are mixed with the nanoparticle-producing reaction mixture and are physically entrapped inside the nanoparticle matrix during synthesis – either by steric constriction or by physical interaction, such as electrostatic interactions, hydrogen bond formation, and hydrophobic interactions, between the photosensitizers and the nanoparticle matrix. Here, the photosensitizers may be inactivated during nanoparticle synthesis and they may form aggregates within the nanoparticle matrix. However, the encapsulated photosensitizers may be leached out from the nanoparticles before the in vivo nanoparticle dose reaches its target area, thus reducing efficiency of treatment and possibly resulting in side effects. In the postloading method, the photosensitizers are added to the already formed nanoparticle, in an appropriate solvent, and the loading mechanism is governed by the physical interactions. It is simple to do, causing no chemical damage to the photosensitizers, and can give a high drug-loading efficiency. However, the nanoparticle agents with postloaded photosensitizers are more prone to premature leaching than those prepared by encapsulation, with potentially bad, or good, consequences for in vivo applications. The covalent linkage method requires a relatively complicated preparation procedure, compared to the above two methods. The photosensitizers are covalently linked either with monomer molecules that are then polymerized into nanoparticles or with already-prepared nanoparticles. This prevents any leaching-related problems during systemic in vivo circulation. Any aggregation of photosensitizers can be avoided by conjugating each photosensitizer to a well-separated location within the nanoparticle matrix. It should be noted that high loading of photosensitizers does not always lead to high PDT efficacy due to the increased probability for drug aggregation (50, 61). This is especially true for agents that are prepared by encapsulation or physical adsorption of the photosensitizers.

The above nanoparticle-based agents have been tested in solution, in cells, and in in vivo animal models, demonstrating their high potential for successful future clinical applications. The solution tests determine if the photochemical and oxygen permeable properties of the photosensitizer-loaded nanoparticles are sufficiently good for PDT, by measuring the singlet oxygen production rate, either by chemical probes (43, 83, 84, 97) or by NIR phosphorescence of singlet oxygen (8, 77). The most commonly used chemical probe is the water-soluble, anthracene-9, 10-dipropionic acid disodium salt (ADPA) (97). However, these chemical probe molecules may enter the nanoparticle matrix, leading to overestimates of the singlet oxygen production by the nanoparticle agents. A nanoparticle-based probe has been made to detect only the singlet oxygen exiting the photodynamic nanoplatforms, which is critical for determining the singlet oxygen production efficiency of the nanoparticle-based PDT agents (98). The in vitro tests can demonstrate the cell-penetration and/or intracellular-localization properties of the PDT agents as well as their effectiveness in killing the target cells. The PDT cell kill is determined typically by a live/dead cell staining assay (61), a clonogenic assay (99), or a MTT assay (100) (see Fig. 5).

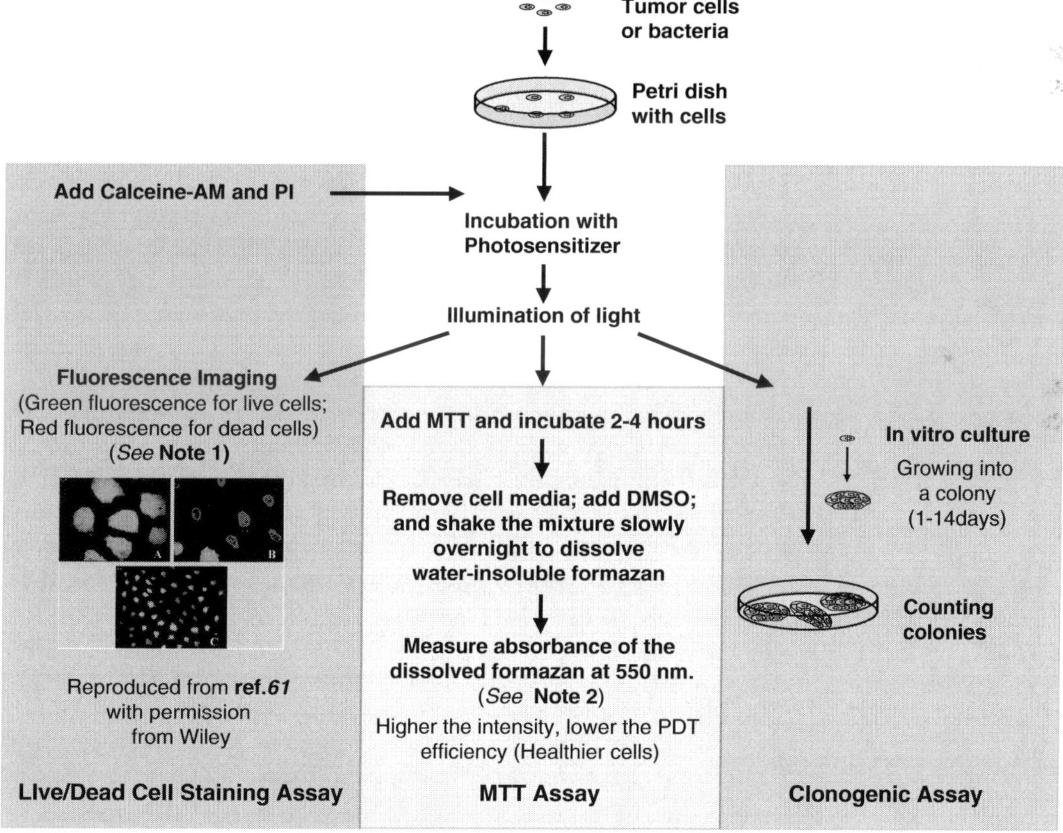

Fig. 5. In vitro tests for PDT efficacy determinations. Note 1: The nonfluorescent Calcein-AM is converted into calcein by esterase in a live cell, emitting strong green fluorescence (excitation: 490 nm, emission: 515 nm). PI enters the dying or dead cells through disrupted membranes and intercalates with nuclear DNA, emitting red fluorescence (excitation: 535 nm, emission: 617 nm). The optimal concentrations of Calcein-AM and PI should be determined for each tested cell line. Note 2: The colored formazan solution has an absorption peak around 500–600 nm. The absorption maximum varies with the solvent employed – for instance, 550 nm for DMSO. The MTT assay depends on reductase enzymes whose activity can be changed upon experimental conditions and therefore sometimes produce erratic results.

The in vivo tests demonstrate the effectiveness of the PDT agents. This efficacy is governed by many additional factors, such as the pharmacokinetics and tissue distribution of the nanoparticles. The in vivo PDT efficiency is determined by monitoring tumor volume reduction, or animal survival, which is typically displayed by a Kaplan–Meier plot (43).

In subsequent sections, recent PDT applications of nanoparticles, especially of polymeric nanoparticles, are shown.

5. Examples of Polymeric Nanoparticle-Based PDT Agents

5.1. Polyacrylamide Nanoparticles

The polyacrylamide (PAA) nanoparticle is a hydrogel that is produced through microemulsion polymerization. Historically, it was first produced as a "nano-PEBBLE" intracellular fluorescence-based sensor, which produces singlet oxygen while sensing (being quenched by) oxygen (101). It can be functionalized with amine or carboxyl groups, for surface modification with a targeting moiety and/or with PEG, and loaded with a combination of actuating molecules that make it suitable for multifunctional tasks, including PDT (102). The PAA nanoparticle is highly soluble in water, offering easy systemic administration without aggregation. The PAA nanoparticle is also biologically inert and nontoxic, which was demonstrated by in vivo toxicity studies, showing no evidence of alterations in histopathology or clinical chemistry for the PAA nanoparticle dose of 10 mg/kg–1 g/kg (29), as well as by a report on safety assessment of PAA (103). It can also be engineered to have controllable biodegradability by introducing different amounts of biodegradable cross-linkers (29).

PAA nanoparticles have been utilized as a platform to produce PDT agents as well as multifunctional agents for tumor-selective PDT and MRI. The PAA nanoparticles typically have the size range of 30–70 nm and have been loaded with methylene blue (47, 61, 66), Photofrin (43, 62, 65), and two-photon absorption photosensitizers including 5,10,15,20-tetrakis(1-methyl 4-pyridinio) porphyrin tetra(p-toluenesulfonate) (TMPyP) (63). The PAA nanoparticles were prepared with either biodegradable (43) or nonbiodegradable cross-linkers (47, 61–65), and both have shown an effective photodynamic activity in cancer cells such as 9L glioma, C6 glioma, and MBA-MD-435 cells. Moreover, targeted PAA nanoparticles with encapsulated Photofrin and surface-conjugated F3 peptide were demonstrated to kill target cells selectively in vitro (43). The same PAA nanoparticles showed successful in vivo PDT therapy in 9L glioma bearing rats when injected in the rat tail vein (43). Treatment of the tumor-bearing rats with F3-targeted nanoparticles, followed by 7.5 min of red light irradiation through an inserted optical fiber, showed a significant improvement in survival rate compared to various sets of control rats: (1) those that received the same light treatment but with nontargeted nanoparticles or Photofrin, (2) those that only received the light treatment, and (3) those that received no treatment. Indeed, 60 days later, 40% of the animals treated with F3-targeted Photofrin nanoparticles were found to be tumor free, and this was still the case 6 months later. In contrast, all control rats were dead within 2 weeks. It should be noted that this demonstrates that the surface-conjugated targeting moieties do make a significant difference in animal survival. These nanoparticles also showed enhanced in vivo MRI tumor contrast due to coencapsulated iron oxide, thus enabling image-guided therapy.

The PAA nanoparticles with encapsulated methylene blue also showed good potential as a PDT agent to eliminate bacterial infections (66). The PDT with the nanoparticles was performed on bacterial cells – of both gram-positive strains (i.e., *Staphylococcus aureus* strain) and gram-negative strains (i.e., *Escherichia coli*, *Pseudomonas aeruginosa*, and *Acinetobacter* sp.) – in suspension as well as in preformed biofilms. The data revealed that the nanoparticles can inhibit biofilm growth and eradicate almost all of the early age biofilms that are formed by all of the bacteria examined. However, the killing efficiency for gram-positive strains was much higher than that for gram-negative strains. PAA nanoparticles were also prepared with ultrasmall size (2–3 nm), for easy renal clearance from the body, and were successfully loaded with a hydrophobic photosensitizer (m-THPC), despite the hydrophilic nature of PAA (64). These ultrasmall nanoparticle agents were as effective in killing cultured cancer cells as free m-THPC photosensitizer molecules.

5.2. PLGA Nanoparticles

Poly (D,L-lactide-co-glycolide) (PLGA) is a biodegradable polymer that is usually prepared using either emulsion solvent evaporation or solvent displacement techniques (104). PLGA particles are made of a mixture of a less hydrophilic lactic acid polymer (PLA) and the more hydrophilic glycolic acid polymer (PGA). They undergo hydrolysis, in the body, to form lactic acid and glycolic acid, which are eventually removed from the body by the citric acid cycle. Owing to their biodegradability and ease of formulation to carry a variety of drugs, these have been the most extensively investigated polymer nanoplatforms for drug delivery (105).

PLGA nanoparticles loaded with a variety of photosensitizers including porphyrins, chlorins, phthalocyanines, and hypericin were studied for PDT efficacy in vitro and in vivo. These PLGA nanoparticles exhibited a higher photoactivity than free PDT drugs in cells such as EMT-6 mammary tumor (54) and NuTu-19 ovarian cancer cells (50), and in an in vivo chick embryo *chorioallantoic membrane* (CAM) model (56).

The meso-tetraphenylporpholactol loaded 50:50 PLGA nanoparticles that were 98 nm in diameter showed complete eradication of subcutaneous implanted U587 gliomas in nude mice at 27 days (191 mW/cm², 57.3 J/cm²) after the light therapy, which was performed at 24 h after i.v. injection of the nanoparticles (60). It should be noted that these nanoparticles themselves have very low fluorescence and singlet oxygen production efficiency probably due to aggregation of the incorporated dye, indicating that the dye released from the nanoparticles contributed to such a PDT outcome.

The PDT efficacy of the PLGA nanoparticle-based agents was found to depend on the lipophilicy of the incorporated photosensitizers, and the properties (the copolymer molar ratio and the size) of the PLGA nanoparticles. A study with PLGA nanoparticles loaded with three photosensitizers of different lipophilicity – meso-tetraphenylporphyrin (TPP), meso-tetra-(4-carboxyphenyl)-porphyrin (TCPP), and chlorin e6 (Ce6) – showed that the higher the photosensitizer's lipophilicity, the higher the incorporation efficiency, the lower the amount of the extravasation (i.e., longer residence time in the blood vessels) and, therefore, the higher their photothrombic efficiency, when tested in an in vivo CAM model (58). Note that the developing chicken embryo is surrounded by a chorioallantoic membrane (CAM), which becomes vascularized as the embryo develops. The CAM is a convenient model for monitoring the modifications of the vasculature as the transparency of its superficial layers allows an examination of structural changes of each blood vessel in real time (56). The molar ratio between PLA and PGA determines the biodegradation rate (106) and the lipophilicity of the particles, affecting the in vitro phototoxicity of incorporated photosensitizers, meso-tetra(hydroxyphenyl)porphyrin (m-THPP), in the order of 50:50 PLA:PGA > 75:25 PLA:PGA > PLA (54). The PLGA particle size seems to play an important role on PDT as demonstrated by a couple of studies. In one study (55), 75:25 PLGA nanoparticles of two different sizes (167 and 370 nm in diameter) were loaded with verteporfin and tested in vitro and in vivo. The smaller nanoparticles exhibited greater photocytotoxicity on EMT-6 mammary tumor cells compared to the larger NPs or to the free drug. The smaller particles also released m-THPP faster than the free photosensitizer in DMSO/PBS. In vivo PDT with mice bearing rhabdomyosarcoma (M1) tumor in the right dorsal area indicated that the i.v. administered PLGA particles (167 nm) with incorporated verteporfin effectively reduced tumor growth for 20 days in mice with short drug-to-light intervals (15 and 30 min). In another study (107), the 50:50 PLGA nanoparticles loaded with m-THPP, with three different mean diameters of 117, 285, and 593 nm, respectively, were prepared. The nanoparticles of 117 nm exhibited the highest rate of reactive oxygen species production and the fastest m-THPP release in vitro. Similarly, the 117-nm-sized nanoparticles exhibited the highest in vivo photodynamic activity in the CAM model. The amount of residual poly(vinyl alcohol) (PVAL) stabilizing agent during the particle synthesis, at the particle surface, were shown to affect both the release and activity of the photosensitizers for the bigger particles (285 and 593 nm) (57). In the other study, four batches of PLA nanoparticles containing meso-tetra(carboxyphenyl)porphyrin (TCPP), with different mean sizes ranging from 121 to 343 nm, were prepared (59). The extravasations of each TCPP-loaded nanoparticle formulation from blood vessels were measured, as well as the extent of photochemically induced vascular occlusion in the CAM model, both revealing size-dependent behaviors. The smallest nanoparticles (121 nm) exhibited the greatest extent of vascular thrombosis as well as the lowest extravasation.

5.3. Silica and Ormosil Nanoparticles

Silica, an inorganic polymer, and organically modified silica (ormosil) nanoparticles are inert but nonbiodegradable. The surface can be easily coated with additional functionalized silica or an ormosil layer for further surface modifications. The size of silica or ormosil nanoparticles used to prepare PDT agents has been reported in the range of 100–200 nm when produced by the Stober process (61, 68), a two-step acid/base-catalyzed procedure, (61, 108) or a surfactant template procedure (76), but the size range has been 20–40 nm when prepared in a micelle (67, 70, 73, 109).

The silica or ormosil nanoparticles have been loaded with various photosensitizers such as protoporphyrin IX (PplX), elsino-chrome A, HPPH, m-THPC, and methylene blue by encapsulation (61, 67–69, 109) and covalent linkage (70–72, 76). The silica/ormosil nanoparticles with both encapsulated and covalently linked photosensitizers produced singlet oxygen and showed in vitro PDT efficiency. However, the leaching of encapsulated photosensitizers was observed for the small (20–30 nm) ormosil nanoparticles containing m-THPC and HPPH (67, 73) in the presence of organic solvent or serum. The ormosil nanoparticles of 20-nm size with covalently linked iodobenzylpyropheophorbide (70) and the mesoporous ormosil nanoparticles of ~100-nm size with PplX (76) showed no leaching but high PDT cell-killing efficiency, indicating high loading of monomeric photosensitizers without aggregation. It also demonstrates the high singlet oxygen permeability of the ormosil matrix.

The ormosil nanoparticles have also been utilized to prepare various PDT agents of synergistic designs by loading another component in addition to the photosensitizers. In one design, the ormosil nanoparticles were loaded with HPPH and an excess of 9,10-*bis* [4′-(4″-aminostyryl)styryl]anthracene (BDSA), a highly two-photon-active molecule (74). HPPH absorption in nanoparticles has significant overlap with the fluorescence of BDSA aggregates, enabling indirect two-photon excitation (850 nm) of HPPH. Drastic changes in the morphology of the cells were observed, which were indicative of impending death as a result of two-photon PDT with these nanoparticles. In another design (75), ormosil nanoparticles containing covalently loaded iodine as a second component that induces the intraparticle external heavy-atom effect on the encapsulated photosensitizer molecules significantly enhanced the efficiency of singlet oxygen generation and, thereby, the in vitro PDT efficacy. Furthermore, the ormosil nanoparticles were loaded with fluorescence dyes (67) and targeting moieties (72) for multitasking (PDT and fluorescent imaging) and for target-specific PDT, respectively.

5.4. Dendrimers

Dendrimers are polymeric materials that are highly branched and monodisperse macromolecules. Dendrimers have been adopted as a platform to deliver 5-Aminolevulinic acid (ALA). ALA is a precursor of protoporphyrin IX (PpIX) that is used as an endogenous photosensitizer and is FDA-approved for topical PDT (see Table 2). However, due to the hydrophilic nature of ALA, ALA–PDT may be limited clinically by the rate of ALA uptake into neoplastic cells and/or penetration into tissue (110). A second-generation dendrimer with conjugated ALA, containing 18 aminolevulinic acid residues attached via ester linkages to a multipodent aromatic core, showed enhancement of porphyrin synthesis in vitro (111) and in vivo (112). In vitro, the dendrimer-ALA was more efficient than ALA for porphyrin synthesis in transformed PAM 212 murine keratinocyte and A431 human epidermoid carcinoma cell lines, up to an optimum concentration of 0.1 mmol/L (111). In an in vivo study with male BALB/c mice, the porphyrin kinetics from ALA exhibited an early peak between 3 and 4 h in most tissues, whereas the dendrimer induced sustained porphyrin production for over 24 h, and basal values were not reached until 48 h after administration (112). Integrated porphyrin accumulation from the dendrimer and that from an equal drug equivalent dose of ALA was comparable, showing that the majority of ALA residues were liberated from the dendrimer. The porphyrin kinetics appears to be governed by the rate of enzymatic cleavage of ALA from the dendrimer, which is consistent with the in vitro results.

5.5. Polymer NPs Based on Natural Polymers

Nanoparticles made of cross-linked natural polymers such as proteins and polysaccharides have been prepared and their PDT efficacy was investigated. The investigated proteins include human serum albumin (HSA) and bovine serum albumin (BSA), a most abundant protein in blood plasma. The polysaccharides include xanthan gum and alginate, both of which have been used as a food additive and rheology modifier.

For example, cross-linked HSA nanoparticles (100–300 nm in diameter) were loaded with photosensitizers, pheophorbide (77). The singlet oxygen quantum yield of the HSA nanoparticles was very low probably due to the attachments of the photosensitizers to proteins and their aggregation within the nanoparticles. When the Jurkat cells were incubated with these nanoparticles, the photosensitizer was released due to nanoparticle decomposition in the cellular lysosomes, which was illustrated by fluorescence lifetime imaging and confocal laser scanning microscopy. After a 24-h incubation period, the nanoparticles showed a higher phototoxicity but lower dark toxicity than free pheophorbide, demonstrating their potential as a PDT agent. BSA nanoparticles containing the photosensitizer silicon (IV) phthalocyanine (NzPc) (452 nm) were reported to show in vitro phototoxicity in human fibroblasts (113). The same nanoplatform containing both magnetic nanoparticles and NzPc were also prepared for potential use in PDT combined with hyperthermia. Both BSA nanoparticles reveal no dark cytotoxicity (113).

Marine atelocollagen/xanthan gum microcapsules with encapsulated photosensitizer, 5,10,15-triphenyl-20-(3-N-methylpyridinium-yl)porphyrin, were prepared (78). The size of these polymeric formulations was relatively large compared to other nanoplatforms (i.e., 100–1,000 nm in diameter even after size reduction by ultrasonic processing in the presence of Tween 20 surfactant). These polymer particles showed about four times more phototoxicity than the respective phosphatidylcholine lipidic emulsion but negligible cytotoxicity toward HeLa cells.

Aerosol OT (AOT)-alginate nanoparticles were also developed into PDT agents by encapsulating methylene blue. These nanoparticles were determined to be ~80 nm in size by dynamic light scattering (DLS) and showed significantly enhanced overall cytotoxicity, following PDT, in comparison to free methylene blue, in two cancer cell lines, MCF-7 and 4T1 (79). Such enhanced in vitro PDT efficiency results from increased ROS and singlet oxygen production by the nanoparticles and significant nuclear localization of the methylene blue released from the nanoparticles. The same nanoplatforms, loaded with both methylene blue and doxorubicine (with measured diameter of 39 nm by AFM and of 62 nm by DLS), were also developed for a combination therapy of chemotherapy and PDT (80). Both methylene blue and doxorubicin were released from the nanoparticles, which resulted in a significantly higher nuclear accumulation of doxorubicin and methylene blue than that with the free drugs. Furthermore, methylene blue serves not only as a PDT drug but also as a P-glycoprotein inhibitor, enhancing doxorubicin-induced cytotoxicity in drug-resistant NCI/ADR-RES cells. Nanoparticle-mediated combination therapy resulted in a significant induction of both apoptosis and necrosis and improved cytotoxicity in drug-resistant tumor cells.

5.6. Polymer-Coated Inorganic Nanoparticles

Two kinds of inorganic nanoparticles, photon upconverting and iron oxide nanoparticles, have been developed as PDT agents after being coated with a polymer layer containing photosensitizers.

Photon upconverting nanoparticles with surface polymer coatings have been developed for deep tissue penetrating PDT, where the upconverting nanoparticles were made of $NaYF_4$:Yb^{3+}, Er^{3+}. Through excitation with multiple photons, these nanoparticles convert lower-energy (NIR) light to higher-energy (visible) light, which then excites the photosensitizing molecules loaded in the surface-coated polymer. The silica-coated upconverting nanoparticles were loaded with photosensitizers, merocyanine, and also functionalized with a monoclonal antibody to target breast cancer cells (87). The resultant nanoparticles were 60–120 nm in diameter and performed efficient PDT after infrared (974 nm) irradiation. The PEI-coated upconverting nanoparticles were loaded with zinc phthalocyanine photosensitizers and conjugated with folic acid (88). These 50-nm nanoparticles produced green/red emission with near-infrared (NIR) excitation, both in vitro and in vivo, demonstrating that these particles could be excited after deep intramuscular injection in rats. The particles showed the ability of significant cell destruction with NIR excitation after targeted binding to cancer cells (HT29 cells).

Iron oxide nanoparticles (typical core size 5–10 nm) have been coated with silica (82–84), dextran (85), and chitosan (86) to produce 20–100 nm nanoparticles. Dextran is a branched polysaccharide and chitosan is a linear polysaccharide. The photosensitizers were loaded into the polymer shell, to serve for PDT, and sometimes for fluorescence imaging, while the iron oxide particle core provides magnetic properties for MRI contrast enhancement. The investigated photosensitizers included an iridium complex (82), a porphyrine (PHPP) (83) and methylene blue (84) encapsulated in the silica shell, PHPP encapsulated in a chitosan shell (86), and a chlorin (TPC) covalently linked to the dextran shell (85). These nanoparticles showed PDT capability in solution (84), in vitro on SW480 carcinoma cells (83, 86), Hela cells (82), murine and human macrophages (85), as well as in vivo (86) on SW480 carcinoma bearing mice. Interestingly, in the in vivo study, magnetic targeting was performed before PDT, which improved tumor accumulation and retention of the nanoparticles (86).

6. Summary and Perspectives

PDT is a rapidly evolving treatment with significant advantages over other therapeutic modalities. PDT has been approved for several cancer applications (see Table 2). However, it has not yet been used for mainstream oncologic practice (i.e., treatment of patients with solid tumors). This results from (a) PDT's dependence on (1) tissue penetration depth by light and (2) oxygen levels in the target tissue, which is usually a problem in solid tumors, as well as (b) inefficient delivery of the PDT drugs to the target tissue, in current clinical practice. Nanoparticles have many advantages as a delivery vehicle, with potential to improve the efficacy of existing imaging and therapy methods, especially for cancer detection and treatments. A variety of nanoparticle matrices, photosensitizers, other active chemicals and targeting components have been utilized to make nanoparticle-based PDT agents that can improve current PDT efficacy. Many of these nanoparticles have excellent optical and tumor-targeting qualities, qualities that are not achievable with molecular PDT drugs. So far, no nanoparticle-based PDT agents have been used in the clinic, although the in vitro and in vivo data from nanoparticle-based PDT agents seem to promise a very high potential for clinical applications of these nanoparticles, provided that several challenging issues are resolved. The bioelimination profile is an important issue to be considered, as it is liable to have a kinetic pattern that is quite different from that of currently used molecular-type drugs and imaging agents. The other challenging issue is the need for a new clinical (FDA) approval procedure. The latter is expected to be "complicated," as these NP-based agents are made of multiple active ingredients. It should be noted that these issues are common to all nanoparticle-based chemotherapeutic agents. In spite of this, several of them have already reached the clinic for chemotherapy (38). Therefore, nanoparticle-based PDT agents are also expected to advance toward clinical applications in the near future. Moreover, with these advanced nanoparticle agents, PDT may become useful in other potential clinical applications.

Acknowledgments

This article is partially supported by funding from NIH grants 1R01EB007977, R33CA125297 03S1, and R21/R33CA125297.

References

1. Siegel, M. M., Tabei, K., Tsao, R., Pastel, M. J., Pandey, R. K., Berkenkamp, S., et al. (1999) Comparative mass spectrometric analyses of photofrin oligomers by fast atom bombardment mass spectrometry, UV and IR matrix-assisted laser desorption/ionization mass spectrometry, electrospray ionization mass spectrometry and laser desorption/jet-cooling photoionization mass spectrometry. *J. Mass Spectrom.* **34**, 661–669.

2. Mitton, D. and Ackroyd, R. (2008) A brief overview of photodynamic therapy in Europe. *Photodiagn. Photodyn. Ther.* **5**, 103–111.

3. Dolmans, D. E. J. G. J., Fukumura, D., and Jain, R. K. (2003) Photodynamic therapy for cancer. *Nat. Rev. Cancer* **3**, 380–387.

4. Brown, S. B., Brown, E. A., and Walker, I. (2004) The present and future role of photodynamic therapy in cancer treatment. *Lancet Oncol.* **5**, 497–508.

5. Wilson, B. C. and Patterson, M. S. (2008) The physics, biophysics and technology of photodynamic therapy. *Phys. Med. Biol.* **53**, R61–R109.

6. Buytaert, E., Dewaele, M., and Agostinis, P. (2007) Molecular effectors of multiple cell death pathways initiated by photodynamic therapy. *Biochim. Biophys. Acta* **1776**, 86–107.

7. Foote, C. S. (1991) Definition of type I and type II photosensitized oxidation. *Photochem. Photobiol.* **54**, 659.

8. Niedre, M., Patterson, M. S., and Wilson, B. C. (2002) Direct near-infrared luminescence detection of singlet oxygen generated by photodynamic therapy in cells in vitro and tissues in vivo. *Photochem. Photobiol.* **75**, 382–391.

9. Tsushima, M., Tokuda, K., and Ohsaka, T. (1994) Use of hydrodynamic chronocoulometry for simultaneous determination of diffusion-coefficients and concentrations of dioxygen in various media. *Anal. Chem.* **66**, 4551–4556.

10. Pryor, W. A. (1986) Oxy-radicals and related species: their formation, lifetimes, and reactions. *Ann. Rev. Physiol.* **48**, 657–667.

11. Morgan, J. and Oseroff, A. R. (2001) Mitochondria-based photodynamic anti-cancer therapy. *Adv. Drug Deliv. Rev.* **49**, 71–86.

12. Eljamel, M. S., Goodman, C., and Moseley, H. (2008) ALA and Photofrin® fluorescence-guided resection and repetitive PDT in glioblastoma multiforme: a single centre Phase III randomised controlled trial. *Lasers Med. Sci.* **23**, 361–367.

13. Vogel, A. and Venugopalan, V. (2003) Mechanisms of pulsed laser ablation of biological tissues. *Chem. Rev.* **103**, 577–644.

14. Stolik, S., Delgado, J. A., Perez, A., and Anasagasti, L. (2000) Measurement of the penetration depths of red and near infrared light in human "ex vivo" tissues. *J. Photochem. Photobiol.* **57**, 90–93.

15. Svaasand, L. O., Gomer, C. J., and Profio, A. E. (1989) Laser-induced hyperthermia of ocular tumors. *Appl. Opt.* **28**, 2280–2287.

16. Muller, P. J. and Wilson, B. C. (1987) Photodynamic therapy of malignant primary braintumours: clinical effects, postoperative ICP, and light penetration in the brain. *Photochem. Photobiol.* **46**, 929–935.

17. Dougherty, T. J., Weishaupt, K. R., and Boyle, D. G. (1985) Photodynamic sensitizers. in: *Cancer: Principles and Practice of Oncology* (DeVita, Jr., V. T. and Rosenberg, S. A., eds.) pp. 2272–2279, J. B. Lippincott Company, Philadelphia.

18. Kavar, B. and Kaye, A. H. (2007) Photodynamic therapy. in: *High-Grade Gliomas: Diagnosis and Treatment* (Barnett, G. E., Ed.) pp. 461–484, Humana Press, Totowa.

19. http://www.accessdata.fda.gov/drugsatfda_docs/label/2008/020451s019lbl.pdf.

20. http://www.photofrin.com/pdf/patient-guide.pdf.

21. Starkey, J. R., Rebane, A. K., Drobizhev, M. A., Meng, F., Gong, A., Elliott, A., et al. (2008) New two-photon activated photodynamic therapy sensitizers induce xenograft tumor regressions after near-IRLaser treatment, through the body of the host mouse. *Clin. Cancer Res.* **14**, 6564–6573.

22. Mir, Y., Houde, D., and van Lier, J. E. (2008) Photodynamic inhibition of acetylcholinesterase after two-photon excitation of copper tetrasulfophthalocyanine. *Lasers Med. Sci.* **23**, 19–25.

23. Josefsen, L. B. and Boyle, R. W. (2008) Photodynamic therapy: novel third-generation photosensitizers one step closer? *Br. J. Pharmacol.* **154**, 1–3.

24. http://www.dusapharma.com/levulan-prescribing-information.html.

25. http://www.biolitecpharma.com/public/smpc.asp?s=foscan.

26. Dolmans, D. E. J. G. J., Kadambi, A., Hill, J. S., Flores, K. R., Gerber, J. N., Walker, J. P., et al. (2002) Targeting tumor vasculature and cancer cells in orthotopic breast tumor by fractionated photosensitizer dosing photodynamic therapy. *Cancer Res.* **62**, 4289–4294.

27. Seshadri, M., Bellnier, D. A., Vaughan, L. A., Spernyak, J. A., Mazurchuk, R., Foster, T. H., et al. (2008) Light delivery over extended time periods enhances the effectiveness of PDT. *Clin. Cancer Res.* **14**, 2796–2805.

28. Henderson, B. W., Busch, T. M., and Snyder, J. W. (2006) Fluence rate as a modulator of PDT mechanism. *Laser Surg. Med.* **18**, 489–493.

29. Koo, Y. L., Reddy, G. R., Bhojani, M., Schneider, R., Philbert, M. A., Rehemtulla, A., et al. (2006) Brain cancer diagnosis and therapy with nano-platforms. *Adv. Drug Deliv. Rev.* **58**, 1556–1577.

30. Redmond, R. W., Land, E. J., and Truscott, T. G. (1985) Aggregation effects on the photophysical properties of porphyrins in relation to mechanisms involved in photodynamic therapy. *Adv. Exp. Med. Biol.* **193**, 293–302.

31. Severino, D., Junqueira, H. C., Gugliotti, M., Gabrielli, D. S., and Baptista, M. S. (2003) Influence of negatively charged interfaces on the ground and excited state properties of methylene blue. *Photochem. Photobiol.* **77**, 459–468.

32. Takakura, Y., Mahato, R. I., and Hashida, M. (1998) Extravasation of macromolecules. *Adv. Drug Deliv. Rev.* **34**, 93–108.

33. Weissleder, R., Bogdanov Jr., A., Tung, C. H., and Weinmann, H. J. (2001) Size optimization of synthetic graft copolymers for in vivo angiogenesis imaging. *Bioconjug. Chem.* **12**, 213–219.

34. Gaumet, M., Vargas, A., Gurny, R., and Delie, F. (2008) Nanoparticles for drug delivery: the need for precision in reporting particle size parameters. *Eur. J. Pharm. Biopharm.* **69**, 1–9.

35. Maeda, H. (2001) The enhanced permeability and retention (EPR) effect in tumor vasculature: the key role of tumor-selective macromolecular drug targeting. *Adv. Enzyme Regul.* **41**, 189–207.

36. Matsumura, Y. and Maeda, H. (1986) A new concept for macromolecular therapeutics in cancer chemotherapy: mechanism of tumoritropic accumulation of proteins and the antitumor agent smancs. *Cancer Res.* **46**, 6387–6392.

37. Vinogradov, S. V., Bronich, T. K., and Kabanov, A. V. (2002) Nanosized cationic hydrogels for drug delivery: preparation, properties and interactions with cells. *Adv. Drug Deliv. Rev.* **54**, 135–147.

38. Davis, M. E., Chen, Z., and Shin, D. M. (2008) Nanoparticle therapeutics: an emerging treatment modality for cancer. *Nat. Rev. Drug Discov.* **7**, 771–782.

39. Jain, R. K. (2001) Delivery of molecular and cellular medicine to solid tumors. *Adv. Drug Deliv. Rev.* **46**, 149–168.

40. Moffat, B. A., Reddy, G. R., McConville, P., Hall, D. E., Chenevert, T. L., Kopelman, R. R., et al. (2003) A novel polyacrylamide magnetic nanoparticle contrast agent for molecular imaging using MRI. *Mol. Imaging* **2**, 324–332.

41. Porkka, K., Laakkonen, P., Hoffman, J. A., Bernasconi, M., and Ruoslahti, E. (2002) A fragment of the HMGN2 protein homes to the nuclei of tumor cells and tumor endothelial cells in vivo. *Proc. Natl. Acad. Sci. U S A* **99**, 7444–7449.

42. Allen, T. M. (2002) Ligand-targeted therapeutics in anticancer therapy. *Nat. Rev. Cancer* **2**, 750–763.

43. Reddy, G. R., Bhojani, M. S., McConville P., Moody, J., Moffat, B. A., Hall, D. E., et al. (2006) Vascular targeted nanoparticles for imaging and treatment of brain tumors. *Clin. Cancer Res.* **12**, 6677–6686.

44. Costantino, L., Gandolfi, F., Tosi, G., Rivasi, F., Vandelli, M. A., and Forni, F. (2005) Peptide-derivatized biodegradable nanoparticles able to cross the blood–brain barrier. *J. Controlled Release* **108**, 84–96.

45. Hong, S., Leroueil, P. R., Majoros, I. J., Orr, B. G., Baker, J. R., and Holl, M. M. B. (2007) The binding avidity of a nanoparticle-based multivalent targeted drug delivery platform. *Chem. Biol.* **14**, 107–115.

46. Montet, X., Funovics, M., Montet-Abou, K., Weissleder, R., and Josephson, L. (2006) Multivalent effects of RGD peptides obtained by nanoparticle display. *J. Med. Chem.* **49**, 6087–6093.

47. Tang, W., Xu, H., Park, E. J., Philbert, M. A., and Kopelman, R. (2008) Encapsulation of methylene blue in polyacrylamide nanoparticle platforms protects its photodynamic effectiveness. *Biochem. Biophys. Res. Commun.* **369**, 579–583.

48. Chen, Y., Gryshuk, A., Achilefu, S., Ohulchansky, T., Potter, W., Zhong, T., et al. (2004) A novel approach to a bifunctional photosensitizer for tumor imaging and phototherapy. *Bioconjug. Chem.* **8**, 1105–1115.

49. Li, G., Slansky, A., Dobhal, M. P., Goswami, L. N., Graham, A., Chen, Y., et al. (2005) Chlorophyll-a analogues conjugated with aminophenyl-DTPA as potential bifunctional agents for magnetic resonance imaging and photodynamic therapy. *Bioconjug. Chem.* **16**, 32–42.

50. Zeisser-Labouebe, M., Lange, N., Gurny, R., and Delie, F. (2006) Hypericin-loaded nanoparticles for the photodynamic treatment of ovarian cancer. *Int. J. Pharm.* **326**, 174–181.

51. Saxena, V., Sadoqi, M., and Shao, J. (2006) Polymeric nanoparticulate delivery system for Indocyanine green: biodistribution in healthy mice. *Int. J. Pharm.* **308**, 200–204.

52. Cheng, Y., Samia, A. C., Meyers, J. D., Panagopoulos, I., Fei, B., and Burda, C. (2008) Highly efficient drug delivery with gold nanoparticle vectors for in vivo photodynamic

therapy of cancer. *J. Am. Chem. Soc.* **130**, 10643–10647.

53. Hu, Z., Pan, Y., Wang, J., Chen, J., Li, J., and Ren, L. (2009) Meso-tetra (carboxyphenyl) porphyrin (TCPP) nanoparticles were internalized by SW480 cells by a clathrin-mediated endocytosis pathway to induce high photocytotoxicity. *Biomed. Pharmacother.* **63**, 155–164.

54. Konan, Y. N., Berton, M., Gurny, R., and Allemann, E. (2003) Enhanced photodynamic activity of meso-tetra(4-hydroxyphenyl)porphyrin by incorporation into sub-200 nm nanoparticles. *Eur. J. Pharm. Sci.* **18**, 241–249.

55. Konan-Kouakou, Y. N., Boch, R., Gurny, R., and Allemann, E. (2005) In vitro and in vivo activities of verteporfin-loaded nanoparticles. *J. Controlled Release* **103**, 83–91.

56. Vargas, A., Pegaz, B., Debefve, E., Konan-Kouakou, Y., Lange, N., Ballini, J.-P., et al. (2004) Improved photodynamic activity of porphyrin loaded into nanoparticles: an in vivo evaluation using chick embryos. *Int. J. Pharm.* **286**, 131–145.

57. Vargas, A., Lange, N., Arvinte, T., Cerny, R., Gurny, R., and Delie, F. (2009) Toward the understanding of the photodynamic activity of m-THPP encapsulated in PLGA nanoparticles: correlation between nanoparticle properties and in vivo activity. *J. Drug Targeting* 17, 599–609.

58. Pegaz, B., Debefve, E., Borle, F., Ballini, J., van den Bergh, H., and Kouakou-Konan, Y. N. (2005) Encapsulation of porphyrins and chlorins in biodegradable nanoparticles: the effect of dye lipophilicity on the extravasation and the photothrombic activity. A comparative study. *J. Photochem. Photobiol. B* **80**, 19–27.

59. Pegaz, B., Debefve, E., Ballini, J., Konan-Kouakou, Y. N., and van den Bergh, H. (2006) Effect of nanoparticle size on the extravasation and the photothrombic activity of meso(p-tetracarboxyphenyl)porphyrin. *J. Photochem. Photobiol. B* **85**, 216–222.

60. McCarthy, J. R., Perez, J. M., Bruckner, C., and Weissleder, R. (2005) Polymeric nanoparticle preparation that eradicates tumors. *Nano Lett.* **5**, 2552–2556.

61. Tang, W., Xu, H., Kopelman, R., and Philbert, M. A. (2005) Photodynamic characterization and in vitro application of methylene blue-containing nanoparticle platforms. *Photochem. Photobiol.* **81**, 242–249.

62. Kopelman, R., Koo, Y. L., Philbert, M., Moffat, B. A., Reddy, G. R., McConville, P., et al. (2005) Multifunctional nanoparticle

platforms for in vivo MRI enhancement and photodynamic therapy of a rat brain cancer. *J. Magn. Magn. Mater.* **293**, 404–410.

63. Gao, D., Agayan, R. R., Xu, H., Philbert, M. A., and Kopelman R. (2006) Nanoparticles for two-photon photodynamic therapy in living cells. *Nano Lett.* **6**, 2383–2386.

64. Gao, D., Xu, H., Philbert, M. A., and Kopelman R. (2007) Ultrafine hydrogel particles: synthetic approach and therapeutic application in living cells. *Angew. Chem. Int. Ed.* **46**, 2224–2227.

65. Ross, B., Rehemtulla, A., Koo, Y. L., Reddy, G. R., Behrend, C., Buck, S., et al. (2004) Photonic and magnetic nanoexplorers for biomedical use: from subcellular imaging to cancer diagnostics and therapy. *Proc. SPIE* **5331**, 76–83.

66. Wu, J. F., Hao, X., Wei, T., Kopelman, R., Philbert, M. A., and Xi, C. (2009) Eradication of bacteria in suspension and biofilms using methylene blue-loaded dynamic nanoplatforms. *Antimicrob. Agents Chemother.* **53**, 3042–3048.

67. Compagnin, C., Bau, L., Mognato, M., Celotti, L., Miotto, G., Arduini, M., et al. (2009) The cellular uptake of meta-tetra (hydroxyphenyl)chlorin entrapped in organically modified silica nanoparticles is mediated by serum proteins. *Nanotechnology* **20**, 345101.

68. Yan, F. and Kopelman, R. (2003) The embedding of metatetra(hydroxyphenyl)-chlorin into silica nanoparticle platforms for photodynamic therapy and their singlet oxygen production and pH dependent optical properties. *Photochem. Photobiol.* **78**, 587–591.

69. Zhou, L., Dong, C., Wei, S. H., Fenga, Y. Y., Zhoua, J. H., and Liu, J. H. (2009) Water-soluble soft nano-colloid for drug delivery. *Mater. Lett.* **63**, 1683–1685.

70. Ohulchanskyy, T. Y., Roy, I., Goswami, L. N., Chen, Y., Bergey, E. J., Pandey, R. K., et al. (2007) Organically modified silica nanoparticles with covalently incorporated photosensitizer for photodynamic therapy of cancer. *Nano Lett.* **7**, 2835–2842.

71. Rossi, L. M., Silva, P. R., Vono, L. L. R., Fernandes, A. U., Tada, D. B., and Baptista, M. S. (2008) Protoporphyrin IX nanoparticle carrier: preparation, optical properties, and singlet oxygen generation. *Langmuir* **24**, 12534–12538.

72. Brevet, D., Gary-Bobo, M., Raehm, L., Richeter, S., Hocine, O., Amro, K., et al. (2009) Mannose-targeted mesoporous silica nanoparticles for photodynamic therapy. *Chem. Commun.* **12**, 1475–1477.

73. Kumar, R., Roy, I., Ohulchanskyy, T. Y., Goswami, L. N., Bonoiu, A. C., Bergey, E. J., et al. (2008) Covalently dye-linked, surface-controlled, and bioconjugated organically modified silica nanoparticles as targeted probes for optical imaging. *ACS Nano* **2**, 449–456.

74. Kim, S., Ohulchanskyy, T. Y., Pudavar, H. E., Pandey R. K., and Prasad, P. N. (2007) Organically modified silica nanoparticles co-encapsulating photosensitizing drug and aggregation-enhanced two-photon absorbing fluorescent dye aggregates for two-photon photodynamic therapy. *J. Am. Chem. Soc.* **129**, 2669–2675.

75. Kim, S., Ohulchanskyy, T. Y., Bharali, D., Chen, Y., Pandey, R. K., and Prasad, P. N. (2009) Organically modified silica nanoparticles with intraparticle heavy-atom effect on the encapsulated photosensitizer for enhanced efficacy of photodynamic therapy. *J. Phys. Chem. C* **113**, 12641–12644.

76. Tu, B. H., Lin, Y., Hung, Y., Lo, L., Chen, Y., and Mou, C. (2009) In vitro studies of functionalized mesoporous silica nanoparticles for photodynamic therapy. *Adv. Mater.* **21**, 172–174.

77. Chen, K., Preuß, A., Hackbarth, S., Wacker, M., Langer, K., and Röder, B. (2009) Novel photosensitizer-protein nanoparticles for photodynamic therapy: photophysical characterization and in vitro investigations. *J. Photochem. Photobiol. B* **96**, 66–74.

78. Deda, D. K., Uchoa, A. F., Carita, E., Baptista, M. S., Toma, H. E., and Araki, K. (2009) A new micro/nanoencapsulated porphyrin formulation for PDT treatment. *Int. J. Pharm.* **376**, 76–83.

79. Khdair, A., Gerard, B., Handa, H., Mao, G., Shekhar, M. P. V., and Panyam, J. (2008) Surfactant-polymer nanoparticles enhance the effectiveness of anticancer photodynamic therapy. *Mol. Pharm.* **5**, 795–807.

80. Khdair, A., Handa, H., Mao, G., and Panyam, J. (2009) Nanoparticle-mediated combination chemotherapy and photodynamic therapy overcomes tumor drug resistance in vitro. *Eur. J. Pharm. Biopharm.* **71**, 214–222.

81. Wieder, M. E., Hone, D. C., Cook, M. J., Handsley, M. M., Gavrilovic, J., and Russell, D. A. (2006) Intracellular photodynamic therapy with photosensitizer-nanoparticle conjugates: cancer therapy using a "Trojan horse." *Photochem. Photobiol. Sci.* **5**, 727–734.

82. Lai, C., Wang, Y., Lai, C., Yang, M.-J., Chen, C.-Y., Chou, P.-T., et al. (2008) Iridium-complex-functionalized Fe_3O_4/SiO_2 core/shell nanoparticles: a facile three-in-one system in magnetic resonance imaging, luminescence imaging, and photodynamic therapy. *Small* **4**, 218–224.

83. Chen, Z., Sun, Y., Huang, P., Yang, X., and Zhou, X. (2009) Studies on preparation of photosensitizer loaded magnetic silica nanoparticles and their anti-tumor effects for targeting photodynamic therapy. *Nanoscale Res. Lett.* **4**, 400–408.

84. Tada, D. B., Vono, L. L. R., Duarte, E. L., Itri, R., Kiyohara, P. K., Baptista, M. S., et al. (2007) Methylene blue-containing silica-coated magnetic particles: a potential magnetic carrier for photodynamic therapy. *Langmuir* **23**, 8194–8199.

85. McCarthy, J. R., Jaffer, F. A., and Weissleder, R. (2006) A macrophage-targeted theranostic nanoparticle for biomedical applications. *Small* **2**, 983–987.

86. Sun, Y., Chen, Z., Yang, X., Huang, P., Zhou, X., and Du, X. (2009) Magnetic chitosan nanoparticles as a drug delivery system for targeting photodynamic therapy. *Nanotechnology* **20**, 135102.

87. Zhang, P., Steelant, W., Kumar, M., and Scholfield, M. (2007) Versatile photosensitizers for photodynamic therapy at infrared excitation. *J. Am. Chem. Soc.* **129**, 4526–4527.

88. Chatterjee, D. K. and Yong, Z. (2008) Upconverting nanoparticles as nanotransducers for photodynamic therapy in cancer cells. *Nanomedicine* **3**, 73–82.

89. Yaghini, E., Seifalian, A. M., and MacRobert, A. J. (2009) Quantum dots and their potential biomedical applications in photosensitization for photodynamic therapy. *Nanomedicine* **4**, 353–363.

90. dos Santos, L. J., Alves, R. B., de Freitas, R. P., Nierengarten, J.-F., Magalhaes, L. E. F., Krambrock, K., et al. (2008) Production of reactive oxygen species induced by a new [60]fullerene derivative bearing a tetrazole unit and its possible biological applications. *J. Photochem. Photobiol. A* **200**, 277–281.

91. Song, L. P., Li, H., Sunar, U., Chen, J., Corbin, I., Yodh, A. G., et al. (2007) Naphthalocyanine-reconstituted LDL nanoparticles for in vivo cancer imaging and treatment. *Int. J. Nanomedicine* **2**, 767–774.

92. Suci, P. A., Varpness, Z., Gillitzer, E., Douglas, T., and Young, M. (2007) Targeting and photodynamic killing of a microbial pathogen using protein cage architectures functionalized with a photosensitizer. *Langmuir* **23**, 12280–12286.

93. Xu, H., Buck, S. M., Kopelman, R., Philbert, M. A., Brasuel, M., Ross, B. D., et al. (2004)

Photo-excitation based nano-explorers: chemical analysis inside live cells and photodynamic therapy. *Isr. J. Chem.* **44**, 317–337.

94. Gu, H., Xu, K., Yang, Z., Changa, C. K., and Xu, B. (2005) Synthesis and cellular uptake of porphyrin decorated iron oxide nanoparticles- a potential candidate for bimodal anticancer therapy. *Chem. Commun.* **34**, 4270–4272.

95. Chen, W. and Zhang, J. (2006) Using nanoparticles to enable simultaneous radiation and photodynamic therapies for cancer treatment. *J. Nanosci. Nanotech.* **6**, 1159–1166.

96. Liu, Y. F., Chen, W., Wang, S. P., and Joly, A. G. (2008) Investigation of water-soluble X-ray luminescence nanoparticles for photodynamic activation. *Appl. Phys. Lett.* **92**, 043901.

97. Moreno, M. J., Monson, E., Reddy, G. R., Rehemtulla, A., Ross, B. D., Philbert, M., et al. (2003) Production of singlet oxygen by Ru(dpp(SO$_3$)$_2$)$_3$ incorporated in polyacrylamide PEBBLEs. *Sens. Act. B: Chem* **90**, 82–89.

98. Cao, Y., Koo, Y.-E. L., Koo, S. M., and Kopelman, R. (2005) Ratiometric singlet oxygen nano-optodes and their use for monitoring photodynamic therapy nanoplatforms. *Photochem. Photobiol.* **81**, 1489–1498.

99. Shinohara, E. T., Cao, C., Niermann, K., Mu, Y., Zeng, F., Hallahan, D. E., et al. (2005) Enhanced radiation damage of tumor vasculature by mTOR inhibitors. *Oncogene* **24**, 5414–5422.

100. Gill, Z. P., Perks, C. M., Newcomb, P. V., and Holly, J. M. P. (1997) Insulin-like growth factor-binding protein (IGFBP-3) predisposes breast cancer cells to programmed cell death in a non-IGF-dependent manner. *J. Biol. Chem.* **272**, 25602–25607.

101. Clark, H. A., Barker, S. L. R., Brasuel, M., Miller, M. T., Monson, E., Parus, S., et al. (1998) Subcellular optochemical nanobiosensors: probes encapsulated by biologically localised embedding (PEBBLEs). *Sens. Act. B* **51**, 12–16.

102. Harrell, J. A. and Kopelman, R. (2000) Biocompatible probes measure intracellular activity. *Biophotonics Int.* **7**, 22–24.

103. Anderson, F. A. (2005) Amended final report on the safety assessment of polyacrylamide and acrylamide residues in cosmetics. *Int. J. Tox.* **24**(**Suppl. 2**), 21–50.

104. Jain, R. A. (2000) The manufacturing techniques of various drug loaded biodegradable poly(lactide-co-glycolide) devices. *Biomaterials* **21**, 2475–2490.

105. Panyama, J. and Labhasetwar, V. (2003) Biodegradable nanoparticles for drug and gene delivery to cells and tissue. *Adv. Drug Deliv. Rev.* **55**, 329–347.

106. Shive, M. S. and Anderson, J. M. (1997) Biodegradation and biocompatibility of PLA and PLGA microspheres. *Adv. Drug Deliv. Rev.* **28**, 5–24.

107. Vargas, A., Eid, M., Fanchaouy, M., Gurny, R., and Delie, F. (2008) In vivo photodynamic activity of photosensitizer-loaded nanoparticles: formulation properties, administration parameters and biological issues involved in PDT outcome. *Eur. J. Pharma. Biopharma.* **69**, 43–53.

108. Hah, H. J., Kim, J. S., Jeon, B. J., Koo, S. M., and Lee, Y. E. (2003) Simple preparation of mono-disperse hollow silica particles without using templates. *Chem. Commun.* **14**, 1712–1713.

109. Qian, J., Gharibi, A., and He, S. L. (2009) Colloidal mesoporous silica nanoparticles with protoporphyrin IX encapsulated for photodynamic therapy. *J. Biomed. Opt.* **14**, 014012.

110. Peng, Q., Warloe, T., Berg, K., Moan, J., Kongshaug, M., Giercksky, K.-E., et al. (1997) 5-Aminolevulinic acid-based photodynamic therapy, clinical research, and future challenges. *Cancer* **79**, 2282–2308.

111. Battah, S., Balaratnam, S., Casas, A., O'Neill, S., Edwards, C., Batlle, A., et al. (2007) Macromolecular delivery of 5-aminolaevulinic acid for photodynamic therapy using dendrimer conjugates. *Mol. Cancer Ther.* **6**, 876–885.

112. Casas, A., Battah, S., Di Venosa, G., Dobbin, P., Rodriguez, L., Fukuda, A., et al. (2009) Sustained and efficient porphyrin generation in vivo using dendrimer conjugates of 5-ALA for photodynamic therapy. *J. Controlled Release* **135**, 136–143.

113. Rodrigues, M. M. A., Simioni, A. R., Primo, F. L., Siqueira-Moura, M. P., Morais, P. C., and Tedesco, A. C. (2009) Preparation, characterization and in vitro cytotoxicity of BSA-based nanospheres containing nanosized magnetic particles and/or photosensitizer. *J. Magn. Magn. Mater.* **321**, 1600–1603.

Chapter 12

Hydrogel Templates for the Fabrication of Homogeneous Polymer Microparticles

Ghanashyam Acharya, Matthew McDermott, Soo Jung Shin, Haesun Park, and Kinam Park

Abstract

Nano/microparticulate drug delivery systems with homogeneous size distribution and predefined shape are important in understanding the influence of the geometry and dimensions of these systems on blood circulation times and cellular uptake. We present a general method using water dissolvable hydrogel templates for the fabrication of homogeneous, shape-specific polymer/drug constructs in the size range of 200 nm to 50 μm. This hydrogel template strategy is mild, inexpensive, and readily scalable for the fabrication of multifunctional drug delivery vehicles.

Key words: Hydrogel, Gelatin, Template, PLGA, Microparticle

1. Introduction

Polymeric nano/microparticulate drug delivery systems have been developed for the delivery of various pharmaceutical agents to targeted areas of the human body (1). The currently available particulate drug delivery systems include liposomes, polymer micelles, and nano/microspheres (2–4). Most nano/microparticles have been prepared by emulsion methods (5, 6). These delivery systems, however, can only be used to prepare particles which carry small amounts of drug compared with the total polymer content. Furthermore, these systems produce particles which are polydisperse in size and so it is difficult to correlate the effect of the size and shape of the particles on biological responses. Recently, various nanofabrication methods have been developed for making nano/microparticles with homogeneous size distribution (7–9). However, nanofabrication of particulate drug delivery systems for drug delivery applications has been a challenging task

Sarah J. Hurst (ed.), *Biomedical Nanotechnology: Methods and Protocols*, Methods in Molecular Biology, vol. 726,
DOI 10.1007/978-1-61779-052-2_12, © Springer Science+Business Media, LLC 2011

mainly due to the lack of general fabrication methods that can produce purified particles in large scale.

Hydrogels are being developed as coating materials on bio-implants to function as interfaces between implants and biological tissues (10, 11) and as reversible thermosensitive templates for biomimetic mineralization (12, 13) and for nanoparticle fabrication. In terms of nanofabrication, the hydrogel template strategy addresses various issues associated with the nanofabrication of particles for use in drug delivery systems. This strategy is broadly applicable and mild and can be used to synthesize nano/microparticles composed of many different polymer/drug combinations ranging in size from 200 nm to greater than 50 µm. Since hydrogels are water soluble, the formed polymer particles (which are not soluble in water) can be easily collected by simply dissolving the template. Further, certain biopolymer hydrogels, such as gelatin, form phase-reversible elastic and mechanically strong hydrogels that can withstand physical manipulation during template preparation and filling and that can maintain the polymer or polymer–drug mixture inside the cavities of the template, preventing its diffusion. Templates for nanoparticle fabrication also can be made from other polymers, such as poly(dimethyl siloxane) (PDMS), however, removing the formed particles from PDMS templates is very difficult since PDMS is not soluble in water.

We herein describe a strategy based on dissolvable hydrogel templates that can be utilized to produce homogeneous polymeric drug delivery systems with a predefined size and shape. Briefly, solutions of hydrogel-forming biopolymer (i.e., gelatin) are used for imprinting nano/microsized features from a PDMS template master. Upon gelation, the hydrogel template retains the exact negative imprint of the features present on the PDMS template (i.e., wells). These wells then are filled with polymer solution and the hydrogel template is dissolved with warm water to release free polymer/drug delivery particles of homogeneous size (14).

2. Materials

2.1. Fabrication of Hydrogel Templates

1. PDMS templates with predefined patterns (Akina, Inc., West Lafayette, IN).
2. Gelatin from porcine skin, Type A, 300 bloom (Sigma, St. Louis, MO).
3. Nanopure water.

2.2. Fabrication of PLGA Microparticles

1. Poly(lactic-*co*-glycolic acid) (PLGA) (MW 65,000, IV 0.82 dL/g).
2. Fluorescent dye: Nile Red.
3. Dichloromethane.

2.3. Collection of PLGA Microparticles

1. Microfilters (Sterlitech Corporation, Kent, WA).

2. Eppendorf Centrifuge 5804, Rotor A-4-44, used at $4,500 \times g$ (Eppendorf, Hauppauge, NY).

3. Olympus Spinning Disk Confocal Imaging Microscope BX61-DSU (Center Valley, PA) equipped with Intelligent Imaging Innovations Slide Book 4.0 software for automated Z-stack and 3D image analysis.

3. Methods

The overall process of the hydrogel template method of making nano/microparticles is described in Fig. 1. Hydrogel templates are prepared by pouring a warm gelatin solution onto a PDMS template containing circular posts (i.e., 20 μm in diameter and 20 μm tall). The PDMS template is then cooled for 10 min in a refrigerator and the formed hydrogel template is carefully peeled off (gelatin undergoes a sol–gel phase transition at 40–45°C). The hydrogel template contains the exact negative imprint of the features present on the PDMS master template (i.e., wells 20 μm in diameter and 20 μm deep). The as-synthesized hydrogel templates then are filled with PLGA polymer solution. Finally, the PLGA-filled, hydrogel template is dissolved in warm water, releasing free homogeneous PLGA microstructures. Using hydrogel templates, homogeneous polymer particles of 200 nm to 50 μm and larger have been prepared (see Fig. 2). Drug loaded PLGA microparticles can be prepared by mixing the PLGA polymer solution with a drug. The drug loaded particles of nanoscale

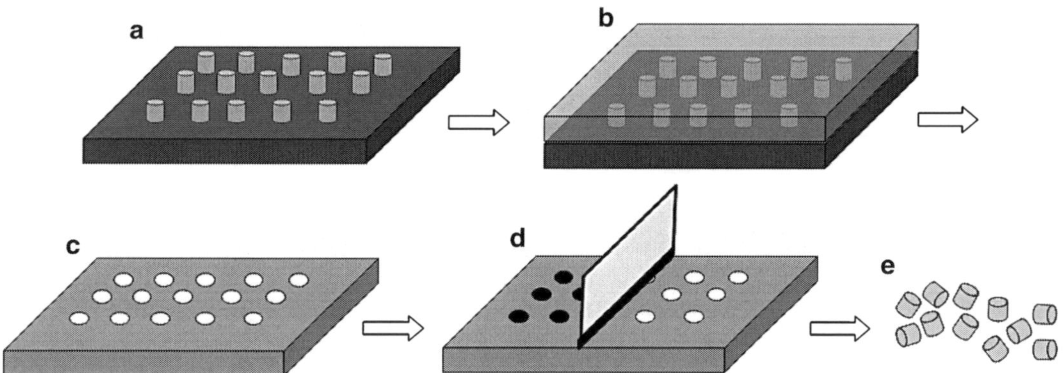

Fig. 1. Schematic diagram for fabrication of particles by the hydrogel template method. (**a**) A PDMS template having vertical posts is prepared from a silicon wafer master template. (**b**) A warm gelatin solution is poured onto the PDMS template. (**c**) The formed hydrogel template is peeled off from the PDMS template (wells are denoted by *white spots*). (**d**) The microwells in the hydrogel template are filled with a PLGA polymer solution (containing a drug) by swiping with a blade. (**e**) Homogeneous free PLGA microstructures are obtained by simply dissolving the hydrogel template in water.

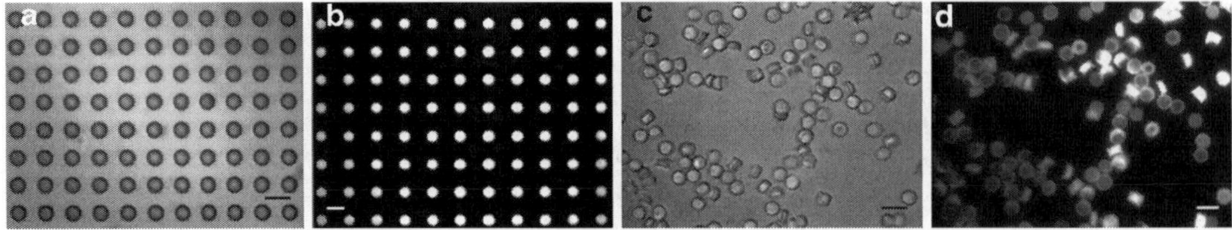

Fig. 2. Fabrication of homogeneous 20 µm PLGA microstructures with a hydrogel template. (**a**) Bright field image of a gelatin hydrogel template. (**b**) A fluorescence image of a hydrogel template filled with PLGA solution containing Nile Red. (**c, d**) Bright field and fluorescence images, respectively, of homogeneous free PLGA microstructures obtained by dissolving the hydrogel templates. Scale bars correspond to 40 µm.

dimension (e.g., 200 nm) are useful for targeting tumor sites by intravenous administration, while those of microscale dimension (e.g., 20 µm) are ideal for long-term delivery ranging from weeks to months after implantation.

3.1. Preparation of Gelatin Solution

1. Weigh gelatin powder (30 g) and transfer it into a 250-mL Pyrex bottle.

2. Add 100 mL Nanopure water to the Pyrex bottle and mix thoroughly.

3. Cap the bottle to prevent evaporation and place in an oven at 65°C for 2 h or until the formation of a clear solution (see Note 1).

4. Use this as-synthesized clear gelatin solution (30% w/v in water) in the fabrication of hydrogel templates (see Note 2).

3.2. Fabrication of Hydrogel Templates

1. Transfer the warm gelatin solution (5 mL) with a pipette onto a PDMS template (3″ diameter) containing circular pillars (i.e., 20 µm diameter and 20 µm height).

2. Spread the gelatin solution evenly to form a thin film completely covering the PDMS template and cool it to 4°C for 10 min by keeping it in a refrigerator (see Note 3).

3. Peel the hydrogel template from the PDMS master template (see Note 4).

4. The hydrogel template thus prepared will be ~3 in. in diameter and contain circular wells (i.e., 20 µm diameter and 20 µm depth).

5. Examine the hydrogel template under a bright field microscope (see Note 5).

3.3. Preparation of PLGA Solution

1. Transfer 200 mg of PLGA polymer into a 5-mL glass vial.

2. Add 1 mL of dichloromethane (CH_2Cl_2) to the vial with a micropipette (see Note 6).

3. Seal the glass vial with Parafilm and place on a flask rocker until all the PLGA is completely dissolved.

4. A clear, thick PLGA solution of 20% w/v concentration will be formed.

5. For fluorescence imaging purposes, the PLGA solution thus prepared may be doped with a fluorescent dye (e.g., 0.001% Nile Red).

3.4. Fabrication of PLGA Microparticles

1. Transfer 200 μL 20% PLGA solution w/v in dichloromethane with a pipette onto a 3 in. diameter hydrogel template containing circular wells 20 μm in diameter and depth (see Note 7).

2. Evenly spread the PLGA solution on the hydrogel template by swiping with a razor blade at a 45° angle (see Note 8).

3. Keep the PLGA-filled gelatin template on the table at room temperature (~25°C) for 5–10 min to evaporate CH_2Cl_2. For the complete removal of dichloromethane, longer drying time may be required.

4. Examine the hydrogel template under a bright field microscope (see Note 9).

3.5. Collection of the PLGA Microparticles

1. Dissolve a batch of ten hydrogel templates in a 100-mL beaker containing 50 mL of water at 40°C.

2. Gently swirl the beaker for 2 min to completely dissolve the hydrogel templates.

3. Upon complete dissolution of the hydrogel templates, free PLGA microparticles will be released into the water.

4. Transfer the solution into 15-mL conical centrifuge tubes and centrifuge at $4,500 \times g$ for 5 min.

5. Discard the supernatant liquid and collect the pellets.

6. Combine all the pellets, redisperse them in water and filter through a 25-mm filter holder equipped with a 25-μm filter (see Note 10).

7. Spot a drop of this dispersion on a glass slide and examine under a bright field microscope (see Note 11).

8. Transfer the solution into a 15-mL conical centrifuge tube and centrifuge at $4,500 \times g$ for 5 min.

9. Freeze dry the pellet obtained upon centrifugation and store it in a refrigerator (see Note 12).

10. This material will be stable for up to 1 year when stored in a freezer.

---

4. Notes

1. Gelatin solution should not be kept hot (>65°C) for long periods of time (>10 h). High temperatures may denature the protein and cause the solution to lose its ability to form hydrogel templates.

2. Elastic and mechanically strong hydrogel templates will be formed with a 30% gelatin solution.

3. Cooling results in the formation of a mechanically stable hydrogel template that can be easily peeled away from the PDMS template.

4. Gelatin templates soften and melt at room temperatures. They need to be stored in the refrigerator.

5. The presence of homogeneous circular wells in the hydrogel template will be clearly visible (see Fig. 2a).

6. PLGA solutions prepared from ethyl acetate and tetrahydrofuran solvents can also be used for hydrogel template filling.

7. Solutions of other polymers, such as polycaprolactone, polystyrene, and poly(vinyl chloride), can also be used for filling the hydrogel templates.

8. Swiping with a razor blade minimizes formation of the PLGA film (i.e., scum layer) on the hydrogel template surface. Gentle pressure has to be applied to force the PLGA solution to completely fill the wells without deforming the hydrogel template. However, avoid pressing the razor blade too hard as it might slough off the template.

9. The circular wells filled with polymer solution in the hydrogel template appear slightly darker compared to an unfilled hydrogel template under a bright field microscope. Alternatively, the filled templates can be observed with a fluorescence microscope if a polymer/dye mixture, such as PLGA/Nile Red, is used (see Fig. 2b).

10. Filtration removes larger aggregates or pieces of polymer film formed during filling of the hydrogel template with polymer solution.

11. Look for the presence of homogeneous circular microstructures dispersed in water (see Fig. 2c). Alternatively, the nanoparticles can be viewed with a fluorescence microscope if a polymer/dye mixture, such as PLGA/Nile Red, is used (see Fig. 2d).

12. This process yields approximately 1 mg of PLGA microstructures. Homogeneous PLGA microparticles containing drugs (e.g., felodipine, progesterone, and paclitaxel) can be prepared by using a solution of PLGA polymer and drug to fill the templates. The stored pellets can be redispersed in water for further use by vortexing.

Acknowledgment

This project was supported in part by the Showalter Trust Fund from Purdue Research Foundation.

References

1. Peer, D., Karp, J. M., Hong, S., Farokhzad, O. C., Margalit, R., and Langer, R. (2007) Nanocarriers as an emerging platform for cancer therapy. *Nat. Nanotechnol.* **2**, 751–760.

2. Torchilin, V. P. (2005) Recent advances with liposomes as pharmaceutical carriers. *Nat. Rev. Drug Discov.* **5**, 145–160.

3. Mundargi, R. C., Babu, V. R., Rangaswamy, V., Patel, P., and Aminabhavi, T. M. (2008) Nano/micro technologies for delivering macromolecular therapeutics using poly (D,L-lactide-co-glycolide) and its derivatives. *J. Control. Release* **125**, 193–209.

4. Blanco, E., Kesssinger, C. W., Sumer, B. D., and Gao, J. (2009) Multifunctional micellar nanomedicine for cancer therapy. *Exp. Biol. Med.* **234**, 123–131.

5. Danhier, F., Lecouturier, N., Vroman, B., Jérôme, C., Marchand-Brynaert, J., Feron, O., et al. (2009) Paclitaxel-loaded PEGylated PLGA-based nanoparticles: In vitro and in vivo evaluation. *J. Control. Release* **133**, 11–17.

6. Heslinga, M. J., Mastria, E. M., and Eniola-Adefeso, O. (2009) Fabrication of biodegradable spheroidal microparticles for drug delivery applications. *J. Control. Release* **138**, 235–242.

7. Euliss, L. E., DuPont, J. A., Gratton, S., and DeSimone, J. (2006) Imparting size, shape, and composition control of materials for nanomedicine. *Chem. Soc. Rev.* **35**, 1095–1104.

8. Glangchai, L. C., Caldorera-Moore, M., Shi, L., and Roy, K. (2008) Nanoimprint lithography based fabrication of shape-specific, enzymatically-triggered smart nanoparticles. *J. Control. Release* **125**, 263–272.

9. Huang, K.-S., Lu, K., Yeh, C.-S., Chung, S.-R., Lin, C.-H., Yang, C.-H., et al. (2009) Microfluidic controlling monodisperse microdroplet for 5-fluorouracil loaded genipin-gelatin microcapsules. *J. Control. Release* **137**, 15–19.

10. Peppas, N. A., Hilt, J. Z., Khademhosseini, A., and Langer, R. (2006) Hydrogels in biology and medicine: From molecular principles to bionanotechnology. *Adv. Mater.* **18**, 1345–1360.

11. Park, K., Shalaby, W. S. W., and Park, H. (1993) *Biodegradable Hydrogels for Drug Delivery.* Technomic Publishing, Lancaster, PA.

12. Liu, G., Zhao, D., Tomsia, A. P., Minor, A. M., Song, X., and Saiz, E. (2009) Three-dimensional biomimetic mineralization of dense hydrogel templates. *J. Am. Chem. Soc.* **131**, 9937–9939.

13. Du, B., Cao, Z., Li, Z., Mei, A., Zhang, X., Nie, J., et al. (2009) One-Pot preparation of hollow silica spheres by using thermosensitive poly(N-isopropylacrylamide) as a reversible template. *Langmuir* **25**, 12367–12373.

14. Acharya, G., Shin, C. S., McDermott, M., Mishra, H., Park, H., Kwon, I. C., et al. (2010) The hydrogel template method for fabrication of homogeneous nano/microparticles. *J. Control. Release* **141**, 314–319.

Chapter 13

Antibacterial Application of Engineered Bacteriophage Nanomedicines: Antibody-Targeted, Chloramphenicol Prodrug Loaded Bacteriophages for Inhibiting the Growth of *Staphylococcus aureus* Bacteria

Lilach Vaks and Itai Benhar

Abstract

The increasing development of bacterial resistance to traditional antibiotics has reached alarming levels, thus there is an urgent need to develop new antimicrobial agents. To be effective, these new antimicrobials should possess novel modes of action and/or different cellular targets compared with existing antibiotics. Bacteriophages (phages) have been used for over a century as tools for the treatment of bacterial infections, for nearly half a century as tools in genetic research, for about two decades as tools for the discovery of specific target-binding proteins and peptides, and for almost a decade as tools for vaccine development. We describe a new application in the area of antibacterial nanomedicines where filamentous phages can be formulated as targeted drug-delivery vehicles of nanometric dimensions (phage nanomedicines) and used for therapeutic purposes. This protocol involves both genetic and chemical engineering of these phages. The genetic engineering of the phage coat, which results in the display of a target-specificity-conferring peptide or protein on the phage coat, can be used to design the drug-release mechanism and is not described herein. However, the methods used to chemically conjugate cytotoxic drugs at high density on the phage coat are described. Further, assays to measure the drug load on the surface of the phage and the potency of the system in the inhibition of growth of target cells as well as assessment of the therapeutic potential of the phages in a mouse disease model are discussed.

Key words: Peptide phage display library, Phage display, Single-chain antibodies, BirA biotin ligase, ZZ domain, IgG, Fc antibody fragment

1. Introduction

The increasing development of bacterial resistance to traditional antibiotics has reached alarming levels (1), forcing scientists to develop new antimicrobial approaches. In both traditional and newly developed antibiotics, the target selectivity lies in the potency

Sarah J. Hurst (ed.), *Biomedical Nanotechnology: Methods and Protocols*, Methods in Molecular Biology, vol. 726,
DOI 10.1007/978-1-61779-052-2_13, © Springer Science+Business Media, LLC 2011

of the drug itself as well as in its ability to affect a mechanism which destroys or hinders the target microorganism and not its host. In fact, a vast number of potentially potent drugs have been excluded from use as therapeutics due to low selectivity (i.e., toxicity to the host as well as to the pathogen) (2). This brings to mind the limited selectivity of anticancer drugs and recent efforts to overcome this low selectivity through the development of novel-targeted, drug-delivery strategies.

In this chapter, we introduce a novel application of filamentous bacteriophage (phage) as targeted drug carriers for the eradication of pathogenic bacteria. In nature, bacteriophages are viruses that selectively invade the specific bacteria cells causing growth inhibition and death. As an efficient immunotherapeutic, it contains the following three components: a drug carrier, a targeting moiety, and a cytotoxic drug, which are responsible for payload, specificity, and efficacy, respectively. In this protocol, the genetically and chemically modified filamentous M13 filamentous bacteriophage takes the role of a nanometric, modular, high-capacity drug-carrying platform, while an antibody provides targeting drug and chloramphenicol serves as a cytotoxic drug. These bacteriophages have dimensions on the nanometer scale (~1 µm in length but less than 10 nm in diameter); hence, we refer to them as phage nanoparticles. The M13 filamentous bacteriophage (see Fig. 1) refers to *Escherichia coli*, male-specific bacteriophage Ff class that is able to infect and replicate in *E. coli* bacteria. However, in this study, the phages are targeted against bacterial pathogens by specific antibodies displayed on its coat; therefore, their natural host specificity is not relevant for the target specificity. We use these phages to deliver a large chloramphenicol payload to pathogenic bacteria such as *Staphylococcus aureus* (3, 4) (see Note 1). In this protocol, chloramphenicol, a hydrophobic drug, is chemically

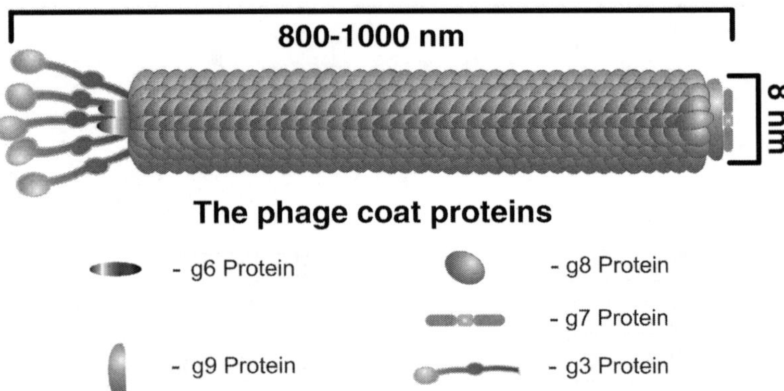

Fig. 1. Structure of the filamentous bacteriophage. The display of proteins and peptides is achieved by in frame fusion of its coding sequence with the sequence of the chosen phage coat protein.

modified to contain an esterase-cleavage linker that enables its slow release from the surface of the phage by serum esterases. The high density of chloramphenicol on the surface of the phage (10^4 drugs/phage) is made possible by linking it to the phage coat through aminoglycoside antibiotics that serve as solubility-enhancing, branched linkers.

This protocol describes genetically engineered fUSE5 phages that display a 15-mer peptide AVITAG (GLNGLNDIFEAQK-IEWHE) (5, 6) on the N-terminus of the p3 minor coat protein (g3p-AVITAG-fUSE5). The AVITAG peptide undergoes efficient biotinylation in vivo by the BirA biotin ligase enzyme (7) which enables coupling with a specific biotinylated antibody via the biotin–streptavidin bridge (see Note 2).

The preparation of drug-carrying phages then can be divided into three major steps:

1. Phage propagation in bacteria and purification by PEG/NaCl precipitation.

2. Prodrug preparation by sequential chemical reactions.

3. Conjugation of the prodrug to the bacteriophage through a hydrophilic aminoglycoside linker via an 1-ethyl-3-(3-dimethy-laminopropyl)-carbodiimide hydrochloride (EDC) coupling reaction.

Further, we describe how to quantify the number of drug molecules on the surface of the phage and how to carry out a bacterial growth inhibition assay to determine the effect that the drug-conjugated phages have on retarding the growth of the target bacteria. Finally, we establish the in vivo therapeutic potential of this system using a disease model in BALB/c mice with a lethal systemic infection of *S. aureus* bacteria. The colony-forming units (CFU) quantification method (8) or radio-labeled phage particles (9) can be used to learn about the targeting ability of drug-carrying phage pharmacokinetics and biodistribution.

2. Materials
(See Note 3)

2.1. Phage Preparation

1. Phage vector g3p-AVITAG-fUSE5 (see Note 4).

2. Bacteria strains: *E. coli* DH5α and DH5αF′ (GibcoBRL, Life Technologies, MD, USA) are used for phage preparation and for phage titration.

3. Growth medium: Yeast extract-tryptone x2 (2YT) and Tryptic Soy Broth (TSB) (see bacteria growth media).

4. Antibiotics: tetracyclin and chloramphenicol (see Sub-heading 2.4).

5. Vectors: pBirAcm (Avidity, LLC, http://www.avidity.com/) is a biotin ligase *birA* expression vector, in which the expression of the *birA* gene is controlled by an isopropyl β-D-1-thiogalactopyranoside (IPTG) inducible promoter.

6. D-biotin: 50 mM D-biotin in water, add 10 N NaOH dropwise to dissolve biotin completely. Filter 0.22 μm to sterilize. Store at 4°C for up to 3 months.

7. IPTG: 1 M of IPTG in sterile double-distilled (MilliQ) water (SDDW) is stored in 1 mL aliquots at –20°C. Use 0.1 mM for birA biotin ligase overexpression.

8. PEG/NaCl: 20% PEG6000 and 2.5 M NaCl in MilliQ water. Sterilize by autoclaving.

9. Vacuum filtration device (0.45 μm) (Amicon, USA).

10. Magnetic beads: Dynabeads M-280 streptavidin (Invitrogen Dynal, http://www.invitrogen.com/site/us/en/home/brands/Dynal.html).

11. Antibodies: biotinylated mouse-anti-*S. aureus* IgG (Abcam, USA, http://www.abcam.com). Any biotinylated IgG that does not bind *S. aureus* as a negative control.

12. Avidin: 10 mg/mL stock solution in DDW. Store at 4°C.

2.2. Prodrug Preparation

Chemicals and solvents were either A.R. grade or purified by standard techniques.

1. Tetrahydrofurane (THF).

2. Glutaric anhydride.

3. Triethylamine (Et_3N).

4. Dimethylaminopyridine.

5. Ethylacetate (EtOAc).

6. Hexane.

7. Hydrochloric acid (HCl).

8. Magnesium sulfate ($MgSO_4$).

9. Silica gel Merck 60 (particle size 0.040–0.063 mm).

10. Dichloromethane (DCM).

11. *N, N'*-Dicyclohexylcarbodiimide (DCC).

12. *N*-Hydroxysuccinimide (NHS).

13. Argon gas.

14. Thin layer chromatography (TLC): silica gel plates Merck $60F_{254}$. Compounds were visualized by irradiation with UV light.

2.3. Drug Conjugation

1. NaHCO$_3$: pH = 8.5 buffer that provides basic conditions for drug conjugation.

2. Dimethyl sulfoxide (DMSO): used for dissolving chloramphenicol–*N*-Hydroxysuccinimide (CAM–NHS) prodrug to a final stock concentration of 44 mg/mL.

3. For high-performance liquid chromatography (HPLC): a reverse phase C-18 column and 80% acetonitrile solution (in water w/w).

4. Citrate buffer: pH = 5.0 solution that is composed of citric acid 1 M and sodium citrate 1 M at appropriate dilution. For 100 μL citrate buffer, pH = 5.0, use 41 μL citric acid and 59 μL sodium citrate.

5. Neomycin: an aminoglycoside antibiotic, which serves as a branched, hydrophilic linker that conjugates the chloramphenicol prodrug to the phage coat proteins via EDC chemistry.

6. EDC, a zero-length crosslinking agent, was used to couple carboxyl groups to primary amines. Store the powder at –20°C. Make a fresh solution in DMSO immediately prior to use.

7. For dialysis: SnakeSkin-Pleated Dialysis tubing (10 kDa cutoff) supplied by PIERCE (Rockford, Illinois, USA).

2.4. General Buffers and Reagents

1. Phosphate-buffered saline (PBS): 8 g NaCl, 0.2 g KCl, 1.44 g Na$_2$HPO$_4$, and 0.24 g KH$_2$PO$_4$ per 1 L, pH = 7.4.

2. Chloramphenicol: 34 mg/mL in 100% ethanol. Store at –20°C.

3. Tetracyclin: 12.5 mg/mL in 50% ethanol. Store at –20°C.

4. Normal rabbit serum. Store at 4°C.

2.5. Bacteria Growth Media

Any supplier of bacterial growth medium components or pre-prepared media. We use products of Becton-Dickinson (http://www.bd.com/).

1. 2YT: 16 g Bacto-Tryptone, 10 g Yeast extract, and 5 g NaCl/L water.

2. TSB: 30 g of Bacto-TBS/L water.

3. To prepare solid media, Bacto-agar at the final concentration of 1.8% was added to the solutions. Following autoclaving, the media were supplemented with 0.4 or 1% glucose and antibiotics. The final concentrations of antibiotics used in this study were as follows: tetracycline (12.5 μg/mL) and chloramphenicol (34 μg/mL).

2.6. Bacterial Strains

In our studies, we used domestic isolates of target bacteria (model pathogens). Such bacterial strains can be obtained from the Global Bioresource Center (ATCC) (http://www.atcc.org). The model bacterial strain used in this protocol is *S. aureus* COL from our laboratory collection (see Note 1).

2.7. Animal Studies

Female BALB/c mice 8–10 weeks old, ~20 g, at least five mice in group. During the experiment, monitor mice weight, behavior, fur condition, and vitality (see Note 5).

---

3. Methods

This protocol provides a detailed description of antibacterial application of engineered bacteriophage nanomedicines as they were carried out in the laboratory of the authors. Ideally, such a targeted drug-delivery system should home and bind to the target cells (bacteria) before the drug release is triggered. Basically, filamentous bacteriophages are first equipped with a targeting moiety that is displayed or linked to a phage coat protein. Variations to the phage display theme can be found in the phage display literature (10–13). We provide examples of peptide-displaying phages that were isolated from a peptide phage display library by affinity selection on *S. aureus* (3). A similar approach can be used to isolate phages that have specificity to any target cell. It is also possible to use phages that display antibody fragments or other target-specificity-conferring proteins, each should be carefully evaluated for how well it tolerates the drug-conjugation chemistry (see Note 6). The biotinylated phage we designed (g3p-AVITAG-fUSE5) is optimal in that it can be complexed with targeting antibodies after completion of the drug-conjugation chemistry.

We provide a description of a chloramphenicol-based prodrug which has an esterase-cleavable linker with a terminal NHS leaving group to facilitate its conjugation to amine groups (see Note 7). Therefore, we began using aminoglycosides (such as neomycin) as solubility-enhancing, branched linkers that provide larger drug-loading capacity, better solubility, longer residence in the blood following i.v. injection into mice, and reduced immunogenicity of the targeted drug-carrying phage nanomedicines (ref. 4 and unpublished data). Based on similar concepts, it is possible to design other means of drug conjugation and release, as we did in a related study, where a protease-based drug-release mechanism was engineered into the phage coat (14). Since in a targeted drug-delivery system, the drug selectivity is replaced with the selectivity conferred by the targeting moiety, a slew of chemically conjugatable toxic compounds can be recruited to serve as potent antimicrobial drugs or as drugs that target other cells that are bearers of disease.

3.1. Phage Preparation

3.1.1. Preparation of Biotinylated g3p-AVITAG-fUSE5 Phage Vector (See Note 8)

Phage g3p-AviTag-fUSE5 displays a 15-amino-acid long peptide (called AVITAG: with the amino-acid sequence (single letter code) GLNDIFEAQKIEWHE) (Avidity, LLC) at the N-terminus of the p3 minor coat protein. The peptide undergoes biotinylation by the enzyme BirA biotin ligase. The plasmid pBirAcm carries an IPTG inducible (see Note 9) copy of *bir*A.

1. G3p-AviTag-fUSE5 should be co-transformed with the pBirAcm plasmid into DH5α *E. coli* cells. As an alternative, a stock of competent cells that already carry pBirAcm may be transformed with phage DNA. Plate the transformed cells on a 2YT-agar plate containing: 12.5 μg/mL tetracycline and 34 μg/mL chloramphenicol. Leave for 16 h at 37°C until colonies of transformed bacteria are clearly visible.

2. Prepare a starter culture by inoculating 3 mL of 2YT medium containing: 12.5 μg/mL tetracycline and 34 μg/mL chloramphenicol in a 13 mL test tube with a single colony of transformed bacteria. Grow in an incubator-shaker for 16 h at 37°C.

3. Transfer the 3 mL starter into 200 mL 2YT containing: 12.5 μg/mL tetracycline, 34 μg/mL chloramphenicol, 50 μM D-biotin, and 0.1 mM IPTG. Under these conditions, and during the subsequent incubation, the phages are produced and the AVITAG peptide is biotinylated before the phages are released from the producing bacteria into the culture medium.

4. Grow for 24 h shaking (250 rpm) at 30°C (see Note 10).

5. Add 0.1 mM IPTG.

6. Grow for 24 h shaking (250 rpm) at 30°C.

7. Precipitate the phage particles with PEG/NaCl (see Subheading 3.1.2 and Note 11).

3.1.2. PEG/NaCl Phage Precipitation

1. To separate the phages from the bacteria, centrifuge the culture (Sorvall centrifuge, GSA rotor, $8,000 \times g$ for 20 min at 4°C) and filter the supernatant (to eliminate remaining bacteria) using a 0.45 μm vacuum filtration device (Amicon, USA) (see Note 12).

2. To precipitate the phages and separate them from the growth medium (that includes components that unless removed, may interfere with the subsequent chemical conjugation of drug to the phages), add one-fifth of the volume of PEG/NaCl to the supernatant, mix well, and incubate for 2 h or more on ice (see Note 13).

3. Collect the phage-precipitates by centrifugation at $8,000 \times g$ for 30 min at 4°C and carefully discard the supernatant.

4. Re-suspend the phages in sterile water, filter at 0.45 μm, and store at 4°C (see Notes 14 and 15).

5. Determinate the phage concentration (see Subheading 3.1.3).

3.1.3. Phage Quantification (See Note 16)

1. Based on optical absorbance: add 50 μL of the phage to 450 μL PBS and measure the absorbance at 269 and 320 nm.

Use special UV-transparent or quartz cuvettes. Quantify the phage concentration according to this formula (15):

$$\frac{\left(\text{O.D.269nm} - \text{O.D.320nm}\right) \times 6 \times 10^{16}}{\text{Phage genome size (b.p.)}} \times 10 \left(\text{Dilution factor}\right) = \text{Phage/ml}$$

2. Live titration (see Note 17):

   (a) Grow DH5αF′ bacteria in 2YT to $A_{600nm} = 0.6\text{–}0.8$.

   (b) Using a multichannel pipette, fill a lane in a sterile 96-well plate with 90 µL of bacteria, add to the first well 10 µL of phage, and make serial dilutions by transferring 10 µL to the next well, making sure to change the tips after each dilution step. Incubate for 1 h at 37°C.

   (c) Plate the 10 µL drops of incubated bacteria onto agar plates supplemented with tetracycline and grown at 37°C overnight to develop colonies of resistant (phage infected) cells.

   (d) Calculate the phage quantity according to the resistant bacteria colonies number multiplied by the dilution factor.

3.1.4. Quantification of Phage Biotinylation Efficiency

Biotinylated phages should be readily captured on streptavidin-coated magnetic beads. The calculation of biotinylation efficiency is based on capturing the phages on such beads followed by determination of the uncaptured (presumably unbiotinylated) phages that are left in the supernatant.

1. Prepare phage stock for reading, by dilution of 33 µL $1\text{–}5 \times 10^{13}$ in 960 µL of PBS.

2. Divide phage solution into two separate tubes.

3. Using a spectrophotometer, read the absorbance of the phage suspension at 269 nm.

4. Place 100 µL of the magnetic streptavidin beads into a new tube.

5. Wash the magnetic beads twice with 500 µL PBS by placing the tube in a magnetic rack for 1 min and then removing the bead-free liquid with a pipette.

6. Block the beads by adding 500 µL of 1% BSA in PBS solution for 1 h at 37°C.

7. Discard the supernatant by pipetting it out of the tube while still on the magnetic rack.

8. Remove the tube from the magnetic rack.

9. Add 500 µL of phages, as prepared earlier.

10. Rotate slowly (10 rpm) on a benchtop tube rotator for 15 min.

11. Place the tube back into the rack for 1 min.

12. Take supernatant into new tube.

13. Read at 269 nm.

14. Compare the concentration of the phages before and after incubation with beads and determine the exact concentration of biotinylated phages (see Note 18).

3.2. Drug Conjugation

3.2.1. Preparation and Conjugation of the Chloramphenicol Prodrug

It is highly recommended that the synthesis of such drugs is carried out by an experienced organic chemist.

1. Dissolve 1 g chloramphenicol (6.2 mmol) in dry THF.

2. Add glutaric anhydride (800 mg, 6.82 mmol), Et$_3$N (1.0 mL, 6.82 mmol), and a catalytic amount of DMAP.

3. Incubate the reaction stirring at room temperature overnight.

4. Check the compound by TLC (EtOAc:Hex = 9:1) to receive acid-like running.

5. Stop the reaction by adding a large volume (~50 mL) of EtOAc. The mixture becomes milky during this step.

6. Add the same volume of 1 N HCl. The mixture becomes partially clear and separates into two layers during this step.

7. Collect the organic layer, dry with magnesium sulfate, and remove the solvent under reduced pressure.

8. Purify the crude product by column chromatography on silica gel (EtOAc:Hex = 4:1). The resulting product is a viscous gel (~2 g weight).

9. Carry out NMR of the product (see Fig. 2): ^{1}H NMR (200 MHz, CD$_3$OD): δ = 8.17 (2H, d, J = 8); 7.65 (2H, d, J = 8); 6.22 (1H, s); 5.08 (1H, d, J = 2); 4.44–4.41 (2H, m); 4.24 (1H, d, J = 2); 2.40–2.32 (4H, m); 1.92 (2H, t, J = 7).

10. If needed, repeat the wash and purification steps 5–8. If there are problems with dissolving a product in EtOAc, then add a small amount of methanol.

11. Dissolve the resulting product (2 g, 4.57 mmol) in DCM.

12. Add DCC (1.4 g, 6.86 mmol) and NHS (790 mg, 6.86 mmol).

13. Incubate the reaction stirring at room temperature overnight.

14. Check the reaction progress by TLC (EtOAc:Hex = 9:1) to receive a polar compound.

15. Filter the reaction and remove the solvent under reduced pressure.

16. Purify the crude product by column chromatography on silica gel (EtOAc:Hex = 4:1) to yield a white solid powder (~1.5 g, 62% yield).

17. Carry out NMR analysis of the CAM–NHS prodrug (see Fig. 2a): ^{1}H NMR (200 MHz, CD$_3$OD): δ = 8.17 (2H, d, J = 8); 7.65 (2H, d, J = 8); 6.22 (1H, s); 5.08 (1H, d, J = 2); 4.44–4.41 (2H, m); 4.24 (1H, d, J = 2); 3.02 (4H, s); 2.91 (2H, t, J = 7); 2.68 (2H, t, J = 7), 2.20 (2H, t, J = 7); 1.43 (1H, t, J = 7).

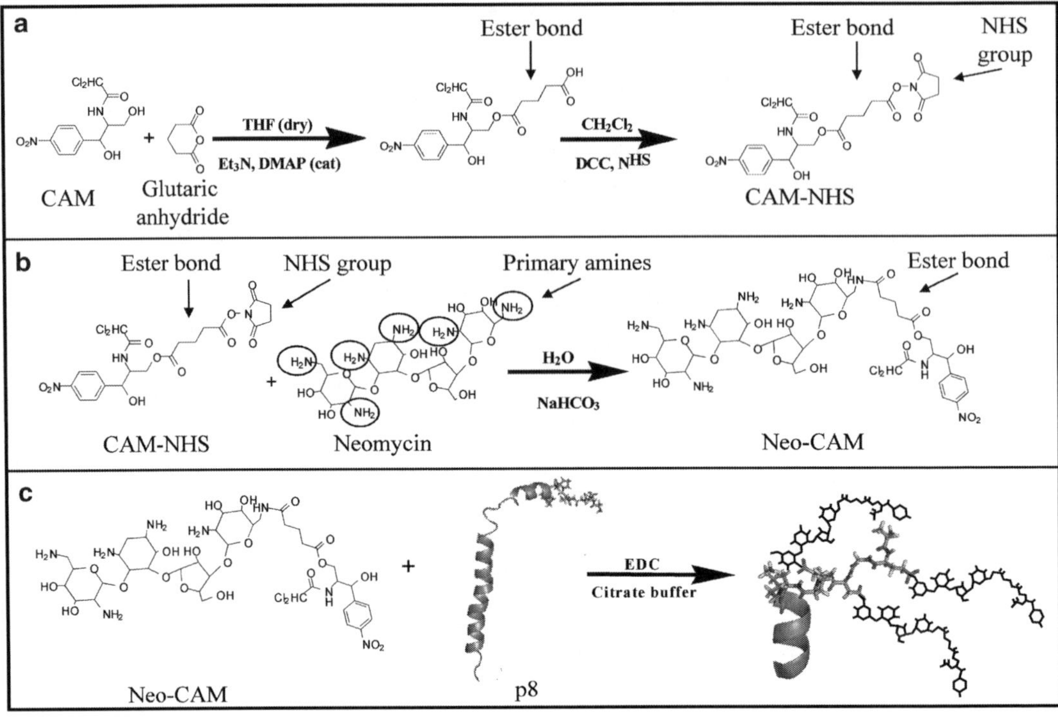

Fig. 2. Schematic representation of the chemical reactions used to prepare neomycin–chloramphenicol adduct for conjugation. (**a**) Two chemical steps were used to modify chloramphenicol (CAM) for conjugation to amine groups. In the first step, the chloramphenicol primary hydroxyl group was reacted with glutaric anhydride to create an ester linkage, resulting in chloramphenicol-linker. In the second step, the free carboxyl group of the chloramphenicol-linker was activated with NHS to allow subsequent linkage to amine groups. (**b**) The chloramphenicol–NHS was reacted with neomycin in a solution of 0.1 M NaHCO$_3$, pH = 8.5, resulting in neomycin–chloramphenicol adduct. The six primary amine groups of neomycin are circled. (**c**) The resulting neomycin–chloramphenicol adduct is conjugated to free carboxyl groups of the phage coat by the EDC procedure.

18. Store as dried powder at –20°C under argon. To dissolve, use DMSO (see Note 19).

3.2.2. Conjugation of Neomycin to the Chloramphenicol Prodrug

1. Mix solid neomycin and 100 µM chloramphenicol prodrug in DMSO within 0.1 M NaHCO$_3$, pH = 8.5, at a molar ratio of 1:2 for the chloramphenicol prodrug:neomycin (see Note 20).

2. Leave on stirrer overnight at room temperature.

3. Determine the prepared Neo–CAM adduct by reverse-phase HPLC. Use reverse phase C-18 column on a Waters machine with a gradient 0–100% of acetonitrile (stock solution of 80% in water w/w) and water (100% water to 0%) in the mobile phase, at 1 mL/min flow rate.

4. The Neo–CAM adduct should elute 18 min after sample injection while the intact CAM prodrug should elute 24 min after sample injection. An example of the results produced is shown in Fig. 3 (see Note 21).

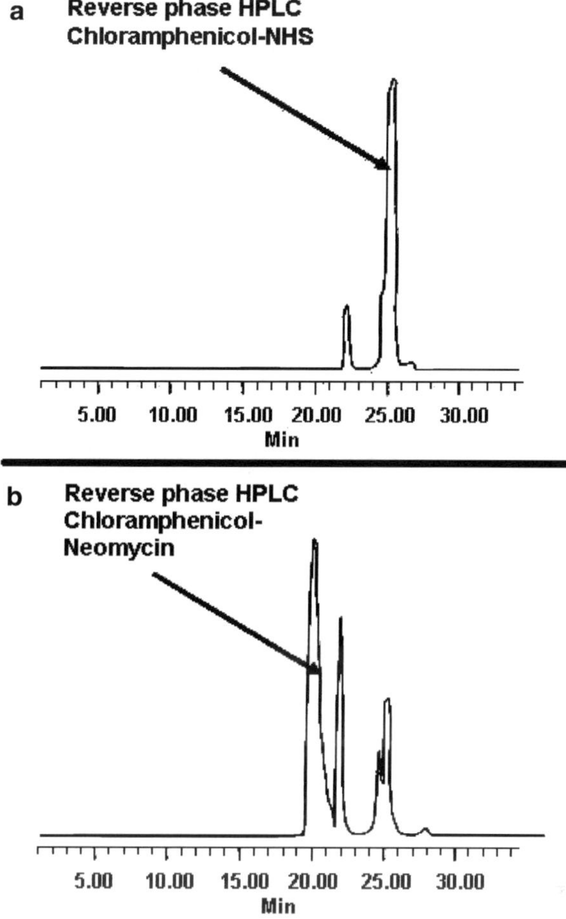

Fig. 3. Reverse-phase HPLC analysis of the Neo–CAM adduct. (**a**) HPLC analysis of chloramphenicol–NHS prior to conjugation to neomycin. The chloramphenicol–NHS prodrug was separated using a gradient of acetonitrile in water on a Waters HPLC machine (RP; C-18 column). CAM–NHS was eluted 25 min postinjection. (**b**) HPLC analysis of Neo–CAM adduct. The Neo–CAM adduct was separated using a gradient of acetonitrile in water on a Waters HPLC machine (RP; C-18 column). The Neo–CAM adduct was eluted at 18–19 min postinjection.

5. Conjugate Neo–CAM to biotinylated or antibody-complexed phage nanoparticles by the EDC procedure.

3.2.3. EDC Conjugation Chemistry

In our system, drug conjugation with EDC is between exposed carboxyl side chains on the phage coat [most of those would be on the major coat protein-p8 that contains four carboxylic amino-acid residues at its exposed N-terminus (Glu2; Asp4; Asp5; Glu12)] and neomycin that contains six primary amines (see Fig. 2).

1. Prepare a conjugation mix in a total volume of 1 mL containing: 0.1 M citrate buffer, pH = 5.0; 0.75 M NaCl; 2.5×10^{-6} mol Neo–CAM; and 5×10^{12} g3p-AVITAG-fUSE5 phage particles (see Note 22).

2. Add 2.5×10^{-6} mol of EDC and leave the reaction at room temperature for 1 h while gently rotating (10 rpm). (see Note 23).

3. Add the same amount of EDC and leave the reaction rotating for 1 h.

4. Perform two-step dialysis against 0.3 M NaCl (see Notes 24 and 25).

3.3. Complexing g3p-AVITAG-fUSE5 Phages with IgG

The biotinylated phages that were prepared according to Subheading 3.1.1 and conjugated to drug according to Subheading 3.2.3 are complexed through an avidin bridge with a biotinylated antibody to confer them with target specificity.

1. Add 0.1 µg avidin to 10^{12} Neo–CAM conjugated phage in 1 mL of 0.3 M NaCl (see Note 26).

2. Incubate for 1 h at room temperature with gentle rotation at 10 rpm.

3. Add 0.3 µg biotinylated IgG to the phage–avidin complex. Prepare a batch of phages in complex with the target-specific IgG and a second batch in complex with a negative control antibody.

4. Incubate for 1 h at room temperature and gentle rotation at 10 rpm on a benchtop tube rotator.

5. Store IgG-complexed phages at 4°C for up to 2 months. A scheme of the complete targeted drug-conjugated phages is shown in Fig. 4.

3.4. Growth Inhibition Experiments (See Note 27)

In the described experiment, *S. aureus* bacteria are treated with chloramphenicol-carrying, antibody-targeted phages (the treatment group is in complex with target-specific IgG and the control group is the phages conjugated to IgG that does not bind the target bacteria) and growth rate is recorded. Normal rabbit serum is added as a source of esterases to facilitate drug release.

1. Grow *S. aureus* bacteria culture overnight at TSB (see Note 1).

2. Collect 0.1–1 mL aliquot of bacteria culture by centrifugation for 1 min at $15,000 \times g$ in a microfuge at 4°C, and wash twice by re-suspension and re-centrifugation in ice-cold PBS.

3. Re-suspend the bacteria in an equal volume of ice-cold PBS.

4. Incubate 10 µL of washed bacteria (~10^7 cells) with 100–300 µL of targeted chloramphenicol-carrying phage nanoparticles (~$1–3 \times 10^{11}$ particles) for 1 h on ice.

5. Add an equal volume (100–300 µL) of normal rabbit serum (see Note 28).

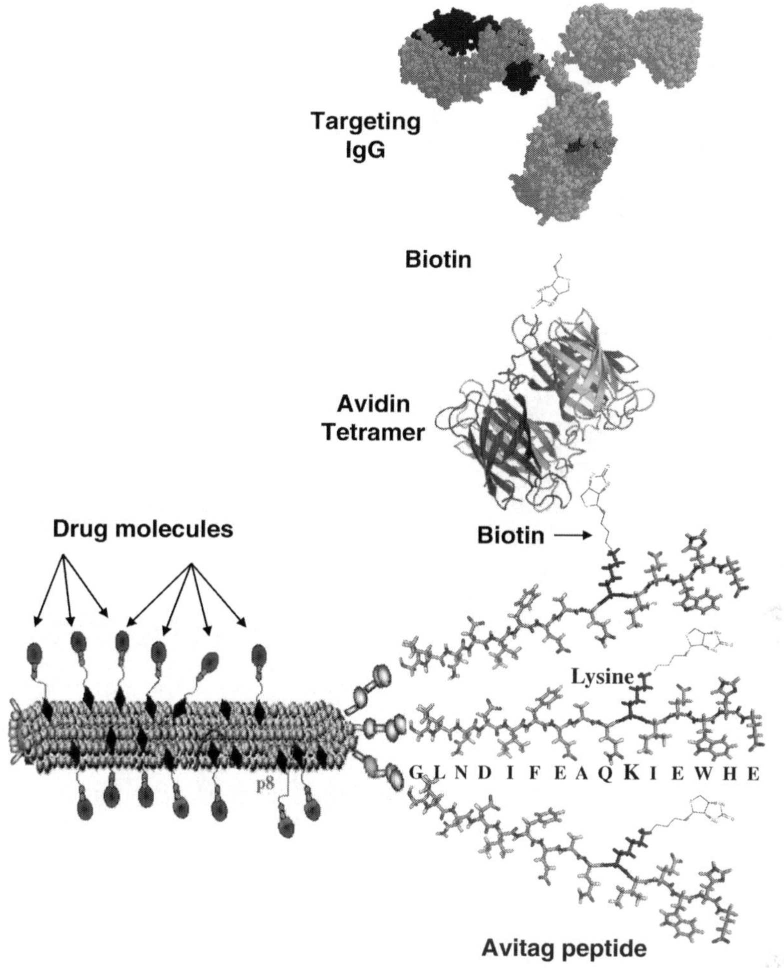

Fig. 4. A scheme of the complete targeted drug-conjugated phages. Each phage is conjugated to about 10,000 drug molecules on its coat. On the phage tip, the AVITAG peptide that undergoes biotinylation is displayed on all (3–5) copies of the g3p minor coat protein. The biotin is bound by an avidin tetramer which through the unoccupied three other biotin binding sites binds biotinylated antibodies (IgG). While for simplicity, the biotin–avidin-biotinylated IgG is shown only once; in theory, every drug-carrying phage may be targeted by many (up to 15) targeting antibodies if all binding sites are occupied. The scheme is not drawn to scale.

6. Incubate for 3 h at 37°C.

7. Dilute 100–300 µL, respectively, of this mixture in 3 mL TSB in 13 mL tubes.

8. Grow at 37°C by shaking at 250 rpm and monitor the absorbance at 600 nm. Plot the OD at 600 nm against time to monitor bacterial growth. An example of the results produced is shown in Fig. 5.

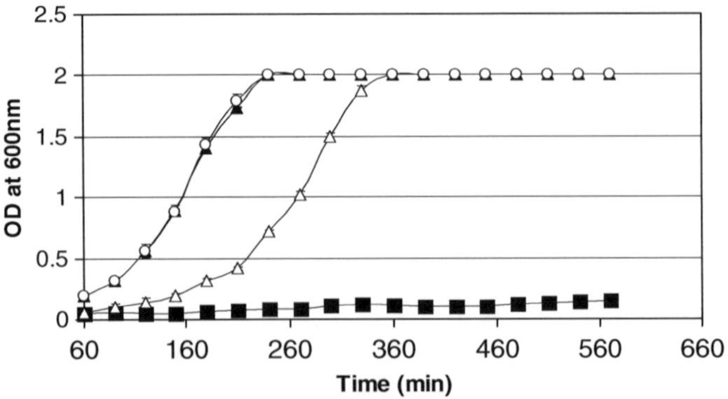

Fig. 5. Effect of drug-carrying antibody-targeted phages on the growth of *S. aureus*. Growth curves of *S. aureus* cells treated with 3×10^{11} Neo–CAM carrying fUSE5-ZZ phages, displaying specific anti-Staphylococcus antibodies bound via ZZ (*filled squares*) or with nontargeted (Fc displaying) Neo–CAM carrying phages (*open triangles*). *Staphylococcus aureus* grown with PBS (*filled triangles*) or with naked fUSE5-ZZ (*open circles*) represents bacteria grown without any inhibitor.

3.5. In Vivo Experiments

3.5.1. Staphylococcus aureus Disease Model

There are no existing "text book" about small animal disease models for pathogenic bacteria. A model has to be designed carefully to allow efficient evaluation of the drug-delivery platform. In the described experiment, mice are injected with *S. aureus* bacteria and are treated with chloramphenicol-carrying, antibody-targeted phages. The mice are followed for signs of toxicity (such as weight loss, apathic behavior, or death). This is a fatal disease model, and untreated mice succumb to death within a few days. Efficacy of the treatment is indicated by delaying of symptoms or prolonging (or preventing) death.

1. Grow a *S. aureus* starter culture overnight in 3 mL of TSB in a 13 mL culture test tube at 37°C by shaking at 250 rpm in a shaking incubator (see Note 29).

2. Dilute the bacteria 1:100 in 200 mL of fresh TSB in 1 L Erlenmeyer flask and grow at 37°C by shaking at 250 rpm in a shaking incubator until the culture reaches $A_{600nm} = 1$ (equivalent to 5×10^8 bacteria/mL).

3. Collect the bacteria by centrifugation and wash the bacteria twice in sterile PBS. Each wash consists of thoroughly resuspending the bacteria in PBS and collecting the cells by centrifugation. Bring the bacteria to a final concentration of 5×10^9/mL in PBS.

4. Inject 10^9 bacteria per mouse into the tail vein of female BALB/c mice (see Notes 30 and 31).

5. Monitor the survival and weight loss for 3 weeks (see Note 32).

3.5.2. In Vivo Therapeutic Activity of Targeted Drug-Carrying Bacteriophages

1. Grow *S. aureus* culture overnight at TSB (follow steps 1–3 of Subheading 3.5.1).

2. Dilute 1:100 in TSB and grow at 37°C until the culture reaches $A_{600nm} = 1$ (equivalent to 5×10^8 bacteria/mL).

3. Wash the bacteria twice in sterile PBS and bring to a final concentration of 5×10^9/mL.

4. Perform a number of comparable injections of *S. aureus* and targeted drug-carrying phage with different time intervals between bacteria and phage injections:

    (a) Incubate 1 mL aliquot of washed bacteria with 1×10^{11} anti-*S. aureus* Neo–CAM carrying bacteriophage for 1 h on ice. Wash in sterile-cold PBS. Re-suspend in 1 mL of cold sterile PBS and inject the mice with 200 µL (=1×10^9 bacteria) into the tail vein.

    (b) Precipitate 1 mL of 5×10^9/mL of *S. aureus* and re-suspend in 1 mL of anti-*S. aureus* Neo–CAM carrying bacteriophage (1×10^{11}). Inject the mice with 200 µL (=1×10^9 bacteria) into the tail vein.

    (c) Inject 10^9 bacteria per mouse into the tail vein. 30 min, 1, 2, 4, 8, and 12 h later inject 1×10^{11} anti-*S. aureus* Neo–CAM carrying bacteriophages into the tail vein.

    (d) Inject 10^9 bacteria per mouse into the tail vein. 30 min, 1, 2, 4, 8, and 12 h later inject the mice with 1×10^{11} anti-*S. aureus* Neo–CAM carrying bacteriophage intraperitonealy.

    (e) Inject 10^9 bacteria per mouse into the tail vein. For 24 and 48 h, monitor the behavior and health of the mice. When the therapy is examined, inject the mice with 1×10^{11} anti-*S. aureus* Neo–CAM carrying bacteriophage into the tail vein/intraperitonealy. In subsequent experiments, you may test different time gaps between the injection of the pathogen and the phages.

5. Monitor the survival and weight loss of the mice for up to for 3 weeks.

4. Notes

1. During the research that was carried out in our laboratory, in addition to *S. aureus* described here in detail, we showed the possibility to target a variety of pathogenic bacteria, such *as Streptococcus pyogenes* and avian pathogen *E. coli* O78. The targeting ability depends exclusively on the targeting moiety displayed on or linked to the phage coat.

2. Different targeting molecules, displayed on the phage coat may be used to confer target specificity to the drug-carrying phages. While the protocol describes in detail phages that display the AVITAG peptide and are linked to targeting antibodies through an avidin–biotin bridge, three alternative targeting moieties have been developed in our laboratory and can be used:

   (a) A specific anti-*S. aureus* 12-mer peptide VHMVAGPGREPT that is displayed as an N-terminal fusion to the p8 (g8p) major coat protein of the fth phage (A12C phage (3)). This *S. aureus* binding peptide was isolated from a disulfide-bond constrained 12-mer phage display library, designed on the fth "type 88" expression vector (16). The library was constructed and kindly provided by Prof. Jonathan Gershoni's group at Tel-Aviv University.

   (b) A single-chain-specific antibacterial antibody (scFv) displayed on the N-terminus of the p3 minor coat protein of the fUSE5 phage (scFv-fUSE5). The scFv display and the two following display methods are based on the fUSE5 vector system developed for polyvalent display on p3 by Smith and co-workers (5) (http://www.biosci.missouri.edu/smithgp/PhageDisplayWebsite/vectors.doc).

   (c) The ZZ domain displayed on the N-terminus of the p3 fUSE5 phage coat protein (17, 18). The ZZ domain is a modified *S. aureus* protein fragment that specifically binds the Fc region of the antibody. The FUSE5-ZZ bacteriophage is able to form a stable complex with target-specific IgGs. The main disadvantage of the above display methods is the fact that the targeting component may lose its binding activity following the drug-conjugation chemistry.

3. Most of the materials and reagents that are listed may be obtained from several vendors. We listed the vendors from whom we routinely purchase, which does not mean that we endorse the products of those particular vendors.

4. The bacteriophages from our laboratory collection: g3p-AVITAG-fUSE5, as well as fUSE5-ZZ, scFv-fUSE5, and A12C can be obtained from the authors upon request (3, 4).

5. The mice behavior and vitality should be monitored daily during the experiment. Healthy animals usually are energetic with shiny fur. One should mention a light increase in body weight (up to 2–3% weekly). However, mice in significant pain or distress typically display a lack of activity, sunken eyes, ruffled fur, and weight decrease.

6. Care should be taken to avoid using phages that have chemically modifiable amino-acid residues at key contact residues of the displayed peptides, as these may lose their target-binding ability upon chemical conjugation of the drug.

7. Early experiments we carried out, in which this drug was linked directly to free amine groups of the phage coat, provided evidence that it is not possible to link a large payload of a hydrophobic drug such as chloramphenicol to the phages that precipitate as a result.

8. The preparation of fUSE5-ZZ, scFv-fUSE4, and A12C bacteriophages is a routine protocol that can be found in ref. (19).

9. The plasmid pBirA carries a copy of birA biotin ligase downstream to the *lac* promoter that enables the birA protein efficient expression following the IPTG addition to the growth media. IPTG is widely used as an inducer for overexpression of the cloned proteins.

10. For efficient growth and biotinylation of g3p-AVITAG-fUSE5, we grow it for approximately 48 h supplied twice with 0.1 mM IPTG. Following the first step of overnight growth, the bacteria culture would not reach growth saturation and would show low optical density. Add an additional dose of IPTG (according to the protocol) and continue the growth for the next 24 h.

11. Phage precipitation step with PEG/NaCl solution is aimed to separate the phages from the bacterial growth media, concentrate them, and re-suspend in the desired solvent.

12. Use only 0.45 μm cutoff filters. Up to 50% of filamentous bacteriophages (that reach 1 μm in length) are trapped and lost in 0.22 μm filter.

13. For large volumes (0.5 L), we recommend to incubate the phage mixture overnight at 4°C.

14. Re-suspend the phage pellet in distilled water (not PBS) because it is preferable for the following chemistry steps.

15. We do not recommend freezing phage stocks because they lose their infectivity probably due to structural instability. It is best to keep it at 4°C.

16. The two approaches are usually used for phage quantification: "live" titration, a process of infecting bacteria with diluted phages followed by counting the resulting infected bacterial colonies that grow on selective plates, and measuring the optical absorbance at 269 nm. The live titration method counts viable infective virus particles, while the optical method quantifies everything that absorbs at 269 nm. In theory, the values should be identical; however, in practice, the absorbance method gives us up to ten times higher value. The difference comes from impurities and noninfective phage particles.

17. Most phages that are used in research laboratories lyse the infected bacteria and form plaques on bacterial lawns which can be counted to calculate their number. However, it is more convenient to count bacterial colonies than to count plaques.

Fortunately, many genetically engineered phages carry antibiotic resistance genes and thus infected cells form colonies on the appropriate selective media. In our protocol, phage quantification by live titration is based on the fact that these specific bacteriophages carry in their genome a tetracycline resistance cassette that provides tetracycline resistance to the infected bacteria. The infection of tetracycline sensitive (tet^s) DH5αF′ bacteria with the above phages will result in tetracycline-resistant (tet^R) bacteria. The number of tet^R bacteria colonies is proportional to phage particle quantity the bacteria were infected with.

18. When you use streptavidin-coated magnetic beads, you also should check PBS only to ensure that the magnetic beads are not broken. Use nonbiotinylated phage as a control to determine the stickiness of the beads.

19. Divide small aliquots of CAM–NHS (for approximately ten reactions) into tubes and keep under argon at –20°C. It is a highly hygroscopic substance! Before use, bring a vial of powder to room temperature and then open. Dissolve in DMSO only before usage.

20. Add CAM–NHS last to the reaction.

21. It is possible to collect purified Neo–CAM using preparative HPLC. However, according to our experience, a small amount of unconjugated chloramphenicol does not interfere with subsequent chemistry steps.

22. You can alternatively use 10^{12} fUSE5-ZZ phages previously complexed with 0.1–0.3 μg IgG (for 1 h at room temperature) (3, 4). The scFv-fUSE5 and A12C phages are already targeted and should be directly used for drug conjugation (10^{12} phage particles) (3, 4). We found that EDC chemistry harms the ZZ domain. Therefore, we recommend to complex targeted IgGs to fUSE5-ZZ before drug conjugation. Biotin, on the other hand, is impervious to EDC chemistry, hence it is possible to conjugate the drug to biotinylated phages before forming a complex with targeting antibodies.

23. EDC conjugation reaction is accompanied with a rise in pressure; therefore, avoid using tubes that can easily be opened.

24. We perform dialysis using 10 kDa snakeskin dialysis tubing (see Subheading 2) while during this step we lose some amount of phage particles. However, if a smaller cutoff is used, some phage particles can precipitate on the snakeskin membrane and clog it.

25. Following the EDC chemistry, bacteriophages partially lose their infectivity and change their absorbance at 269 nm; therefore, it is impossible to determine the exact drug-carrying phage concentration. According to our calculation, the final dialyzed drug-conjugated bacteriophage is $\sim 1 \times 10^{12}$/mL.

26. It is important not to add a too high concentration of avidin, because free avidin can trap the biotinylated IgG added at the next step and can eliminate its binding to phages.

27. According to our calculation, we can conjugate up to 10,000 chloramphenicol molecules per phage.

28. Pay attention that serum is clean and noncontaminated. Filter-sterilize it if needed.

29. Bacteria dosage for injection may differ among the strains. We worked with fairly pathogenic bacteria; thus, the amounts that were used for lethal model were very high. When highly pathogenic bacteria are available, a lower concentration of bacteria may be injected with the same drug-carrying phage dosage. This can extremely improve the therapeutic results.

30. We injected the bacteria and the drug directly to the tail vein of the mice. As an alternative, intraperitoneal administration can be performed. It allows using larger injection volumes, and more easily performed by inexperienced experimenters.

31. It is highly recommended to warm the animal in an incubator or under an incandescent light. This procedure makes the tail veins more visible for injection. Use 28 gauge needles. Be sure there are no air bubbles in the solution to be injected, as this can harm the mice. Before injection, wipe the injection site clean with a disinfecting gauze to avoid unintended infection.

32. In this disease model, expect a massive decrease in body weight of infected animals leading to at least 80% mortality within 1 week from injection of the *S. aureus* bacteria.

Acknowledgments

Studies of targeted drug-carrying phage nanomedicines at the author's laboratory received a grant from the Israel Public Committee for Allocation of Estate Funds, Ministry of Justice, Israel and by the Israel Cancer Association.

References

1. Yoshikawa, T. T. (2002) Antimicrobial resistance and aging: beginning of the end of the antibiotic era? *J. Am. Geriatr. Soc.* **50**, S226–S229.

2. Forget, E. J. and Menzies, D. (2006) Adverse reactions to first-line antituberculosis drugs. *Expert Opin. Drug Saf.* **5**, 231–249.

3. Yacoby, I., Shamis, M., Bar, H., Shabat, D., and Benhar, I. (2006) Targeting antibacterial agents by using drug-carrying filamentous bacteriophages. *Antimicrob. Agents Chemother.* **50**, 2087–2097.

4. Yacoby, I., Bar, H., and Benhar, I. (2007) Targeted drug-carrying bacteriophages as antibacterial nanomedicines. *Antimicrob. Agents Chemother.* **51**, 2156–2163.

5. Smith, G. P. (1985) Filamentous fusion phage: novel expression vectors that display cloned

antigens on the virion surface. *Science* **228**, 1315–1317.

6. Beckett, D., Kovaleva, E., and Schatz, P. J. (1999) A minimal peptide substrate in biotin holoenzyme synthetase-catalyzed biotinylation. *Protein Sci.* **8**, 921–929.

7. Scholle, M. D., Kriplani, U., Pabon, A., Sishtla, K., Glucksman, M. J., and Kay, B. K. (2006) Mapping protease substrates by using a biotinylated phage substrate library. *Chembiochem* 7, 834–838.

8. Zou, J., Dickerson, M. T., Owen, N. K., Landon, L. A., and Deutscher, S. L. (2004) Biodistribution of filamentous phage peptide libraries in mice. *Mol. Biol. Rep.* **31**, 121–129.

9. Molenaar, T. J., Michon, I., de Haas, S. A., van Berkel, T. J., Kuiper, J., and Biessen, E. A. (2002) Uptake and processing of modified bacteriophage M13 in mice: implications for phage display. *Virology* **293**, 182–191.

10. Benhar, I. (2001) Biotechnological applications of phage and cell display. *Biotechnol. Adv.* **19**, 1–33.

11. Cortese, R., Felici, F., Galfre, G., Luzzago, A., Monaci, P., and Nicosia, A. (1994) Epitope discovery using peptide libraries displayed on phage. *Trends Biotechnol.* **12**, 262–267.

12. Hoogenboom, H. R., de Bruine, A. P., Hufton, S. E., Hoet, R. M., Arends, J. W., and Roovers, R. C. (1998) Antibody phage display technology and its applications. *Immunotechnology* **4**, 1–20.

13. Sidhu, S. S., Weiss, G. A., and Wells, J. A. (2000) High copy display of large proteins on phage for functional selections. *J. Mol. Biol.* **296**, 487–495.

14. Bar, H., Yacoby, I., and Benhar, I. (2008) Killing cancer cells by targeted drug-carrying phage nanomedicines. *BMC Biotechnol.* **8**, 37.

15. Berkowitz, S. A. and Day, L. A. (1976) Mass, length, composition and structure of the filamentous bacterial virus fd. *J. Mol. Biol.* **102**, 531–547.

16. Enshell-Seijffers, D., Smelyanski, L., and Gershoni, J. M. (2001) The rational design of a "type 88" genetically stable peptide display vector in the filamentous bacteriophage fd. *Nucleic Acids Res.* **29**, E50–E60.

17. Nilsson, B., Moks, T., Jansson, B., Abrahmsen, L., Elmblad, A., Holmgren, E., et al. (1987) A synthetic IgG-binding domain based on staphylococcal protein A. *Protein Eng.* **1**, 107–113.

18. Nilsson, J., Larsson, M., Stahl, S., Nygren, P. A., and Uhlen, M. (1996) Multiple affinity domains for the detection, purification and immobilization of recombinant proteins. *J. Mol. Recognit.* **9**, 585–594.

19. Enshell-Seijffers, D. and Gershoni, J. M. (2002) Phage display selection and analysis of Ab-binding epitopes. in: *Current Protocols in Immunology* (Coligan, J. E., Bierer, B. E., Margulies, D. H., Shevach, E. M., and Strober, W., eds) pp. 9.8.1-9.8.27. John Wiley & Sons, Inc., USA.

Chapter 14

Viruses as Nanomaterials for Drug Delivery

Dustin Lockney, Stefan Franzen, and Steven Lommel

Abstract

Virus delivery vectors are one among the many nanomaterials that are being developed as drug delivery materials. This chapter focuses on methods utilizing plant virus nanoparticles (PVNs) synthesized from the *Red clover necrotic mosaic virus* (RCNMV). A successful vector must be able to effectively carry and subsequently deliver a drug cargo to a specific target. In the case of the PVNs, we describe two types of ways cargo can be loaded within these structures: encapsidation and infusion. Several targeting approaches have been used for PVNs based on bioconjugate chemistry. Herein, examples of such approaches will be given that have been used for RCNMV as well as for other PVNs in the literature. Further, we describe characterization of PVNs, in vitro cell studies that can be used to test the efficacy of a targeting vector, and potential routes for animal administration.

Key words: Bioconjugation, Capsid, Drug delivery, Encapsidation, Infusion, Nanomaterial, Peptides, Plant virus nanoparticle, Targeting

1. Introduction

The methods and strategies we use to synthesize a plant virus nanoparticle (PVN), as a drug delivery platform, are related to the structure and function of *Red clover necrotic mosaic virus* (RCNMV). Therefore, it is important to review the basic structure of RCNMV and how it functions as a virus before we discuss the methods for PVN synthesis.

RCNMV has an icosahedral protein capsid that contains its highly organized bipartite genome, named RNA-1 and RNA-2 (Fig. 1). RCNMV is ~36 nm in diameter and is constructed from 180 chemically equivalent 37 kDa capsid proteins (CPs) (1). These CPs assume one of three conformations in an ABC trimer, called an icosahedral asymmetric unit (IAU) (Fig. 1). Sixty of these IAU's are needed to form the $T=3$ capsid (1). The $T=3$

Sarah J. Hurst (ed.), *Biomedical Nanotechnology: Methods and Protocols*, Methods in Molecular Biology, vol. 726,
DOI 10.1007/978-1-61779-052-2_14, © Springer Science+Business Media, LLC 2011

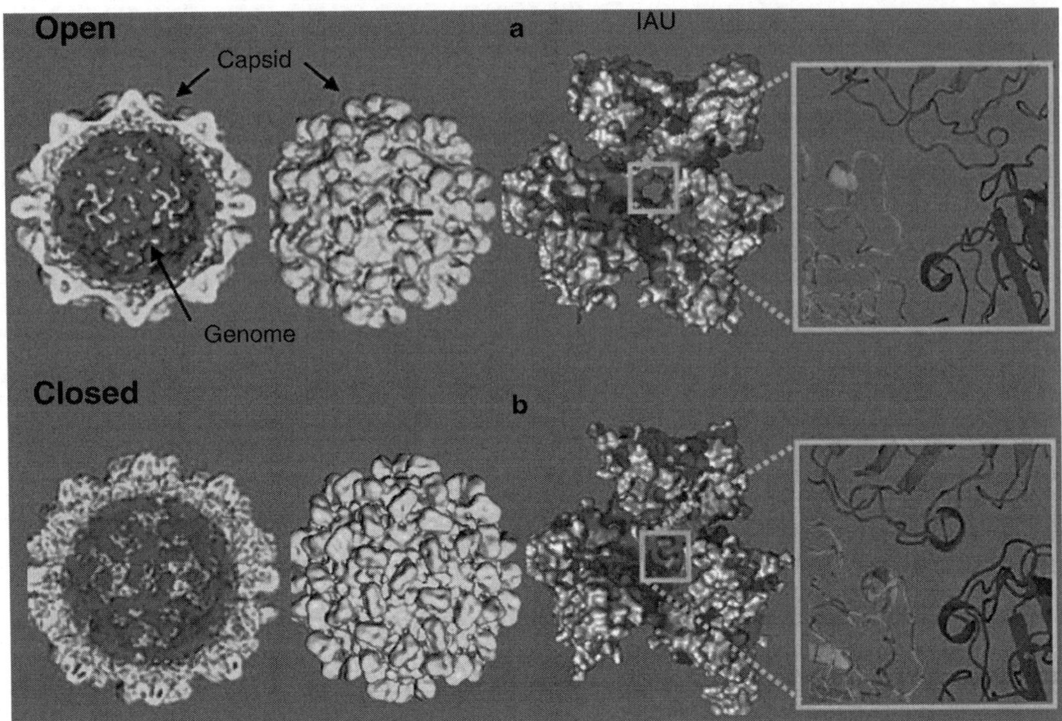

Fig. 1. Cryoelectron microscopy image of RCNMV and the quasi-threefold axis of the IAU with and without divalent cations. The capsid shows the icosahedral symmetry. *Open*: RCNMV free of Ca^{2+} and Mg^{2+} (**a**) and *closed*: RCNMV in the presence of Ca^{2+} and Mg^{2+} (**b**).

value is the triangulation number (T) and refers to the number of CPs in the IAU. The size of the virus can also be related to the T number. A $T=1$ virus has a single protein as the subunit for viral assembly and is the smallest virus. Depending on the size of the capsid subunit, $T=1$ viruses can be as small as 20 nm in diameter. The most common structure in the field of plant virus nanoparticles (NPs) is the $T=3$. *Brome mosaic virus* (BMV), *Cowpea chlorotic mottle virus* (CCMV), *Cowpea mosaic virus* (CPMV), *Hibiscus chlorotic ringspot virus* (HCRSV), *Tomato bushy stunt virus* (TBSV), and RCNMV are all $T=3$ viruses which have diameters in the range from 29 to 37 nm (2–5).

RCNMV, like other plant viruses, does not use an endosomal mechanism for entry into its host. Plant viruses must use a vector, such as an insect, fungus, or animal, to create a mechanical opening in the cell for virus entry. Once the virus has entered the cell, the virus disassembles and delivers its genome. The mechanisms of virus assembly and disassembly comprise an entire field of research. It is important to realize that capsid assembly and disassembly must be energetically favorable in the same cell. In addition, viral proteins and nucleic acids are able to assemble into virions containing only viral proteins and/or nucleic acids. Investigation into these mechanisms has elucidated some features of RCNMV that are important in PVN synthesis.

Investigation into the possible disassembly mechanism (also known as uncoating) of RCNMV, using cryoelectron microscopy, revealed conformational changes that are sensitive to the presence of Ca^{2+} and Mg^{2+} (1). Figure 1a shows that when Ca^{2+} and Mg^{2+} are removed, the CPs rotate and move away from the center of the quasi-threefold axis (1). This conformational transition leads to the opening of a 10–13 Å channel at the center of each trimer axis. This indicates that when RCNMV is in a Ca^{2+} and Mg^{2+} rich environment (e.g., soil), it will be in a closed conformation (Fig. 1b). When RCNMV is in a cytosolic environment, where the concentration of Ca^{2+} is ~100 nM, the capsid will be in an open conformation.

Further work on understanding how RCNMV functions has led to the identification of genomic sequences/structures that are important for viral replication and genome packaging. In particular the transactivator (TA), located on RNA-2, is a hairpin structure that hybridizes to the transactivator binding sequence (TABS) on RNA-1. The TA is necessary for CP expression and is hypothesized to be the origin of assembly (OAS). The OAS is thought to act as a nucleation site for capsid assembly.

It is most convenient to formulate strategies and methods, when engineering a plant virus nanoparticle that exploit the intrinsic properties of the plant virus. Figure 2 shows an overview

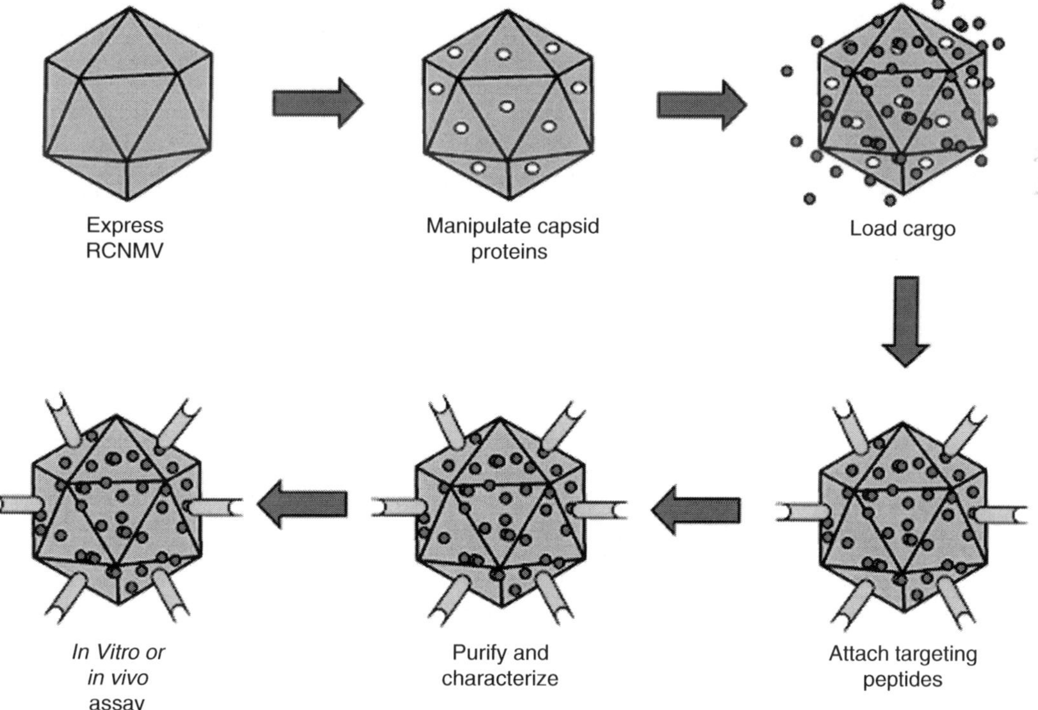

Fig. 2. Representation of the general protocol for synthesis of a PVN. The sequence involves isolation of purified plant viruses followed by capsid protein manipulation. Then a therapeutic or other molecule of interest is infused. Targeting peptides are added to the surface using bioconjugate chemistry methods. This is followed by purification and characterization. Finally the PVN is tested in vitro or in vivo.

of the procedures used to synthesize PVNs for testing both in vitro and in vivo. The first step is expression and purification of RCNMV from a host. Once we have our plant virus we manipulate the capsid proteins in preparation for cargo loading. Cargo loading can be done by either encapsidation or infusion. For encapsidation, the virus is completely disassembled and then reassembled around the cargo. Infusion exploits the swelling properties of RCNMV and is the method used to infuse cargo (e.g., doxorubicin) into the virus. Next, the virus is functionalized with the desired targeting moieties; we use small peptides. Once the nanoparticle is synthesized it is very important to purify and properly characterize the product. Finally, the PVN is tested in vitro and then in vivo. Herein, we will discuss these procedures.

2. Materials

2.1. Purification of Red Clover Necrotic Mosaic Virus

1. 10 mM sodium phosphate buffer, pH = 7.0.
2. 200 mM sodium acetate buffer, pH = 5.3.
3. Carborundum.
4. Cheesecloth.
5. DNA plasmids: RC-169 and RC-2.
6. Homogenizing buffer: 200 mM sodium acetate pH = 5.3, 1,000-fold dilution of β-mercaptoethanol.
7. High capacity ultracentrifuge.
8. *Nicotiana clevelandii* plants.
9. Miracloth.
10. Mortar and pestle.
11. Polyethylene glycol (PEG) 8000.
12. Sucrose.
13. MEGAshortscript™ (Applied Biosciences).
14. Viral RNA transcripts: RNA-1 and RNA-2.
15. Deionized water.

2.2. Encapsidation

1. Appropriate NPs less than 15 nm in diameter. This can include Au, CdSe, Fe_3O_4, etc. We will use Au NPs coated with bis-sulfonatophenyl phenylphosphine (BSPP) as an example.
2. An oligonucleotide with an appropriate chemical group for attachment to the NP (e.g., alkane thiolate linker for Au).
3. RCNMV, prepared as discussed in Subheading 3.1.
4. RNA-1 transcript (1,539 bp).
5. Micro Bio-Spin P-30 columns (Biorad).
6. 100 mM dithiothreitol (DTT).

7. 10 mM sodium phosphate buffer, pH = 7.0.

8. 0.1 M NaCl, 10 mM sodium phosphate, pH = 7.0.

9. 50 mM Tris–HCl buffer, pH = 5.5.

10. 200 mM sodium ethylenediaminetetraacetic acid (EDTA), pH = 10.

11. Sephadex G75 columns.

12. Slide-A-Lyzer dialysis cassette (10 kDa molecular weight cut-off (MWCO)).

13. Sucrose cushion.

14. 100 mM glycine–NaOH, pH = 10.

2.3. Infusion

1. 1× Dulbecco's phosphate buffered saline (1× DPBS).

2. 6,000–8,000 MWCO dialysis tubing. A higher MWCO (100 kDa) may be substituted to remove possible proteolytic enzymes or degraded coat proteins that are present from the purification process.

3. Conjugation buffer 1: 50 mM sodium phosphate buffer, pH = 7.25.

4. Conjugation buffer 2: 50 mM 4-(2-hydroxyethyl)-1-piperazineethanesulfonic acid (HEPES), 50 mM NaCl, pH = 7.25.

5. Dimethyl sulfoxide (DMSO, cell grade).

6. Infusion buffer: 50 mM Tris base, 50 mM sodium ethylenediaminetetraacetic acid (EDTA), 50 mM sodium acetate, pH = 7.0.

7. Infusion dye.

8. NAP 25 column.

9. RCNMV, prepared as discussed in Subheading 3.1.

10. Sephadex G25 column (30 cm L × 1.1 cm I.D.). The sephadex G25 may be replaced by a higher molecular weight exclusion resin (sephadex G200 is very common) if it provides better resolution between the eluted virus peak and infusion dye.

2.4. Decoration of the PVN Surface with Targeting Peptides: Conjugation Protocols

1. 1× Dulbecco's phosphate buffered saline.

2. Dimethyl sulfoxide (DMSO, cell grade).

3. NAP 25 column.

4. Peptides.

5. Sephadex G25 column (30 cm × 1.1 cm I.D.). A higher molecular weight exclusion resin may be used if it provides better resolution of the virus and excess peptides.

6. Sulfosuccinimidyl 4-[*N*-maleimidomethyl]cyclohexane-1-carboxylate (Sulfo-SMCC). There are derivatives containing polyethylene glycol spacer arms that can be substituted.

7. 50 mM sodium phosphate, 50 mM NaCl, pH = 7.0.

3. Methods

3.1. Purification of Red Clover Necrotic Mosaic Virus

Plant virus nanoparticle production begins with the simple mechanical inoculation of the virus or a clone (or transcript) of the virus to the leaf of a susceptible plant. The virus infection spreads from the inoculated cells through the plasmodesmata (cell-to-cell junctions) and the vascular system of the plant, thus resulting in a large percentage of the cells comprising the plant becoming infected. After 1 week, the plant reaches its maximum capacity for production of virus and can be harvested. The biomass of the virus can be as high as 1% of the plant tissue. Plant virus yields vary depending on the species of plant virus and the host combination, but on the high end can yield 2–10 mg of virus per gram of wet weight tissue. Plant bioreactors represent well developed and robust technology that can be employed for the production of large scale quantities of the virus NPs under pharmaceutical and good manufacturing practice (GMP) conditions. Plant bioreactors as compared to cell bioreactors offer both distinct advantages and drawbacks. The high yields in plants are not always reproducible in cells in a bioreactor. However, plant tissue presents greater problems for purification.

1. Transcribe the genome of RCNMV from DNA templates RC-169 and RC2 using MEGAshortscript™ (T7-polymerase kit).

2. Mix 1 μL of RNA-1 and RNA-2 transcripts together and dilute to 110 μL with 10 mM phosphate buffer, pH = 7.0.

3. Gently rub carborundum, an abrasive powder, on the leaves of *Nicotiana clevelandii* plants and then wash using dH_2O.

4. Inoculate the leaves with 27 μL RNA transcript mixture by supporting the underside of the leaf with one hand and gently rubbing the drops of inoculum over the surface of the leaf with the other. Wash away the inoculum with dH_2O and allow the plants to incubate at 18–26°C for 7–10 days.

5. Combine three infected leaves with 5 mL dH_2O and grind with a mortar and pestle.

6. For a large preparation of RCNMV, rub this pulverized plant material on carborundum treated plants using a sponge. Gently rinse off inoculum and carborundum with a water hose.

7. Incubate these plants in a greenhouse 7–10 days. Symptoms should appear in 4–5 days as little ring spots.

8. Harvest the whole plants by cutting the main stems and weigh the tissue.

9. Combine plant tissue with homogenizing buffer (one part tissue: two parts buffer) and homogenize using a blender

(blender capacity: 350 g tissue, buffer capacity: 700 mL). Blend for 30 s (three times) at low speed.

10. Gradually pour slurry through four layers of cheesecloth into a beaker to remove plant debris.

11. Twist cheesecloth into a ball and squeeze through remaining liquid.

12. Incubate on ice for 10 min with stirring.

13. Pour above slurry into 250-mL centrifuge bottles and spin at $6,600 \times g$ for 25 min.

14. Pour supernatant through a sheet of miracloth into a graduated cylinder.

15. Measure the filtrate volume and pour it into a beaker.

16. Add ¼ the volume of 40% PEG 8000/NaCl to clarified filtrate.

17. Incubate filtrate on ice for 1 h with stirring at low speed.

18. Pour precipitated samples into centrifuge bottles and centrifuge at $6,600 \times g$ for 20 min at 4°C.

19. Discard supernatant to waste container and keep pellet by draining inverted tubes on a paper towel.

20. Resuspend each pellet in centrifuge tube with 30 mL 200 mM sodium acetate buffer, pH = 5.3, and transfer to 30-mL centrifuge tubes. Centrifuge at $7,800 \times g$ for 20 min at 4°C.

21. Save supernatant at 4°C for later use or continue purification.

22. Add 3 mL 20 % sucrose to supernatant.

23. Ultracentrifuge above supernatant at $163,000 \times g$ for 2 h at 5°C.

24. Discard supernatant.

25. Add 500 mL 200 mM sodium acetate buffer, pH = 5.3 to surface of pellet and incubate at 4°C overnight.

26. Collect suspension and portion to a microcentrifuge tube. Centrifuge at $9,800 \times g$ for 5 min, three times, each time collecting supernatant and moving it to a fresh microcentrifuge tube.

27. Determine A_{260} and A_{280}. Calculate the concentration of RCNMV using $\varepsilon_{260} = 6.46$ mL/mg/cm, and assess purity compared to a standard value of $A_{260}/A_{280} = 1.69$.

28. Viruses can be stored at 4°C or frozen.

3.2. Encapsidation

Encapsidation refers to inclusion inside the virus capsid protein shell using the viral RNA OAS as scaffolding. If the RNA is not present, the protein shell can still be used to encapsulate NPs by various routes (6–8) with a strong contribution due to electrostatic interactions (9, 10). Several groups have explored the use of plant virus capsid proteins to encapsidate gold NPs up to 15 nm in diameter (11, 12), collections of smaller particles including quantum dots (13), or metal oxide particles (9, 14).

The encapsidation protocol consists of packaging NPs and proteins with radii ranging from 4 to 15 nm. This protocol is demonstrated using a particle bioconjugated with a short hairpin, DNA-2, a thiolated DNA analog of the RCNMV OAS (Fig. 3a) (9, 12, 15). An RNA-1 fragment, containing the TABS, is hybridized with the OAS (Fig. 3b) to facilitate in vitro self-assembly of the virus (Fig. 3c). Based on these studies, it is evident that many plant viruses have a cargo capacity equivalent to a sphere of 10 nm or greater.

1. Synthesize 5′-thiol deoxyuridine modified DNA oligonucleotides, DNA-2, with the sequence: 5′-SH-AGAGGUAUCG CCCCGCCUCU-3′. In order to deprotect the thiol group, add 1 mL 100 mM DTT to DNA-2 and allow to react for 30 min at room temperature.

2. Remove excess DTT using a Micro Bio-Spin P-30 column.

3. Synthesize Au NPs coated with bissulfonatophenyl phenylphosphine (BSPP).

4. Incubate a 1:300 mole ratio of Au NPs with DNA-2 at 37°C for 8 h in 100 μL 10 mM phosphate buffer, pH = 7. Dilute to 500 μL with 0.1 M NaCl, 10 mM phosphate buffer, pH = 7 and incubate for an additional 40 h. Unattached DNA should be removed by centrifugation at 14,000 rpm in a SS-34 rotor (23,000 × g) for 25 min. Remove the supernatant, which consists of the unreacted DNA. Resuspend the precipitate in 500 μL of 10 mM phosphate buffer, pH = 7 and recentrifuge. This step should be repeated twice. Finally, suspend the DNA/Au conjugates in 10 mM phosphate buffer, pH = 7.

5. Obtain full-length RCNMV RNA-1 used for encapsidation by transcription from a Sma1 linearized plasmid vector in

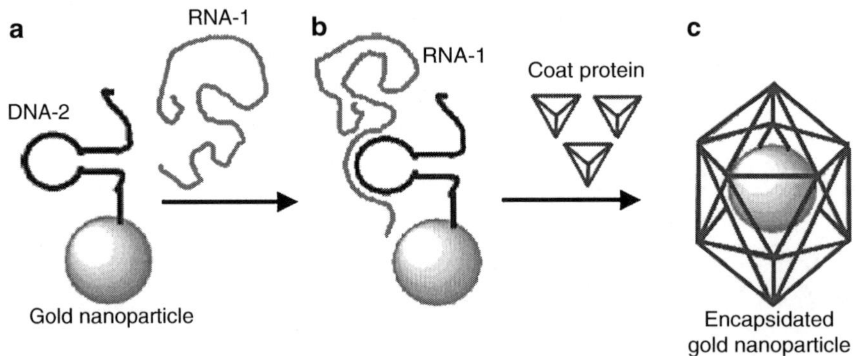

Fig. 3. Schematic representation of encapsidation strategy using DNA-2, an analog of the TA of RNA-2, as the origin of assembly, combined with synthetic RNA-1 to capture CPs.

which the bacteriophage T7 RNA polymerase promoter was used to transcribe full-length RCNMV cDNA clones.

6. Prepare RCNMV CPs by suspending RCNMV at a final concentration of 0.5 mg/mL in 200 mM EDTA, pH = 10.0 at room temperature. Pellet aggregates at $10,000 \times g$ for 10 min and collect supernatant. EDTA causes the opening of the 60 pores in the capsid.

7. Separate the CP from the genome by size-exclusion chromatography (Sephadex G 75). The concentration of CP can be estimated using a Bradford assay.

8. Dissolve 20 μL of RCNMV CP (0.5 mg/mL) in a total volume of 100 μL with glycine–NaOH, pH 10 at room temperature. Use of the glycine buffer minimizes capsid protein aggregation.

9. Dialyze using Slide-A-Lyzer Dialysis Cassette (10 kDa molecular weight cutoff) against 50 mM Tris buffer at three different pH's (5.5, 6.0, and 6.5) overnight at room temperature. A sample at each pH should be analyzed by dynamic light scattering (DLS) to observe appropriate PVN diameter of ~34 nm.

10. Add DNA/nanoparticle conjugates to 1 μL T7 RCNMV RNA-1 transcripts (4 mg/mL). Incubate this mixture for 10 min, and then add 5 μL purified RCNMV CP (10 mg/mL) (see steps 6–9).

11. Carry out the encapsidation reaction by dialysis of the sample against 50 mM Tris–HCl, pH = 5.5 overnight at room temperature, using a Slide-A-Lyzer Dialysis Cassette (10 kDa MWCO). Separate unencapsidated virus by ultra centrifugation through a sucrose cushion at $218,000 \times g$ for 20 min using a SW-55 rotor.

3.3. Infusion

The swelling and pore opening of the PVN can be used to incorporate small molecules, peptides, or oligonucleotides. The infusion process exploits a natural mechanism employed by the virus to release its genome upon entry into a newly infected cell. First, the virus is opened using ethylenediamine tetraacetic acid (EDTA) which removes Ca^{2+} and Mg^{2+} from the solution. Second, molecules are infused into the opened virions. Initially, this infusion protocol was demonstrated for rhodamine (positive charge), luminarosine (neutral) fluorescein (negative charge), and doxorubicin (positive charge) (15). The PVN was exposed to a 1,000-fold excess of the molecule. After an incubation period, the PVNs were closed by addition of Ca^{2+} and Mg^{2+}. Subsequent experience has shown that 900–1,000 doxorubicin molecules can reproducibly be infused into the RCNMV capsid by these methods.

1. Prepare dialysis tubing by soaking for 15 min and washing thoroughly with dH_2O, inside and out, to remove sodium azide preservative.

2. Dilute RCNMV stock sample to 5 mg/mL with distilled water. Use $\varepsilon_{260} = 6.46$ mL/mg/cm when determining concentration.

3. Place the desired amount of RCNMV in dialysis tubing and seal using clips. Allow to dialyze over night at 4°C.

4. Dissolve dye/chemotherapeutic in 100% DMSO to a concentration ≥ 5mM.

5. Remove RCNMV from dialysis tubing and aliquot 1 mL volumes into 1.5-mL microcentrifuge tubes. Add dye/chemotherapeutic to the RCNMV aliquots at a dye:virus mole ratio of 1,000:1 (see Note 1). The infusion buffer should be at a final concentration of ≤10% DMSO, 50 mM Tris base, 50 mM sodium acetate, 50 mM EDTA, pH 7.1. A higher concentration of DMSO may be used if it is shown that the virus is stable.

6. Wrap the dye/chemotherapeutic with virus in aluminum foil and incubate at room temperature for a minimum of 4 h and maximum of 24 h.

7. Remove excess dye/chemotherapeutic from the suspension/ solution using a NAP 25 column preequilibrated with buffer of choice (see Note 2). If proceeding to conjugation of peptides, equilibrate the NAP 25 column with 30 mL of Conjugation buffer 1 or 2 (see Note 3).

8. Trace Ca^{2+} in buffers used for gel filtration is sufficient to close the virus.

9. Pellet aggregates at $10,000 \times g$ for 10 min at room temperature.

3.4. Decoration of the PVN Surface with Targeting Peptides: Conjugation Protocols

The virus capsid provides a highly organized array of amino acids on the icosahedral structure of the capsid. Common methods of achieving structural modifications in PVN synthesis use a combination of bioconjugation techniques and/or mutagenesis.

Bioconjugation strategies are used for attaching fluorophores, biotin, folic acid, chemotherapeutic drugs, polyethylene glycol (PEG), antibodies, and peptides. The most useful amino acid residues for chemical modification are lysines and cysteines, and to a lesser extent aspartic and glutamic acids (16). Lysines are abundant in proteins and will react with N-hydroxysuccinimidyl esters (NHS-esters). Glutamic and aspartic acids are also abundant and can be modified using 1-ethyl-3-(3-dimethylaminopropyl) carbodiimide (EDC) to form a reactive species that will undergo an amidation reaction with a primary amine (17). Reaction of aspartic and glutamic acids may decrease the colloidal stability of the viral suspension and is an uncommon practice in viral nanotechnology. Cysteines are the least abundant amino acid on the surface, but provide a point for selective conjugation using maleimides. In our lab, we use succinimidyl-4-(*N*-maleimidomethyl) cyclohexane-1-carboxylate (SMCC) for the orthogonal conjugation of cysteine terminated peptides to the surface lysines of the RCNMV (Fig. 4).

1. Prepare RCNMV (1–5 mg/mL) in 50 mM sodium phosphate, 50 mM sodium chloride buffer, pH = 7.0. The phosphate buffer may be substituted with HEPES or MES, if necessary.

2. Portion RCNMV into 1 mL aliquots.

3. Dissolve 1–2 mg of sulfo-SMCC in 100 μL 100% DMSO.

4. Slowly add the sulfo-SMCC solution to the 1 mL virus aliquot, vortex immediately.

5. Check pH. The pH should be 7.2 ± 0.1. Ideally, the ionic strength of the buffer should be strong enough such that the pH does not need to be adjusted (see Note 4).

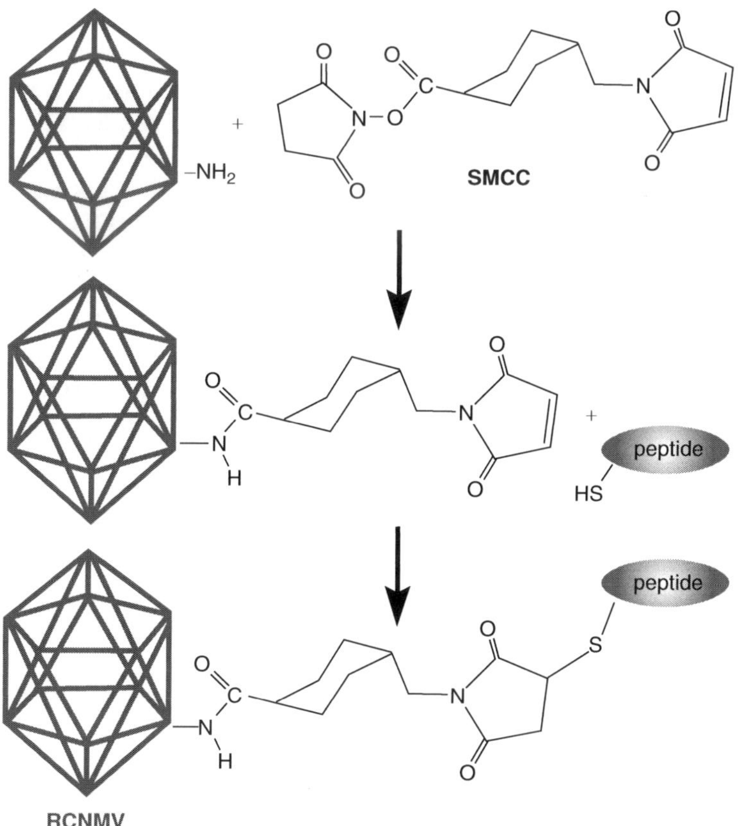

Fig. 4. Bioconjugate coupling by SMCC is shown schematically. In this scheme a surface lysine on the CP is conjugated to a terminal cysteine of a targeting peptide.

6. Allow the reaction to incubate at room temperature on a rocker plate for 30–45 min. This is enough time to functionalize the surface of RCNMV with maleimides.

7. Remove excess sulfo-SMCC using a NAP-25 column pre-equilibrated with 50 mM sodium phosphate buffer, 50 mM sodium chloride, pH = 7.0 (or HEPES).

8. Dissolve peptides at a concentration of 1–2 mg/mL in ~50 μL 100% DMSO. Make sure the solution is clear, indicating the peptide has fully dissolved.

9. Slowly add the peptide solution to the maleimide activated virus and check the pH. The pH should be between 6.8 and 7.3.

10. Incubate for at least 6 h to overnight at room temperature. If stored at lower temperature (i.e., 4°C), the reaction must be allowed to continue for 24 h.

11. Pellet aggregates at 10,000×g for 10 min at room temperature.

12. Remove excess peptide using size-exclusion chromatography (i.e., 30 × 1.1 cm sephadex G 25, 50, 75, or 200 column) (see Note 5).

13. Concentrate samples using centrifugal filters (100 kDa MWCO is best) at ~7,000×g, if necessary. Do not use polystyrene filters as this can cause severe aggregation.

3.5. Characterization

One major objective of PVN synthesis is to specifically deliver a chemotherapeutic to cancerous tissue. However, success depends upon accurate understanding of the mechanism in which a PVN targets and enters a cell. Any testable hypothesis that is made of the mechanism is highly dependant on purity and careful characterization of the nanomaterial. For the most part, purification of PVNs is predominantly done by size-exclusion chromatography, but anion exchange can be used as well. Density gradient centrifugation, sucrose or iodixanol, can also be used to determine whether infusion is successful. DLS and transmission electron microscopy (TEM) can be used to verify if the synthesized particle is intact and has the correct dimensions.

It is important to realize that any manipulation of the PVN will result in some level of aggregation. The amount of aggregation varies depending on the procedure performed and is not always noticeable by eye. Aggregation that is not noticeable by eye can be detected using DLS. DLS is best measured at a concentration of 0.5 mg/mL. Removal of these aggregates can be done by low speed centrifugation (~10,000×g for 10 min).

For TEM analysis, we use either carbon type A or B copper grids from Ted Pella with a 2% uranyl acetate stain. It is important to not overload the grid. We have found that a 20 μL drop of

10–20 μg/mL sample, placed on a grid for 30 s, is sufficient for particle detection. The excess drop can be wicked away using a paper towel. The virus that remains can then be stained with 20 μL of 2% uranyl acetate for 30 s and the excess solution can be wicked away in the same manner. Allow the virus sample to dry before placing it into the grid holder.

3.6. In Vitro Cell Studies of PVN Delivery

The PVN platform can be tested for efficacy using in vitro cell culture. The cells of interest are grown to 80% confluence in appropriate plates (12- or 96-well depending on the signal-to-noise ratio needed in a plate reader measurement). Subsequently, the PVN formulation is introduced to the plated cells in serial dilutions. The PVN formulation at the end of the purification process is suspended in buffer (1× DPBS, 10 mM HEPES pH = 7.1, or 10 mM phosphate buffer pH = 7.1), and not in growth media. For delivery, the sample must be filter sterilized. We use 0.2-μm syringe filters. It is acceptable to add the formulation directly to the well; however this will dilute the growth media. Control lanes for mock deliveries should be included to test for alteration of cell survival. To minimize buffer effects, one should avoid dilutions greater than ~10% of the volume.

The efficacy of the delivery can be determined using a survivability assay. There are several standard kits that can be used to determine the percent survival in a 12- or 96-well plate format. The adenosine triphosphate (ATP) and 3-(4,5-dimethylthiazol-2-yl)-2,5-diphenyltetrazolium bromide (MTT) assays are most widely used. The ATP assay monitors the level of ATP, which is a measure of cell viability. The assay is based on the ATP-dependent reaction of luciferin with oxygen catalyzed by the enzyme luciferase. The chemiluminescent signal from the reaction is proportional to the number of living cells. This value is determined at an initial time and then the value is used as a reference for subsequent measurements as a function of time. The MTT assay is a colorimetric test that measures the enzymatic activity of the cell. This test also relies on the cell density as a reporter of the cell viability. These assays can be used to determine whether chemotherapeutic agents such as a doxorubicin have been delivered to cells.

3.7. Routes of Administration of PVNs

Animal trials testing the immunogenicity and clearance rates of PVNs use a parenteral route of administration. Injection into the saphenous vein is the common route in a murine model. Rapid clearance is often observed in studies of PVNs and other NPs. As mentioned above, surface attached PEG provides a method to prolong circulation, presumably by avoiding uptake by the reticuloendothelial system (RES). This has been demonstrated in diverse nanoparticle platforms including liposomes, Au NPs, polymers, and PVNs.

4. Notes

1. To increase loading, a higher dye:virus mole ratio can be used. However, the stability of the virus can be compromised by overloading.

2. The choice of buffer greatly depends on the solubility of the dye/chemotherapeutic agent. Otherwise, the conjugation chemistry works best in a phosphate buffer.

3. If one wishes to calculate quantum yields or use sample in a drug delivery experiment, then extra purification methods must be used to remove the excess cargo. Fluorophores such as rhodamine or doxorubicin self-quench when infused.

4. If the pH exceeds 8 during this step, the sample should be discarded. RCNMV is not stable at pH > 8.0 and hydrolysis of the NHS-ester is faster at higher pH.

5. Peptide conjugation can be validated using several methods. The most accurate method for quantization and verification of peptide conjugation is mass spectrometry. Gel electrophoresis can be used to verify peptide conjugation and spectroscopy can be used to quantify the number of peptides attached if they have a spectroscopic handle (fluorophore or chromophore). However, the extinction coefficient of the spectroscopic handle can change significantly when attached to a peptide on the surface of a virus. Moreover, the folding of peptides can be disrupted when attached to the surface of a virus. Validation of peptide conjugation by these methods does not necessarily indicate that a peptide will retain its targeting capabilities after conjugation. Molecular dynamics simulations may give insight into the nature of peptide folding on virus surfaces. However, retention of targeting function can only be validated in a cell based assay.

Acknowledgments

We thank NanoVector, Inc. for funding that supported D. L.

References

1. Sherman, M. B., Guenther, R. H., Tama, F., Sit, T. L., Brooks, C. L., Mikhailov, A. M., et al. (2006) Removal of divalent cations induces structural transitions in Red clover necrotic mosaic virus, revealing a potential mechanism for RNA release. *J. Virol.* **80**, 10395–10406.

2. Speir, J. A., Munshi, S., Wang, G., Baker, T. S., and Johnson, J. E. (1995) Structures of the native and swollen forms of cowpea chlorotic mottle virus determined by X-ray crystallography and cryoelectron microscopy. *Structure* **3**, 63–78.

3. Lucas, R. W., Larson, S. B., and McPherson, A. (2002) The crystallographic structure of brome mosaic virus. *J. Mol. Biol.* **317**, 95–108.

4. Lin, T. W., Chen, Z., Usha, R., Stauffacher, C. V., Dai, J. B., Schmidt, T., et al. (1999) The refined crystal structure of cowpea mosaic virus at 2.8 angstrom resolution. *Virology* **265**, 20–34.

5. Hopper, P., Harrison, S. C., and Sauer. R. T. (1984) Structure of tomato bushy stunt virus. V. Coat protein sequence determination and its structural implications. *J. Mol. Biol.* **177**, 701–713.

6. Aniagyei, S. E., DuFort, C., Kao, C. C., and Dragnea, B. (2008) Self-assembly approaches to nanomaterial encapsulation in viral protein cages. *J. Mater. Chem.* **18**, 3763–3774.

7. Douglas, T. and Young, M. (1998) Host-guest encapsulation of materials by assembled virus protein cages. *Nature* **393**, 152–155.

8. Dragnea, B., Chen, C., Kwak, E. -S., Stein, B., and Kao, C. C. (2003) Gold nanoparticles as spectroscopic enhancers for *in vitro* studies on single viruses. *J. Am. Chem. Soc.* **125**, 6374–6375.

9. Loo, L., Guenther, R. H., Lommel, S. A., and Franzen, S. (2007) Encapsidation of nanoparticles by red clover necrotic mosaic virus. *J. Am. Chem. Soc.* **129**, 11111–11117.

10. Ren, Y. P., Wong, S. M., and Lim, L.Y. (2006) *In vitro*-reassembled plant virus-like particles

for loading of polyacids. *J. Gen. Virol.* **87**, 2749–2754.

11. Chen, C., Kwak, E. S., Stein, B., Kao, C. C., and Dragnea, B. (2005) Packaging of gold particles in viral capsids. *J. Nanosci. Nanotechnol.* **5**, 2029–2033.

12. Loo, L., Guenther, R. H., Basnayake, V. R., Lommel, S. A., and Franzen, S. (2006) Controlled encapsidation of gold nanoparticles by a viral protein shell. *J. Am. Chem. Soc.* **128**, 4502–4503.

13. Dixit, S. K., Goicochea, N. L., Daniel, M. -C., Murali, A., Bronstein, L., De, M., et al. (2006) Quantum dot encapsulation in viral capsids. *Nano Lett.* **6**, 1993–1999.

14. Shtykova, E.V., Huang, X., Gao, X., Dyke, J. C., Schmucker, A. L., Dragnea, B., et al. (2008) Hydrophilic monodisperse magnetic nanoparticles protected by an amphiphilic alternating copolymer. *J. Phys. Chem. C* **112**, 16809–16817.

15. Loo, L., Guenther, R. H., Lommel, S. A., and Franzen, S. (2008) Infusion of dye molecules into Red clover necrotic mosaic virus. *Chem. Commun. (Camb)* (1), 88–90.

16. Strable, E. and Finn, M. G. (2009) Chemical Modification of Viruses and Virus-Like Particles. In: *Viruses and Nanotechnology* (Manchester, M. and Steinmetz, N. F., eds.) pp. 1–21, Springer, Berlin, Heidelberg.

17. Hermanson, G. T. (2008) *Bioconjugate Techniques*. Academic Press, San Diego.

Chapter 15

Applications of Carbon Nanotubes in Biomedical Studies

Hongwei Liao, Bhavna Paratala, Balaji Sitharaman, and Yuhuang Wang

Abstract

Carbon nanotubes (CNTs) are novel, one-dimensional nanomaterials with many unique physical and chemical properties that have been increasingly explored for biological and biomedical applications. In this chapter, we briefly summarize the intrinsic properties of single-walled carbon nanotubes (SWNTs), a special class of CNTs, and their corresponding applications in these fields. SWNTs have been utilized for the ultrasensitive detection of biological species, providing a label-free approach. SWNT-Raman tags have achieved detection sensitivity down to 1 fmol/L. SWNT-based drug delivery systems have shown promising potential based on preliminary in vitro and in vivo studies. Also, the remarkable optical properties of SWNTs have made them promising candidates as contrast agents for imaging in cells and animals. Moreover, due to their excellent mechanical strength, SWNTs have been used to improve the mechanical properties of solid polymeric nanocomposites and porous scaffolds. Sample preparation procedures for the use of SWNTs as fluorescent imaging labels and in biological composites will be discussed.

Key words: Single-walled carbon nanotubes, Surface functionalization, Biomedical applications, Drug delivery, Biomedical imaging, Nanocomposite

1. Introduction

Carbon nanotubes (CNTs) are seamless cylinders of graphene sheets, exhibiting a wide variety of remarkable chemical and physical properties, which have drawn tremendous interest in the past decade (1, 2). Depending on the number of graphene layers, CNTs are classified as single-walled carbon nanotubes (SWNTs) or multiwalled carbon nanotubes (MWNTs). Applications of CNTs span many fields; they can be used as nanoelectronics (2), field-effect emitters (3), and composite materials (4), for example. In recent years, due to their interesting size, shape, structure, and their unique physical properties, efforts have been devoted to

Sarah J. Hurst (ed.), *Biomedical Nanotechnology: Methods and Protocols*, Methods in Molecular Biology, vol. 726,
DOI 10.1007/978-1-61779-052-2_15, © Springer Science+Business Media, LLC 2011

exploring the potential biological applications of CNTs (5–9). In this chapter, we will give a brief review of the applications of SWNTs in biomedical studies, followed by protocols describing the use of SWNTs in fluorescent imaging and biocomposites as specific examples.

1.1. What Properties Make SWNTs Useful in Biological Applications?

SWNTs are one-dimensional (1D) hollow nanomaterials with diameters of 0.4–4 nm and lengths ranging from 50 nm up to 1 cm. With all atoms on the surface, SWNTs have ultrahigh surface area (1,300 m^2/g), permitting the loading of multiple molecules onto the nanotube sidewall. These functionalized nanotubes can bend to interact with one cell at multiple binding sites, resulting in improved binding affinity.

Due to quantum confinement of the electronic density of states (DOS) along the circumference, SWNTs exhibit sharp electronic DOS at the van Hove singularities, giving rise to fascinating optical properties uniquely associated with each SWNT structure (see Fig. 1) (10). Depending on the structure, SWNT can be metallic or semiconducting. Semiconducting SWNTs exhibit strong optical absorption and photoluminescence in the NIR range with emission between 800 and 1,600 nm (11). Since biological tissues are transparent within part of this range, SWNTs are therefore suitable for use in biological imaging. SWNTs also have several distinctive Raman scattering features including the radial breathing mode (RBM) and tangential mode (G-band) (15), which are sharp, strong peaks that can be easily distinguished from fluorescence background and are thus suitable for Raman detection/imaging (16). Other examples of SWNT applications include photothermal therapy (12, 13) and photoacoustic imaging (14).

SWNTs due to their excellent mechanical strength (Young's modulus ~1 TPa) are also being investigated for use as reinforcing agents, to enhance the mechanical properties of solid polymeric nanocomposites and porous scaffolds for biomedical applications. The properties of SWNT and the corresponding biomedical applications are schematically shown in Fig. 2.

1.2. Examples of Biological Applications

Motivated by various chemical and physical properties of SWNTs, many efforts have been devoted to apply SWNTs in biomedical applications. SWNT-based sensors have been developed to detect biological species including proteins and DNA (6, 17, 18). SWNTs can be utilized as optical tags or contrast agents in biological imaging techniques (9, 14, 16, 19). Properly functionalized SWNTs are able to deliver biological or molecular cargo into cells (7, 8, 20–22). Recently, SWNTs have shown promise for in vivo cancer treatment in a mouse model (23). In this section, examples of applications of SWNTs in biomedical studies are briefly reviewed.

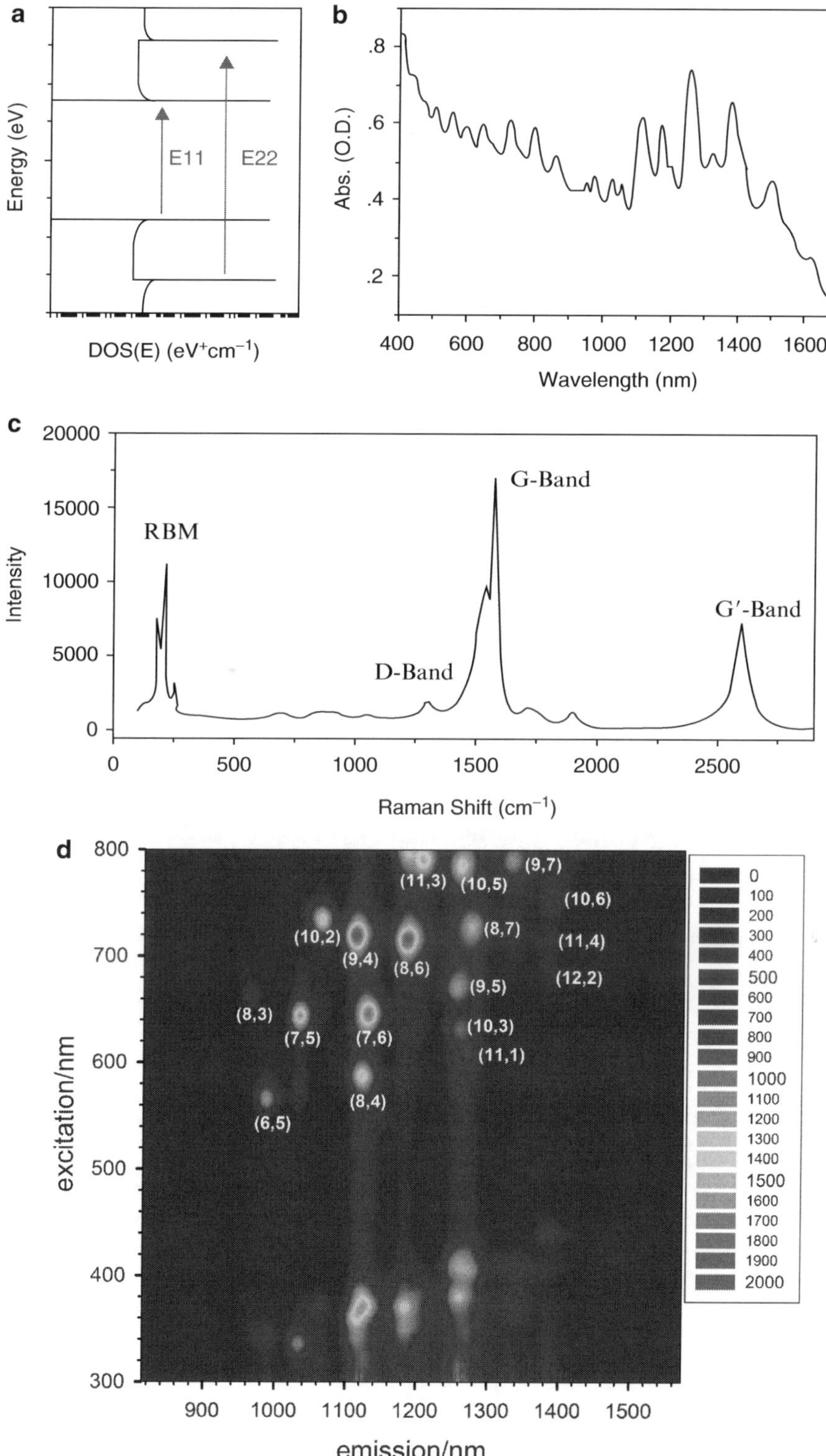

Fig. 1. Optical properties of SWNTs. (**a**) Schematic density of electronic states of a single SWNT structure. The sharp features of the DOS are attributed to van Hove singularities. $E11$ and $E22$ are optical transitions correspond to photon absorption in the NIR and visible (vis) ranges, respectively. (**b**) Absorption spectrum of an aqueous solution of SWNTs. Peaks in the spectrum are due to SWNTs with different structures. (**c**) Raman spectrum of SWNTs. The peaks at 200–300, ~1,340, ~1,590, and ~2,700 cm^{-1} are the radial breathing modes (RBM), D-band mode, G-band mode, and G′-band mode, respectively. (**d**) Contour plot of fluorescence intensity versus excitation and emission wavelengths for a sample of semiconducting HiPCo SWNTs. SWNTs with different structures emit at different wavelengths under different excitations.

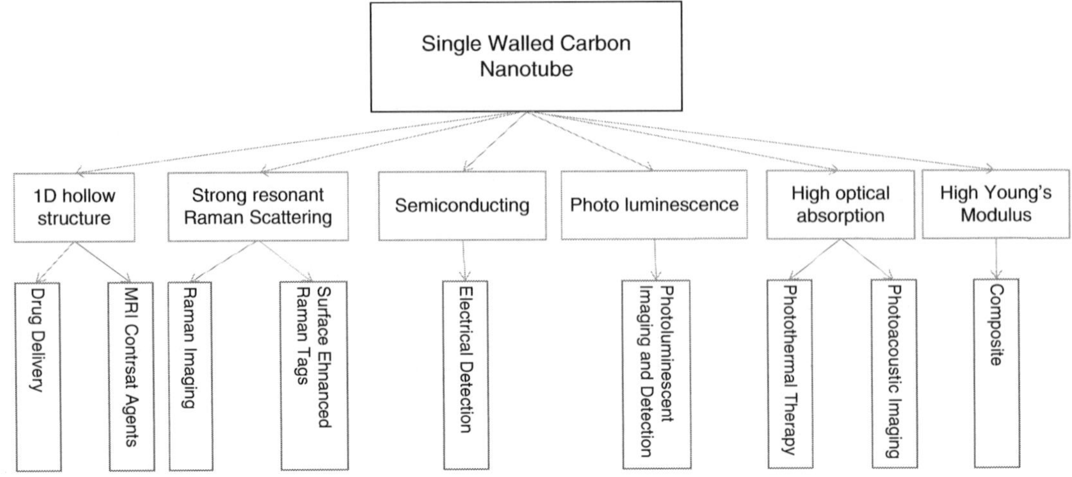

Fig. 2. SWNT properties and the corresponding biological applications.

1.2.1. Electrical Detection

Field-effect transistors (FET) based on semiconducting SWNTs have been utilized for biomolecule detection. SWNTs conjugated with biotin, Staphylococcal protein A, and U1A antigen (6, 24) have been reported to impart specific binding of streptavidin, immunoglobulin G, and the monoclonal mouse antibody 10E6, respectively. This technique has achieved in situ direct detection of these analytes in the nmol/L range via electrical read out. A variety of SWNT devices have been demonstrated for selective detection of oxidase and dehydrogenase activity, as well as for other biomolecules of interest, in a label-free fashion (25, 26).

1.2.2. Photoluminescent Detection on Semiconducting SWNT

The interesting optical properties of semiconducting SWNTs open another route for sensitive and selective detection of biomolecules. Two mechanisms, charge transfer and fluorescence quenching (27, 28), have been used for biomolecule detection based on SWNT band-gap fluorescence. The band-gap fluorescence is sensitive to the local dielectric environment around the SWNT and this property can be exploited in chemical sensing. DNA conformational polymorphism, induced by divalent metal cations that bind to DNA and stabilize the Z-form, has been detected by monitoring the SWNT band-gap fluorescence red shift (29). This strategy has also been utilized to detect the metal ions. Strano and co-workers have developed an array of SWNT sensors for detecting H_2O_2 molecules, generated upon growth factor stimulation in living A431 human epidermal carcinoma cells, which stochastically absorb and quench the SWNT fluorescence with spatial and temporal resolution (30). SWNT NIR fluorescence does not photobleach, has negligible autofluorescence from other assay components in the NIR range, and demonstrates a large Stokes shift compared to that of traditional fluorophores. These properties make SWNTs a novel class of fluorophores that allow the use of a range of excitation energies and the real time tracking of biological processes. This chapter highlights one method that can be used to synthesize highly fluorescent SWNTs.

1.2.3. SWNT-Raman Tags

SWNTs show intense Raman scattering cross-sections and high scattering efficiencies (16). The Raman scattering spectra of SWNTs are simple, with strong, well-defined Lorentzian peaks which are easily distinguishable from noise (18). SWNT-Raman tags do not photobleach even under high laser powers.

Surface-enhanced Raman spectroscopy (SERS) (31) can be used to increase the intensity of Raman active molecules in proximity to surface plasmons associated with gold, silver, and copper nanostructures (32). By coupling the intense Raman scattering efficiency of SWNTs with SERS substrates, the limit of detection of traditional fluorescence assays can be extended from approximately 1 pmol/L (33) to the femtomolar level or below.

1.2.4. In Vitro Delivery of Biomolecules

Properly functionalized SWNTs are able to enter cells by endocytosis without obvious toxicity (7) and they are chemically stable in biological environments. Owing to these properties, functionalized SWNTs have been used to efficiently deliver various biological cargos such as drugs, proteins, and DNA/RNA into cells. Once taken up by cells via endocytosis, functionalized SWNTs are able to exit cells through exocytosis (28).

Drug molecules can be covalently or noncovalently conjugated to SWNTs for in vitro delivery. SWNTs covalently tethered with the platinum (IV) complexes are taken into cancer cells. The platinum (II) core complex is released in reducing pH environments, thus killing the cancer cells. When attached to SWNTs, the cytotoxicity of the free platinum (IV) complex increases 100-fold (34). Aromatic molecules can be noncovalently loaded onto functionalized SWNTs via π–π stacking. Doxorubicin, a commonly used cancer chemotherapy drug, has been noncovalently loaded in high amounts onto the surface of PEGylated SWNTs (up to 4 g drug/1 g nanotube). The loading/binding is pH dependent and favorable for drug release in tumor microenvironments with acidic pH (35).

1.2.5. In Vivo Tumor Targeting and Cancer Therapy

In addition, SWNTs can be used for in vivo tumor targeting and cancer therapy. Dai and co-workers have conjugated paclitaxel (PTX), a commonly used chemotherapy drug, to branched PEG functionalized SWNTs via a cleavable ester bond (23). The SWNT–PTX conjugate exhibits improved treatment efficacy over the clinical Cremophor-based PTX formulation, Taxol®, in a 4T1 murine breast cancer model in mice.

1.2.6. Biological Imaging Using Carbon Nanotubes

The intrinsic optical properties of SWNTs make them useful as optical probes. Owing to their unique 1D structure, SWNTs exhibit strong resonance Raman scattering, high optical absorption, and photoluminescence in the NIR range. All of these properties can be utilized for in vitro and in vivo imaging in biological systems. Individual semiconducting SWNTs have photoluminescence in the NIR range between 800 and 1,600 nm, depending on their structure. This is useful for biological imaging, due to the high optical transparency of biological tissue near 800–1,000 nm and the inherently low autofluorescence from tissue in the NIR range (36). Biological imaging using SWNTs also benefits from the reduced background from autofluorescence. Using the intrinsic NIR photoluminescence of SWNTs, Jin et al. can track endocytosis and exocytosis of SWNTs in NIH-3T3 cells in real time (28).

SWNTs exhibit strong resonance Raman scattering that can be easily distinguished from fluorescence background. Raman microscopy has been utilized to image SWNTs in liver cells and tissue slices, using either their RBM or G-band peaks (16, 37–39). Also, Raman signals of SWNTs do not photobleach. They can be used for long-term imaging and tracking (16, 39).

SWNTs have strong optical absorption in the visible and NIR range which can be utilized in photoacoustic imaging. Photoacoustic imaging has higher spatial resolution than traditional ultrasound and deeper tissue penetration than fluorescence imaging (40). RGD-conjugated SWNTs have been used as the contrast agent for photoacoustic molecular imaging of cancer in living mice (14). The RGD-SWNT conjugate showed eight times greater photoacoustic signal than nontargeted SWNTs.

Recently, SWNTs conjugated with Gd^{3+} have been used in magnetic resonance imaging (MRI) (5). The aquated Gd^{3+} ion clusters within ultrashort SWNTs were found to be superparamagnetic, with a MRI efficacy of 40 times greater than the Gd^{3+}-based contrast agents in current clinical use.

1.2.7. SWNT-Based Composite and Porous Scaffolds for Tissue Engineering Applications

The field of tissue engineering and regenerative medicine requires the development of biomaterials with superior mechanical and bioactive properties. Metallic- or ceramic-based materials used in the development of implants and devices can support significant functional loads. However, their limitations include a weak tissue interface, potential for corrosion and fatigue, and poor bioactivity. The past decade has seen a significant increase in research on polymers as materials to overcome some or all of the above limitations. The mechanical properties of polymers can be improved by modifying the processing conditions or composition and by incorporating nanomaterials as reinforcing agents.

Recently, CNTs (SWNTs and MWNTs) have been demonstrated to substantially improve the mechanical and structural properties of polymer composites (4, 41–43). These porous bioscaffolds provide structural support, guide cell growth, and transport nutrients and waste products, essential for tissue regeneration. The presence of less than 0.5 wt% of SWNTs in a polypropylene fumarate (PPF, a linear biodegradable, biocompatible polyester) matrix has been shown to significantly enhance the compressive and flexural properties of the matrix up to threefold against the polymer alone (43, 44). Recent studies also show that the presence of small nontoxic amounts of SWNTs in polymeric scaffolds enhances cellular adhesion and proliferation and may induce bioactive properties into the scaffolds (41, 42, 44). The present chapter provides protocols for preparing SWNT-reinforced polymer nanocomposites and porous scaffolds. In addition, it elaborates on the techniques and methodologies that are used to characterize the structural and mechanical properties of reinforced polymer nanocomposites and porous scaffolds.

1.3. Five Degrees of Control to Nontoxic Nanotubes

Similar to other nanomaterials, the intrinsic toxicity level of CNTs depends on their physicochemical characteristics (e.g., size distribution, metal catalyst residual, and surface modifications). There are a wide variety of end products with different physicochemical characteristics given the combination of different synthesis methods and purification processes that can be used. A SWNT sample suitable for biomedical applications should be: (a) metal free, (b) water soluble, (c) length controlled, (d) structurally sorted, and (e) architecturally controlled (surface chemistry, defect density, and 3D assembly). For most applications, (a) and (b) are essential; for the most demanding applications, one may need all five degrees of control.

For instance, only semiconducting SWNTs are useful for photoluminescent and electrical detection; the existence of metal particles, metallic SWNTs, and other carbonaceous particles in the raw SWNTs may significantly affect the sensitivity of using nanotubes as fluorescent imaging tags and cause toxic side effects to the biological system. Purification, separation, and isolation of these semiconducting SWNTs from other by-products are therefore required for biomedical applications. Because pristine SWNTs are hydrophobic, surface functionalization is required in order to improve their aqueous solubility. Recent results have shown that while nonfunctionalized, hydrophobic SWNTs can be toxic (45–48), those with biocompatible coatings (9, 12, 34, 35, 49–54) are harmless to cells in vitro and in vivo, at least to mice within tested dose ranges (38, 39).

1.4. Materials Chemistry Toward Five Degrees of Control

1.4.1. Synthesis of SWNTs

Since their discovery, synthesis has been the main challenge in the basic and applied research of CNTs. A variety of techniques have been developed and improved for SWNT production, and commercial SWNT materials are now available. Typically, SWNTs are grown by heating a carbon-containing feedstock at elevated temperature in the presence of a transition metal catalyst such as iron or cobalt. Carbon feedstocks used to date include bulk graphite, hydrocarbons, and carbon monoxide. The most common processes are arc discharge (55), laser ablation (56), and chemical vapor deposition (CVD) (57). Among the CVD methods, high-pressure CO (HiPco) developed at Rice University (57) and CoMoCat (Southwest Nanotechnologies, Inc.) (58) are two major commercial approaches used to produce large-scale, high-quality SWNTs. The comparison of those methods is summarized in Table 1.

1.4.2. Functionalization of SWNTs

As-grown SWNTs are insoluble in organic solvents. For biomedical applications, surface chemistry or functionalization is required to solubilize SWNTs and to render biocompatibility and low toxicity. Surface functionalization of SWNTs can be covalent or noncovalent. Covalent methods involve addition of functional groups to SWNT sidewalls by chemical bonds. In noncovalent approaches, the SWNT sidewalls are functionalized by, for example, aromatic compounds, surfactants, and polymers, employing π–π stacking or hydrophobic interactions. Because noncovalent modifications of SWNTs can preserve their desired properties while imparting high water solubilities, this approach to synthesizing aqueous soluble nanotubes is widely used.

1.4.3. Noncovalent Functionalization of SWNTs

Noncovalent functionalization of SWNTs can be carried out by coating SWNTs with aromatic small molecules, biomacromolecules, polymers, and amphiphilic surfactant molecules. Since the chemical structure of the π-network of the CNTs remains, the physical properties of CNTs are essentially preserved. Consequently, aqueous solutions of noncovalently functionalized SWNTs are promising for multiple biomedical studies including imaging.

The polyaromatic graphitic surface of SWNTs is accessible to the binding of aromatic molecules, such as pyrene, porphyrin, and their derivatives, via π–π stacking (24, 59, 60). Chen et al. showed that proteins can be tethered on SWNTs functionalized

Table 1
Comparison of synthesis methods of SWNT

Methods	Quantity	Quality	Yield
Arc discharge	Grams	Good	Up to 30%
Laser ablation	Grams	Good	Around 70%
Chemical vapor deposition	Large scale	Excellent	Up to 95%

Source: http://en.wikipedia.org/wiki/Carbon_nanotube#Synthesis

with an amine-reactive pyrene derivative (24). Dai and co-workers have shown that fluorescein (FITC)-terminated PEG chains are able to solubilize SWNTs through the aromatic FITC domain π–π stacked on the nanotube surface. The obtained SWNT conjugates have visible fluorescence which is useful for biological detection and imaging (61).

Amphiphiles can also be used to solublize SWNTs in aqueous solutions, with hydrophobic domains attached to the SWNT surface via van der Waals forces and hydrophobic effects (62–64). Cherukuri et al. used Pluronic tri-block polymer to solubilize SWNTs for in vivo experiments (62). Surfactants, such as sodium dodecyl sulfate (SDS), have also been used to suspend SWNTs in water (11). PEGylated phospholipids (PL-PEG) have been developed to noncovalently functionalize SWNTs (12, 23, 35, 65). The hydrocarbon chains of the lipid strongly anchor onto the nanotube surface while the hydrophilic PEG chain imparts water solubility and biocompatibility. Biological molecules can be conjugated onto PEGylated SWNTs by using a functional group (e.g., amine) at the PEG terminal.

1.4.4. Covalent Functionalization of SWNTs

Various methods have been developed to covalently functionalize SWNTs. Oxidation is one of the most common (66). During the oxidation process, carboxyl groups are formed at the ends of tubes as well as on the sidewalls. After oxidation, sp^2 carbon atoms can be turned into sp^3 which can be covalently conjugated with amino acids (67).

Cycloaddition reactions are widely used to covalently functionalize SWNTs. The cycloaddition reaction occurs on the aromatic sidewalls, instead of the nanotube ends and defect sites as with oxidation. A 1,3-dipolar cycloaddition reaction on SWNTs developed by Prato et al. is now a commonly used reaction (68, 69). An azomethine-ylide generated by condensation of an α-amino acid and an aldehyde is added to the graphitic surface, forming a pyrrolidine ring coupled to the SWNT sidewall. Functional groups introduced via a modified α-amino acid can be used for further conjugation of biological molecules (22, 70).

Billups and co-workers applied the Birch reduction (71) to functionalize SWNTs (72, 73). The Billups reaction is particularly effective as lithium and other alkali metals can intercalate nanotube ropes in liquid ammonia, permitting the exfoliated nanotubes to react homogeneously with alkyl or aryl radicals produced from dissociation of corresponding halide precursors (72, 73). Various functional groups, including carboxylic acids (74), can be covalently added to SWNT sidewalls via this chemistry to afford predominantly individual nanotubes in water.

After chemical reactions, the intrinsic physical properties of CNTs such as photoluminescence and Raman scattering are likely destroyed due to the disrupted nanotube structure. The intensities of Raman scattering and photoluminescence of SWNTs are drastically decreased after covalent functionalization, reducing their potential for their use in optical applications in biomedical studies. Recently, the Wang group developed an approach to selectively oxidize the outer wall of double-walled carbon nanotubes using a combination of oleum and nitric acid (75). This double-wall chemistry enabled high water solubility through carboxylic acid functional groups introduced to the outer wall, while leaving the inner tube intact. This provides the opportunity to covalently tether molecules of interest onto the outer wall of the double-walled nanotube while using the optical properties of the inner wall of the nanotube for detection and imaging.

2. Materials

2.1. Preparation of Water-Soluble, Brightly Fluorescent SWNT Conjugates

1. High-pressure CO converted (HiPco) SWNTs (see Note 1).
2. Sodium cholate.
3. 1,2-Distearoyl-sn-glycero-3-phosphoethanolamine–N-[methoxy(polyethylene glycol) 5000] (DSPE-mPEG$_{5k}$).
4. 3,500 Molecular weight cut-off (MWCO) membranes.

2.2. Preparation of SWNT-Reinforced Polymers and Porous Scaffolds

1. PPF, prepared and purified as described previously (76), and propylene fumarate di-acrylate (PPF-DA).
2. Purified HiPco SWNTs (iron content approximately 2%) (77).
3. Benzoyl peroxide (BP), diethyl fumarate, and N,N-dimethyl-p-toluidine (DMT) NaCl (300–500 μm crystal size) sieved with USA Standard Testing Sieves.

2.3. Characterization

1. Gold wire.
2. Methylene chloride.

3. Methods

3.1. Preparation of Water-Soluble, Brightly Fluorescent SWNT Conjugates (Adapted from Ref. 19)

1. Add 1 mg raw HiPco SWNTs and 40 mg sodium cholate to 4 mL of water.
2. Bath sonicate the mixture for 1–6 h.
3. Ultracentrifuge the resulting black suspension at $300,000 \times g$ for 1 h to remove large aggregates and bundled nanotubes, leaving a dark supernatant of predominantly individual SWNTs.

4. Add 1 mg/mL DSPE–mPEG$_{5k}$ to the supernatant and sonicate briefly (<1 min) to ensure that the DSPE–mPEG$_{5k}$ is fully dissolved.

5. Dialyze the solution against water using a 3,500 MWCO membrane over a period of 4–5 days with multiple water changes per day. This process slowly removes the sodium cholate from the solution, allowing the DSPE–mPEG$_{5k}$ to coordinate to the surface of the nanotubes.

6. Following dialysis, ultracentrifuge the solution again at 300,000 $\times g$ for 1 h to remove any bundles that may have formed during the exchange process.

3.2. Preparation of SWNT-Reinforced Polymers

1. Disperse SWNTs in chloroform by high shear mixing for 5 min and then sonication for 15 min.

2. Mix PPF and PPF-DA (cross-linking agent of PPF) in chloroform (1 g PPF and 2 g PPF-DA per 3 mL chloroform).

3. Add the SWNTs (from step 1) immediately to the PPF/PPF-DA mixture to achieve SWNT concentrations ranging between 0 and 2 wt% (see Note 2).

4. Sonicate this mixture for 15 min, then remove the chloroform by rotary evaporation and vacuum-drying to obtain the uncross-linked, SWNT-reinforced polymer nanocomposites.

5. To achieve cross-linking, trigger the thermal polymerization reaction by adding 1 wt% BP (a free radical initiator, 0.1 g/mL dissolved in diethyl fumarate) and then 0.15 wt% DMT (accelerator) to the SWNTs under vigorous stirring.

6. Centrifuge the specimens at 721 $\times g$ for 5 min to remove any air bubbles.

7. Add the polymeric mixture into cylindrical glass vials. Vials 6.5 mm in diameter and 40 mm in length or 3 mm in diameter and 150 mm in length work best for compressive and flexural testing, respectively.

8. Cure them at 60°C for 24 h.

9. Recover the specimens by breaking the glass container. Specimens can also be cut into appropriate shapes and lengths with a diamond saw.

3.3. Preparation of SWNT-Reinforced Porous Scaffolds ("Thermal Cross-Linking Particulate Leaching")

1. Perform steps 1–4 in Subheading 3.2.

2. Mix the uncross-linked nanocomposites with 1 wt% free radical initiator (BP), then add the appropriate amount of NaCl to achieve the desired porosity (see Note 3). NaCl is used as the water-soluble porogen (78).

3. Cast the mixtures and thermally cross-link them at 100°C (see Note 4) for 24 h in cylindrical Teflon molds (4 mm diameter

and 8 mm height) or cylindrical glass molds (6.5 mm diameter and 40 mm in length).

4. Soak the cross-linked samples in water (change the water every 8 h) on a shaker table (80 rpm) at room temperature for 3 days to leach out the NaCl porogen. Blot them with absorbent paper and dry them under vacuum for 24 h.

3.4. Characterization of Structural Properties

3.4.1. SWNT-Polymer Interactions and Scaffold Pore Structure Analysis Using SEM

1. Sputter coat cross-sections of cut disks with gold using a sputter coating system.

2. Examine the characteristics, such as the dispersion of SWNTs, the pore structure of the scaffold, the pore size, the morphology, and the interconnectivity, by observing the samples above under a field emission scanning electron microscope at an accelerating voltage of 15 kV.

3.4.2. Sol Fraction Analysis

In order to assess the influence of the SWNTs on cross-linking density in the PPF polymer matrix, sol fraction analysis can be carried out on uncross-linked, SWNT-reinforced polymers.

1. Weigh 0.5 g of the sample (W_i, accuracy = 0.001 g) into a vial with 20 mL of methylene chloride.

2. Seal the vial and place it on a shaker table (80 rpm) at room temperature for 7 days.

3. Filter the solid sample with a weighed filter paper (W_p). Dry the retained material on the filter paper at 60°C for 1 h and keep it at room temperature for another 1 h. Then, weigh it again (W_{p+s}).

4. Calculate the sol fraction using the following equation for each group ($n \geq 5$):

$$\text{Sol fraction} = \frac{W_i - (W_{p+s} - W_p)}{W_i} \times 100\%. \tag{1}$$

3.4.3. Porosity Measurements Using Micro-CT and Mercury Porosimetry

3.4.3.1. Micro-CT Analysis

Micro-CT can be used to nondestructively and quantitatively measure the three-dimensional (3D) porosity and porous interconnectivity of SWNT-reinforced scaffolds.

1. Scan 4 mm × 8 mm cylindrical samples of each scaffold type with a micro-CT imaging system at 10 mm resolution using a voltage of 40 kV and a current of 250 mA.

2. Conduct image reconstruction and analysis. First, reconstruct the raw images of scaffolds to serial coronal-oriented tomograms using a 3D cone beam reconstruction algorithm.

3. Perform a threshold analysis to determine the threshold value for which grayscale tomograms of scaffolds are most accurately represented by their binarized counterparts in terms of porosity. Apply the optimal threshold value for all 3D reconstructions and quantitative analysis.

4. Generate representative 3D reconstructions (top and side views at a camera viewing angle of 10°) of porous scaffolds based on binarized tomograms.

5. In order to eliminate potential edge effects, select a cylindrical volume of interest (VOI) with a diameter of 3 mm and a height of 6 mm in the center of a scaffold. Perform a shrink-wrap process between two 3D measurements to shrink the outside boundary of the VOI in a scaffold through any openings whose size is equal to or larger than the threshold value.

6. Calculate scaffold porosity as:

$$\text{Porosity} = 100\% - \text{Vol\% of binarized object.} \qquad (2)$$

7. Interconnectivity is quantified as the fraction of the pore volume in a scaffold that is accessible from the outside through openings of a certain minimum size (79). Calculate interconnectivity as follows:

$$\text{Interconnectivity} = \frac{V - V_{shrink\text{-}wrap}}{V - V_m} \times 100\%, \qquad (3)$$

where V is the total volume of the VOI, $V_{shrink\text{-}wrap}$ is the VOI volume after shrink-wrap processing, and V_m is the volume of scaffold material.

3.4.3.2. Mercury Intrusion Porosimetry

1. Evacuate the sample chamber of a mercury intrusion porosimeter and fill it with mercury until an initial pressure of ~0.6 psi.

2. Weigh each 4 mm × 8 mm cylindrical samples for each scaffold type ($n \geq 3$) and place it into the sample chamber.

3. Increase the chamber pressure at a rate of 0.01 psi/s to 50 psi, and record the intruded volume of the mercury. The intruded mercury volume per gram of the sample is assumed to be equal to the pore volume (V_{pore}).

4. Calculate the porosity, ε, using the formula:

$$\varepsilon = \frac{V_{pore}}{V_{pore} + (1/\rho)} \times 100\%, \qquad (4)$$

where ρ is the density of the nanocomposites (see Note 3).

5. Make pore size measurements using the Washburn equation

$$D = \frac{4\gamma |\cos\theta|}{P}, \qquad (5)$$

where D is the pore diameter, γ is the surface tension of mercury, θ is the contact angle between mercury and the scaffold material (see Note 5), and P is the pressure.

3.5. Characterization of Bulk Properties

3.5.1. Viscoelastic Testing Using Rheometer

1. Test viscoelastic properties of the uncross-linked, SWNT-polymer nanocomposites with a rheometer in oscillatory shear mode at 25°C.

2. Melt the nanocomposite samples (weight percentages up to 0.2 wt%) (see Note 6) and place them between the base plate and cone geometry (60 mm diameter, 59 min cone angle, and 26 μm truncation).

3. Perform the measurements as a function of the oscillatory strain frequency (ω) of 0.001–30 Hz using 0.01–0.1 strain amplitude (see Note 7).

4. Record the complex viscosity magnitude, storage modulus, and loss modulus as measures of viscoelastic properties of the polymer nanocomposites.

3.5.2. Mechanical Testing: Compressive and Flexural Properties (See Note 8)

Mechanical properties under compression and flexion are tested at room temperature using a mechanical testing machine.

3.5.2.1. Compressive Testing

1. Compress the prepared specimens (4 mm × 8 mm) along their long axis until failure, and record the force and displacement throughout the compression.

2. Generate stress–strain curves based on the initial specimen dimensions (see Note 9).

3.5.2.2. Flexural Testing

1. The testing specimens are placed on a three-point bending apparatus with two supports ~40 mm from each other.

2. Load a nose midway between the supports until the specimen fails.

3. Record the force and displacement and convert to a stress–strain curve (see Note 9).

4. Notes

1. As-grown HiPco materials contain both iron catalysts and carbon nanoparticles. The majority of the iron can be removed in the ultracentrifugation step. Alternatively, these impurities can be removed by wet chemistry (80) or other purification methods.

2. The same basic procedure can be followed to disperse SWNTs into other hydrophobic polymers.

3.
$$\varepsilon = \frac{V_{NaCl}}{V_{NaCl} + V_{nano}} \times 100\%, \tag{6}$$

$$W_{NaCl} = \frac{\varepsilon}{1-\varepsilon} \times \frac{\rho_{NaCl}}{\rho_{Nano}} \times W_{Nano}, \tag{7}$$

where ε is the apparent porosity (volume percent of porogen in a scaffold), V_{NaCl} and V_{Nano} are the volumes of NaCl and the nanocomposite in a scaffold, W_{NaCl} and W_{Nano} are the weights of NaCl and the nanocomposite in a scaffold, and ρ_{NaCl} is the density of NaCl (2.17 g/mL). The density of the nanocomposite (ρ_{Nano}) is calculated by measuring the mass and volume of five solid, cross-linked nanocomposite cylinders; found here to be 1.25 g/mL (41).

4. The curing temperature of 100°C is applied to ensure complete cross-linking of the scaffold materials (81).

5. 140° is the reported contact angle (θ) between mercury and this scaffold material (82).

6. Rheological measurements are performed only with composites up to 0.2 wt% CNT since previous studies (4, 83) already confirm that solid-like behavior commences at very low SWNT weight percentages (0.05–0.2 wt%).

7. The 0.01–0.1 strain amplitude is chosen as it allows for rheological measurement in the linear dynamic range (4). Use the low end of the reported strain amplitude range for the nanocomposite melts; use the high end for the uncross-linked polymer melts.

8. Compressive and flexural testing should follow the American Society of Testing Materials (ASTM) Standard D695-02a and ASTM Standard D790-03, respectively.

9. The slope of the initial linear portion of the curve gives the compressive modulus and a line drawn parallel to the curve defining the modulus, beginning at 1.0% strain (offset) gives the offset compressive yield strength (the stress at which the stress–strain curve intersects the line). The flexural modulus can be calculated from the stress–strain curve using similar methods. The compressive and flexural strength are defined as the maximum stress carried by the specimen during compression or flexural testing, respectively

Acknowledgments

The authors would like to thank Jarrett Leeds for helping in the preparation of soluble SWNT. This work was supported by Office of the Vice President of Research at Stony Brook University (SB).

References

1. Liu, Z., Tabakman, S., Welsher, K., and Dai, H. (2009) Carbon nanotubes in biology and medicine: *in vitro* and *in vivo* detection, imaging and drug delivery. *Nano Res.* **2**, 85–120.

2. Cao, Q. and Rogers, J. A. (2008) Random networks and aligned arrays of single-walled carbon nanotubes for electronic device applications. *Nano Res.* **1**, 259–272.

3. Fan, S. S., Chapline, M. G., Franklin, N. R., Tombler, T. W., Cassell, A. M., and Dai, H. J. (1999) Self-oriented regular arrays of carbon nanotubes and their field emission properties. *Science* **283**, 512–514.

4. Shi, X., Hudson, J., Spicer, P., Tour, J., Krishnamoorti, R., and Mikos, A. (2005) Rheological behaviour and mechanical characterization of injectable poly (propylene fumarate)/single-walled carbon nanotube composites for bone tissue engineering. *Nanotechnology* **16**, 531.

5. Sitharaman, B., Kissell, K. R., Hartman, K. B., Tran, L. A., Baikalov, A., Rusakova, I., et al. (2005) Superparamagnetic gadonanotubes are high-performance MRI contrast agents. *Chem. Commun.* **31**, 3915.

6. Chen, R. J., Bangsaruntip, S., Drouvalakis, K. A., Kam, N. W. S., Shim, M., Li, Y. M., et al. (2003) Non-covalent functionalization of carbon nanotubes for highly specific electronic biosensors. *Proc. Natl. Acad. Sci.USA* **100**, 4984–4989.

7. Kam, N. W. S., Jessop, T. C., Wender, P. A., and Dai, H. (2004) Nanotube molecular transporters: internalization of carbon nanotube-protein conjugates into mammalian cells. *J. Am. Chem. Soc.* **126**, 6850–6851.

8. Bianco, A., Kostarelos, K., Partidos, C. D., and Prato, M. (2005) Biomedical applications of functionalised carbon nanotubes. *Chem. Commun.* 571–577.

9. Cherukuri, P., Bachilo, S. M., Litovsky, S. H., and Weisman, R. B. (2004) Near-infrared fluorescence microscopy of single-walled carbon nanotubes in phagocytic cells. *J. Am. Chem. Soc.* **126**, 15638–15639.

10. Tans, S. J., Devoret, M. H., Dai, H., Thess, A., Smalley, R. E., Geerligs, L. J., et al. (1997) Individual single-wall carbon nanotubes as quantum wires. *Nature* **386**, 474–477.

11. O'Connell, M. J., Bachilo, S. M., Huffman, C. B., Moore, V. C., Strano, M. S., Haroz, E. H., et al. (2002) Band gap fluorescence from individual single-walled carbon nanotubes. *Science* **297**, 593–596.

12. Kam, N. W. S., O'Connell, M., Wisdom, J. A., and Dai, H. (2005) Carbon nanotubes as multifunctional biological transporters and near-infrared agents for selective cancer cell destruction. *Proc. Natl. Acad. Sci. USA* **102**, 11600–11605.

13. Chakravarty, P., Marches, R., Zimmerman, N. S., Swafford, A. D. E., Bajaj, P., Musselman, I. H., et al. (2008) Thermal ablation of tumor cells with anti body-functionalized single-walled carbon nanotubes. *Proc. Natl Acad. Sci. USA* **105**, 8697–8702.

14. De la Zerda, A., Zavaleta, C., Keren, S., Vaithilingam, S., Bodapati, S., Liu, Z., et al. (2008) Photoacoustic molecular imaging in living mice utilizing targeted carbon nanotubes. *Nat. Nanotechnol.* **3**, 557–562.

15. Rao, A. M., Richter, E., Bandow, S., Chase, B., Eklund, P. C., Williams, K. A., et al. (1997) Diameter-selective Raman scattering from vibrational modes in carbon nanotubes. *Science* **275**, 187–191.

16. Heller, D. A., Baik, S., Eurell, T. E., and Strano, M. S. (2005) Single walled carbon nanotube spectroscopy in live cells: towards long-term labels and optical sensors. *Adv. Mater.* **17**, 2793–2799.

17. Tang, X. W., Bansaruntip, S., Nakayama, N., Yenilmez, E., Chang, Y. L., and Wang, Q. (2006) Carbon nanotube DNA sensor and sensing mechanism. *Nano Lett.* **6**, 1632–1636.

18. Chen, Z., Tabakman, S. M., Goodwin, A. P., Kattah, M. G., Daranciang, D., Wang, X., et al. (2008) Protein microarrays with carbon nanotubes as multi-color Raman labels. *Nat. Biotechnol.* **26**, 1285–1292.

19. Welsher, K., Liu, Z., Sherlock, S., Robinson, J., Chen, Z., Daranciang, D., et al. (2009) A route to brightly fluorescent carbon nanotubes for near-infrared imaging in mice. *Nat. Nanotechnol.* **4**, 773–780.

20. Singh, R., Pantarotto, D., McCarthy, D., Chaloin, O., Hoebeke, J., Partidos, C. D., et al. (2005) Binding and condensation of plasmid DNA onto functionalized carbon nanotubes: toward the construction of nanotube-based gene delivery vectors. *J. Am. Chem. Soc.* **127**, 4388–4396.

21. Kam, N. W. S. and Dai, H. (2005) Carbon nanotubes as intracellular protein transporters: generality and biological functionality. *J. Am. Chem. Soc.* **127**, 6021–6026.

22. Pantarotto, D., Briand, J. P., Prato, M., and Bianco, A. (2004) Translocation of bioactive peptides across cell membranes by carbon nanotubes. *Chem. Commun.* 16–17.

23. Liu, Z., Chen, K., Davis, C., Sherlock, S., Cao, Q., Chen, X., et al. (2008) Drug delivery

with carbon nanotubes for *in vivo* cancer treatment. *Cancer Res.* **68**, 6652–6660.

24. Chen, R. J., Zhang, Y. G., Wang, D. W., and Dai, H. (2001) Non-covalent sidewall functionalization of single-walled carbon nanotubes for protein immobilization. *J. Am. Chem. Soc.* **123**, 3838–3839.

25. Kim, S. N., Rusling, J. F., and Papadimitrakopoulos, F. (2007) Carbon nanotubes for electronic and electrochemical detection of biomolecules. *Adv. Mater.* **19**, 3214–3228.

26. Wang, J. (2005) Carbon-nanotube-based electrochemical biosensors: a review. *Electroanalysis* **17**, 7–14.

27. Barone, P. W., Parker, R. S., and Strano, M. S. (2005) *In vivo* fluorescence detection of glucose using a single-walled carbon nanotube optical sensor: design, fluorophore properties, advantages, and disadvantages. *Anal. Chem.* **77**, 7556–7562.

28. Barone, P. W., Baik, S., Heller, D. A., and Strano, M. S. (2004) Near-infrared optical sensors based on single-walled carbon nanotubes. *Nat. Mater.* **4**, 86–92.

29. Heller, D., Jeng, E., Yeung, T., Martinez, B., Moll, A., Gastala, J., et al. (2006) Optical detection of DNA conformational polymorphism on single-walled carbon nanotubes. *Science* **311**, 508–511.

30. Jin, H., Heller, D., Kalbacova, M., Kim, J., Zhang, J., Boghossian, A., et al. (2010) Detection of single-molecule H_2O_2 signaling from epidermal growth factor receptor using fluorescent single-walled carbon nanotubes. *Nat. Nanotech.* **5**, 302–309.

31. Nie, S. and Emory, S. R. (1997) Probing single molecules and single nanoparticles by surface-enhanced Raman scattering. *Science* **275**, 1102–1106.

32. Jeanmaire, D. L. and Van Duyne, R. P. (1977) Surface Raman spectroelectrochemistry part 1. Heterocyclic, aromatic, and aliphatic amines adsorbed on anodized silver electrode. *J. Electroanal. Chem.* **84**, 1–20.

33. Espina, V., Woodhouse, E. C., Wulfkuhle, J., Asmussen, H. D., Petricoin, E. F., and Liotta, L. A. (2004) Protein microarray detection strategies: focus on direct detection technologies. *J. Immunol. Methods* **290**, 121–133.

34. Feazell, R. P., Nakayama-Ratchford, N., Dai, H., and Lippard, S. J. (2007) Soluble single-walled carbon nanotubes as longboat delivery systems for platinum(IV) anticancer drug design. *J. Am. Chem. Soc.* **129**, 8438–8349.

35. Liu, Z., Sun, X., Nakayama, N., and Dai, H. (2007) Supramolecular chemistry on water-soluble carbon nanotubes for drug loading and delivery. *ACS Nano* **1**, 50–56.

36. Aubin, J. E. (1979) Autofluorescence of viable cultured mammalian cells. *J. Histochem. Cytochem.* **27**, 36–43.

37. Liu, Z., Winters, M., Holodniy, M., and Dai, H. (2007) siRNA delivery into human T cells and primary cells with carbon nanotube transporters. *Angew. Chem. Int. Ed.* **46**, 2023–2027.

38. Schipper, M. L., Nakayama-Ratchford, N., Davis, C. R., Kam, N. W. S., Chu, P., Liu, Z., et al. (2008) A pilot toxicology study of single-walled carbon nanotubes in a small sample of mice. *Nat. Nanotechnol.* **3**, 216–221.

39. Liu, Z., Davis, C., Cai, W., He, L., Chen, X., and Dai, H. (2008) Circulation and long-term fate of functionalized, biocompatible single-walled carbon nanotubes in mice probed by Raman spectroscopy. *Proc. Natl Acad. Sci. USA* **105**, 1410–1415.

40. Xu, M. H. and Wang, L. H. V. (2006) Photoacoustic imaging in biomedicine. *Rev. Sci. Instrum.* **77**, 041101.

41. Shi, X., Sitharaman, B., Pham, Q., Liang, F., Wu, K., Billups, W. E., et al. (2007) Fabrication of porous ultra-short single-walled carbon nanotube nanocomposite scaffolds for bone tissue engineering. *Biomaterials* **28**, 4078–4090.

42. Shi, X., Sitharaman, B., Pham, Q., Spicer, P., Hudson, J., Wilson, L., et al. (2008) *In vitro* cytotoxicity of single-walled carbon nanotube/biodegradable polymer nanocomposites. *J. Biomed. Mater. Res. A* **86**, 813–823.

43. Sitharaman, B., Shi, X., Tran, L., Spicer, P., Rusakova, I., Wilson, L., et al. (2007) Injectable *in situ* cross-linkable nanocomposites of biodegradable polymers and carbon nanostructures for bone tissue engineering. *J. Biomater. Sci. Polymer Ed.* **18**, 655–671.

44. Sitharaman, B., Shi, X., Walboomers, X. F., Liao, H., Cuijpers, V., Wilson, L. J., et al. (2008) *In vivo* biocompatibility of ultra-short single-walled carbon nanotube/biodegradable polymer nanocomposites for bone tissue engineering. *Bone* **43**, 362–370.

45. Cui, D. X., Tian, F. R., Ozkan, C. S., Wang, M., and Gao, H. J. (2005) Effect of single wall carbon nanotubes on human HEK293 cells. *Toxicol. Lett.* **155**, 73–85.

46. Lam, C. W., James, J. T., McCluskey, R., and Hunter, R. L. (2004) Pulmonary toxicity of single-wall carbon nanotubes in mice 7 and 90 days after intratracheal instillation. *Toxicol. Lett.* **77**, 126–134.

47. Warheit, D. B., Laurence, B. R., Reed, K. L., Roach, D. H., Reynolds, G. A. M., and Webb, T. R. (2004) Comparative pulmonary toxicity assessment of single-wall carbon nanotubes in rats. *Toxicol. Lett.* **77**, 117–125.

48. Poland, C. A., Duffin, R., Kinloch, I., Maynard, A., Wallace, W. A. H., Seaton, A.,

et al. (2008) Carbon nanotubes introduced into the abdominal cavity of mice show asbestos-like pathogenicity in a pilot study. *Nat. Nanotechnol.* **3**, 423–428.

49. Wu, W., Wieckowski, S., Pastorin, G., Benincasa, M., Klumpp, C., Briand, J. P., et al. (2005) Targeted delivery of amphotericin B to cells by using functionalized carbon nanotubes. *Angew. Chem. Int. Ed.* **44**, 6358–6362.

50. Dumortier, H., Lacotte, S., Pastorin, G., Marega, R., Wu, W., Bonifazi, D., et al. (2006) Functionalized carbon nanotubes are noncytotoxic and preserve the functionality of primary immune cells. *Nano Lett.* **6**, 1522–1528.

51. Chen, X., Lee, G. S., Zettl, A., and Bertozzi, C. R. (2004) Biomimetic engineering of carbon nanotubes by using cell surface mucin mimics. *Angew. Chem. Int. Ed.* **43**, 6111–6116.

52. Chen, X., Tam, U. C., Czlapinski, J. L., Lee, G. S., Rabuka, D., Zettl, A., et al. (2006) Interfacing carbon nanotubes with living cells. *J. Am. Chem. Soc.* **128**, 6292–6293.

53. Chin, S. F., Baughman, R. H., Dalton, A. B., Dieckmann, G. R., Draper, R. K., Mikoryak, C., et al. (2007) Amphiphilic helical peptide enhances the uptake of single-walled carbon nanotubes by living cells. *Exp. Biol. Med.* **232**, 1236–1244.

54. Yehia, H. N., Draper, R. K., Mikoryak, C., Walker, E. K., Bajaj, P., Musselman, I. H., et al. (2007) Single-walled carbon nanotube interactions with HeLa cells. *J. Nanobiotech.* **5**, 8.

55. Journet, C., Maser, W. K., Bernier, P., Loiseau, A., Delachapelle, M. L., Lefrant, S., et al. (1997) Large-scale production of single walled carbon nanotubes by the electric-arc technique. *Nature* **388**, 756–758.

56. Thess, A., Lee, R., Nikolaev, P., Dai, H., Petit, P., Robert, J., et al. (1996) Crystalline ropes of metallic carbon nanotubes. *Science* **273**, 483–487.

57. Nikolaev, P., Bronikowshi, M. J., Bradley, R. K., Rohmund, F., Colbert, D. T., Smith, K. A., et al. (1999) Gas-phase catalytic growth of single-walled carbon nanotubes from carbon monoxide. *Chem. Phys. Lett.* **313**, 91–97.

58. Resasco, D. E., Alvarez, W. E., Pompeo, F., Balzano, L., Herrera, J. E., Bitiyanan, B., et al. (2002) A scalable process for production of single-walled carbon nanotubes (SWNT) by catalytic disproportionation of CO on a solid catalyst. *J. Nanopart. Res.* **4**, 131–136.

59. Guldi, D. M., Taieb, H., Rahman, G. M. A., Tagmatarchis, N., and Prato, M. (2005) Novel photoactive single-walled carbon nanotube-porphyrin polymer wraps: efficient and longlived intracomplex charge separation. *Adv. Mater.* **17**, 871–875.

60. Chen, J., Liu, H. Y., Weimer, W. A., Halls, M. D., Waldeck, D. H., and Walker, G. C. (2002) Non-covalent engineering of carbon nanotube surfaces by rigid, functional conjugated polymers. *J. Am. Chem. Soc.* **124**, 9034–9035.

61. Nakayama-Ratchford, N., Bangsaruntip, S., Sun, X. M., Welsher, K., and Dai, H. (2007) Non-covalent functionalization of carbon nanotubes by fluorescein-polyethylene glycol: supramolecular conjugates with pH-dependent absorbance and fluorescence. *J. Am. Chem. Soc.* **129**, 2448–2449.

62. Cherukuri, P., Gannon, C. J., Leeuw, T. K., Schmidt, H. K., Smalley, R. E., Curley, S. A., et al. (2006) Mammalian pharmacokinetics of carbon nanotubes using intrinsic near-infrared fluorescence. *Proc. Natl Acad. Sci. USA* **103**, 18882–18886.

63. Richard, C., Balavoine, F., Schultz, P., Ebbesen, T. W., and Mioskowski, C. (2003) Supramolecular self-assembly of lipid derivatives on carbon nanotubes. *Science* **300**, 775–778.

64. Wang, H., Zhou, W., Ho, D. L., Winey, K. I., Fischer, J. E., Glinka, C. J., et al. (2004) Dispersing single-walled carbon nanotubes with surfactants: a small angle neutron scattering study. *Nano Lett.* **4**, 1789–1793.

65. Liu, Z., Cai, W. B., He, L. N., Nakayama, N., Chen, K., Sun, X. M., et al. (2007) *In vivo* biodistribution and highly efficient tumour targeting of carbon nanotubes in mice. *Nat. Nanotechnol.* **2**, 4–52.

66. Niyogi, S., Hamon, M. A., Hu, H., Zhao, B., Bhowmik, P., Sen, R., et al. (2002) Chemistry of single-walled carbon nanotubes. *Acc. Chem. Res.* **35**, 1105–1113.

67. Zeng, L., Alemany, L. B., Edwards, C. L., and Barron, A. R. (2008) Demonstration of covalent sidewall functionalization of single wall carbon nanotubes by NMR spectroscopy: side chain length dependence on the observation of the sidewall sp³ carbons. *Nano. Res.* **1**, 72–88.

68. Georgakilas, V., Kordatos, K., Prato, M., Guldi, D. M., Holzinger, M., and Hirsch, A. (2002) Organic functionalization of carbon nanotubes. *J. Am. Chem. Soc.* **124**, 760–761.

69. Tagmatarchis, N. and Prato, M. (2004) Functionalization of carbon nanotubes via 1,3-dipolar cycloadditions. *J. Mater. Chem.* **14**, 437–439.

70. Pastorin, G., Wu, W., Wieckowski, S., Briand, J. P., Kostarelos, K., Prato, M., et al. (2006) Double functionalisation of carbon nanotubes for multimodal drug delivery. *Chem. Commun.* 1182–1184.

71. Birch, A. J. and Smith, H. (1958) Reduction by metal–amine solutions: applications in synthesis and determination of structure. *Q. Rev. Chem. Soc.* **12**, 17–33.

72. Liang, F., Sadana, A. K., Peera, A., Chattopadhyay, J., Gu, Z., Hauge, R. H., et al. (2004) A convenient route to functionalized carbon nanotubes. *Nano Lett.* **4**, 1257–1260.

73. Liang, F., Alemany, L. B., Beach, J. M., and Billups, W. E. (2005) Structure analyses of dodecylated single-walled carbon nanotubes. *J. Am. Chem. Soc.* **127**, 13941–13948.

74. Deng, S. L., Brozena, A. H., Zhang, Y., Piao, Y.-M., Wang, Y. H. (2011) "Diameter-Dependent, Progressive Alkylcarboxylation of Single-Walled Carbon Nanotubes." *Chem. Comm.* **47**, 758–760.

75. Brozena, A., Moskowitz, J., Shao, B., Deng, S., Liao, H., Gaskell, K., et al. (2010) Outer wall selectively oxidized, water-soluble double-walled carbon nanotubes. *J. Am. Chem. Soc.* **132**, 3932–3938.

76. Shung, A., Timmer, M., Jo, S., Engel, P., and Mikos, A. (2002) Kinetics of poly (propylene fumarate) synthesis by step polymerization of diethyl fumarate and propylene glycol using zinc chloride as a catalyst. *J. Biomater. Sci. Polym. Ed.* **13**, 95–108.

77. Chiang, I., Brinson, B., Huang, A., Willis, P., Bronikowski, M., Margrave, J., et al. (2001) Purification and characterization of single-wall carbon nanotubes (SWNTs) obtained from the gas-phase decomposition of CO (HiPco process). *J. Phys. Chem. B* **105**, 8297–8301.

78. Fisher, J. P., Holland, T., Dean, D., Engel, P., and Mikos, A. (2001) Synthesis and properties of photocross-linked poly (propylene fumarate) scaffolds. *J. Biomater. Sci. Poly. Ed.* **12**, 673–687.

79. Moore, M., Jabbari, E., Ritman, E., Lu, L., Currier, B., Windebank, A., et al. (2004) Quantitative analysis of interconnectivity of porous biodegradable scaffolds with micro-computed tomography. *J. Biomed. Mater. Res.* **71**, 258–267.

80. Wang, Y., Shan, H., Hauge, R., Pasquali, M., and Smalley, R. E. (2007) A highly selective, one-pot, green chemistry for carbon nanotube purification. *J. Phys. Chem. B* **111**, 1249–1252.

81. Timmer, M., Carter, C., Ambrose, C., and Mikos, A. (2003) Fabrication of poly (propylene fumarate)-based orthopaedic implants by photo-crosslinking through transparent silicone molds. *Biomaterials* **24**, 4707–4714.

82. Lowell, S., Shields, J., Thomas, M., and Thommes, M. (2004) *Characterization of porous solids and powders: surface area, pore size, and density.* Springer, The Netherlands.

83. Shi, X., Hudson, J., Spicer, P., Tour, J., Krishnamoorti, R., and Mikos, A. (2006) Injectable nanocomposites of single-walled carbon nanotubes and biodegradable polymers for bone tissue engineering. *Biomacromolecules* **7**, 2237–2242.

Chapter 16

Electrospun Nanofibrous Scaffolds for Engineering Soft Connective Tissues

Roshan James, Udaya S. Toti, Cato T. Laurencin, and Sangamesh G. Kumbar

Abstract

Tissue-engineered medical implants, such as polymeric nanofiber scaffolds, are potential alternatives to autografts and allografts, which are short in supply and carry risks of disease transmission. These scaffolds have been used to engineer various soft connective tissues such as skin, ligament, muscle, and tendon, as well as vascular and neural tissue. Bioactive versions of these materials have been produced by encapsulating molecules such as drugs and growth factors during fabrication. The fibers comprising these scaffolds can be designed to match the structure of the native extracellular matrix (ECM) closely by mimicking the dimensions of the collagen fiber bundles evident in soft connective tissues. These nanostructured implants show improved biological performance over the bulk materials in aspects of cellular infiltration and in vivo integration, and the topography of such scaffolds has been shown to dictate cellular attachment, migration, proliferation, and differentiation, which are critical steps in engineering complex functional tissues and crucial to improved biocompatibility and functional performance. Nanofiber matrices can be fabricated using a variety of techniques, including drawing, molecular self-assembly, freeze-drying, phase separation, and electrospinning. Among these processes, electrospinning has emerged as a simple, elegant, scalable, continuous, and reproducible technique to produce polymeric nanofiber matrices from solutions and their melts. We have shown the ability of this technique to be used to fabricate matrices composed of fibers from a few hundred nanometers to several microns in diameter by simply altering the polymer solution concentration. This chapter will discuss the use of the electrospinning technique in the fabrication of ECM-mimicking scaffolds. Furthermore, selected scaffolds will be seeded with primary adipose-derived stromal cells, imaged using scanning electron microscopy and confocal microscopy, and evaluated in terms of their capacity toward supporting cellular proliferation over time.

Key words: Nanofiber scaffolds, Electrospinning, Tissue engineering, Soft tissue, Skin, Tendon, Extracellular matrix, Biodegradable scaffold, Cell behavior, Stem cells

1. Introduction

The increasing demand for biologically compatible donor tissue and organ transplants (allografts) far outstrips the availability, leading to an acute shortage. The available allografts have the

Sarah J. Hurst (ed.), *Biomedical Nanotechnology: Methods and Protocols,* Methods in Molecular Biology, vol. 726, DOI 10.1007/978-1-61779-052-2_16, © Springer Science+Business Media, LLC 2011

potential to elicit an immune response and carry risk of disease transmission. Autografts from an individual are associated with issues such as donor site morbidity and very limited availability. Tissue-engineered implants, such as biodegradable three-dimensional (3D) porous scaffolds, have emerged as a viable alternative to these materials to repair/regenerate damaged tissues and restore functionality. Isolated cells alone cannot reassemble into complex 3D functional tissues; however, these materials can be used to fill the tissue void and provide anchorage for cells to attach, infiltrate, populate, and differentiate to create functional tissue (1, 2). These scaffolds of both natural and/or synthetic origin are being designed to mimic the structure and functions of the native tissue (i.e., extracellular matrix (ECM)) morphologically, mechanically, and dimensionally (3, 4). By virtue of their small fiber size, these materials show exceptionally high surface area and porosity and superior mechanical and degradation properties compared to micro-sized porous materials, and as a result have shown improved tissue regeneration capabilities (5). Additionally, biodegradable nanofiber scaffolds can be used to deliver drugs, proteins, and growth factors locally, which can further accelerate or modulate the in vivo response (6, 7). For these reasons, nanofiber scaffolds are a popular choice to repair and regenerate various soft tissues such as skin, blood vessel, nerve, tendon, and cartilage (5, 8–10).

Nanofiber matrices can be fabricated using a variety of techniques including electrospinning (7), molecular self-assembly (11), phase separation (12), and drawing. The choice of the fabrication method is largely based on the properties of the chosen matrix material. For instance, molecular self-assembly is a preferred technique to produce peptide nanofiber scaffolds (13), while temperature-induced phase separation is the popular choice to fabricate highly crystalline poly(L-lactic acid) (PLLA) nanofiber scaffolds. All these processes have advantages and disadvantages in terms of their fabrication, controllability, reproducibility, and desired end application (Table 1). Drawing is a simple process where a micropipette, a few micrometers in diameter, is dipped into a polymer liquid and withdrawn at a fixed speed to produce fibers. This labor-intensive process can produce only fibers with diameters in the micrometer size regime, and an additional step such as weaving is needed to produce scaffolds for tissue engineering applications. This process is also limited by the cohesive properties of the material which must be able to support the stress produced under deformation/pulling. The phase separation technique utilizes the physical incompatibility of two materials and their tendency to separate into two phases to fabricate nanofiber scaffolds. In brief, for example, flash-frozen highly crystalline PLLA solution in dimethylformamide will undergo gelation and is washed with water followed by freeze-drying to remove the solvent phase to obtain highly porous

Table 1
Comparison of nanofiber fabrication techniques that can be used to synthesize tissue-engineered implants

Fabrication technique	Advantages	Disadvantages
Drawing	• Simple equipment	• Discontinuous process • Not scalable • No control on fiber dimensions
Temperature-induced phase separation	• Simple equipment • Convenient fabrication process • Mechanical properties of the fiber matrices can be varied by changing polymer composition	• Limited to specific polymers • Not scalable • No control on fiber dimensions
Molecular self-assembly	• Only smaller nanofibers of few nanometer in diameter and few microns in length can be fabricated • Complex functional structures	• Complex process involving intermolecular forces • Not scalable • Weak nanofiber strength
Electrospinning	• Simple instrument • Continuous process • Cost effective compared to other existing methods • Scalable • Ability to fabricate fiber diameters few nanometer to several microns • Aligned and random-oriented fibers	• Jet instability • Toxic solvents • Packaging, shipping, and handling

nanofiber scaffolds (14). Though the technique is simple and requires minimum instrumentation, it can be used only for certain specific polymer–solvent combinations. Nanofiber scaffolds fabricated from phase separation processes are highly porous and lack the mechanical properties suitable for load-bearing applications. Molecular self-assembly is based on spontaneous organization of molecules and components into patterns or structure without human intervention. This technique involves building nanoscale fibers using small molecules which interact by intermolecular forces such as van der Waals forces, hydrogen bonds, and electrostatic forces. The system is highly flexible with the possibility of self-assembling innumerable structural shapes by modifying the structure of the molecule. Self-assembly is ideally suited to design short structures composed of ECM peptide or bioactive groups.

Electrospinning, or the electrostatic spinning process, is a versatile platform that can be used to fabricate nanofiber scaffolds from polymer solutions. Polymers of both natural and synthetic origin as well as their blends have been fabricated into nanofiber scaffolds for a variety of biomedical applications (6, 15, 16). This process requires relatively simple instrumentation and is a continuous, scalable, and highly reproducible technique. Electrospinning is mediated by the application of high electric potential to a polymer solution. A typical electrospinning apparatus has three essential components including a power source, a polymer solution delivery system, and a collector (see Fig. 1). In brief, a programmable syringe pump delivers polymer solution at the desired rate

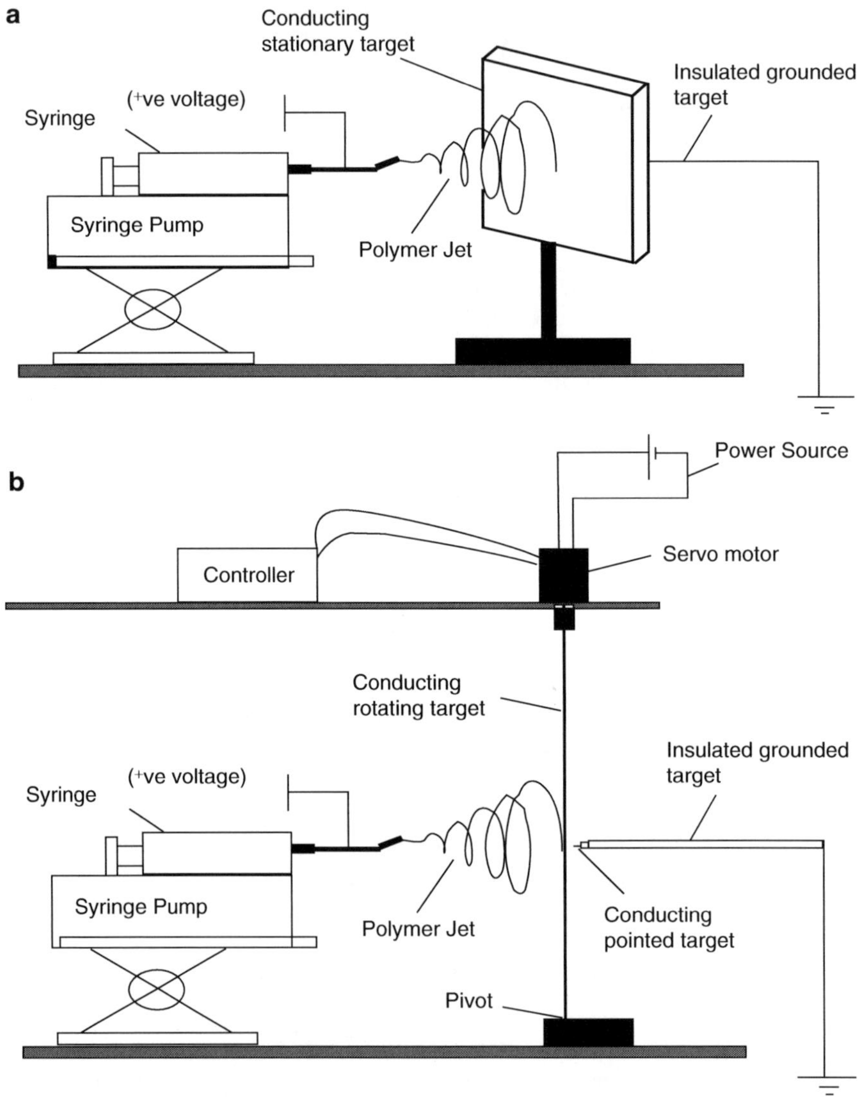

Fig. 1. Schematics of the electrospinning process using (**a**) a stationary target such as aluminum foil and (**b**) moving target such as a rotating mandrel. A high-voltage power source is used to apply an electric potential of a few kV to a pendant polymer droplet at the end of a blunt needle. The polymer solution is pumped out of the syringe at a controlled flow rate. The charged polymer solution undergoes a series of bending and stretching instabilities across the air-gap distance, moving toward the grounded target. The solvent rapidly evaporates and ultra-thin polymeric fibers are deposited.

and, when the applied electric potential exceeds the forces of surface tension acting in the opposite direction, a thin polymer jet ejects from the needle tip of the syringe and travels toward the grounded target. During this journey, the jet undergoes a series of electrically driven bending and stretching instabilities which result in looping and spiraling motions. The jet elongates and stretches to minimize this instability due to repulsive electrostatic forces, and ultra thin fibers are deposited on the target. Various process parameters, such as needle diameter, solution flow rate, applied electric potential, and working distance (distance between needle and target), and system parameters, including polymer molecular weight and solution viscosity, surface tension, and conductivity, affect the electrospinning and hence need to be optimized to produce continuous fibers of desired morphology and mechanical properties. The type of collector target also plays an important role in fiber orientation. For instance, the use of a stationary target results in random nanofiber deposition, while the use of a rotating mandrel-like target results in oriented nanofiber deposition (5). It is possible to achieve any nanofiber orientation by manipulating rotation speed and the local electrical field at the target. Several efforts are also being made to align nanofibers by using special collector configurations where fibers experience a local electrical field that forces them to align.

This chapter will only emphasize electrospinning of nanofiber matrices and their related soft tissue regeneration applications in skin and tendons. In this chapter, electrospun nanofiber morphology, pore structure, and diameters are analyzed by scanning electron microscopy (SEM). SEM micrographs are also used to study cell morphology, infiltration, and behavior following in vitro and in vivo experimentation. Cell proliferation and differentiation on nanofiber scaffolds are determined using a calorimetric assay to quantify cellular metabolic activity; a standard curve correlating absorbance of cell metabolic activity with cell number was used to calculate cell number. Confocal microscopy analysis was used to analyze fluorescently stained cells and provides valuable information on cell infiltration and cell viability. Detection of live and dead cells on the nanofiber scaffolds may not be possible using light microscopy due to scaffold opacity.

2. Materials

2.1. Electrospinning Polymer Sheets

1. Organic solvents: Tetrahydrofuran (THF) and N,N-dimethylformamide (DMF).

2. Synthetic polymer: poly(lactic-co-glycolic acid) 65:35 (PLAGA 65:35) (SurModics Pharmaceuticals, Birmingham, AL). Store at –20°C (see Note 1).

3. Vial: Glass vial (28 × 95 mm) with screw thread.

4. Paraffin film.

5. Shaker: Vortex mixer.

6. Syringe: 10-mL Luer-Lok tip.

7. Dispensing needle: 18-gauge, 1.0″ length, blunt-end, stainless needle with threaded cap.

8. DC power: HV power supply (Gamma High Voltage Research, Ormond Beach, FL).

9. Pump: Aladdin-1000 syringe.

10. Lab jack.

11. Grounded target: heavy duty aluminum foil.

12. Fine tip tweezers.

2.2. Scanning Electron Microscopy

1. Double-sided carbon tape (8 mm × 20 mm).

2. Aluminum SEM specimen mount stubs.

3. Sharp blade or scalpel.

4. Sputter coater with gold foil.

2.3. Absorbance Assay for Proliferation Using Primary Adipose-Derived Stromal Cells

1. Ethyl alcohol 200 proof. Working solution is prepared by diluting to 70% ethyl alcohol using sterile double-distilled deionized water.

2. UV light source (cell culture hood).

3. Sterile gauze.

4. Sterile fine tip tweezers.

5. Dulbecco's phosphate-buffered saline 1× (PBS).

6. Dulbecco's modified Eagle's medium (DMEM) low glucose 1×.

7. Fetal bovine serum (FBS).

8. Penicillin streptomycin (P/S) having 10,000 units/mL penicillin and 10,000 μg/mL streptomycin.

9. 0.5% Tryspin–EDTA (10×).

10. Sterile T-75 cell culture flasks.

11. Aseptic cell culture hood.

12. Aseptic cell culture incubator.

13. Sterile non-tissue culture-treated plate, 24-well (Non-TCP).

14. CellTiter 96 AQ$_{ueous}$ One Solution Cell Proliferation Assay (Promega, Madison, WI).

15. 10% Sodium dodecyl sulfate (SDS).

16. 96-Well cell culture plate.

2.4. Confocal Microscopy of Live and Dead Cells

1. Dulbecco's PBS 1×.

2. Live/Dead Viability/Cytotoxicity kit for mammalian cells (Molecular Probes, Eugene, OR).

3. Fine tip tweezers.

4. Lab-Tek two-well glass chamber slide (Nalge Nunc, Naperville, IL).

---

3. Methods

Electrospinning system and process parameters vary with the use of different polymers, which have different physical and chemical properties. Solution viscosity is one property that we have found to affect the morphology of the nanofiber scaffolds greatly. For instance, a lower molecular weight copolymer solution (low viscosity) will result in electrospray (i.e., bead formation resulting from breakage of the polymer jet as it is deposited on the target), while a higher molecular weight copolymer solution will result in electrospun fibers (17). Furthermore, as the viscosity increases, the fiber diameter increases. This methodology exhibits very high reliability and reproducibility as evidenced by the consistent fiber diameter distribution, porosity, pore size, and degradation properties. Herein, electrospun scaffold morphology (e.g., bead formation and fiber diameter) is evaluated as a function of varying process parameters. Fiber diameter is quantified using imaging software.

Cell proliferation response to nanofiber scaffolds of various fiber diameters is determined by evaluating cellular metabolic activity or DNA content. A cell proliferation assay that can be used to quantify cell number as a function of metabolic activity is discussed. This assay gives results that are indicative of the inhibitory or stimulatory effect on cell growth over the time course of cell culture. Furthermore, it is reported in the literature that cells exhibit recognition of the nanofiber dimensions and orientation (18). Fibroblastic cells are known to exhibit a round or a flat well-spread morphology depending on the surface they are seeded onto (19). We use confocal microscopy to determine the localization of live and dead cells on the scaffold and the shape of the cells present on the nanofibers.

3.1. Electrospun PLAGA Sheets

1. Prepare a polymeric solution of PLAGA 65:35, by weighing accurately 2.2 g of PLAGA 65:35 polymer pellets in a glass vial and then adding 10 mL of an organic solvent composed of 3:1 THF:DMF. Tightly cap the vial immediately and wrap in paraffin film. Shake the vial containing the polymer and organic solvent vigorously overnight at room temperature to dissolve the polymer completely, which will yield a 22% w/v polymer solution (see Note 2).

2. Place aluminum foil of dimension 10 cm × 10 cm in the center of a plastic sheet (¼ in. thickness and such that it fits inside the plastic chamber) with a ½ in. diameter hole in the middle.

Place the plastic sheet upright inside the electrospinning chamber and fix its position. Pass the grounded cable through the hole in the middle of the plastic sheet and attach it to the now vertically oriented aluminum foil target (see Note 3). A schematic of the electrospinning setup is shown in Fig. 1.

3. Place the syringe pump on the lab jack placed inside the chamber. Place an empty 10-mL syringe with an 18-G needle attached onto the syringe pump with the blunt end of the needle facing the vertically oriented aluminum foil. Adjust the height of the lab jack such that the needle tip is aligned with the center of the plastic sheet. Adjust the horizontal distance between the needle tip and the grounded aluminum foil to be 30 cm (this is referred to as the air-gap distance). Remove the empty syringe from the syringe pump and detach the needle.

4. Fill the 10-mL syringe with 8 mL of the polymer solution and attach an 18-G blunt end needle to it. Remove all air bubbles from the polymer solution contained in the syringe (see Note 4).

5. Place the syringe–needle assembly containing the polymer solution onto the syringe pump and set the flow rate to 4.0 mL/h.

6. Attach the DC voltage source to the tip of the stainless steel blunt needle. Turn on the DC voltage and adjust it to 20 kV (see Note 5). Switch off the power supply and the syringe pump when all the polymer solution has been electrospun.

7. Remove the aluminum foil containing the electrospun scaffold from the plastic sheet. Peel off the aluminum foil using tweezers to separate the scaffold, and store at least overnight under vacuum or until further use. The vacuum will remove any residual solvent. An example of the structural changes resulting from the use of varying polymer concentrations is shown in Fig. 2. Scaffolds can be fabricated into various shapes and size as shown in Fig. 3 (see Note 6).

3.2. SEM

1. Place double-sided carbon tape onto the aluminum specimen mount stubs.

2. Cut the electrospun scaffolds using a sharp blade into squares of side 0.5 cm. Attach the scaffolds to the carbon tape and sputter-coat with gold for 150 s at 60 mA current and below 10^{-1} mbar vacuum. Sputter-coating deposits a conductive metal on the scaffold to enable imaging using the electron beam current (see Note 7).

3. Prepare SEM images of the scaffolds and determine fiber diameter using NIH Image J software available freely at http://rsbweb.nih.gov/ij/.

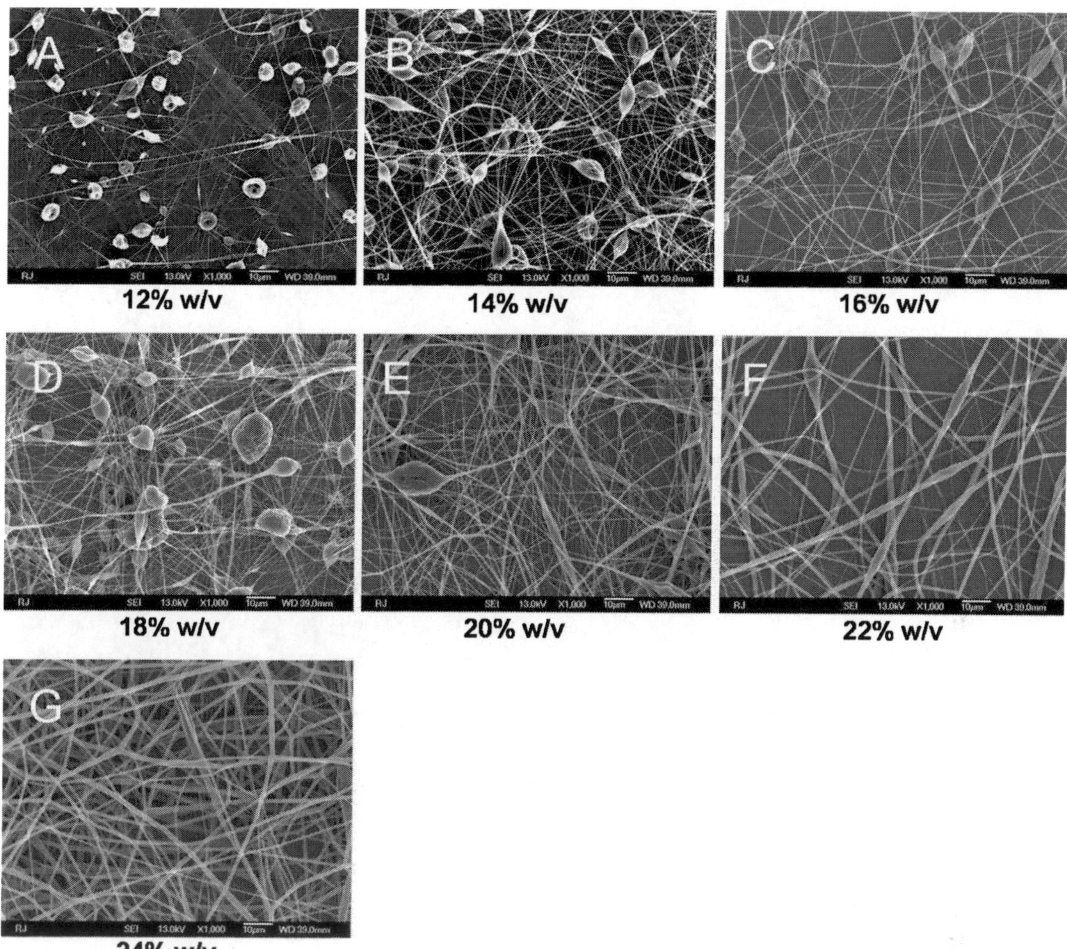

Fig. 2. SEM image of electrospun scaffolds fabricated from (**a**) 12%, (**b**) 14%, (**c**) 16%, (**d**) 18%, (**e**) 20%, (**f**) 22%, and (**g**) 24% w/v PLAGA 65:35 polymer in 3:1 THF:DMF pumped at a flow rate of 4 mL/h from a 18-G blunt-end needle with an applied electric potential of 20 kV across an air-gap distance of 30 cm. At 12% w/v, the SEM image shows beads of polymer attached by a thin fiber of polymer. This is referred to as beads-on-a-string morphology. With increasing concentration/viscosity of the polymer solution, the rounded beads begin to flatten out with more fibers being fabricated. At concentrations of 20, 22, and 24% w/v polymer, beads are minimal or not present at all, and the fabricated scaffold is composed completely of nano-diameter fibers.

3.3. Cell Proliferation Assay

1. The following instructions assume familiarity with primary cell isolation and aseptic cell culture techniques. Isolate adipose-derived stromal cells from the inguinal fat pad of wild-type Fischer 344 rats and maintain in culture using low glucose DMEM supplemented with 10% FBS and 1% P/S at 37°C in a humidified incubator. DMEM provides the nutrient media for the cells, in which they survive, grow, and divide. FBS contains a rich variety of proteins which helps maintain cultured cells in media. P/S is an antibiotic to prevent bacterial

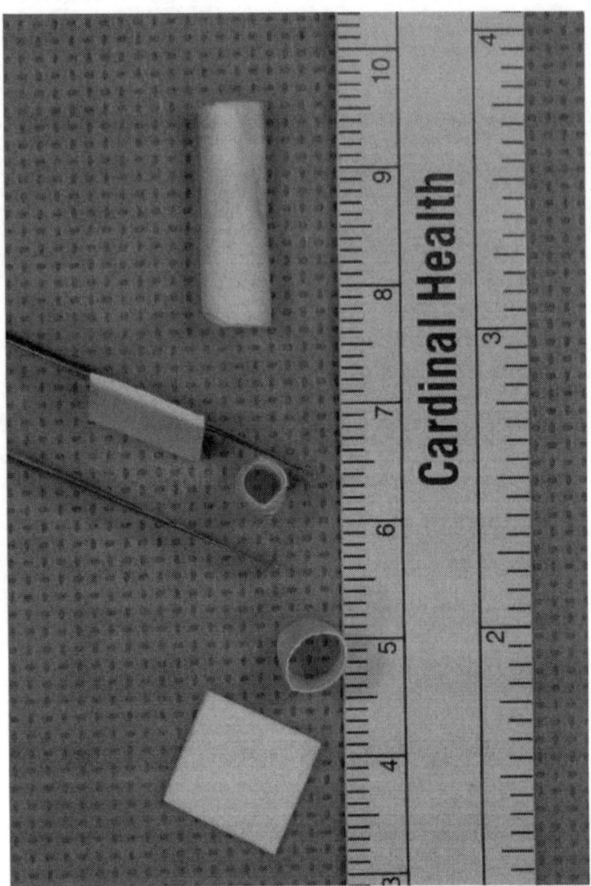

Fig. 3. Electrospun PLAGA 65:35 nanofiber scaffolds fabricated as sheets and tubes of varying diameter. The electrospinning technique is highly versatile and can be readily modified to fabricate scaffolds of different shapes and sizes. Based on the type of collector/target and its motion, it is possible to shape the scaffold and align the deposited fibers. The use of a rotating mandrel/drum results in tubular scaffolds as shown.

contamination during culture. Passage the cells when they are 70% confluent. Detach the cells using 0.05% trypsin to provide new maintenance cultures in T-75 flasks.

2. Cut the electrospun scaffolds into squares of side 1 cm and sterilize them for 5 min in 70% ethanol and then through UV sterilization in a cell culture hood for 30 min each on both sides. Soak the sterile scaffolds overnight in low glucose DMEM at 37°C in a humidified incubator. Incubation overnight in media will wet the scaffolds and allow the media to infiltrate into the porous structure (see Note 8).

3. Ensure that the cells have reached 70% confluence in the T-75 flasks. Rinse the flasks gently with warmed PBS and lift the monolayer of cells using 0.05% trypsin for 5 min. Neutralize trypsin activity with low glucose DMEM supplemented with 10% FBS and gently pellet the cells for 5 min at $500 \times g$.

Resuspend the cell pellet in DMEM and dilute using media to 30,000 cells/100 µL.

4. Place the wetted scaffolds into wells of a 24-well non-TCP plate.

5. Place 100 µL of the cell suspension onto each scaffold and incubate for 2 h at 37°C in a humidified chamber.

6. Add 0.5 mL of supplemented DMEM into each well and incubate at 37°C in a humidified chamber. Replace with fresh supplemented low glucose DMEM every 3 days (see Note 9).

7. The following instructions assume the availability of and familiarity with an absorbance spectrophotometer. Aspirate the media present in the wells. Gently rinse the cell-seeded scaffolds three times with 1 mL of warmed PBS solution. Transfer the scaffold using forceps to a new well plate, incubate with 1 mL of DMEM, and add 200 µL of MTS reagent (provided in the cell proliferation assay kit). MTS reagent will be metabolized in the mitochondria of live cells into a colored product, which is then quantified by absorbance spectroscopy.

8. Incubate the scaffolds for 2 h in a humidified chamber at 37°C. Stop the reaction with the addition of 250 µL of 10% SDS.

9. Transfer 250 µL of the incubated solution from each scaffold into a single well of a 96-well plate. Measure the absorbance at 490 nm (see Note 10). At 490 nm, the absorbance of the metabolized product of the MTS reagent is quantified.

10. Create a standard curve using samples with known numbers of primary adipose-derived stromal cells to correlate absorbance to cell number. An example result of cellular proliferation of MSCs on a 22% w/v PLAGA 65:35 electrospun nanofiber scaffold is shown in Fig. 4. Proliferation of human skin fibroblasts on PLAGA 50:50 electrospun nanofiber scaffolds composed of varying fiber diameters is depicted in Fig. 5.

3.4. Confocal Microscopy

1. Prepare PBS solution of 4 µM ethidium homodimer-1 (EthD-1) and 2 µM calcein AM (EthD-1 and calcein AM are reagents supplied in the assay kit) using the stock solution reagents. Cover and protect from light as the reagents are sensitive to light (see Note 11).

2. Rinse the cell–scaffold constructs with warmed PBS three times and then transfer them to a glass chamber slide.

3. Add PBS reagent solution to the wells to cover the cell–scaffold constructs sufficiently. Incubate at room temperature and protected from light for up to 45 min.

4. Wash the scaffolds with warmed PBS and transfer to clean glass chamber slide. Keep the scaffolds wet using PBS.

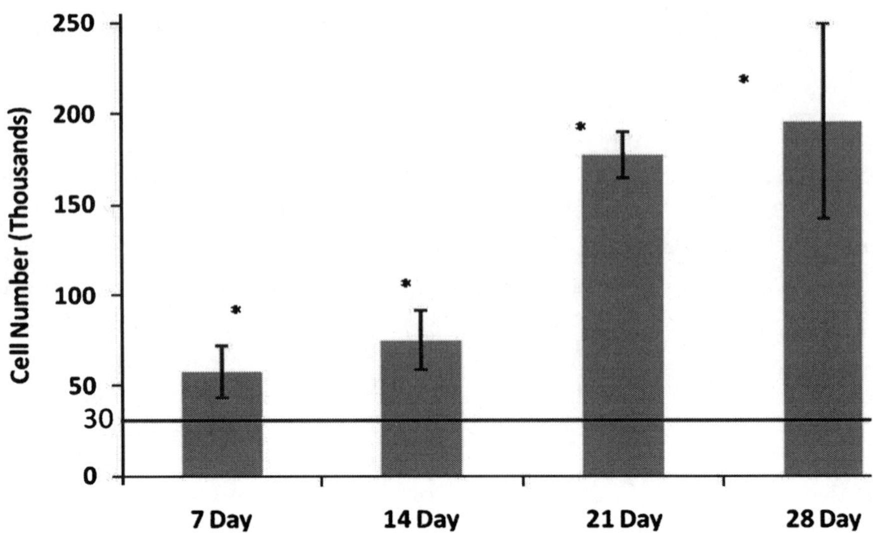

Fig. 4. Cellular proliferation of primary adipose stromal cells over 28 days in vitro culture on PLAGA 65:35 electrospun scaffold sheets fabricated from a 24% w/v polymer solution in 3:1 THF:DMF using process parameters of 20 kV potential, 30 cm air gap, 4.0 mL/h flow rate, and 18-G blunt needle. Cellular proliferation was significantly upregulated at 7 days and at 21 days in culture. Initial seeding density was 30,000 cells/scaffold ($n = 4$, $p < 0.05$, *asterisk* indicates significant difference). Stromal cells seeded on two-dimensional PLAGA 65:35 flat sheets proliferated very rapidly, but easily detached during media changes and washing with PBS (data not shown).

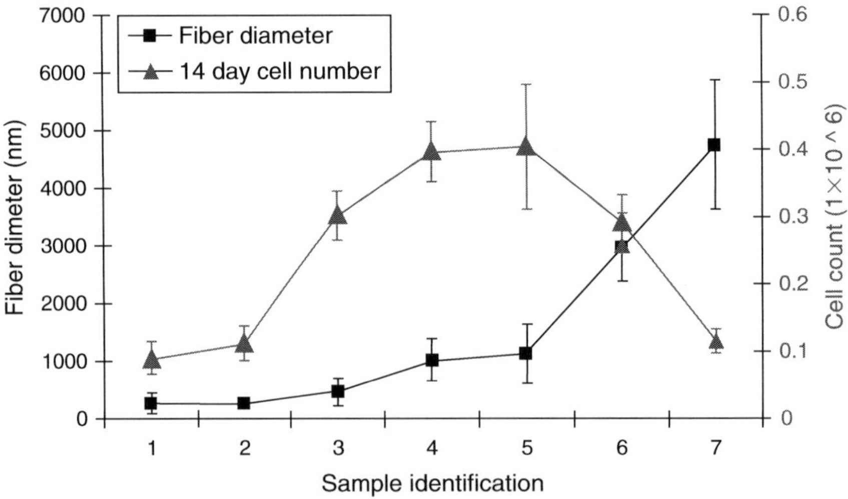

Fig. 5. Cellular proliferation of human skin fibroblasts seeded onto PLAGA 50:50 electrospun scaffolds at 14 days. Scaffolds of varying polymer concentrations were electrospun to fabricate meshes of different fiber diameters. Cell numbers were significantly higher on scaffolds composed of fibers 350–1,100 nm in diameter (i.e., Matrix 3–5). The proliferation trend was consistent throughout all the time points studied up to 28 days in culture. Fiber diameter range for the electrospun scaffolds is measured from SEM images: Matrix 1 and Matrix 2: 200–300 nm (same diameter range), Matrix 3: 250–467 nm, Matrix 4: 500–900 nm, Matrix 5: 600–1,200 nm, Matrix 6: 2,500–3,000 nm, and Matrix 7: 3,250–6,000 nm. Reprinted from ref. 19 with permission from Elsevier.

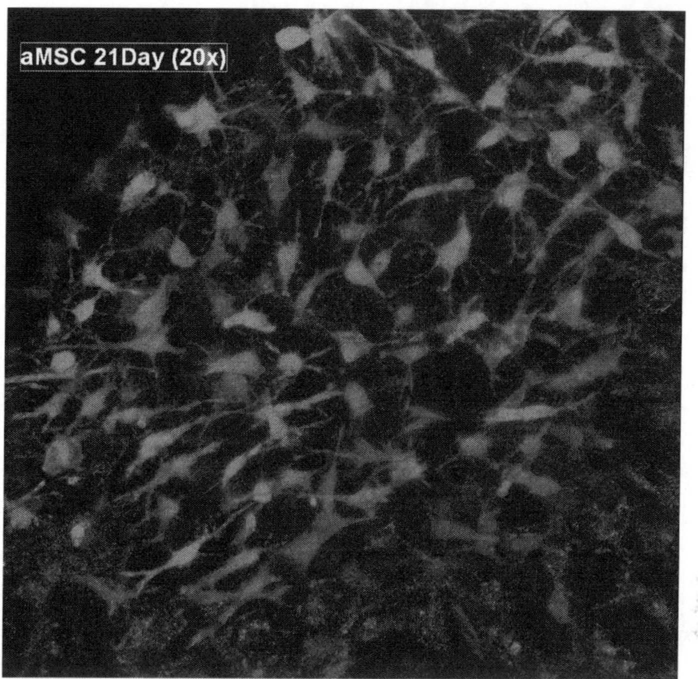

Fig. 6. Confocal microscopy image (×20) of aMSCs on a PLAGA 65:35 electrospun nano-fiber scaffold sheet at 21 days in vitro culture. Live cells have been stained using the Live/Dead Viability/Cytotoxicity kit for mammalian cells. The aMSCs show a well-spread morphology with extending cellular processes both at the surface and at various depths into the PLAGA 65:35 nanofiber scaffold.

5. View the constructs using confocal microscopy. Excitation at 515 nm induces calcein AM emission (green emission, em=495 nm) for the detection of live cells present on the scaffold, while excitation at 635 nm induces EthD-1 (red emission, em=495 nm) for the detection of dead cells on the scaffold. Software can be used to overlay the fluorescence images. Examples of primary adipose stromal cells present on the electrospun PLAGA 65:35 nanofiber scaffolds are shown in Fig. 6.

4. Notes

1. Long-term storage of PLAGA is best at –20°C. Additionally, it is best to aliquot smaller amounts of the polymer pellets soon after the sealed pouch is opened. Prepare the aliquots and wrap the capped end of the tubes or vials in paraffin film. The polymer can then be stored long term until needed without repeated opening, thus avoiding moisture condensation. Additionally, allow the vials/tubes containing the stored polymer to come to room temperature before opening.

When possible, use the same batch of the polymer. There may be variations in the polydispersity index between different manufactured batches. These differences may affect the electrospinning processing parameters, which then may require modification.

2. Always use adequate protection (organic solvent-resistant gloves and protective eye wear) and work with organic solvents inside a chemical fume hood. Prepare all the pipettes and glass vials, and weigh out the polymer before measuring the organic solvent since the solvent evaporates rapidly. Delay in capping the vials will lead to solvent evaporation altering the final polymer solution concentration. Always prepare a fresh solvent (i.e., THF:DMF 3:1) just before adding it to the polymer pellets. Store protected with similar organic solvents in a fireproof cabinet. Ideally, use the polymer solution as soon as it is dissolved. Do not keep the polymer solution in glass vials for long durations as the solvent may evaporate in small amounts from the capped vials.

3. Electrospinning setup must be housed inside a transparent plastic-walled chamber exhausted to a chemical fume hood. The inside of the plastic box or chamber must be accessible as necessary and the chamber designed to be completely closed to prevent solvent fumes from leaking out during the fabrication process and to prevent accidental contact with the high-voltage power source.

4. Prepare the electrospinning setup in advance. Transferring the polymer solution to the syringe, removing air bubbles from the syringe, and placing the syringe onto the syringe pump should be the penultimate step before attaching the high-voltage power source, closing the chamber, and turning the power on.

5. When charged, ensure that the polymer droplet forming at the needle end is moving toward the target. When sufficient electric potential is applied, the rounded polymer droplet coming out of the needle end becomes conical in shape. This phenomenon is referred to as a Taylor cone. Due to solvent evaporation occurring at the needle end, some polymer may deposit and clog the needle orifice. In such a scenario, turn off the voltage source and then open the chamber to access the needle. Stop the syringe pump and clean the needle tip using Kimwipes.

6. When about 10–20 mL of polymer solution is deposited onto a 10 cm × 10 cm aluminum foil target, the polymer sheet can be readily separated without tearing. The electrospinning parameters can be readily modified to accommodate other polymer–solvent systems. Additionally, the target can be modified to include different shapes and rotational movement

such as a rotating mandrel. For any new polymer, it is best to prepare different polymer concentrations in suitable solvents. Electrospinning small volumes onto smaller targets will enable one to narrow down a working range for the system parameters.

7. Quantification of the fiber diameter data is done by taking measurements of 100 individual fibers within each image and averaging the results.

8. Immersion in ethanol may cause shrinkage of some electrospun scaffolds. Thicker and bigger scaffolds may accommodate these dimensional changes.

9. Media should be added gently along the sides of the well following incubation of the wetted scaffold with the cell suspension.

10. Make sure no bubbles are present in the incubated solution transferred to the 96-well plate prior to absorbance measurement. Gently pop any bubbles using a sharp needle. Additionally run a blank sample, which is a wetted scaffold without any cells. This control sample will determine any interference of the polymer with the assay reagents.

11. For economy, only prepare enough PBS solution with 4 µM EthD-1 and 2 µM calcein AM to just cover the scaffold. Staining is done in chamber wells where the gasket around the samples will retain the solutions.

Acknowledgments

The authors gratefully acknowledge funding from the NIH (R01 EB004051 and R01 AR052536), the NSF-(EFRI-0736002), and the Coulter Foundation (526203-FY10-01).

References

1. Almarza, A. J., Yang, G., Woo, S. L., Nguyen, T., and Abramowitch, S. D. (2008) Positive Changes in Bone Marrow-Derived Cells in Response to Culture on an Aligned Bioscaffold. *Tissue Eng. Part A* **14**, 1489–1495.

2. Li, S., Patel, S., Hashi, C., Huang, N. F., and Kurpinski, K. (2007) *Biomimetic Scaffolds made from Biodegradable Nanofibers Covalently Linked with Extracellular Matrix and/or Animal Growth Factors and Entrapped Cells.* PCT Designated States: Designated States W: AE, AG, AL, AM, AT, AU, AZ, BA, BB, BG, BR, BW, BY, BZ, CA, CH, CN, CO, CR, CU, CZ, DE, DK, DM, DZ, EC, EE, EG, ES, FI, GB, GD, GE, GH, GM, GT, HN, HR, HU, ID, IL, IN, IS, JP, KE, KG, KM, KN, KP, KR, K (TRUNCATED).

3. Laurencin, C. T., Nair, L. S., Bhattacharyya, S., Allcock, H. R., Bender, J. D., Brown, P. W., and Greish, Y. E. (2005) *Polymeric Nanofibers for Tissue Engineering and Drug Delivery.* Main IPC: A61L027-18; Secondary IPC: A61L027-50, PCT Designated States: Designated States W: AE, AG, AL, AM, AT, AU, AZ, BA, BB, BG, BR, BW, BY, BZ, CA, CH, CN, CO, CR, CU, CZ, DE, DK, DM, DZ, EC, EE, EG, ES, FI, GB, GD, GE, GH,

GM, HR, HU, ID, IL, IN, IS, JP, KE, KG, KP, KR, KZ, LC, LK, LR, L (TRUNCATED).

4. James, R., Kesturu, G., Balian, G., and Chhabra, A. B. (2008) Tendon: Biology, Biomechanics, Repair, Growth Factors, and Evolving Treatment Options. *J. Hand Surg. [Am]* **33**, 102–112.

5. Kumbar, S. G., James, R., Nukavarapu, S. P., and Laurencin, C. T. (2008) Electrospun Nanofiber Scaffolds: Engineering Soft Tissues. *Biomed. Mater.* **3**, 034002.

6. Kumbar, S. G., Nair, L. S., Bhattacharyya, S., and Laurencin, C. T. (2006) Polymeric Nanofibers as Novel Carriers for the Delivery of Therapeutic Molecules. *J. Nanosci. Nanotechnol.* **6**, 2591–2607.

7. Kumbar, S. G., Nukavarapu, S. P., James, R., Hogan, M. V., and Laurencin, C. T. (2008) Recent Patents on Electrospun Biomedical Nanostructures: An Overview. *Recent Pat. Biomed. Eng.* **1**, 68–78.

8. He, W., Yong, T., Teo, W. E., Ma, Z., and Ramakrishna, S. (2005) Fabrication and Endothelialization of Collagen-Blended Biodegradable Polymer Nanofibers: Potential Vascular Graft for Blood Vessel Tissue Engineering. *Tissue Eng.* **11**, 1574–1588.

9. Janjanin, S., Li, W. J., Morgan, M. T., Shanti, R. M., and Tuan, R. S. (2008) Mold-Shaped, Nanofiber Scaffold-Based Cartilage Engineering using Human Mesenchymal Stem Cells and Bioreactor. *J. Surg. Res.* **149**, 47–56.

10. Koh, H. S., Yong, T., Chan, C. K., and Ramakrishna, S. (2008) Enhancement of Neurite Outgrowth using Nano-Structured Scaffolds Coupled with Laminin. *Biomaterials* **29**, 3574–3582.

11. Park, Y. and Choo, J. E. (2007) *Self-Assembly Nanocomposites Comprising Hydrophilic Bioactive Peptides and Hydrophobic Materials for Surface Treatment of Biomaterials and Tissue Regeneration Therapy.* PCT Designated States: Designated States W: AE, AG, AL, AM, AT, AU, AZ, BA, BB, BG, BR, BW, BY, BZ, CA, CH, CN, CO, CR, CU, CZ, DE, DK, DM, DZ, EC, EE, EG, ES, FI, GB, GD, GE, GH, GM, GT, HN, HR, HU, ID, IL, IN, IS, JP, KE, KG, KM, KN, KP, KZ, L (TRUNCATED).

12. Li, X. T., Zhang, Y., and Chen, G. Q. (2008) Nanofibrous Polyhydroxyalkanoate Matrices as Cell Growth Supporting Materials. *Biomaterials* **29**, 3720–3728.

13. Stupp, S. I., Hartgerink, J. D., and Niece, K. L. (2004) *Self-Assembling Peptide-Amphiphiles and Self-Assembled Peptide Nanofiber Networks for Tissue Engineering.* Main IPC: C07K, PCT Designated States: Designated States W: AE, AG, AL, AM, AT, AU, AZ, BA, BB, BG, BR, BY, BZ, CA, CH, CN, CO, CR, CU, CZ, DE, DK, DM, DZ, EC, EE, EG, ES, FI, GB, GD, GE, GH, GM, HR, HU, ID, IL, IN, IS, JP, KE, KG, KP, KR, KZ, LC, LK, LR, LS, L (TRUNCATED).

14. Liu, X. and Ma, P. X. (2009) Phase Separation, Pore Structure, and Properties of Nanofibrous Gelatin Scaffolds. *Biomaterials* **30**, 4094–4103.

15. Jiang, H., Hu, Y., Li, Y., Zhao, P., and Zhu, K. (2005) *Preparation of Core-Shell Nano/Micro-fiber or Capsule for Drug Sustained Release.* Main IPC: A61K009-00, Patent Application Country: Application: CN; Patent Country: CN.

16. Andersen, E., Smith, D., and Reneker, D. (2005) *A Medical Device with Nanofiber Outer Surface Layer Incorporating Nitric Oxide and Poly(Ethylenimine) Diazeniumdiolate for Insertion in to Vascular System.* Main IPC: A61L029-00, PCT Designated States: Designated States W: AE, AG, AL, AM, AT, AU, AZ, BA, BB, BG, BR, BW, BY, BZ, CA, CH, CN, CO, CR, CU, CZ, DE, DK, DM, DZ, EC, EE, EG, ES, FI, GB, GD, GE, GH, GM, HR, HU, ID, IL, IN, IS, JP, KE, KG, KP, KR, KZ, LC, LK, LR, L (TRUNCATED).

17. Kumbar, S. G., Bhattacharyya, S., Sethuraman, S., and Laurencin, C. T. (2007) A Preliminary Report on a Novel Electrospray Technique for Nanoparticle Based Biomedical Implants Coating: Precision Electrospraying. *J. Biomed. Mater. Res. B* **81**, 91–103.

18. Li, W. J., Jiang, Y. J., and Tuan, R. S. (2006) Chondrocyte Phenotype in Engineered Fibrous Matrix Is Regulated by Fiber Size. *Tissue Eng.* **12**, 1775–1785.

19. Kumbar, S. G., Nukavarapu, S. P., James, R., Nair, L. S., and Laurencin, C. T. (2008) Electrospun Poly(Lactic Acid-Co-Glycolic Acid) Scaffolds for Skin Tissue Engineering. *Biomaterials* **29**, 4100–4107.

<h1>Chapter 17</h1>

Peptide Amphiphiles and Porous Biodegradable Scaffolds for Tissue Regeneration in the Brain and Spinal Cord

Rutledge G. Ellis-Behnke and Gerald E. Schneider

Abstract

Many promising strategies have been developed for controlling the release of drugs from scaffolds, yet there are still challenges that need to be addressed in order for these scaffolds to serve as successful treatments. The RADA4 self-assembling peptide spontaneously forms nanofibers, creating a scaffold-like tissue-bridging structure that provides a three-dimensional environment for the migration of living cells. We have found that RADA4: (1) facilitates the regeneration of axons in the brain of young and adult hamsters, leading to functional return of behavior and (2) demonstrates robust migration of host cells and growth of blood vessels and axons, leading to the repair of injured spinal cords in rats.

Key words: CNS regeneration, Spinal cord injury, Tissue repair, Self-assembling peptide, Nanofiber scaffold, Schwann cell, Neural progenitor cell, Surgery, Trauma, Nanomedicine

1. Introduction

Nanotechnology has ushered in a new era of possibilities in tissue and organ reconstruction. The ability of researchers to establish fine control of the nanodomain is making way for increased targeting of cell placement and therapeutic delivery, amplified by cell encapsulation and implantation. In particular, scaffolds play a central role in organ regeneration (1) and repair of central nervous system (CNS) tissues (2, 3). Also, drug delivering scaffolds may need to be combined with cells to obtain functional recovery in treating traumatic brain injury (TBI) or spinal cord injury (SCI) (4).

Acting as a template, scaffolds can guide cell proliferation, cell differentiation, and tissue growth, influencing the survival of transplanted cells or the invasion of cells from the surrounding tissue, by providing a surface for cell adhesion and migration. In addition, they can be used to deliver drugs at a rate designed to

Sarah J. Hurst (ed.), *Biomedical Nanotechnology: Methods and Protocols*, Methods in Molecular Biology, vol. 726,
DOI 10.1007/978-1-61779-052-2_17, © Springer Science+Business Media, LLC 2011

match the physiological need of the tissue (5). Yet, there are still challenges to be addressed in order for these scaffolds to serve as successful treatments (2, 4, 6–8). First, the cells must be precisely placed into the scaffold before it is implanted to prevent their migration. Second, in order for the tissue to be reconstituted from the surrounding area, the scaffold must be able to allow cells to migrate into it. Third, it is important to prevent the acidic breakdown of the cell scaffold, which, if allowed, results in an adverse environment for cell growth.

Scaffolds can be synthesized from a variety of different materials and each type of material has distinct advantages and disadvantages. Natural scaffolds, including alginate (9–11), chitosan (12–14), collagen (15, 16), fibrin (17–23), and hyaluronan (6, 24–26), impart intrinsic signals within the structure that can enhance tissue formation (27), contain sites for cell adhesion, and allow for cell attachment; they also exhibit similar properties to the tissues they are replacing (28). Some, but not all, natural materials also allow for cell infiltration (28). However, since these materials are obtained from natural sources, homogeneity of the product between batch preparations can be an issue (28), and purification is essential to prevent a foreign body response after implantation. For example, chitosan can cause an allergic reaction (29) while fibrins from blood products and collagen from animal products have been known to cause an immune response as well as transfer infectious agents from donor to recipient (28). In addition, the modulus of these natural scaffolds may be very different from that of the tissue in which they are implanted (30). This difference can cause expression patterns of genes in the cell that are not conducive to repair and that could lead to the development of tissues unable to function like the original tissues (31). In a pulsatile environment, these materials could shear away from the surrounding tissue, causing physical damage or even increased rates of cellular compression (30). Many flexible scaffold systems have been developed, utilizing both a natural (i.e., collagen) and a synthetic component that can degrade by both hydrolysis and collagenase degradation pathways, as well as support cell growth. This type of scaffold could possibly be used in soft tissue applications (32).

Unlike natural scaffolding materials, synthetic materials, including PEG [poly (ethylene glycol)] (33–36), PLA [poly (lactic acid)]/PGA [poly (glycolic acid)]/PLGA [poly (lactic-co-glycolic) acid] (37–41), and pHEMA-MMA [poly (2-hydroxyethyl methacrylate-co-methyl methacrylate)] (42–46), have known compositions and can be custom designed with specific mechanical or degradation properties or to minimize the immune response (27). Also, these polymers can be conjugated to produce materials with properties that exhibit only the beneficial aspects of the individual components (28). PLAs, PLGAs, and MMAs have been used to form scaffolds pre-impregnated with cells (47) which then can be implanted in the damaged tissue area (48). Unfortunately, the by-products of these degraded synthetic materials can be absorbed by the body and may cause pH changes around the implantation site, leading to necrosis, delayed apoptosis, and, in rare cases, pain (28).

Designed self-assembling peptides (SAPs) that spontaneously form nanofibers are the next generation of cell scaffolds to emerge (2, 6, 8, 49, 50). These peptides create a scaffold-like tissue-bridging structure that provides a three-dimensional (3D) environment for cell growth and migration, similar to the native extracellular matrix (6, 49, 51, 52). The scaffolds are assembled through the weak ionic bonds and van der Waals forces of the self-complementary peptides (2, 6, 8, 49, 50). Typically, these peptides are composed of alternating positive and negative L-amino acids that assemble into highly hydrated structures in the presence of physiological concentrations of salts, tissue culture media, and human body fluids, such as cerebrospinal fluid (2, 8, 53). The benefits of SAPs over some natural and/or synthetic scaffold materials are that they: (1) pose a minimized risk of carrying biological pathogens or contaminants (53, 54), (2) elicit no immune response, and (3) show excellent physiological compatibility as well as minimal cytotoxicity (55, 56). In addition, SAPs allow for high cell implantation densities and enable cells to migrate freely in and out of the scaffold and the surrounding tissue. Further, these materials can be designed to match the modulus of the surrounding tissue (30) and they can either be prebuffered or allowed to be buffered during implantation by any physiological fluid available from the surrounding sites (30). Finally, the pH in the local environment does not decrease during breakdown of this type of scaffold.

The RADA4 peptide is one family of amphiphilic molecules that contain both a hydrophobic and a hydrophilic face. These peptides spontaneously self-assemble into nanofibers about 10 nm in diameter when the ionic strength of the solvent is increased to above its acidity level or pH values are raised to neutrality (e.g., physiological salt concentrations, culture media, and buffers). The nanofibers are highly hydrated, with greater than 99% water content (16, 34, 35, 53); however, they can be mixed at higher concentrations to reduce their water content and change their physical properties. We have utilized this peptide scaffold to create tissue-bridging structures that provide a 3D environment for the migration of living cells that can be used to: (1) regenerate axons in the brain of hamsters, leading to functional return of behavior (55) and (2) repair the injured spinal cord of rats, demonstrating robust migration of host cells, and growth of blood vessels and axons into the scaffolds (57). This chapter outlines in detail the materials and methods that are needed to: (1) repair the brain and (2) repair the spinal cord using this self-assembling nanopeptide scaffold system.

2. Materials

2.1. Preparation of the RADA4 Solution

1. 1% RADA4 solution: 10 mg RADA4 powder (obtained from the Massachusetts Institute of Technology Center for Cancer Research Biopolymers Laboratory, Cambridge, MA) in 1 mL deionized (DI) water (Millipore Corp, Billerica, MA) (see Fig. 1).

2.2. Animals – Brain

1. 53 P2 Syrian hamster pups and adult Syrian hamsters (*Mesocricetus aurotus*) (see Note 1).
2. Sodium pentobarbital (50 mg/kg).
3. Gelfoam.
4. Wound clips (or suturing materials).
5. Knife.
6. Head holder.
7. Isotonic saline (typically used in any surgical procedure).

2.3. Animal Behavior Testing – Brain

1. Sunflower seeds.
2. Small black rubber ball or block, 1–1.5 cm in diameter.
3. White wire.
4. Cages (26 cm × 43 cm).
5. Video camera.

2.4. Preparation for Tracing Regenerated Axons – Brain

1. Sodium pentobarbital (50 mg/kg).
2. Glass micropipette with a tip diameter (~10 μm) attached to a Pico Spritzer (General Valve, Fairfield, NJ).

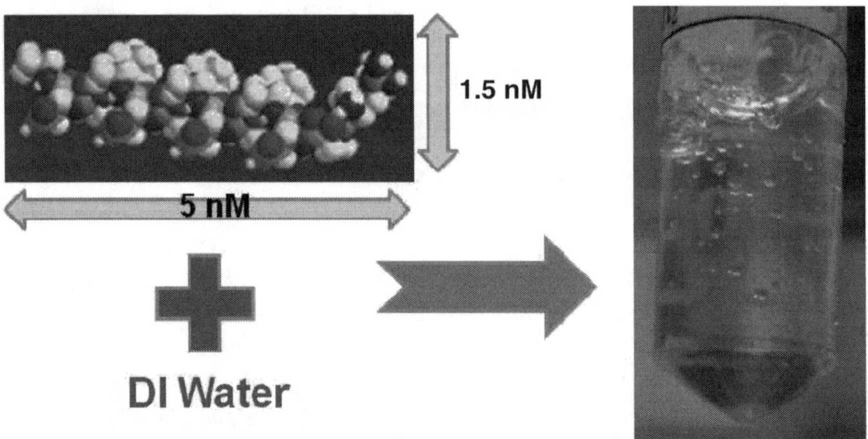

Fig. 1. Self-assembling peptide nanofiber scaffold RADA4. (*Left*) Molecular model of the RADA4 (arginine, alanine, aspartate, and alanine) building block. (*Right*) This peptide is a liquid when dissolved in deionized (DI) water (shown in a vial). When applied to a physiological environment, the material becomes a gel (data not shown).

3. 1% Cholera-toxin subunit B conjugated with FITC (CTB-FITC).

4. 0.9% NaCl.

5. 0.25% NaNO$_3$.

6. 2% Paraformaldehyde in phosphate-buffered saline (PBS) (pH=7.4).

7. 30% Sucrose.

8. Gelatin-coated slides.

2.5. Immunolabeling of the Optic Tract Axons – Brain

1. 0.1 M PBS (pH=7.4).

2. 2% Triton 100, 2% normal rabbit serum, 2.5% bovine serum albumin (BSA) in 0.1 M PBS (pH=7.4).

3. Goat anticholeragenoid (List Biological Laboratories, Campbell, CA) (1:8,000 dilution), 2% Triton 100, 2% normal rabbit serum, 2.5% BSA in 0.1 M PBS (pH=7.4).

4. Fluorescent donkey anti-IgG antibodies Alexa-488 (secondary antibody from Invitrogen – Molecular Probes) (1:200 dilution).

5. DAKO mounting medium (DAKO, Carpinteria, CA).

6. Fluorescence microscope.

7. Digital camera.

2.6. Animals – Spinal Cord (See Note 1)

1. Adult female wild-type Sprague–Dawley rats (220–250 g).

2. Adult female green fluorescent protein (GFP)-transgenic Sprague–Dawley rats ["green rat CZ-004" SD TgN (act-EGFP) OsbCZ-004] (220–250 g).

2.7. Pretreatment for RADA4 – Spinal Cord

1. Dulbecco's Modified Eagle Medium: Nutrient Mixture F-12 (DMEM/F12) with 10% fetal bovine serum (FBS) (Gibco, Grand Island, NY).

2.8. Isolation and Culture of Schwann Cells (ScCs) – Spinal Cord

1. 10% FBS in DMEM/F12.

2. 1.25 U/mL Dispase.

3. 0.05% Collagenase.

4. 15% FBS in DMEM/F12.

5. Poly-L-lysine (0.01%, Sigma, St. Louis, MO).

6. 20 µg/mL Pituitary extract (Sigma) and 2 µM forskolin (Sigma).

7. Ca^{2+}- and Mg^{2+}-free Hanks balanced salt solution (CMF-HBSS, Gibco).

8. 0.05% Trypsin in CMF-HBSS (Gibco).

9. 0.02% Ethylenediaminetetraacetic acid (EDTA) in CMF-HBSS (Gibco).

2.9. Isolation and Culture of Neural Precursor Cells (NPCs) – Spinal Cord

1. Ca^{2+}- and Mg^{2+}-free Hanks balanced salt solution (CMF-HBSS, Gibco).
2. DMEM/F12 supplemented with B27 (2%, Gibco).
3. N_2 (1%, Gibco).
4. Epidermal growth factor (EGF, 20 ng/mL, Gibco).
5. Basic fibroblast growth factor (bFGF, 20 ng/mL, Sigma).
6. Penicillin (100 U/mL).
7. Streptomycin (100 µg/mL).
8. Poly-L-lysine (0.01%, Sigma).

2.10. ScCs and NPCs in 3D Culture Within the RADA4 Scaffold – Spinal Cord

1. 10% FBS in DMEM/F12.
2. Two-photon confocal microscope (Zeiss LSM510 META, Jena, Germany).

2.11. Immunocyto-chemistry – Spinal Cord

1. 4% Paraformaldehyde.
2. Poly-L-lysine (0.01%, Sigma).
3. 30% Sucrose.
4. Gelatin-coated slides.
5. 1% BSA, 10% normal goat serum, 0.3% Triton X-100 in 0.1 M PBS (pH = 7.4).
6. Rabbit antip75 (1:200 dilution, Promega, Madison, WI).
7. Mouse anti-nestin (1:2,000 dilution, BD Biosciences, Cambridge, MA).
8. Rabbit anti-glial fibrillary acidic protein (anti-GFAP, 1:1,000 dilution, Chemicon, Temecula, CA).
9. Mouse anti-Rip (1:50 dilution, BD Biosciences, Cambridge, MA).
10. Mouse anti-β-tubulin type III (1:1,000 dilution, Sigma).
11. Fluorescent Alexa 568 goat anti-mouse or anti-rabbit secondary antibody (1:400 dilution, Molecular Probes, Eugene, OR).
12. DAKO mounting medium containing 4',6-diamidino-2-phenylindole (DAPI, 2 µg/mL).

2.12. Surgical Procedures – Spinal Cord

1. Ketamine (80 mg/kg).
2. Xylazine (10 mg/kg).
3. 11-0 suture.

4. Microscoop (suction can also be used).

5. Isotonic saline.

2.13. Preparation of Tissue Sections – Spinal Cord

1. 4% Paraformaldehyde in 0.1 M PBS (pH = 7.4).

2. Pentobarbital anesthesia.

3. 10% Sucrose and 4% paraformaldehyde in 0.1 M PBS (pH = 7.4).

4. 30% Sucrose in 0.1 M PBS (pH = 7.4).

5. Gelatin-coated slides.

6. Optimum cutting temperature compound (OCT).

2.14. Immunohisto-chemistry on Tissue Sections – Spinal Cord

1. 0.1 M PBS (pH = 7.4).

2. 1% BSA, 10% normal goat serum, and 0.3% Triton X-100 in 0.1 M PBS (pH = 7.4).

3. Mouse anti-nestin (1:2,000 dilution).

4. Rabbit anti-GFAP (1:1,000 dilution).

5. Mouse anti-Rip (1:50 dilution).

6. Mouse anti-β-tubulin III (1:500 dilution).

7. Mouse anti-NF200 (1:400 dilution, Sigma).

8. Rabbit anti-5HT (serotonin, 1:200 dilution, Sigma).

9. Rabbit anticalcitonin gene-related peptide (CGRP, 1:200 dilution, Sigma).

10. Mouse anti-ED1 (1:1,000 dilution, Serotec, Raleigh, NC).

11. Rabbit antip75 (1:200 dilution, Promega) for SCs.

12. Mouse anti-myelin basic protein (anti-MBP, Ipswich, MA, 1:1,000 dilution).

13. Fluorescent Alexa 568 goat anti-mouse or anti-rabbit secondary antibody (1:400 dilution).

14. DAKO mounting medium containing DAPI (2 μg/mL).

15. Two-photon confocal microscope.

2.15. Hematoxylin and Eosin (H&E) Staining and Alkaline Phosphatase (AP) Histochemical Staining – Spinal Cord

1. H&E and native endothelial alkaline phosphatase (AP).

2. Collagen-coated or -charged slides.

3. Distilled water.

4. Alum hematoxylin.

5. 0.3% Acid alcohol.

6. Scott's tap water substitute.

3. Methods

3.1. Preparation of the RADA4 Solution – Brain

1. Mix 10 mg of RADA4 powder in 1 mL of DI water, sonicate for 30 s, and filter (see Note 2).

3.2. Young Animals – Brain

1. Anesthetize 53 P2 Syrian hamster pups hypothermically using whole-body cooling. Open the scalp and sever the optic tract within the superior colliculus (SC) with a sharp surgical knife through a slot cut in the cartilaginous skull, extending 1.5 mm below the surface, from the midline to a point beyond the lateral margin of the SC.

2. Treat animals by injection into the brain wound with 10 μL of 1% RADA4 or 10 μL isotonic saline (control) (see Fig. 2). The cut and untreated controls receive no injection.

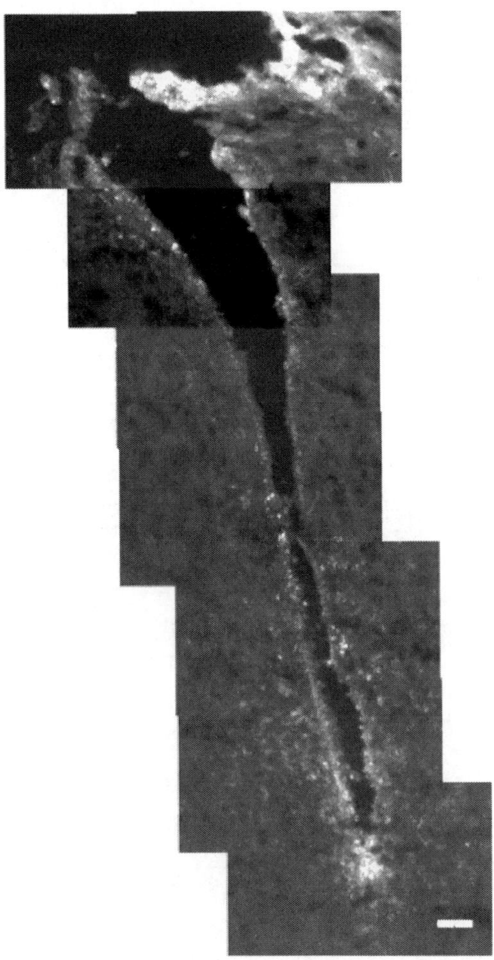

Fig. 2. Montage of parasagittal sections of a cut filled with saline solution (control), 72 h survival time postlesion. Scale bars = 100 μm.

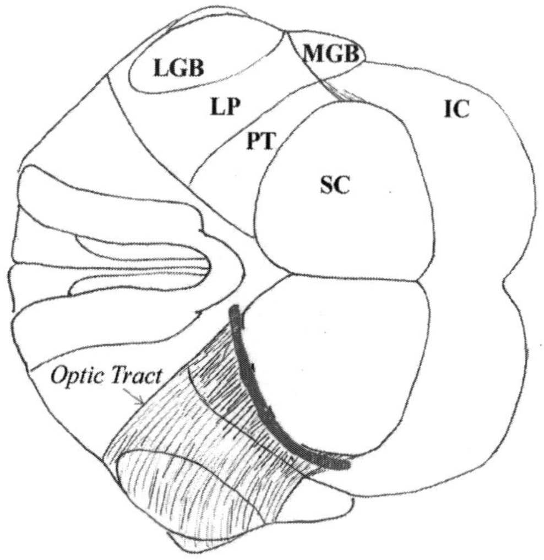

Fig. 3. Dorsal view reconstruction of the hamster brain with cortex removed. Rostral is to the left and caudal is to the right. The heavy black line depicts the location of the transection of the optic tract (brachium of the SC). The locations of the superior colliculus (SC), pretectal area (PT), lateral posterior nucleus (LP), medial geniculate body (MGB), lateral geniculate body (LGB), and inferior colliculus (IC) are shown.

3.3. Adult Animals – Brain (See Note 3 and Ref. 58)

1. Anesthetize adult Syrian hamsters with an intraperitoneal injection of sodium pentobarbital (50 mg/kg) and then fit them in a head holder. Expose and aspirate the overlying cortex to reveal the rostral edge of the SC and the brachium of the SC on the left side. Completely transect the brachium of the SC of each animal (see Fig. 3 and Note 4).

2. With the aid of a sterile glass micropipette, inject either 30 μL of 1% RADA4 solution or 30 μL isotonic saline (control) into the site of the lesion. The cut and untreated control group receives no injection.

3. Fill the surgical opening in the skull with saline-soaked gelfoam and close the overlying scalp with wound clips.

4. Perform behavior testing at several time points after the date of surgery (see Subheading 3.4).

3.4. Animal Behavior Testing – Brain (See Note 5)

1. Video-record animal behavior testing trials for analysis.

2. Allow the animal to wait, relatively motionless, on the platform for several seconds (see Note 6).

3. Test visually elicit orienting movements by presenting sunflower seeds (see Note 7) to part of the hamster's visual field (temporal or nasal, upper or lower), avoiding the nasal-most 45° (see Fig. 4) first by hand, and later with the aid of a white metal wire, on the end of which there is a small black rubber ball, slotted for holding a seed (see Note 8).

4. Make a minimum of ten presentations on each side, with random choice of side. Animals should be tested two to three times per week by two different investigators independently.

3.5. Preparation for Tracing Regenerated Axons – Brain

1. Anesthetize animals with an intraperitoneal injection of sodium pentobarbital (50 mg/kg).

2. Using a glass micropipette (tip diameter, ~10 µm) attached to a Pico Spritzer, deliver intraocular injections of 1 µL of 1% cholera-toxin subunit B conjugated with FITC into the vitreous humor of the right eye of each animal (see Fig. 5).

3. Return the animals to their cages, place them under a heat lamp, and monitor them until they recover.

4. Four days after intraocular injection, sacrifice the animals with an overdose of anesthetic (see Note 9) and perfuse transcardially with 0.9% NaCl and 0.25% $NaNO_3$ (pH = 7.4), followed by 2% paraformaldehyde in 0.1 M PBS (pH = 7.4).

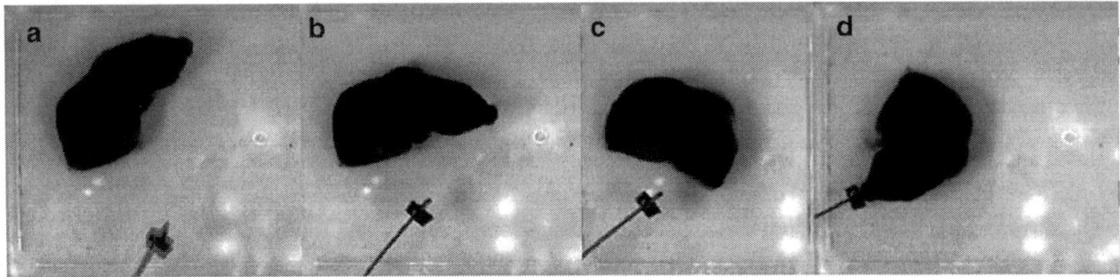

Fig. 4. Behavior. This adult animal turns toward the stimulus in the affected left visual field in small steps, prolonged here by movements of the stimulus away from him. Each frame is taken from a single turning movement, at times (**a**) 0, (**b**) 0.27, (**c**) 0.53, and (**d**) 0.80 s from movement initiation. The animal reached the stimulus just after the last frame. This is about 0.20 s slower than most turns by a normal animal.

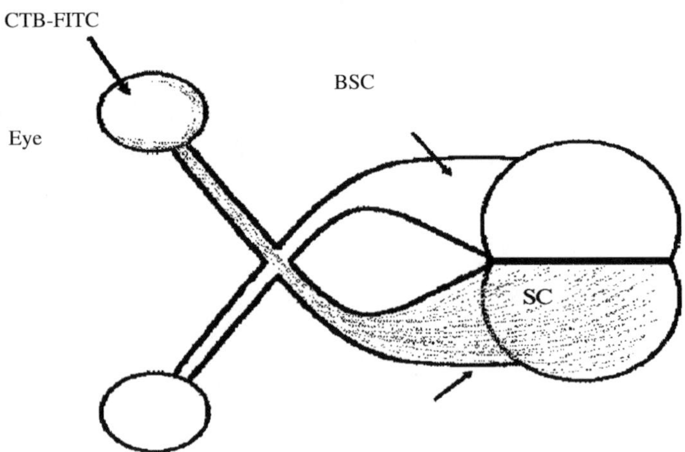

Fig. 5. The location of the eyes and location of the injection of CTB-FITC. This figure also illustrates how the eyes are connected to the SC.

5. Remove the brains and eyes and postfix in 2% paraformaldehyde at 4°C for 4 days. For cryoprotection, place brains in 30% sucrose at 4°C until they sink. Cut 30 μm parasagittal sections on a cryostat and mount them directly on gelatin-coated slides.

3.6. Immunolabeling of the Optic Tract Axons – Brain

1. Air-dry the mounted sections, wash three times with 0.1 M PBS (pH = 7.4) at 10 min intervals and preblock in 0.1 M PBS (pH = 7.4) containing 2% Triton 100, 2% normal rabbit serum, and 2.5% BSA for 30 min at room temperature (see Note 10).

2. Incubate the slides with goat anticholeragenoid (1:8,000 dilution), 2% Triton 100, 2% normal rabbit serum, 2.5% BSA for 48 h at room temperature (goat anticholergenoid labels CTB-FITC, previously injected into the eye to trace the optic pathway).

3. Wash the slides again three times in 0.1 M PBS (pH = 7.4) and incubate with fluorescent donkey anti-IgG antibodies Alexa-488 (secondary antibody) (1:200 dilution) for 1.5 h at room temperature in a light-protected chamber.

4. Wash the slides four to five times in 0.1 M PBS (pH = 7.4) at 5 min intervals and coverslip with DAKO to mount the sections.

5. Examine the sections under a fluorescence microscope and take pictures with a digital camera (see Fig. 6).

6. Analyze serial sections to reconstruct the locations of regenerated axons in the SC.

3.7. Pretreatment of RADA4 – Spinal Cord (See Note 11)

1. Because untreated RADA4 peptide has a very low pH (about 3–4) and will damage the host tissue, either in the dorsal column-transected or right-hemisected spinal cord, neutralize the RADA4 in culture medium before transplantation. Gently and quickly plate 1% RADA4 peptide to a dish in DMEM/F12 with 10% FBS. The self-assembly of the peptide into a scaffold material will occur as soon as the material contacts the medium.

2. Change the medium twice at 1, 10, and 30 min after plating, and once every 3 days during the following days.

3. On day 7, use the medium-treated RADA4 material for transplantation.

3.8. Isolation and Culture of Schwann Cells (ScCs) – Spinal Cord

1. Isolate the ScCs by cutting the sciatic nerves from adult GFP-transgenic Sprague–Dawley rats into 1-mm³ explants and place in culture dishes with DMEM/F12 supplemented with 10% FBS.

2. When the outgrowth of migratory cells (predominantly fibroblasts) reach a near-confluent monolayer around the explants (about 7 days), transfer the explants to new culture dishes with fresh medium.

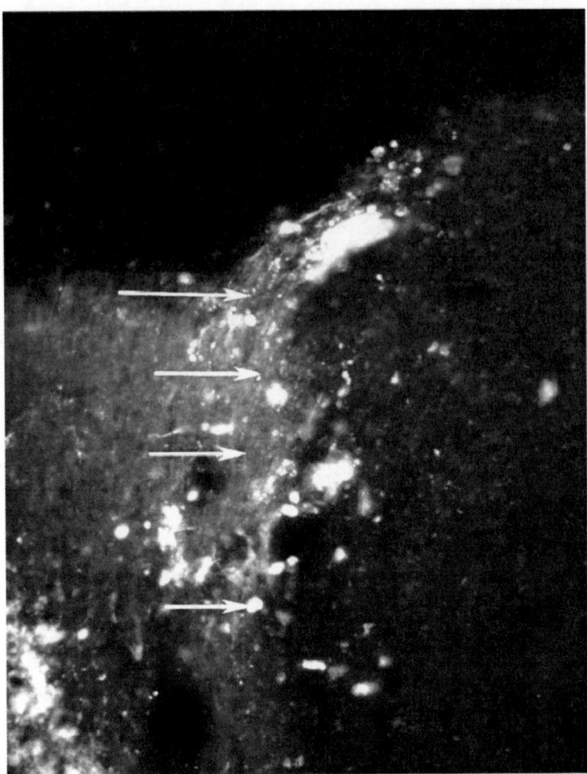

Fig. 6. Brain of 8-month old hamster. Dark-field photo of a parasagittal section from the brain treated with 1% RADA4 at the time of surgery in the lesion site. Rostral is *left* and caudal is *right*. Arrows show the location of the lesion. The axons have grown through the site of lesion and are reinnervating the SC. Note the lack of tissue disruption.

3. After three to five such passages (3–5 weeks), the cells that emerge from the explants will be primarily ScCs. Transfer the explants to a 35 mm dish containing 1.25 U/mL dispase, 0.05% collagenase, and 15% FBS in DMEM/F12 for incubation overnight at 37°C in 5% CO_2.

4. On the following day, dissociate the explants and plate the cells onto poly-L-lysine (0.01%)-coated dishes in DMEM/F12 with 10% FBS.

5. Later, re-feed the cultures with the same medium supplemented with 20 µg/mL pituitary extract and 2 µM forskolin for dividing.

6. When the ScCs reach confluence rinse in Ca^{2+}- and Mg^{2+}-free Hanks balanced salt solution (CMF-HBSS) and briefly treat with 0.05% trypsin and 0.02% EDTA in CMF-HBSS.

7. Wash cells twice in DMEM/F12 with 10% FBS and transfer into new dishes at a density of 2×10^6 cells per 100 mm dish.

8. When the cells reach confluence again, collect for transplantation.

3.9. Isolation and Culture of NPCs – Spinal Cord

1. Dissect the hippocampi of embryonic day 16 (E16) embryos of GFP-transgenic Sprague–Dawley rats in cooled CMF-HBSS and dissociate mechanically.

2. Collect the cells after centrifugation and re-suspend in DMEM/F12-B27 supplement (2%) N2 (1%), EGF (20 ng/mL), bFGF (20 ng/mL), penicillin (100 U/mL), and streptomycin (100 μg/mL).

3. Adjust the cells to 1×10^5 cells/mL and plant into culture flasks.

4. Replace the medium by one-half every 3 days.

5. Typically, the cells grow in suspending neurospheres (individual clones of spheres derived from neural stem/progenitor cells); mechanically dissociated them approximately once each week and re-seed at approximately 1×10^5 cells/mL.

6. Use the second neurospheres for transplantation. To assess the purification of the NPCs, dissociate some of the neurospheres and seed onto poly-L-lysine-coated plates with the same medium as above.

3.10. ScCs or NPCs in 3D Culture Within the RADA4 – Spinal Cord (See Note 12)

1. Before transplantation, culture the ScCs or NPCs within the RADA4 scaffold.

2. Collect the ScCs or NPCs and finally adjust to 5×10^5 cells/μL.

3. Then, mix 1 μL cell suspension with 9 μL RADA4 peptide; gently and quickly plate the mixture to a dish in DMEM/F12 with 10% FBS.

4. Change the medium twice at 1, 10, and 30 min after plating, and once every 3 days during the following days.

5. On day 7, the cultures can be used for transplantation (see Subheading 3.12).

6. Maintain some of the cultures for 4 weeks, take images of living cells using a two-photon confocal microscope, to determine cell-material viability and perform immunostaining.

3.11. Immunocytochemistry – Spinal Cord (See Note 13)

1. For immunostaining, directly fix the ScCs or dissociated cultured NPCs with 4% paraformaldehyde.

2. Plate the neurospheres of the NPCs on poly-L-lysine-coated coverslips and grow in culture medium. After the neurospheres are attached on the coverslips (about 1 h), fix with 4% paraformaldehyde.

3. Cryoprotect the fixed cultures of RADA4 with ScCs or NPCs in 30% sucrose, then cut the cryostat sections (10 μm) and mount on gelatin-coated slides.

4. Prior to exposure to antibodies, preblock all samples with 1% BSA, 10% normal goat serum, and 0.3% Triton X-100 in 0.1 M PBS (pH = 7.4) for 1 h at room temperature (25°C);

then apply the following primary antibodies: (1) rabbit anti-p75 (1:200) for identifying ScCs; (2) mouse anti-nestin (1:2,000) for NPCs; (3) rabbit anti-glial fibrillary acidic protein (anti-GFAP, 1:1,000) for astrocytes; (4) mouse anti-Rip (1:50) for oligodendrocytes; and (5) mouse anti-β-tubulin type III (1:1,000; Sigma) for neurons.

5. Incubate the sections with the primary antibody in 0.1 M PBS (pH = 7.4) plus 1% BSA and 0.3% Triton X-100 overnight at 4°C and treat with fluorescent Alexa 568 goat anti-mouse or anti-rabbit secondary antibody (1:400), respectively, for 2 h at room temperature.

6. Finally, mount the sections with mounting medium containing DAPI to counterstain the nuclei (see Figs. 7 and 8).

3.12. Surgical Procedures – Spinal Cord

1. Anesthetize the wild-type Sprague–Dawley rats with ketamine (80 mg/kg) and xylazine (10 mg/kg).

2. Perform a dorsal laminectomy on the sixth (C6) and seventh cervical (C7) segments. Incise the dura longitudinally and pull laterally. Make a spinal cord dorsal column between C6 and C7, followed by removing 1 mm dorsal column tissue (see Fig. 9).

3. With the assistance of a microscoop, transfer 5 µL precultured RADA4 scaffold or 5 µL cultured mixture of RADA4 with ScCs or NPCs (see Subheading 3.10) from the culture dish into the lesion cavity.

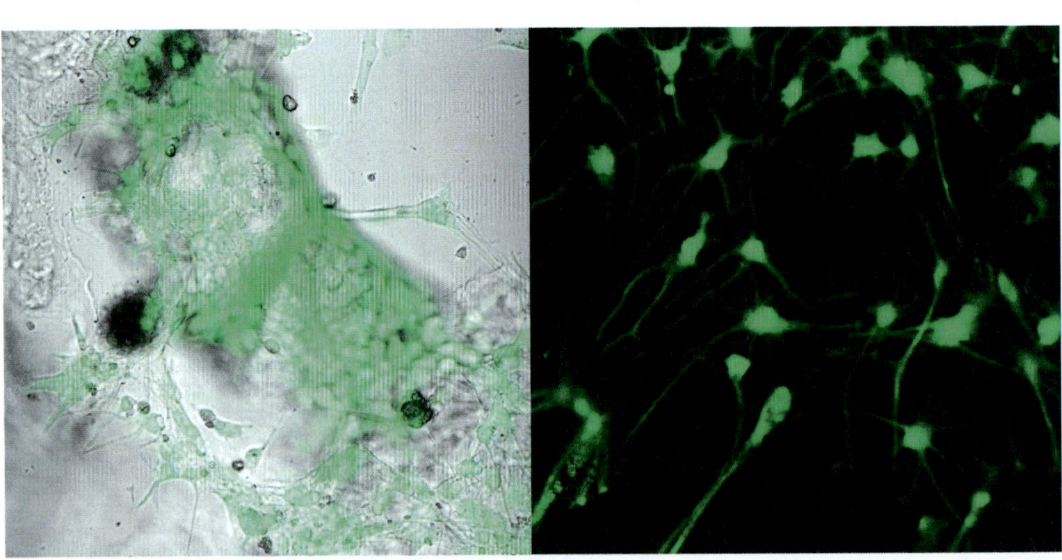

Fig. 7. Images of live neural precursor cells (NPCs) in 1% RADA4 that survived in cell culture for 1 month. (*Left*) Merged bright field and fluorescent images showing that the cells only grow where the RADA4 is and not in other parts of the culture well. (*Right*) Two-photon microscope picture 4 weeks after the cells were cultured within RADA4 in vitro.

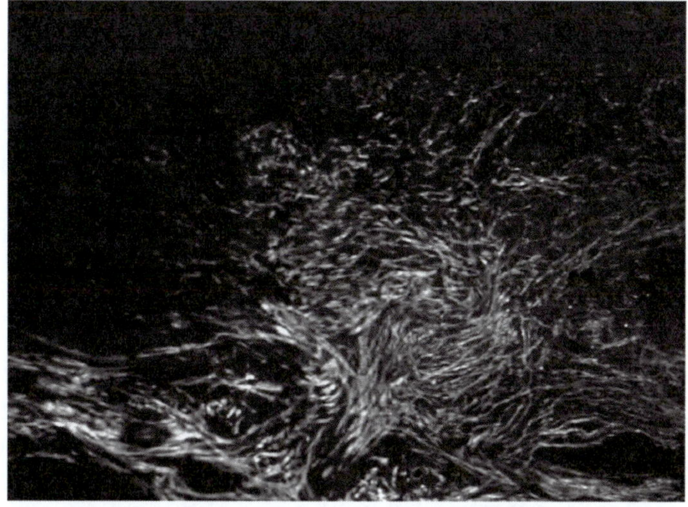

Fig. 8. RADA4 pretreated with Schwann cells (ScCs) added by culture medium before transplantation. The implants of RADA4 integrate very well with host tissue, and no obvious cavities or gaps are observed between the implants and host. Moreover, many cells migrated into the surrounding tissue from the implants, as shown by the GFP expressing cells (white) that have migrated beyond the boundaries of the original injury site. With this type of pretreatment, the cells survive very well in the implants. There is no evidence of RADA4 material in the implants after 8 weeks. Also, there is no sign of a barrier at the tissue implant interface.

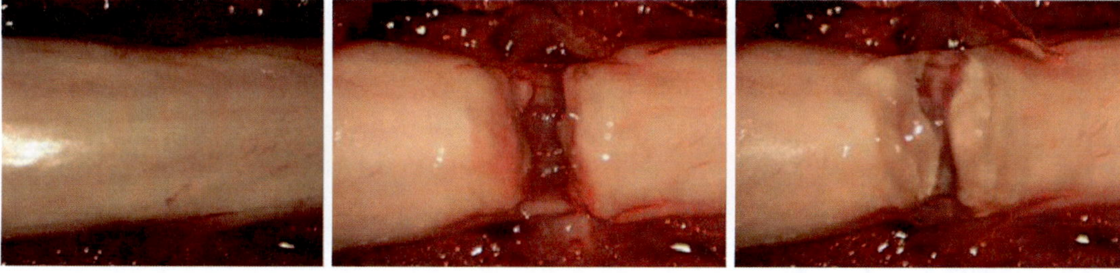

Fig. 9. Rat spinal cord. (*Left*) Intact rat spinal cord. (*Center*) Complete transection of the spinal cord. (*Right*) Transected spinal cord treated with RADA4 only. The molecule is 5 nm in length; the spinal cord is 2 mm.

4. As a control, place 5 µL uncultured RADA4 or saline in the lesion cavity.

5. After transplantation, close the dura by suturing with an 11-0 suture. The muscle layers and skin should also be closed with a suture.

3.13. Immunohisto-chemistry on Tissue Sections – Spinal Cord (See Note 14)

1. Perfuse the animals transcardially with 4% paraformaldehyde in 0.1 M PBS (pH = 7.4) after an overdose of pentobarbital anesthesia at a certain time point after transplantation.

2. Postfix the spinal cords overnight in the perfusing fixative plus 10% sucrose at 4°C.

3. Cryoprotect the tissues in 30% sucrose in 0.1 M PBS (pH = 7.4) for 12 h at 4°C.

4. Separate a 1 cm length of the spinal cord, centered at the injury site and embed it in OCT. Cut horizontal cryostat sections (30 μm) and mount onto gelatin-subbed slides (see Note 15). Store at –20°C.

5. Air-dry the frozen slides at room temperature for 30 min and wash with 0.1 M PBS (pH = 7.4) for 10 min. Then, preblock with 1% BSA, 10% normal goat serum, and 0.3% Triton X-100 in PBS for 1 h at room temperature.

6. Afterward, apply the following primary antibodies overnight at 4°C: mouse anti-nestin (1:2,000) for NPCs; rabbit anti-GFAP (1:1,000) for astrocytes; mouse anti-Rip (1:50) for oligodendrocytes; mouse anti-β-tubulin III (1:500) for neurons; mouse anti-NF200 (1:400) for axons; rabbit anti-5HT (serotonin, 1:200) for raphespinal axons; rabbit CGRP (1:200) for primary sensory axons; mouse anti-ED1 (1:1,000) for macrophages; rabbit antip75 (1:200) for SCs; and mouse anti-myelin basic protein (anti-MBP 1:1,000) for myelin.

7. Wash the slides in 0.1 M PBS (pH = 7.4) three times and incubate with fluorescent Alexa 568 goat anti-mouse or anti-rabbit secondary antibody (1:400) for 2 h at room temperature.

8. Coverslip the slides with DAKO mounting medium containing DAPI to counterstain the nuclei.

9. Take images using a confocal microscope.

3.14. Hematoxylin and Eosin (H&E) Staining and Alkaline Phosphatase (AP) Histochemical Staining – Spinal Cord (See Note 16)

1. Mount fixed sections at 10–50 μm on collagen-coated or -charged slides.

2. To show the morphology and the angioregeneration around the grafts, use H&E and native endothelial alkaline phosphatase (AP).

3. Bring sections to distilled water.

4. Stain nuclei with the alum hematoxylin. The level and type of fixation may affect the duration required for reaction. Four to five minutes is a good start.

5. Rinse in running tap water.

6. Differentiate with 0.3% acid alcohol. Differentiation will take between 3 and 5 min and could be longer depending on the thickness of the tissue.

7. Rinse in running tap water.

8. Rinse in Scott's tap water substitute, used for blueing, to obtain optimal contrast for cell differentiation.

9. Rinse in tap water.

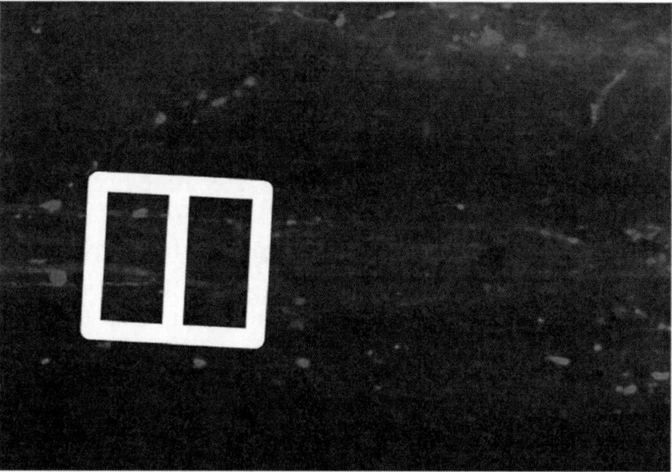

Fig. 10. Spinal cord with quantification grid. The grid is 200 by 200 μm with a line bisecting the square at the center and the grid is placed in the center of the lesion on every fourth section. The neurofilaments (white) are passing through the center of the grid; each fiber is counted whenever it crosses any of the white lines. The data is compared to a nonoperated control; the nonoperated control is a series of age-matched controls where the average number of filaments is set at 100%. Even though 100% is not needed for behavioral return, this is a way to measure the amount of reinnveration and the continuity of the fibers reconnecting the disconnected area.

10. Stain with eosin 2 min. If the staining is too intense, it can be reduced by rinsing.

11. Dehydrate, clear, and mount.

3.15. Quantification of Axons – Spinal Cord

1. Quantify NF-200-, CGRP-, and 5HT-positive axons that regenerated within the graft on immunostaining sections by using a fluorescent microscope. Only immunolabeled cells of interest need to be quantified.

2. Superimpose three lines at intervals of 0.5 mm onto the graft perpendicularly to the longitudinal axis with the middle one through the center of the graft (see Fig. 10).

3. Count the axons intercepted with the superimposed lines. The mean number of axons obtained from five sections of each animal is defined as the number of axons.

4. Notes

1. In this note, we discuss the number of animals needed to perform the experiments described herein.

   In all of the experiments, the majority of the animals are allowed to survive the maximum amount of time. For example, we usually sacrifice 10, 10, 10, 20, and 50% of the animals at each subsequent time point, for an experiment including five time points; for experiments with three time points, we sacrifice 10, 30, and 60% of the animals at each subsequent time point. This allows for a comparison of each of the time points, and during these experiments the most regeneration is usually seen at the later time points. Typical time points are: 30, 45, 60, and 90 days from the date of surgery.

In the case of the hamster pups, four groups of animals are necessary for a valid comparison: (1) uncut controls, (2) cut and untreated controls, (3) cut and saline-treated controls, and (4) cut and treated controls. There should be six animals in each group, for each time point. For example: 4 groups × 6 animals/group = 24 animals plus 5 time points × 6 animals for each time point = 30 animals, for a total of 54 animals needed for an experiment with 5 time points. For the adult hamsters, five groups of animals are necessary for a valid comparison: (1) uncut controls, (2) enucleated controls (where one eye is removed to determine the rate of spontaneous turning toward the blind side), (3) cut and untreated controls, (4) cut and saline-treated controls, and (5) cut and treated controls. In addition, behavior controls will be needed to determine spontaneous turning. Again, there should be six animals in each group, for each time point. The group identity of an animal should be unknown to the investigator during testing periods. The uncut controls should be injected with the same carrier fluid used to mix the peptide material.

For the wild-type rats, five groups of animals are necessary for a valid comparison: (1) a saline control group, (2) an uncultured RADA4 group, (3) a precultured RADA4 only group, (4) a precultured RADA4 seeded with NPCs group, and (5) a precultured RADA4 seeded with ScCs group. There should be six animals in each group, for each time point.

We note that these are the minimum numbers of animals needed and it is always better to have a few extras, especially as experiments can last for up to a year. If your animal handling and surgery skills are good and consistent, fewer animals could be used; if you have not handled animals very often, you may want to increase the number of extra animals by up to 50% of the total.

2. Take extra care to ensure that the material is pure by mixing the powder into the liquid and letting it sit at room temperature for 1 month. An indication of its purity is that it remains clear and odorless. The material can also be analyzed using HPLC to assess purity.

3. It is important to use both young and adult animals. In young animals, the optic tract is still growing (our intention was to see if a permissive environment for growth could be created). Further, with young animals, there is no need for growth factors.

4. Pay special attention to ensure that there is a complete transection of the brachium of the SC from the lateral edge to the midline. In many CNS regeneration models, the lesions are shallow. To drive regeneration through the center of the lesion, use a 2 mm deep cut. The increased depth reduces the possibility of axons growing around the bottom of the cut. Look specifically for regeneration growing through the center of the lesion location since that shows the creation of a permissive environment.

5. The purpose of animal behavior testing is to show that there is functional return of vision driven by behavior. Many regeneration papers refer to action potentials in the axons but if there is no return of functional behavior then regeneration is useless. Only test adult animals for behavior.

6. The cage bottom is a convenient and useful test platform, because the odors are very familiar to the animal. Investigators should wear clean laboratory gowns and gloves (disposable ones work well) because novel odors will interrupt the hamster's movements.

7. Sunflower seeds are a favorite food of hamsters, and under the right conditions they will respond to visual presentations of a seed, turn toward it, take it into their mouth, and transfer it into a cheek pouch.

8. Count trials where the response occurs within 2 s of stimulus presentation. In both normal and blind animals, a turning response can also be elicited by touching the whiskers; therefore, care should be taken to avoid whisker contact during visual presentations. Do not count a trial if the animal turns before the visual stimulus presentation commences or if the seed comes into contact with the whiskers. The completion of a turn is signaled by the animal's head coming to a stationary position within 5° of the stimulus for at least one-third of a second. If the animals orient in the direction of the stimulus more than 70% of the time, vision is deemed as successfully restored. Because an enucleated animal generally fails to respond when the stimulus was placed outside the intact visual field, its whiskers were often touched on the blind side to elicit turns in that direction.

9. An overdose typically requires more than 60 mg/kg, but the exact amount is dependent upon local animal handling requirements.

10. The preblocking solutions are used to reduce nonspecific binding of the primary and secondary antibodies and are specific for each primary antibody used. Care must be taken when using multiple primary antibodies in the same section that the preblock eliminates all nonspecific binding.

11. Neutralization is necessary in the spinal cord for culturing ScCs prior to implantation into the lesion site because, unlike

the brain, the spinal cord has less cerebral spinal fluid that allows for instantaneous buffering of the material.

12. ScCs are used because they are myelinating cells and can provide growth factor support; NPCs are used to replace the lost neuronal connections.

13. These steps are used to determine if the scaffold material causes the cells to be transformed before they are implanted (i.e., if the cells are healthy and viable after they have been mixed with the material).

14. These steps are used to assess the regeneration in the spinal cord and reconnection of the axons after lesion. Look for everything from inflammation to numbers of astrocytes, neurons, oligodendrocytes, raphespinal axons, macrophages, and ScCs.

15. Take care to maintain the orientation of the tissue during embedding and sectioning. Section the tissue longitudinally and mount directly on coated slides.

16. The tissue structure is being labeled to look at the overall distribution of collagen and cell structure. Immunostaining reveals various cell types in the sections; H&E is a general stain used to reveal the general appearance of the tissue so the distribution of immunostained cells can be precisely located. AP is a standard used for the visualization of vasculature.

Acknowledgments

The authors gratefully acknowledge Dr. David K. C. Tay in the Department of Anatomy at the University of Hong Kong Faculty of Medicine for reviewing the chapter and for taking pictures.

References

1. Ellis-Behnke, R. G., Teather, L. A., Schneider, G. E., and So, K. F. (2007) Using nanotechnology to design potential therapies for CNS regeneration. *Curr. Pharm. Des.* **13**, 2519–2528.

2. Zhang, S., Holmes, T., Lockshin, C., and Rich, A. (1993) Spontaneous assembly of a self-complementary oligopeptide to form a stable macroscopic membrane. *Proc. Natl Acad. Sci. USA* **90**, 3334–3338.

3. Takenaga, M., Ohta, Y., Tokura, Y., Hamaguchi, A., Suzuki, N., Nakamura, M., et al. (2007) Plasma as a scaffold for regeneration of neural precursor cells after transplantation into rats with spinal cord injury. *Cell Transplant.* **16**, 57–65.

4. Willerth, S. M. and Sakiyama-Elbert, S. E. (2008) Cell therapy for spinal cord regeneration. *Adv. Drug Deliv. Rev.* **60**, 263–276.

5. Langer, R. (1998) Drug delivery and targeting. *Nature* **392**, 5–10.

6. Holmes, T. C., de Lacalle, S., Su, X., Liu, G., Rich, A., and Zhang, S. (2000) Extensive neurite outgrowth and active synapse formation on self-assembling peptide scaffolds. *Proc. Natl Acad. Sci. USA* **97**, 6728–6733.

7. Schmidt, C. E. and Leach, J. B. (2003) Neural tissue engineering: strategies for repair and regeneration. *Annu. Rev. Biomed. Eng.* **5**, 293–347.

8. Zhang, S., Holmes, T. C., DiPersio, C. M., Hynes, R. O., Su, X., and Rich, A. (1995) Self-complementary oligopeptide matrices support mammalian cell attachment. *Biomaterials* **16**, 1385–1393.

9. Kataoka, K., Suzuki, Y., Kitada, M., Hashimoto, T., Chou, H., Bai, H., et al. (2004) Alginate enhances elongation of early regenerating axons in spinal cord of young rats. *Tissue Eng.* **10**, 493–504.

10. Kataoka, K., Suzuki, Y., Kitada, M., Ohnishi, K., Suzuki, K., Tanihara, M., et al. (2001) Alginate, a bioresorbable material derived from brown seaweed, enhances elongation of amputated axons of spinal cord in infant rats. *J. Biomed. Mater. Res.* **54**, 373–384.

11. Prang, P., Muller, R., Eljaouhari, A., Heckmann, K., Kunz, W., Weber, T., et al. (2006) The promotion of oriented axonal regrowth in the injured spinal cord by alginate-based anisotropic capillary hydrogels. *Biomaterials* **27**, 3560–3569.

12. Crompton, K. E., Goud, J. D., Bellamkonda, R. V., Gengenbach, T. R., Finkelstein, D. I., Horne, M. K., et al. (2007) Polylysine-functionalised thermoresponsive chitosan hydrogel for neural tissue engineering. *Biomaterials* **28**, 441–449.

13. Freier, T., Koh, H. S., Kazazian, K., and Shoichet, M. S. (2005) Controlling cell adhesion and degradation of chitosan films by *N*-acetylation. *Biomaterials* **26**, 5872–5878.

14. Freier, T., Montenegro, R., Shan Koh, H., and Shoichet, M. S. (2005) Chitin-based tubes for tissue engineering in the nervous system. *Biomaterials* **26**, 4624–4632.

15. Archibald, S. J., Krarup, C., Shefner, J., Li, S. T., and Madison, R. D. (1991) A collagen-based nerve guide conduit for peripheral nerve repair: an electrophysiological study of nerve regeneration in rodents and nonhuman primates. *J. Comp. Neurol.* **306**, 685–696.

16. Mahoney, M. J., Krewson, C., Miller, J., and Saltzman, W. M. (2006) Impact of cell type and density on nerve growth factor distribution and bioactivity in 3-dimensional collagen gel cultures. *Tissue Eng.* **12**, 1915–1927.

17. Herbert, C. B., Bittner, G. D., and Hubbell, J. A. (1996) Effects of fibinolysis on neurite growth from dorsal root ganglia cultured in two- and three-dimensional fibrin gels. *J. Comp. Neurol.* **365**, 380–391.

18. Sakiyama, S. E., Schense, J. C., and Hubbell, J. A. (1999) Incorporation of heparin-binding peptides into fibrin gels enhances neurite extension: an example of designer matrices in tissue engineering. *FASEB J.* **13**, 2214–2224.

19. Sakiyama-Elbert, S. E. and Hubbell, J. A. (2000) Controlled release of nerve growth factor from a heparin-containing fibrin-based cell ingrowth matrix. *J. Control. Release* **69**, 149–158.

20. Sakiyama-Elbert, S. E. and Hubbell, J. A. (2000) Development of fibrin derivatives for controlled release of heparin-binding growth factors. *J. Control. Release* **65**, 389–402.

21. Taylor, S. J., McDonald III, J. W., and Sakiyama-Elbert, S. E. (2004) Controlled release of neurotrophin-3 from fibrin gels for spinal cord injury. *J. Control. Release* **98**, 281–294.

22. Taylor, S. J., Rosenzweig, E. S., McDonald III, J. W., and Sakiyama-Elbert, S. E. (2006) Delivery of neurotrophin-3 from fibrin enhances neuronal fiber sprouting after spinal cord injury. *J. Control. Release* **113**, 226–235.

23. Taylor, S. J. and Sakiyama-Elbert, S. E. (2006) Effect of controlled delivery of neurotrophin-3 from fibrin on spinal cord injury in a long term model. *J. Control. Release* **116**, 204–210.

24. Gupta, D., Tator, C. H., and Shoichet, M. S. (2006) Fast-gelling injectable blend of hyaluronan and methylcellulose for intrathecal, localized delivery to the injured spinal cord. *Biomaterials* **27**, 2370–2379.

25. Hahn, S. K., Jelacic, S., Maier, R. V., Stayton, P. S., and Hoffman, A. S. (2004) Anti-inflammatory drug delivery from hyaluronic acid hydrogels. *J. Biomater. Sci. Polym. Ed.* **15**, 1111–1119.

26. Tian, W. M., Hou, S. P., Ma, J., Zhang, C. L., Xu, Q. Y., Lee, I. S., et al. (2005) Hyaluronic acid-poly-d-lysine-based three-dimensional hydrogel for traumatic brain injury. *Tissue Eng.* **11**, 513–525.

27. Segura, T., Anderson, B. C., Chung, P. H., Webber, R. E., Shull, K. R., and Shea, L. D. (2005) Crosslinked hyaluronic acid hydrogels: a strategy to functionalize and pattern. *Biomaterials* **26**, 359–371.

28. Willerth, S. M. and Sakiyama-Elbert, S. E. (2007) Approaches to neural tissue engineering using scaffolds for drug delivery. *Adv. Drug Deliv. Rev.* **59**, 325–338.

29. Retrieved from HemCon Medical Technologies, Inc. on 1/20/2011: HemCon bandage FAQ/Shellfish allergy study at www.hemcon.com/Products/HemConBandageOverview.aspx.

30. Ellis-Behnke, R. G., Liang, Y. X., Tay, D. K., Kau, P. W., Schneider, G. E., Zhang, S., et al. (2006) Nano hemostat solution: immediate hemostasis at the nanoscale. *Nanomedicine* **2**, 207–215.

31. Leach, J. B., Brown, X. Q., Jacot, J. G., Dimilla, P. A., and Wong, J. Y. (2007) Neurite outgrowth and branching of PC12 cells on very soft

substrates sharply decreases below a threshold of substrate rigidity. *J. Neural Eng.* **4**, 26–34.

32. Guan, J., Stankus, J. J., and Wagner, W. R. (2006) Development of composite porous scaffolds based on collagen and biodegradable poly(ester urethane)urea. *Cell Transpl.* 15 Suppl. **1**, S17–S27.

33. Burdick, J. A., Ward, M., Liang, E., Young, M. J., and Langer, R. (2006) Stimulation of neurite outgrowth by neurotrophins delivered from degradable hydrogels. *Biomaterials* **27**, 452–459.

34. Krause, T. L. and Bittner, G. D. (1990) Rapid morphological fusion of severed myelinated axons by polyethylene glycol. *Proc. Natl Acad. Sci. USA* **87**, 1471–1475.

35. Mahoney, M. J. and Anseth, K. S. (2006) Three-dimensional growth and function of neural tissue in degradable polyethylene glycol hydrogels. *Biomaterials* **27**, 2265–2274.

36. Piantino, J., Burdick, J. A., Goldberg, D., Langer, R., and Benowitz, L. I. (2006) An injectable, biodegradable hydrogel for trophic factor delivery enhances axonal rewiring and improves performance after spinal cord injury. *Exp. Neurol.* **201**, 359–367.

37. Aubert-Pouessel, A., Venier-Julienne, M. C., Clavreul, A., Sergent, M., Jollivet, C., Montero-Menei, C. N., et al. (2004) *In vitro* study of GDNF release from biodegradable PLGA microspheres. *J. Control. Release* **95**, 463–475.

38. Kim, D. H. and Martin, D. C. (2006) Sustained release of dexamethasone from hydrophilic matrices using PLGA nanoparticles for neural drug delivery. *Biomaterials* **27**, 3031–3037.

39. Lam, X. M., Duenas, E. T., and Cleland, J. L. (2001) Encapsulation and stabilization of nerve growth factor into poly(lactic-co-glycolic) acid microspheres. *J. Pharm. Sci.* **90**, 1356–1365.

40. Patist, C. M., Mulder, M. B., Gautier, S. E., Maquet, V., Jerome, R., and Oudega, M. (2004) Freeze-dried poly(D,L-lactic acid) macroporous guidance scaffolds impregnated with brain-derived neurotrophic factor in the transected adult rat thoracic spinal cord. *Biomaterials* **25**, 1569–1582.

41. Teng, Y. D., Lavik, E. B., Qu, X., Park, K. I., Ourednik, J., Zurakowski, D., et al. (2002) Functional recovery following traumatic spinal cord injury mediated by a unique polymer scaffold seeded with neural stem cells. *Proc. Natl Acad. Sci. USA* **99**, 3024–3029.

42. Belkas, J. S., Munro, C. A., Shoichet, M. S., Johnston, M., and Midha, R. (2005) Long-term *in vivo* biomechanical properties and biocompatibility of poly(2-hydroxyethyl methacrylate-co-methyl methacrylate) nerve conduits. *Biomaterials* **26**, 1741–1749.

43. Carone, T. W. and Hasenwinkel, J. M. (2006) Mechanical and morphological characterization of homogeneous and bilayered poly(2-hydroxyethyl methacrylate) scaffolds for use in CNS nerve regeneration. *J. Biomed. Mater. Res. B, Appl. Biomater.* **78**, 274–282.

44. Nomura, H., Katayama, Y., Shoichet, M. S., and Tator, C. H. (2006) Complete spinal cord transection treated by implantation of a reinforced synthetic hydrogel channel results in syringomyelia and caudal migration of the rostral stump. *Neurosurgery* **59**, 183–192; discussion 183–192.

45. Tsai, E. C., Dalton, P. D., Shoichet, M. S., and Tator, C. H. (2004) Synthetic hydrogel guidance channels facilitate regeneration of adult rat brainstem motor axons after complete spinal cord transection. *J. Neurotrauma* **21**, 789–804.

46. Tsai, E. C., Dalton, P. D., Shoichet, M. S., and Tator, C. H. (2006) Matrix inclusion within synthetic hydrogel guidance channels improves specific supraspinal and local axonal regeneration after complete spinal cord transection. *Biomaterials* **27**, 519–533.

47. Newman, K. D. and McBurney, M. W. (2004) Poly(D,L lactic-co-glycolic acid) microspheres as biodegradable microcarriers for pluripotent stem cells. *Biomaterials* **25**, 5763–5771.

48. Lloyd, D. A., Ansari, T. I., Gundabolu, P., Shurey, S., Maquet, V., Sibbons, P. D., et al. (2006) A pilot study investigating a novel subcutaneously implanted pre-cellularised scaffold for tissue engineering of intestinal mucosa. *Eur. Cell Mater.* **11**, 27–33; discussion 34.

49. Zhang, S. (2003) Fabrication of novel biomaterials through molecular self-assembly. *Nat. Biotechnol.* **21**, 1171–1178.

50. Zhang, S., Marini, D. M., Hwang, W., and Santoso, S. (2002) Design of nanostructured biological materials through self-assembly of peptides and proteins. *Curr. Opin. Chem. Biol.* **6**, 865–871.

51. Garreta, E., Genove, E., Borros, S., and Semino, C. E. (2006) Osteogenic differentiation of mouse embryonic stem cells and mouse embryonic fibroblasts in a three-dimensional self-assembling peptide scaffold. *Tissue Eng.* **12**, 2215–2227.

52. Zhang, S., Gelain, F., and Zhao, X. (2005) Designer self-assembling peptide nanofiber scaffolds for 3D tissue cell cultures. *Semin. Cancer Biol.* **15**, 413–420.

53. Holmes, T. C. (2002) Novel peptide-based biomaterial scaffolds for tissue engineering. *Trends Biotechnol.* **20**, 16–21.

54. Lupi, O., Madkan, V., and Tyring, S. K. (2006) Tropical dermatology: bacterial tropical diseases. *J. Am. Acad. Dermatol.* **54**, 559–578.

55. Ellis-Behnke, R. G., Liang, Y. X., You, S. W., Tay, D. K., Zhang, S., So, K. F., et al. (2006) Nano neuro knitting: peptide nanofiber scaffold for brain repair and axon regeneration with functional return of vision. *Proc. Natl Acad. Sci. USA* **103**, 5054–5059.

56. Davis, M. E., Hsieh, P. C., Takahashi, T., Song, Q., Zhang, S., Kamm, R. D., et al. (2006) Local myocardial insulin-like growth factor 1 (IGF-1) delivery with biotinylated peptide nanofibers improves cell therapy for myocardial infarction. *Proc. Natl Acad. Sci. USA* **103**, 8155–8160.

57. Guo, J., Su, H., Zeng, Y., Liang, Y. X., Wong, W. M., Ellis-Behnke, R. G., et al. (2007) Reknitting the injured spinal cord by self-assembling peptide nanofiber scaffold. *Nanomedicine* **3**, 311–321.

58. Schneider, G. E., Ellis-Behnke, R. G., Liang, Y. X., Kau, P. W., Tay, D. K., and So, K. F. (2006) Behavioral testing and preliminary analysis of the hamster visual system. *Nat. Protoc.* **1**, 1898–1905.

Chapter 18

Computational Simulations of the Interaction of Lipid Membranes with DNA-Functionalized Gold Nanoparticles

One-Sun Lee and George C. Schatz

Abstract

We develop a shape-based coarse-grained (SBCG) model for DNA-functionalized gold nanoparticles (DNA-Au NPs) and use this to study the interaction of this potential antisense therapeutic with a lipid bilayer model of a cell membrane that is also represented using a coarse-grained model. Molecular dynamics simulations of the SBCG model of the DNA-Au NP show structural properties which coincide with our previous atomistic models of this system. The lipid membrane is composed of 30% negatively charged lipid (1,2-dioleoyl-sn-glycero-3-phosphoserine, DOPS) and 70% neutral lipid (1,2-dioleoyl-sn-glycero-3-phosphocholine, DOPC) in 0.15 M sodium chloride solution. Molecular dynamics (MD) simulations of the DNA-Au NP near to the lipid bilayer show that there is a higher density of DOPS than DOPC near to the DNA-Au NP since sodium counterions are able to have strong electrostatic interactions with DOPS and the DNA-Au NP at the same time. Using a steered MD simulation, we show that this counterion-mediated electrostatic interaction between DNA-Au NP and DOPS stabilizes the DNA-Au NP in direct contact with the lipid. This provides a model for interaction of DNA-Au NPs with cell membranes that does not require protein mediation.

Key words: DNA, Gold, Nanoparticle, Lipid, DOPS, DOPC, Molecular dynamics simulation, Charge–charge interaction, Sodium ion

---

1. Introduction

The delivery of inorganic nanoparticles into cells has been one of the most exciting recent developments in therapeutics or imaging (1–4). Among the many different nanoparticles being considered, gold nanoparticles have emerged as an attractive candidate for the delivery of various drug molecules into targeted cells (5–9). In particular, Mirkin and coworkers have developed a method for delivering DNA or RNA-functionalized gold nanoparticles into cells for therapeutic purposes such as antisense treatment (10).

Sarah J. Hurst (ed.), *Biomedical Nanotechnology: Methods and Protocols*, Methods in Molecular Biology, vol. 726, DOI 10.1007/978-1-61779-052-2_18, © Springer Science+Business Media, LLC 2011

According to recent work in the Mirkin lab, these DNA-functionalized gold nanoparticles (DNA-Au NPs) are readily taken up into cells through endocytosis. However, this is somewhat surprising since the surface of the gold nanoparticle is densely covered by negatively charged ss-DNA ($\sim$19 pmol/cm^2), and the cell membrane is also negatively charged. According to a study by Giljohann et al., the DNA-Au NPs adsorb a large number of positively charged serum proteins on the nanoparticle surface (7), resulting in a zeta potential for the DNA-Au NP which is changed from -21 ± 4 to -13 ± 1 mV after adsorption. Based on these experiments, they proposed that the interaction of DNA-Au NP with proteins is a possible mechanism for recognition of the nanomaterial by the cell and subsequent cellular internalization. However, even though this explains many aspects of cellular internalization of the DNA-Au NPs, there still remain many questions about the interaction between these negatively charged nanoparticles and the cell membrane. Indeed, even after the adsorption of positively charged serum proteins, the DNA-Au NP still has a negative charge. Moreover, according to recent experiments, DNA-Au NPs are internalized by cells without serum proteins, albeit a smaller amount of internalized nanoparticles (11). Therefore, the fundamental nature of the interaction between these negatively charged nanoparticles and a cell membrane has yet to be fully elucidated, and this has proven to be a hindrance in understanding how cells interact with nanoparticles.

Computer simulation is an emerging and promising tool for investigating large systems composed of bio- and nanomaterials as a result of recent progress in the development of molecular theories and computer technologies. In our earlier work, our group has developed an atomistic model of DNA-Au NPs which has provided deeper understanding concerning a variety of structural features including the effective radius of the DNA-Au NP, the interaction between DNA strands on the surfaces of gold particles, and their local salt concentration (12–14). However, even though our atomistic model correctly describes many details of this system, this model is not appropriate for studying the interactions between a nanoparticle and a cell membrane due to its high demand for computational resources. Therefore, a new computational approach such as coarse-grained model is needed for the study of DNA-Au NPs and their interactions with membranes. Recently, the Schulten group has developed the shape-based coarse-grained (SBCG) model of cell membranes for the study of membrane curvature that is induced by membrane proteins. Such models are essential for understanding endocytosis pathways that are important for many processes in cell biology (15). The SBCG model is simpler than previous CG lipid models in terms of the number of beads, but it still describes the properties of membranes effectively. This suggests that the SBCG lipid model will be appropriate for studying the interaction of DNA-Au NPs with membranes, and since the required computational resources are reasonable, it is a logical place to start such studies.

In Subheading 2, we develop a SBCG model for the DNA-Au NP based on an atomistic model we previously reported (12). In addition, we show that the properties of this newly developed SBCG model are very close to what we obtained with our atomistic modeling. In Subheading 3, molecular dynamics (MD) simulations are performed to study the interaction between the DNA-Au NPs and a lipid bilayer model of a cell membrane. In addition, we perform steered MD simulations to provide a deeper understanding of the interactions between the negatively charged DNA-Au NP and the lipid membrane.

2. Development of a SBCG Model for DNA-Functionalized Gold Nanoparticles

Force field parameters for the SBCG model of the DNA-Au NP (see Fig. 1) are developed based on atomistic simulations described in our previous report (12). In this work, we considered a 2-nm gold nanoparticle functionalized with four single stranded oligonucleotides as representation of the smallest DNA-Au NP system that has been studied in the Mirkin group. In the present application, this is approximated by one bead for the gold nanoparticle and two for each DNA strand (see Fig. 1c).

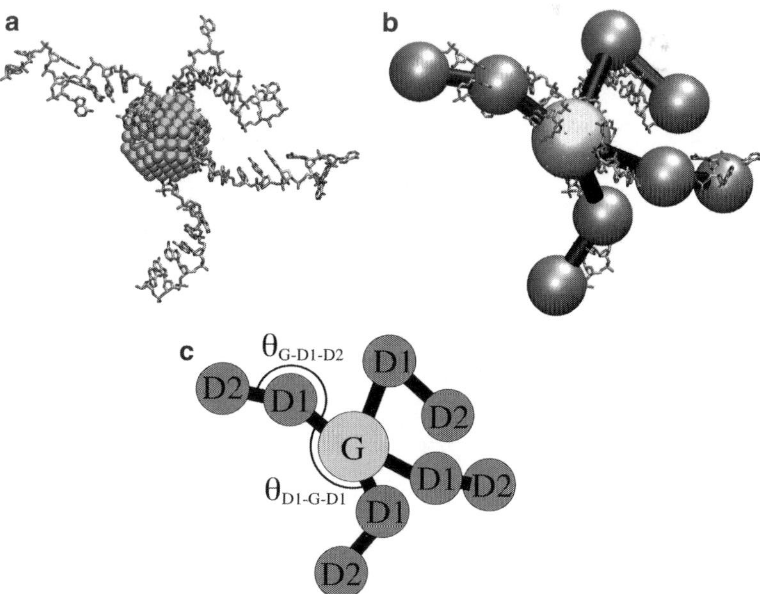

Fig. 1. (a) Atomistic model of a DNA-functionalized gold nanoparticle. (b) Coarse-grained model of the same structure, with the atomistic model superimposed. (c) Name assignments of the CG beads.

Equations 1–3 are used for the bond, angle, and van der Waals interaction energy calculations, respectively, the parameters for which are listed in Table 1. Each bead for DNA (D1 and D2) in Fig. 1 corresponds to 5.5 adenine bases, and the bead for gold nanoparticle corresponds to 201 gold atoms. Each bead that comprises the DNA strand has –5.5 charges, whereas the bead for the gold nanoparticle is electrically neutral.

$$E_{bond} = k_d (d - d_0)^2. \tag{1}$$

$$E_{angle} = k_\theta (\theta - \theta_0)^2. \tag{2}$$

$$E_{LJ} = \varepsilon \left[\left(\frac{R}{r} \right)^{12} - 2 \left(\frac{R}{r} \right)^{6} \right]. \tag{3}$$

To compare the dynamic properties of the SBCG model with the atomistic model, we performed MD simulations of the SBCG

Table 1
Force field parameters for the CG model of DNA-Au NP and membrane

Bond	k_d (kcal/mol/Å^2)	d_0 (Å)
G–D1	10.0	28.0
D1–D2	10.0	25.0
PCH–DOT*	0.2	12.0
PSH–DOT*	0.2	12.0

Angle	k_θ (kcal/mol/rad^2)	θ_0 (°)
D1–G–D1	200.0	109.5
G–D1–D2	100.0	140.0

Atom (LJ)	ε (kcal/mol)	R (Å)	Charge (−e)
G	20.0	18.0	0.0
D1	0.1	13.6	−5.5
D2	0.1	13.6	−5.5
PCH*	0.1	13.6	0.0
PSH*	0.1	13.6	−2.2
DOT*	10.0	13.6	0.0
NAB*	0.1	14.0	2.2
CLB*	0.1	14.0	−2.2

See Figs. 1 and 3 for the name assignment of the CG beads. *Adapted from ref. 15

model. Periodic boundary conditions were used corresponding to a box of dimensions of $100 \times 100 \times 100\,\text{Å}^3$ (16). To neutralize the system, 20 NAB ions are added. The NAB ion corresponds to 2.2 sodium ions and has +2.2 charges. In addition to these sodium ions, another 123 NAB and CLB ions (CLB ion corresponds to 2.2 chlorine ions, and has –2.2 charge) are added to make the concentration of sodium ions in the box to be 0.52 M. This choice matches what has been used previously in the atomistic simulations.

The system is simulated for 12 ns using the NVT ensemble (17) and Langevin dynamics at a temperature of 300 K with a damping coefficient of $\gamma = 2/\text{ps}$ (18, 19). According to the work of Arkhipov et al. (15), a value of 2/ps for the damping coefficient in Langevin dynamics reproduces the viscosity of water for the coarse-graining level used in our study. No atomic coordinates were constrained during the production period. Atomic coordinates were saved every 1 ps for the trajectory analysis. MD simulations were carried out using NAMD2 (20).

To compare the flexibility of the SBCG DNA-Au NP model with the atomistic results, the distribution of the angles $\theta_{\text{G-D1-D2}}$ and $\theta_{\text{D1-G-D1}}$ (*see* Fig. 1) obtained from MD are compared with our previous atomistic MD simulations. As shown in Table 2, the average value of $\theta_{\text{D1-G-D1}}$ obtained from the SBCG model is $109 \pm 8°$, whereas it is $108 \pm 24°$ for the atomistic model. Also, the average $\theta_{\text{G-D1-D2}}$ in SBCG is $123 \pm 12°$ and it is $127 \pm 20°$ for atomistic model.

To calculate the effective radius of the DNA-NPs, the radius of gyration (R_{G}) is introduced. The radius of gyration is defined in Eq. 4, where N_D is the number of beads in the DNA, $\langle \cdots \rangle$ denotes a time average, r_D is the position vector of the Dth DNA bead, and r_{G} is the position vector of the center of the gold particle bead.

$$R_{\text{G}}^2 = \frac{1}{N_{\text{D}} + 1}\left\langle \sum_{D=0}^{N_{\text{D}}} \left(\vec{r_D} - \vec{r_{\text{G}}}\right)^2 \right\rangle. \tag{4}$$

Table 2

Average angle $\theta_{\text{D1-G-D1}}$ and $\theta_{\text{G-D1-D2}}$ obtained from a 12-ns MD simulation for the CG model of DNA-Au NP

	$\theta_{\text{D1-G-D1}}$ (°)	$\theta_{\text{G-D1-D2}}$ (°)
CG model	109 ± 8	123 ± 12
Atomistic simulation	108 ± 24	127 ± 20

The average angle from the atomistic simulation is also shown for comparison

During the last 2 ns of the simulation, the value of R_{G} is $31 \pm 1\,\text{Å}$, whereas it is $28.4 \pm 2.3\,\text{Å}$ in our previous simulation at the atomistic level. All statistical uncertainties are $\pm 1\sigma$ (one standard deviation).

We calculated the number density of sodium ions within $r = 31\,\text{Å}$ from the center of the gold particle. This distance was chosen to be close to R_{G}, so the number density refers to ions that are in the volume occupied by the DNA. The number density of sodium ions $\rho(r)$ is calculated using Eq. 5.

$$\rho(r) = \frac{\langle n_i(r_i) \rangle}{v}.\tag{5}$$

Here v is the volume of a sphere with a radius $31\,\text{Å}$ minus the volume of the gold particle. $\langle n_i(r_i) \rangle$ is the number of particles averaged over time as shown in Eq. 6, where T is the total time for the sampling.

$$\langle n_i(r_i) \rangle = \frac{1}{T} \sum_{\tau=1}^{T} n_i(r_i(\tau)).\tag{6}$$

The distribution of number density of sodium atoms (normalized to its bulk value) during the last 2 ns of the MD simulation is shown in Fig. 2. Here, the number density of sodium ions around the gold nanoparticle is normalized relative to that for the 0.52 M bulk concentration. Figure 2 shows that the concentration of sodium ions around the gold nanoparticle is about 22% higher than the bulk concentration. This is consistent with our previous atomistic simulation results (20%).

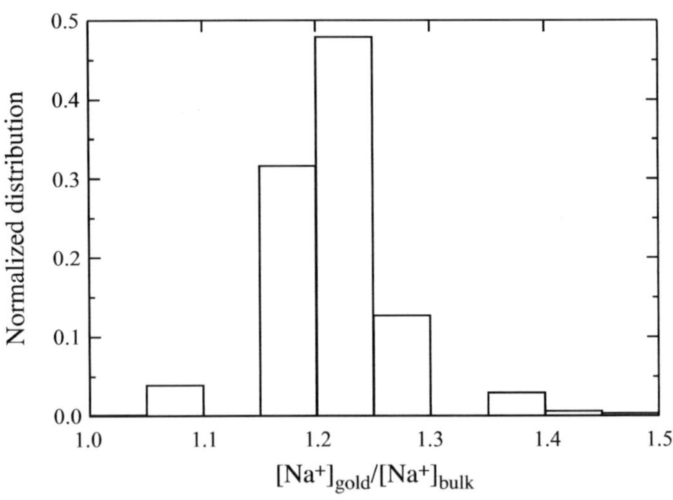

Fig. 2. Distribution of the relative concentration of sodium ions within 31 Å of the gold nanoparticle in the DNA-Au NP complex. The Na$^+$ concentration around the gold particle is about 22% higher than the bulk concentration.

3. Nanoparticle–Membrane Interaction

We performed 1 μs MD simulations for the SBCG DNA-Au NP interacting with a lipid bilayer to study the nanoparticle/membrane interaction. Force field parameters for the chosen lipids are listed in Table 1 as adapted from the work of Arkhipov et al. (15). Two kinds of lipid molecules, 1,2-dioleoyl-sn-glycero-3-phosphoserine (DOPS) and 1,2-dioleoyl-sn-glycero-3-phosphocholine (DOPC) (21), are used in the simulations. As shown in Fig. 3, both DOPS and DOPC are composed of two beads: one for the head group and one for the tail group. Note that the DOPS or DOPC correspond to 2.2 lipid molecules, and the head group of DOPS has a −2.2 charge, whereas DOPC has zero charge (15). The total number of lipid molecules is 2,738 (37×37×2) and consists of 820 DOPS and 1,918 DOPC (see Fig. 3). Therefore, ~30% of our lipids have a negative charge. Periodic boundary conditions are used, corresponding to a box of dimensions of $464 \times 464 \times 440 \,\text{Å}^3$. To neutralize the system, 840 NAB ions are

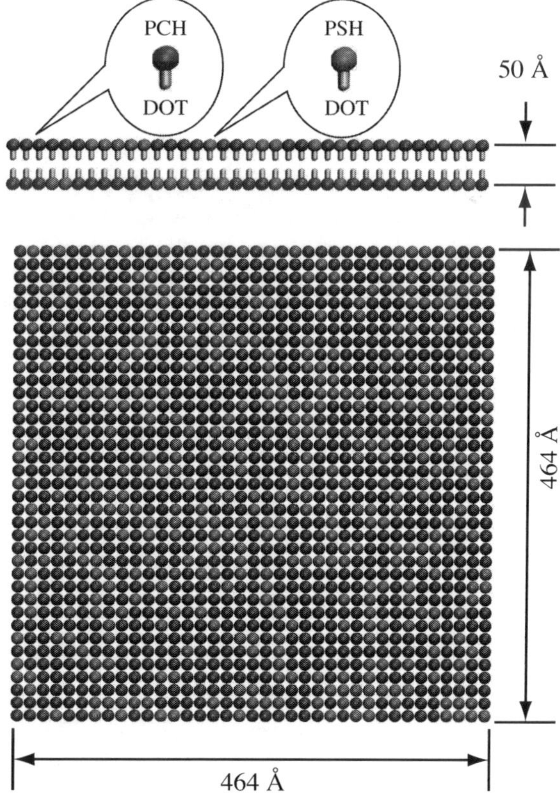

Fig. 3. Side and top views of coarse-grained model of a lipid bilayer composed of 70% DOPC (PCH) and 30% DOPS (PSH). The surface of the lipid is composed of a 37 × 37 × 2 array.

added. In addition to these sodium ions, another 2,714 NAB and CLB ions are added to make the concentration of sodium ions to be 0.15 M.

The distance between the gold nanoparticle and the surface of the lipid is taken to be ~75 Å in the starting structure. A 1-ns MD simulation at 5,000 K with a NVT ensemble is performed to equilibrate the system. During the equilibration, the position of the gold and lipid is fixed with harmonic constraints. In the production period, the system is simulated for 1 μs using the NVT ensemble and Langevin dynamics at a temperature of 310 K with a damping coefficient $\gamma = 2/ps$ (18, 19). A time step of 100 fs is used for the simulation. The long-range interaction cutoff is taken to be 35 Å. No atomic coordinates were constrained during the production period. Atomic coordinates were saved every 1 ns for the trajectory analysis. MD simulations were carried out using NAMD2 (20).

A snapshot of the system after a 1-μs MD simulation is shown in Fig. 4. Even though the distance between the gold nanoparticle and the surface of the membrane is about 75 Å at the starting position, the DNA-Au NP then approaches to the membrane

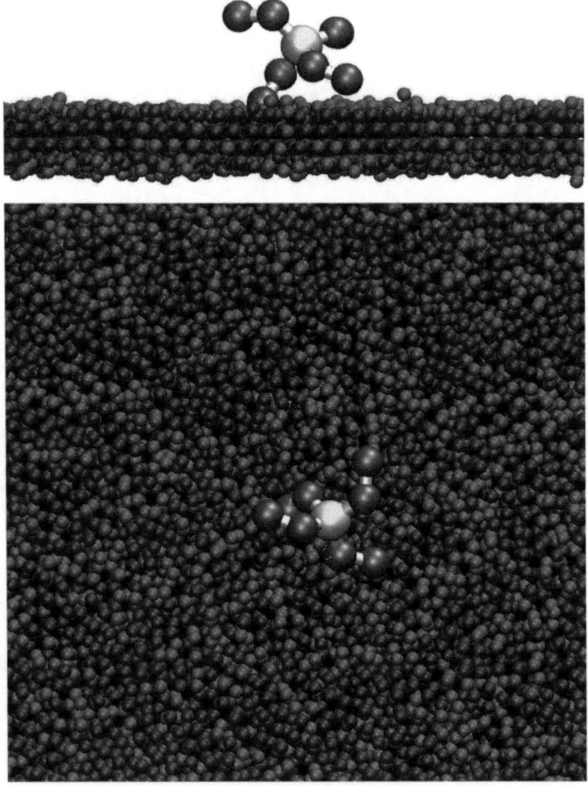

Fig. 4. Side and top views of a DNA-Au NP on a membrane after a 1-μs MD simulation. The DNA-Au NP is adsorbed on the membrane, but penetration of the membrane is not observed.

such that the distance after 200 ns is ~40 Å. At this distance, the surfaces of the DNA-Au NP and the membrane are almost touching each other as shown in Fig. 4. During the 1-μs MD simulation, penetration of the DNA-Au NP through the membrane (leading to endocytosis) is not observed.

Radial distribution functions (RDFs) between sodium ions and the head group of DOPS or DOPC are presented in Fig. 5. The first peak in $g_{DOPS-Na^+}(r)$ appears at ~10 Å and is very intense, whereas the first peak in $g_{DOPC-Na^+}(r)$ appears at ~12 Å and is much weaker. This indicates that sodium ions are localized to DOPS, as it makes sense based on electrostatic effects. A snapshot of the membrane, shown in Fig. 5c, shows the distribution of DOPS and DOPC, while Fig. 5d shows the distribution of the sodium ions superimposed on the data of Fig. 5c. Both figures show distinct dark gray patches, indicating that most of the sodium ions are closely associated with DOPS.

RDFs between the gold nanoparticle in the DNA-Au NP and DOPS or DOPC have been generated in order to scrutinize the position of the DNA-Au NP during the MD simulation. Figure 6 shows the results, and we see that the first peak in $g_{Au-DOPS}(r)$ appears at 47 Å, whereas the first peak in $g_{Au-DOPC}(r)$ appears at 52 Å (and is less intense). Therefore, the DNA-Au NP is predominantly

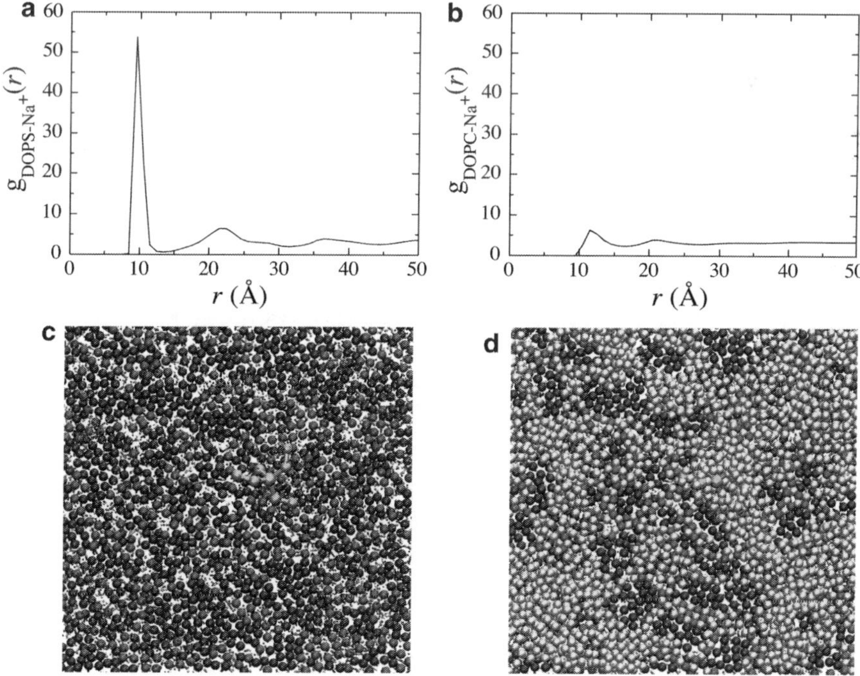

Fig. 5. RDFs between (**a**) DOPS and sodium ion and (**b**) DOPC and sodium ion obtained from a 1-μs MD simulation. Sodium ions are localized around DOPS even though DOPS is only 30% of the lipid. (**c**) Snapshot of a DNA-Au NP on a membrane after 1 μs MD simulation. DOPS is shown in *light gray* and DOPC is shown in *dark gray*. (**d**) The distribution of sodium ions is shown. Sodium ions are preferentially localized around the DOPS and the DNA-Au NP.

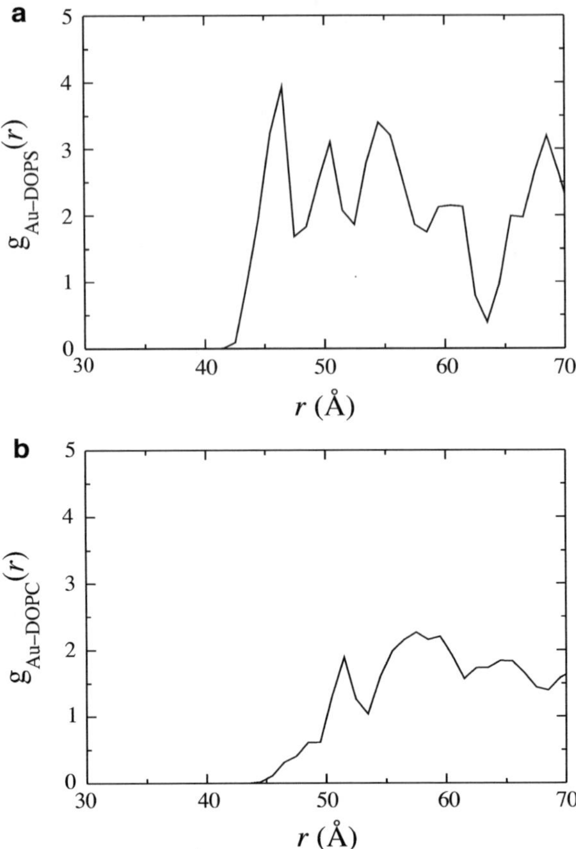

Fig. 6. RDFs between the gold nanoparticle of the DNA-Au NP and (**a**) DOPS and (**b**) DOPC during a 1-µs MD simulation. This shows that the DNA-Au NP is in closer contact with DOPS than DOPC.

localized near DOPS even though both the DNA-Au NP and DOPS are negatively charged and DOPS represents only 30% of the lipid composition (15). Therefore, we conclude that sodium ions interacting with both the DNA-Au NP and DOPS induce the DNA-Au NP to be closer to DOPS than to DOPC.

To examine the interaction between the DNA-Au NP and DOPS or DOPC, we performed a steered MD simulation. A force is exerted on the gold nanoparticle of the DNA-Au NP along the path shown in Fig. 7a. Note that the path is parallel to the bilayer, so the effect of the steered MD is to pull the NP through regions of varying lipid composition. The pulling velocity is 150 Å/µs, and a time step of 50 fs is used. The interaction energies of the DNA-Au NP with the membrane, and with the sodium and chloride ions along the path of the steered MD simulation are shown in Fig. 7b. The total interaction energy of the DNA-Au NP is the sum of these individual interaction energies.

$$E_{total} = E_{NP-lipid} + E_{NP-Na^+} + E_{NP-Cl^-}. \tag{7}$$

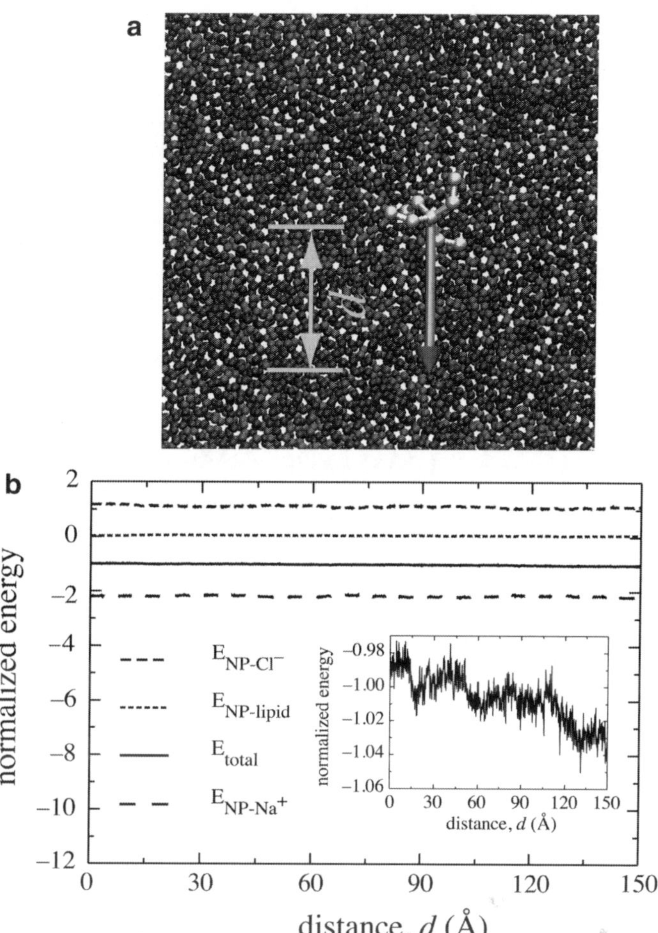

Fig. 7. (a) The trajectory used for the DNA-Au NP during the steered MD simulation is shown with an *arrow*. A force is exerted on the gold nanoparticle of the DNA-Au NP and the total distance of the steered MD is 150 Å. (b) Interaction energies of the DNA-Au NP with sodium ion, chlorine ion, and membrane during the steered MD simulation. The total energy that is the sum of the individual interaction energies is shown in the *inset*.

This energy is shown in the inset of Fig. 7b. Note that all energies in Fig. 7b are normalized relative to the absolute value of the interaction energy of the DNA-Au NP with its environment in 0.15 M sodium chloride solution. To obtain the normalization factor, we performed a MD simulation for the DNA-Au NP in 0.15 M sodium chloride solution for 0.5 μs with periodic boundary conditions of $406 \times 406 \times 406 \, \text{Å}^3$. The average interaction energy of the DNA-Au NP with the sodium and chloride ions is calculated during the last 0.1 μs of this simulation.

As shown in Fig. 7b, the interaction energy between the DNA-Au NP and lipid ($E_{\text{NP–lipid}}$) is negligible compared with that of $E_{\text{NP–Na}^+}$ and $E_{\text{NP–Cl}^-}$. E_{total} is overall negative, meaning that interaction of the functionalized nanoparticle with Na^+ is more important than with Cl$^-$. Its value for $d=0$ is close to -1.00, meaning that the energy at that point matches that from bulk solution conditions and the particle is not strongly bound to the surface. However, we see that E_{total} decreases to -1.04 as the DNA-Au NP is dragged across the surface. This means that the DNA-Au NP is stabilized by 4% of the bulk interaction energy as a result of moving it across the surface and exposing it to more DOPS. For a deeper understanding of this stabilization of the DNA-Au NP close to the membrane, the number of DOPS molecules near the DNA-Au NP (within 35 Å) has been calculated during the steered MD calculation. This number is plotted in Fig. 8a, and we see that this number starts at 15 but increases to ~20 at $d=130$ Å. This happens while the interaction energy in Fig. 7 is decreasing, so it is apparent that more DOPS

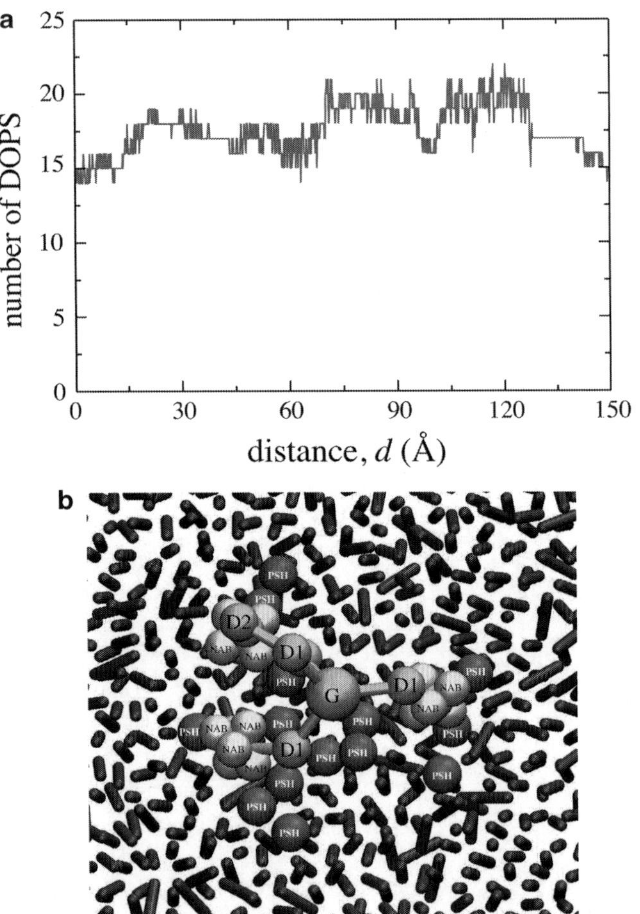

Fig. 8. (a) The number of DOPS within 35 Å of the DNA-Au NP during the steered MD simulation is shown. (b) Snapshot from the steered MD simulation. Sodium ions (NAB) are intercalated between the DNA-Au NP and the head of DOPS (PSH). All other lipids are shown with a stick model for clarity.

bind to the cluster as the steered MD proceeds (somewhat like the function of a vacuum cleaner), and this causes the functionalized nanoparticle to be attracted to the surface. A snapshot of the DNA-Au NP during the steered MD simulation is shown in Fig. 8b. We see that the sodium ions are interposed between DOPS and bead D2 of the DNA-Au NP, essentially providing the "glue" to bind the nanoparticle to the DOPS-coated surface.

4. Conclusion

An important accomplishment of this chapter is the development of a SBCG model for DNA-Au NP, and the demonstration of its use for simulating the interaction of this potential therapeutic material with a lipid bilayer composed of 30% DOPS and 70% DOPC in 0.15 M sodium chloride solution. MD simulations for several microseconds are possible for this system, which makes it possible to determine the thermodynamic and structural properties, providing an explanation for why the negatively charged functionalized nanoparticle is attracted to the negatively charged membrane. Indeed, we find that counterion-mediated electrostatic attractive interactions are sufficient to enable the particle to strongly bind to the negatively charged surface. This indicates why protein mediation is not essential for getting the DNA-Au NP to stick to the surface, but it leaves open the question of how endocytosis works for this system. This latter point will be the subject of future studies.

Acknowledgments

This research was supported by National Science Foundation (grant CHE-0843832), and by the Northwestern Center for Cancer Nanobiotechnology Excellence (1 U54 CA119341-01).

References

1. Alemdaroglu, F. E., Alemdaroglu, N. C., Langguth, P., and Hermann, A. (2008) DNA block copolymer micelles – A combinatorial tool for cancer nanotechnology. *Adv. Mater.* **20**, 899–902.

2. Ghosh, P., Han, G., De, M., Kim, C. K., and Rotello, V. M. (2008) Gold nanoparticles in delivery applications. *Adv. Drug Deliv. Rev.* **60**, 1307–1315.

3. Nel, A. E., Madler, L., Velegol, D., Xia, T., Hoek, E. M. V., Somasundaran, P., et al. (2009) Understanding biophysicochemical interactions at the nano-bio interface. *Nat. Mater.* **8**, 543–557.

4. Thurn, K. T., Brown, E. M. B., Wu, A., Vogt, S., Lai, B., Maser, J., et al. (2007) Nanoparticles for applications in cellular imaging. *Nanoscale Res. Lett.* **2**, 430–441.

5. Chithrani, B. D. and Chan, W. C. W. (2007) Elucidating the mechanism of cellular uptake and removal of protein-coated gold nanoparticles of different sizes and shapes. *Nano Lett.* 7, 1542–1550.

6. Chithrani, B. D., Ghazani, A. A., and Chan, W. C. W. (2006) Determining the size and shape dependence of gold nanoparticle uptake into mammalian cells. *Nano Lett.* 6, 662–668.

7. Giljohann, D. A., Seferos, D. S., Patel, P. C., Millstone, J. E., Rosi, N. L., and Mirkin, C. A. (2007) Oligonucleotide loading determines cellular update of DNA-modified gold nanoparticles. *Nano Lett.* 7, 3818–3821.

8. Giljohann, D. A., Seferos, D. W., Prigodich, A. E., Patel, P. C., and Mirkin, C. A. (2009) Gene regulation with polyvalent siRNA-nanoparticle conjugates. *J. Am. Chem. Soc.* 131, 2072–2073.

9. Rosi, N. L., Giljohann, D. A., Thaxton, C. S., Lytton-Jean, A. K. R., Han, M. S., and Mirkin, C. A. (2006) Oligonucleotide-modified gold nanoparticles for intracellular gene regulation. *Science* 312, 1027–1030.

10. Patel, P. C., Giljohann, D. A., Seferos, D. S., and Mirkin, C. A. (2008) Peptide antisense nanoparticles. *Proc. Natl. Acad. Sci. U.S.A.* 105, 17222–17226.

11. Patel, P.C. at Northwestern University (2009) *Personal communication.*

12. Lee, O.-S. and Schatz, G. C. (2009) Molecular dynamics simulation of DNA-functionalized gold nanoparticles. *J. Phys. Chem. C* 113, 2316–2321.

13. Lee, O.-S. and Schatz, G. C. (2009) Interaction between DNAs on a gold surface. *J. Phys. Chem. C* 113, 15941–15947.

14. Lee, O.-S. and Schatz, G. C. (2009) Molecular dynamics simulation of ds-DNA on a gold surface at low surface coverage. *MRS Proc.* 1177, 209–233.

15. Arkhipov, A., Yin, Y., and Schulten, K. (2008) Four-scale description of membrane sculpting by BAR domains. *Biophys. J.* 95, 2806–2821.

16. Allen, M. P. and Tildesley, D. J. (1987) *Computer simulation of liquids.* Oxford University Press, Inc., New York.

17. McQuarrie, D. A. (1976) *Statistical mechanics.* Harper Collins Publishers, New York.

18. Feller, S. E., Zhang, Y. H., Pastor, R. W., and Brooks, B. R. (1995) Constant-pressure molecular-dynamics simulation – The Langevin piston method. *J. Chem. Phys.* 103, 4613–4621.

19. Martyna, G. J., Tobias, D. J., and Klein, M. L. (1994) Constant-pressure molecular-dynamics algorithms. *J. Chem. Phys.* 101, 4177–4189.

20. Kale, L., Skeel, R., Bhandarkar, M., Brunner, R., Gursoy, A., Krawetz, N., et al. (1999) NAMD2: Greater scalability for parallel molecular dynamics. *J. Comput. Phys.* 151, 283–312.

21. Becker, W. M., Reece, J. B., and Poenie, M. F. (1996) *The world of the cell.* The Benjamin/Cummings Publishing Company, Inc., Menlo Park.

Part II

Translating Nanoscience and Technology from the Lab to the Clinic

Chapter 19

Cytotoxic Assessment of Carbon Nanotube Interaction with Cell Cultures

Hanene Ali-Boucetta, Khuloud T. Al-Jamal, and Kostas Kostarelos

Abstract

The field of nanotoxicology recently has emerged out of the need to systematically study the biocompatibility and potential adverse effects of novel nanomaterials. Carbon nanotubes (CNT) are one of the most interesting types of nanomaterials, and recently, their use in applications has dramatically increased. Their potential adverse impact on human health and the environment, however, have caused them to be viewed with apprehension in certain cases so further studies into their toxicology are justified. Current methodologies using cell culture (in vitro) models are unreliable and are not yet able to offer conclusive results about the toxicity profile of CNT. The need for reliable and rapid toxicity assays that will allow high throughput screening of nanotube materials is a prerequisite for the valid assessment of CNT toxicity. The assay described here was developed based on the pitfalls and drawbacks of traditionally used cytotoxicity assays. A methodological description of the main problems associated with the MTT and the LDH assays is offered to illustrate the advantages of this novel assay for the study and determination of the cytotoxic profile of CNT. Most importantly, a thorough account of this novel assay which is considered to be rapid, reliable, and suitable for broad-spectrum cytotoxicity screening of different types of CNT is described.

Key words: Nanotechnology, Nanotoxicology, MTT, LDH, Fluorescence, Cell death, Apoptosis

1. Introduction

Carbon nanotubes (CNT), novel cylindrical nanostructures, have already exceeded many expectations in terms of widespread usage and large-scale manufacturing due to their extraordinary properties which include high electrical and thermal conductivity and robust mechanical properties (1). CNT also can be utilized as components in a variety of biomedical applications ranging from probes in biosensing strategies to drug delivery vectors in therapeutic schemes (1–6). However, before such applications are used

Sarah J. Hurst (ed.), *Biomedical Nanotechnology: Methods and Protocols*, Methods in Molecular Biology, vol. 726,
DOI 10.1007/978-1-61779-052-2_19, © Springer Science+Business Media, LLC 2011

in a clinical setting systematic toxicological assessment of CNT is needed and warranted.

Despite the increased usage of CNT, general conclusions about their in vitro cytotoxicity have proven difficult because of the wide variety of available CNT materials (e.g., in terms of their manufacturing, colloidal dispersion, and chemical purity) and the variability in the assays used to determine their toxicity. The interaction between CNT and the molecules used in traditional and well-established toxicity assays is one key factor attributing to this variability (7–12). Most traditional assays are based on colorimetry and fluorescence, and the CNT strongly interact with the chromophore molecules used. CNT also intrinsically interact through supramolecular stacking and assembly with species, such as macromolecules (e.g., polymers, proteins, and nucleic acids) and small molecules (e.g., doxorubicin).

Worle-Knirsch et al. (7) have previously indicated that the MTT assay is unreliable to use with CNT due to the commonly occurring false-positive results caused by the strong interaction between the CNT and the insoluble formazan crystals. They suggested using alternative cytotoxicity assays (such as LDH and flow cytometry with Annexin V/PI staining) and other tetrazolium-based assays (WST-1, INT, XTT). Casey et al. confirmed through spectroscopic analysis that single-walled carbon nanotubes (SWNT) interact with the dyes (Coomassie Blue, Alamar Blue, Neutral Red, MTT, and WST-1) used in cytotoxicity assays and concluded that most are not suitable for the quantitative toxicity assessment of CNT (8). In an attempt to propose a reliable cytotoxicity assay that would not rely on light absorbance, the same group described a novel approach using the clonogenic assay (13). Although reliable this assay is too time-consuming to allow for rapid screening. No other report has used the clonogenic assay since this work was published, and despite the reported inaccuracies most in vitro toxicity studies are still carried out using colorimetry-based methodologies. More recently, Monteiro-Riviere et al. (11) studied the reliability of a range of widely used viability and cytotoxicity assays with many types of nanoparticles including SWNT. They also found that SWNT interfered to varying degrees with the results of most established toxicity assays. The need for reliable toxicity assays that would allow rapid screening of nanomaterials has now become a serious obstacle toward safety validation of the myriad types of CNT as well as other types of nanoparticles.

Here, we propose a modified version of one of the most widely used and established cytotoxicity assays, the lactate dehydrogenase (LDH) assay, that circumvents all interactions between CNT and the fluorophore molecules leading to a reliable and technically straightforward methodological solution. A comparison with other established assays like the MTT and the original

LDH assay is also provided to illustrate the pitfalls and problems with those methodologies compared to the proposed *modified LDH* (mLDH) method that offers a trustworthy and reproducible determination of cellular toxicity following their interaction with CNT.

2. Materials

2.1. CNT Preparation

1. Pristine multiwalled carbon nanotubes (MWNT) (Nanocyl, Belgium) (see Note 1).
2. Pluronic F127 copolymer (Sigma, UK).
3. Sterile deionized water.
4. Glass vials.
5. Water bath.
6. Bath sonicator (Ultrasonic cleaner, VWR).

2.2. Cell Culture

1. Adherent cells, such as the lung epithelial cell line A549 (CCL-185, ATCC, UK), or others.
2. 0.05% Trypsin with 0.53 mM ethylenediaminetetraacetic acid (EDTA) tetrasodium salt (Gibco, Invitrogen, UK).
3. Culture media appropriate for the cell line being studied. F12 Ham media supplemented with 10% fetal bovine serum (FBS), 50 U/mL penicillin, and 50 µg/mL streptomycin (all from Gibco, Invitrogen, UK) was used with A549 cells.
4. 96-Well flat bottom plate (Corning Costar Corporation®, USA).
5. 10% v/v Dimethyl sulfoxide (DMSO) (>99.7%, Hybri-Max™, sterile filtered, hybridoma tested) in complete cell culture medium.
6. 1, 5, and 25 mL serological pipettes.
7. Incubate at 37°C with 5% CO_2.
8. Trypan blue dye exclusion assay kit.

2.3. In Vitro Cytotoxicity Assays

2.3.1. MTT Assay

1. MTT (3-(4, 5-dimethylthiazol-2-yl)-2,5-diphenyltetrazolium bromide) powder (Sigma, UK). Stored at 4°C before reconstitution.
2. Sterile phosphate-buffered saline (PBS) (1×) (Gibco, Invitrogen, UK).
3. Sterile filter (0.22 µm).
4. DMSO, 100%.
5. Plate reader.

2.3.2. LDH Assay (Original and Modified)

1. LDH kit: CytoTox 96 non-radioactive cytotoxicity assay (Promega, UK) containing substrate mix (five vials), assay buffer (60 mL), LDH positive control (25 µL), lysis solution (3 mL) (see Note 2), and stop solution (65 mL). Store substrate mix and assay buffer at –20°C, protected from light until use. Store LDH positive control, lysis solution, and stop solution at 4°C.

2. 96-Well flat bottom plate (Corning Costar Corporation®, USA).

3. Phenol-free media (e.g., RPMI media) (Gibco, Invitrogen, UK).

4. 9% v/v Triton X-100.

5. Plate reader.

3. Methods

3.1. Preparation of CNT Dispersions

1. In a glass vial, hydrate the Pluronic F127 to a final concentration of 1% w/v with sterile deionized water.

2. Place the vial in a water bath (37°C) for 30–45 min or until the Pluronic F127 flocculates disappear.

3. Disperse 1 mg/mL pristine MWNT powder in 1% (10 mg/mL) Pluronic F127 by bath sonication for 30–45 min (see Note 3).

4. Store the stock MWNT:F127 dispersion at 4°C until further use. When ready to use, sonicate for 15 min.

5. Use the 1% F127 stock solution for the controls in the toxicological assessments. Store at 4°C until further use.

3.2. Cell Culture

The following protocol describes the incubation of the MWNT:F127 dispersions and control samples with A549 lung epithelial cell lines. If desired, the cells can be treated with various inhibitors and the incubation times of the samples with the cells and CNT concentration in the samples can be varied. DMSO is used as a positive control for cytotoxicity.

1. Passage A549 cells when they reach 70–80% confluency to maintain exponential growth. Use for a maximum of ten passages.

2. To trypsinize the monolayer, rinse with 1× PBS then incubate with trypsin–EDTA at 37°C for 5 min. Detach the cells by vigorous up and down pipetting.

3. Centrifuge the cells at $240 \times g$ for 5 min at 4°C and resuspend in complete media.

4. Count cells and determine cell viability by Trypan blue dye exclusion assay.

5. Seed 10,000 cells per well (150 µL/well) in a 96-well plate and incubate for 24 h at 37°C in a humidified atmosphere (5% CO_2) incubator.

6. Dilute the stock MWNT:F127 dispersion or stock 1% F127 in complete cell culture medium to reach the desired concentrations (see Note 4).

7. Sonicate the diluted MWNT:F127 dispersion in media for 2 min prior to its addition to cells.

8. Incubate the cells with the CNT dispersions or control samples (1% F127 or 10% DMSO) for 24 h at 37°C in a humidified atmosphere (5% CO_2) incubator. Also, leave some cells untreated as controls.

9. Proceed with the cytotoxicity assessments (see Subheading 3.3).

3.3. In Vitro Cytotoxicity Assays

3.3.1. MTT Assay

The colorimetric MTT assay is used to measure cell viability (14). The yellow tetrazolium salt (MTT) is reduced by mitochondrial reductase in living and metabolically active cells to purple, water-insoluble formazan crystals, which can then be dispersed using DMSO or other detergents. A decrease in absorbance at 570 nm compared to untreated control cells is then a measure of the cell viability or the amount of apoptosis or necrosis that has been caused by the test material (see Fig. 1).

1. Aspirate the media after the incubation period is over.

2. Prepare the MTT solution by reconstituting the MTT powder in 1× sterile PBS to a final concentration of 5 mg/mL and

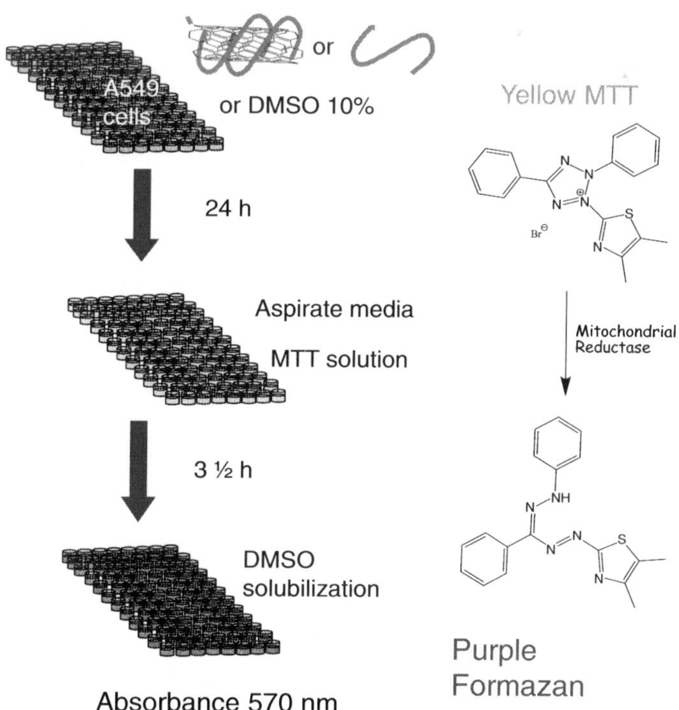

Fig. 1. Schematic of the MTT assay.

subsequently filter and sterilize this solution using a 0.22-μm sterile filter. Store in 2 mL aliquots at –20°C protected from light until use (stable for at least 6 months after reconstitution).

3. Dilute the MTT solution with complete media containing FBS (at a ratio of 1:6) and add 120 μL of this solution to each well.

4. Incubate the cells for 3.5 h at 37°C in a humidified atmosphere (5% CO_2) incubator.

5. Remove the MTT solution by gently inverting the 96-well plate into a paper tissue.

6. Add 150 μL DMSO (100%) to solubilize the formazan crystals and incubate the plate for 15 min at 37°C to remove air bubbles.

7. Read the absorbance at 570 nm in a plate reader and express the results as the percentage cell viability ($n = 8 \pm$ S.D.) compared to untreated control cells (see Fig. 2 and Note 5). The percentage cell viability is calculated using this formula:

$$\% \text{ Cell viability} = \frac{A_{570\,nm} \text{ of treated cells}}{A_{570\,nm} \text{ of untreated cells}} \times 100.$$

3.3.2. Original LDH Assay (See Note 6)

LDH is a stable cytosolic enzyme that is released from the cell upon cell lysis. The LDH assay is based on quantitatively measuring released LDH using a coupled enzymatic assay, in which LDH plays a role in the conversion of a tetrazolium salt (INT) into a red soluble formazan product which then can be measured colorimetrically. The amount of LDH released is proportional to the number of lysed cells (15) (see Figs. 3 and 4).

1. Transfer 50 μL media containing released LDH from all wells into a fresh 96-well plate.

2. For maximum LDH release: Add 10 μL lysis solution (10×) for every 100 μL of fresh media. Keep the cells after treatment at 37°C for 45–60 min.

3. Centrifuge the plate at $240 \times g$ for 4 min and supernatant into the fresh 96-well plate.

4. Thaw the assay buffer and warm to room temperature, while keeping it protected from light.

5. Transfer 12 mL assay buffer into one vial of substrate mix. Gently mix to dissolve the substrate mix, while keeping it protected from light. Both the assay buffer and non-used reconstituted substrate mix can be stored again at –20°C (see Note 7).

6. Add 50 μL reconstituted substrate mix to each well containing the transferred aliquots. Cover the plate with foil and incubate for 30 min at room temperature.

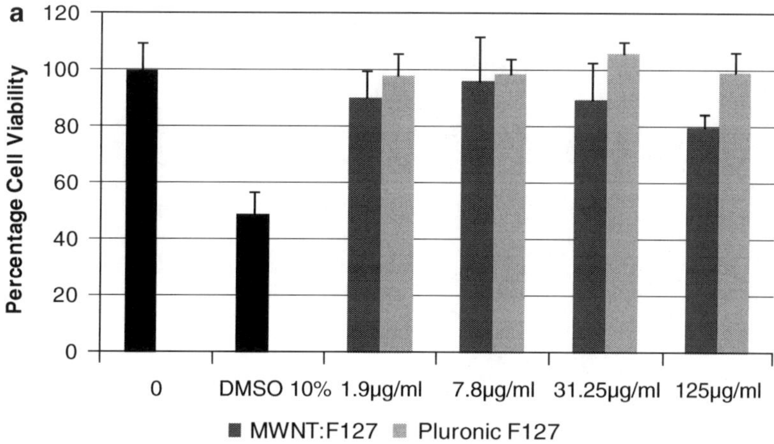

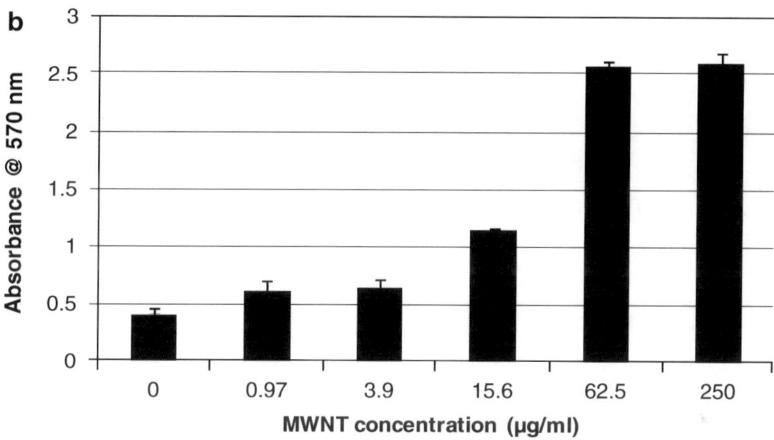

Fig. 2. (a) Percentage cell viability of A549 assessed by the MTT assay for varying MWNT and Pluronic F127 concentrations. DMSO is used as a positive control; untreated cells are the negative control (0). The MWNT:F127 dispersions show concentration-dependent toxicity after 24 h exposure. The values listed are the concentrations of MWNT in solution, the corresponding F127 control for each concentration is at a concentration ten times higher (e.g., 1.9 μg/mL MWNT, 19 μg/mL F127) (see Note 4). (b) Absorbance (at 570 nm) of insoluble formazan mixed with MWNT:F127 dispersions. The spiking experiment shows that the intrinsic absorbance of MWNT can interfere with the results of the MTT assay.

7. Add 50 μL stop solution to each well and pop any large bubbles using a syringe needle.

8. Read the absorbance of the solutions at 490 nm in a plate reader and express the results as the percentage LDH released ($n = 4 \pm$ S.D.) compared to maximum LDH released from the untreated control cells (see Fig. 5). The percentage LDH released (% cytotoxicity) is calculated using this formula:

$$\% \text{ LDH released} = \frac{A_{490\,nm} \text{ of treated and untreated cells} - A_{490\,nm} \text{ of media alone}}{A_{490\,nm} \text{ of maximum of untreated cells} - A_{490\,nm} \text{ of media alone}} \times 100.$$

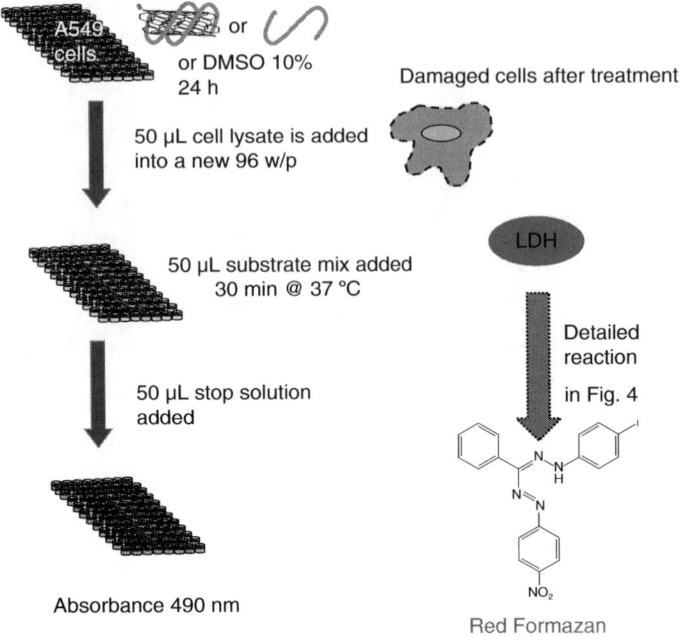

Fig. 3. Schematic of the original LDH assay.

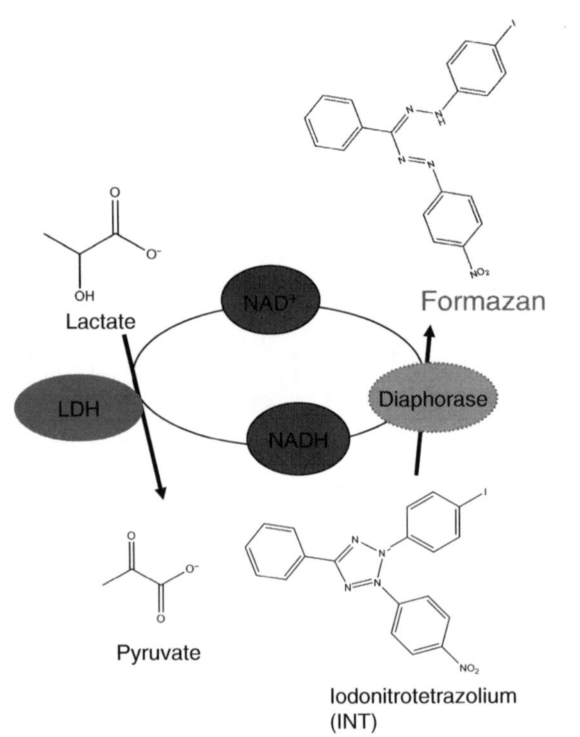

Fig. 4. LDH-mediated conversion of the INT salt into formazan.

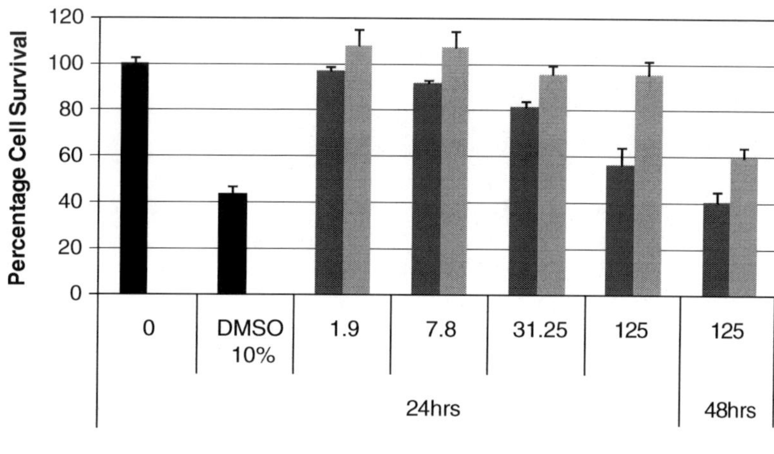

Fig. 5. Percentage cell survival with varying concentrations of MWNT:F127 and Pluronic F127. The values listed are the concentrations of MWNT in solution (in μg/mL), the corresponding F127 control for each concentration is at a concentration ten times higher (e.g., 1.9 μg/mL MWNT, 19 μg/mL F127) (see Note 4). DMSO was used as a positive control; untreated cells (0) were the negative control. MWNT:F127 showed a clear dose-dependent toxicity after 24 h of exposure, which was potentiated after 48 h due to Pluronic F127 toxicity. The Pluronic F127 did not, however, cause any cytotoxicity after 24 h incubation at the concentrations used.

The absorbance data at 490 nm also can be shown without converting it into percentage LDH release to highlight the interference of CNT with the results of the assay (see Fig. 6 and Note 8).

3.3.3. The "Modified LDH" Assay

The original colorimetric LDH assay was modified to avoid interference of the components used in the assay with the CNT. The survived cells after treatment are artificially lysed with Triton X-100, and the cell lysate is centrifuged in order to precipitate the CNT. The released LDH is therefore an indication of the number of viable cells that survived treatment with CNT (see Figs. 4 and 7).

1. Replace the media with 100 μL per well phenol and serum-free media (RPMI, phenol-free media) (see Note 6).

2. Add 10 μL 9% v/v Triton X-100 per 100 μL added phenol and serum-free media.

3. Incubate the plate at 37°C for 45–60 min (see Note 9).

4. Transfer the cell lysate into tubes and centrifuge at $16,000 \times g$ for 5 min to pellet the uptaken CNT (see Note 10).

5. Transfer 50 μL cell lysate, avoiding the CNT pellet, into a fresh 96-well plate.

6. Add 50 μL reconstituted substrate mix to each well containing the transferred and centrifuged cell lysate. Cover the plate

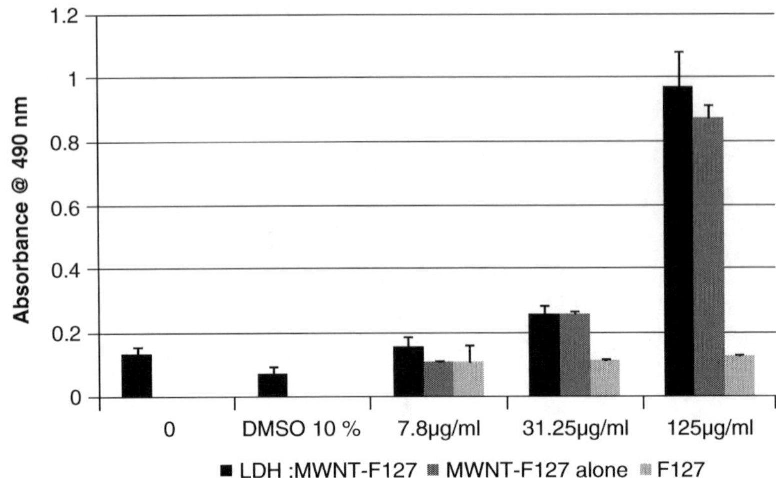

Fig. 6. Percentage LDH release (absorbance at 490 nm) after treatment of cells with different concentrations of MWNT:F127 and Pluronic F127. The values listed are the concentrations of MWNT in solution (in µg/mL), the corresponding F127 control for each concentration is at a concentration ten times higher (e.g., 7.8 µg/mL MWNT, 78 µg/mL F127) (see Note 4). MWNT:F127 dispersions (no assay) were used as controls (no assay). The absorbance of the released LDH in the MWNT-treated wells (LDH:MWNT:F127) is identical to the intrinsic absorbance of the MWNT:F127 which indicates that the observed LDH readings are attributed in part to the intrinsic absorbance of CNT. The LDH enzyme might also be inhibited by the presence of the positive control (DMSO 10%) as it shows low absorbance compared to the untreated control.

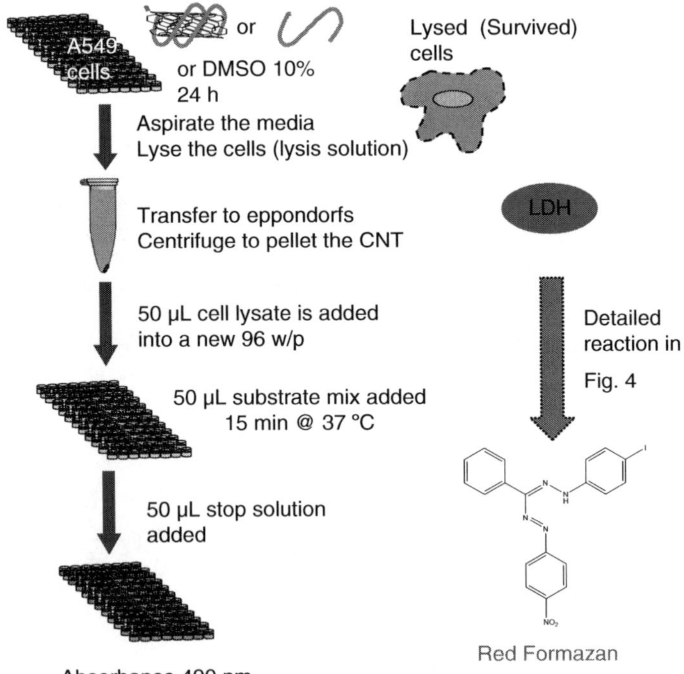

Fig. 7. Schematic of the modified LDH assay.

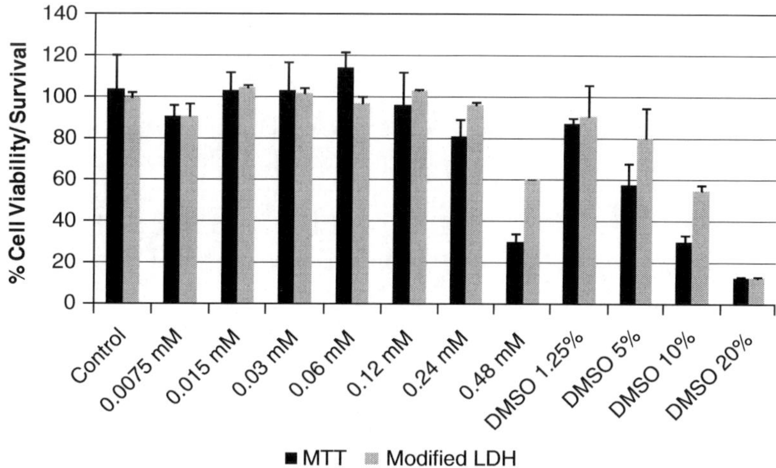

Fig. 8. Different concentrations of cationic liposomes (0.0075–0.48 mM) and DMSO (1.25–20%) were used to assess the reliability of the modified LDH version compared to the MTT assay. Very similar toxicity patterns were observed with both assays, emphasizing the reliability of using the modified LDH assay for the in vitro cytotoxicity assessment of CNT.

with foil and incubate for 15 min at room temperature. (Follow steps 4 and 5 from Subheading 3.3.2 for the preparation of the reconstituted substrate mix).

7. Add 50 µL stop solution to each well and pop any large bubbles using a syringe needle.

8. Read the absorbance at 490 nm in a plate reader and express the results as the percentage cell survival ($n = 4 \pm$ S.D.) compared to untreated control cells (see Fig. 8 and Note 11). The percentage cell survival is calculated using this formula:

$$\% \text{ Cell survival} = \frac{A_{490\,nm} \text{ of treated cells}}{A_{490\,nm} \text{ of untreated cells}} \times 100.$$

4. Notes

1. Noncovalently functionalized CNT are used as an example. If desired, different dispersing agents or CNT that are modified with different surface molecules also can be used. If covalently functionalized CNT are used, disperse them in 5% dextrose by bath sonication for 30–45 min and store at 4°C until further use.

2. 9% v/v Triton X-100 in deionized water can also be used as a lysis solution.

3. A well-dispersed sample of MWNT should contain no precipitates or aggregates. If desired, SWNT can be used, however,

they are not as easily dispersed in F127 as MWNT; longer sonication times may be required. Concentrations higher than 1 mg/mL MWNT are not easily dispersed.

4. We used MWNT:F127 dispersions with MWNT concentrations ranging from 0 to 125 µg/mL. As you dilute the MWNT:F127 stock in cell culture media, the concentration of the Pluronic F127 is also diluted. So, in these samples, the F127 concentration varies as the CNT concentration is varied. As a result, the corresponding F127 controls should have concentrations in the range between 0 and 1,250 µg/mL.

5. The concentration-dependent decrease in cell viability (see Fig. 2a) should be viewed with apprehension. According to published data, the MWNT might adsorb to the formazan crystals (through a strong, π–π stacking interaction) and not allow them to dissolve when the solubilizing agent (DMSO) is added. This would cause a falsely low reading in the absorbance at 570 nm and falsely low percentage cell viability. Further, we believe that the intrinsic absorbance of CNT can alter the results of the assay when high cellular uptake and internalization of CNT occurs. We spiked the insoluble formazan with different concentrations of the MWNT:F127 dispersion and read the absorbance at 570 nm (see Fig. 2b). An increase in absorbance was observed as the concentration of MWNT increased indicating that the intrinsic absorbance of the MWNT is also contributing significantly to the absorbance at 570 nm. This effect results in falsely high cell viability, explaining why during many assays cell viability over 100% is obtained. The systematic effects of these two different types of interference can have the opposite effect on the results of the assay causing unreliable and irreproducible results. In conclusion, the use of the MTT assay for the assessment of CNT cytotoxicity should be avoided due to this unpredictable balance between false-positive and -negative readings.

6. Media containing phenol and serum (FBS) can contribute to background absorbance. This background should be subtracted from all results before calculating the percentage LDH released. In order to reduce this background without affecting cell viability, use phenol-free media with a reduced serum concentration (5%). If a positive control is desired for this LDH assay, gently vortex the LDH positive control and mix 2 µL into 10 mL PBS + 1% bovine serum albumin (BSA) (1:5,000 dilution). This stock should be prepared fresh before each use and triplicate or quadruplicate wells are recommended. This control should have an absorbance of 1.39 ± 25% at 490 nm.

7. Reconstituted substrate mix can be stored for 6–8 weeks at − 20°C without loss of activity. Upon storage, a precipitate

may occur in the assay buffer which can be removed by centrifugation at $300 \times g$ for 5 min and does not affect the assay performance (15).

8. The original LDH assay, similar to the MTT assay, is a colorimetry-based method therefore significant CNT interference is possible. As can be seen from Fig. 6, the media containing the released LDH showed exactly the same absorbance at 490 nm as MWNT:F127 dispersions diluted in media. Therefore, it is not possible to attribute the trend seen in Fig. 6 (as the concentration of the MWNT was increased, there was an increase in the LDH released) to actual cytotoxicity. Note that the positive control (DMSO 10%) showed low LDH release (low absorbance) compared to the other controls. This could be due to the inhibition of LDH enzyme by the DMSO in the media.

9. This step can be replaced by a freeze–thawing cycle. Incubate the plate at $-70°C$ for approximately 30 min followed by thawing at $37°C$ for 15 min and then proceed to step 4.

10. It can be difficult to precipitate the CNT from the media due to their high dispersability. If necessary, the centrifugation time can be increased depending on the amount of CNT uptaken by the cells. A discernible pellet should be observed at the end of the centrifugation step. The centrifugation speed does not seem to affect the release of LDH over a range of speed from 300 to $16,000 \times g$. Centrifugation at $4°C$ is preferable since the LDH enzyme is stable at $4°C$.

11. The toxicity of cationic liposomes, which do not interact with the chemicals used in such assays, was analyzed using this modified protocol and the MTT assay. Results of both assays showed the same trends, proving that the modified version of the LDH assay is in fact promising and reliable (see Fig. 8).

Acknowledgments

This work was partially supported by the European Union FP7 ANTICARB (HEALTH-2007-201587) program. H.A.-B. wishes to acknowledge the Ministére de l'Enseignement Supèrieur et de la Recherche Scientifique (Algeria) for a full Ph.D. scholarship.

References

1. Kostarelos, K., Bianco, A., and Prato, M. (2009) Promises, facts and challenges for carbon nanotubes in imaging and therapeutics. *Nat. Nanotechnol.* **4**, 627–633.

2. Bianco, A. and Prato, M. (2003) Can carbon nanotubes be considered useful tools for biological applications? *Adv. Mater.* **15**, 1765–1768.

3. Prato, M., Kostarelos, K., and Bianco, A. (2008) Functionalized carbon nanotubes in drug design and discovery. *Acc. Chem. Res.* **41**, 60–68.

4. Ali-Boucetta, H., Al-Jamal, K. T., McCarthy, D., Prato, M., Bianco, A., and Kostarelos, K. (2008) Multiwalled carbon nanotube-doxorubicin supramolecular complexes for cancer therapeutics. *Chem. Commun.* **4**, 459–461.

5. Podesta, J. E., Al-Jamal, K. T., Herrero, M. A., Tian, B., Ali-Boucetta, H., Hegde, V., et al. (2009) Antitumor activity and prolonged survival by carbon-nanotube-mediated therapeutic siRNA silencing in a human lung xenograft model. *Small* **5**, 1176–1185.

6. Pastorin, G., Wu, W., Wieckowski, S., Briand, J. -P., Kostarelos, K., Prato, M., et al. (2006) Double functionalization of carbon nanotubes for multimodal drug delivery. *Chem. Commun.* **11**, 1182–1184.

7. Worle-Knirsch, J. M., Pulskamp, K., and Krug, H. F. (2006) Oops they did it again! Carbon nanotubes hoax scientists in viability assays. *Nano Lett.* **6**, 1261–1268.

8. Casey, A., Herzog, E., Davoren, M., Lyng, F. M., Byrne, H. J., and Chambers, G. (2007) Spectroscopic analysis confirms the interactions between single walled carbon nanotubes and various dyes commonly used to assess cytotoxicity. *Carbon* **45**, 1425–1432.

9. Casey, A., Davoren, M., Herzog, E., Lyng, F. M., Byrne, H. J., and Chambers, G. (2007) Probing the interaction of single walled carbon nanotubes within cell culture medium as a precursor to toxicity testing. *Carbon* **45**, 34–40.

10. Monteiro-Riviere, N. A. and Inman, A. O. (2006) Challenges for assessing carbon nanomaterial toxicity to the skin. *Carbon* **44**, 1070–1078.

11. Monteiro-Riviere, N. A., Inman, A. O., and Zhang, L. W. (2009) Limitations and relative utility of screening assays to assess engineered nanoparticle toxicity in a human cell line. *Toxicol. Appl. Pharmacol.* **234**, 222–235.

12. Davoren, M., Herzog, E., Casey, A., Cottineau, B., Chambers, G., Bryne, H. J., et al. (2007) In vitro toxicity evaluation of single walled carbon nanotubes on human A549 lung cells. *Toxicol. In Vitro* **21**, 438–448.

13. Herzog, E., Casey, A., Lyng, F. M., Chambers, G., Byrne, H. J., and Davoren, M. (2007) A new approach to the toxicity testing of carbon-based nanomaterials – The clonogenic assay. *Toxicol. Lett.* **174**, 49–60.

14. Mosmann, T. (1983) Rapid colorimetric assay for cellular growth and survival: Application to proliferation and cytotoxicity assays. *J. Immunol. Methods* **65**, 55–63.

15. *CytoTox 96 Non-Radioactive Cytotoxicity Assay-Technical Bulletin.* http://www.promega.com/catalog/catalogproducts.aspx?categoryname=productleaf_405. Accessed 2010.

Chapter 20

Nanoparticle Toxicology: Measurements of Pulmonary Hazard Effects Following Exposures to Nanoparticles

Christie M. Sayes, Kenneth L. Reed, and David B. Warheit

Abstract

Health risks following exposures to nanoparticle types are dependent upon two primary factors, namely, hazard and exposure potential. This chapter describes a pulmonary bioassay methodology for assessing the hazardous effects of nanoparticulates in rats following intratracheal instillation exposures; these pulmonary exposures are utilized as surrogates for the more physiologically relevant inhalation route of exposure. The fundamental features of this pulmonary bioassay are dose–response evaluations and time-course assessments to determine the sustainability of any observed effect. Thus, the major endpoints of this assay are the following: (1) time course and dose–response intensity of pulmonary inflammation and cytotoxicity, (2) airway and lung parenchymal cell proliferation, and (3) histopathological evaluation of lung tissue. This assay can be performed using particles in the fine (pigmentary) or ultrafine (nano) size regimes.

In this assay, rats are exposed to selected concentrations of particle solutions or suspensions and lung effects are evaluated at 24 h, 1 week, 1 month, and 3 months postinstillation exposure. Cells and fluids from groups of particle-exposed animals and control animals are recovered by bronchoalveolar lavage (BAL) and evaluated for inflammatory and cytotoxic endpoints. This protocol also describes the lung tissue preparation and histopathological analysis of the lung tissue of particle-instilled rats. This assay demonstrates that instillation exposures of particles produce effects similar to those previously measured in inhalation studies of the same particulates.

Key words: Pulmonary toxicity, Particulate materials, Particles, Fine particles, Ultrafine particles, Nanoparticles, Nanomaterials, In vivo, Rat, Lung, Intratracheal instillation, Lung hazards, Pulmonary bioassay

1. Introduction

There is a great need for the development of rapid and reliable short-term bioassays to evaluate the pulmonary toxicity of novel nanoparticles and materials. Although inhalation is the only relevant physiological route of administration that simulates human

Sarah J. Hurst (ed.), *Biomedical Nanotechnology: Methods and Protocols*, Methods in Molecular Biology, vol. 726, DOI 10.1007/978-1-61779-052-2_20, © Springer Science+Business Media, LLC 2011

exposure, it is in most cases not a practical method for conducting lung hazard studies. One critical requirement of an inhalation bioassay is the use of adequate exposures to ensure that sufficient amounts of material deposit in the lung so that pulmonary effects can be appropriately assessed. If sufficient amounts of material are not available, then predictive studies via inhalation are of little value. Also, it is not a practical way to examine the effects of multiple particulates simultaneously.

To this end, several investigators have proposed methods of administration using intratracheal instillations of particles into the lungs of experimental animals to simulate high-dose exposures (1–4). Intratracheal instillation procedures allow for a direct route of entry and exposure to the lung using a small amount of material in the liquid phase. This route of exposure has some practical advantages relative to inhalation studies because the methodology is (1) relatively inexpensive, (2) simple to implement, and (3) easily permits the instillation of a sufficient, quantifiable bolus of particle suspension to reach target tissues in the respiratory tract. This third advantage is of considerable importance when testing particles or materials in the nanoscale-sized regime. Researchers must employ a method of exposing animals to small amounts of nanoparticles, most often in suspension or solution because synthesizing and producing new nanoparticle types on a large scale requires a nontrivial amount of effort.

This chapter describes an intratracheal instillation bioassay performed in rats that can be used to predict the potential for inhaled particles to produce lung hazard effects, such as pulmonary fibrosis and sustained inflammation in exposed humans. We postulate that the mechanisms related to particle-induced pulmonary disease are dependent upon three interdependent general factors: (1) the tendency of inhaled particles or materials to cause lung cell injury, measured by general cytotoxicity, (2) the affinity for inhaled particles to produce ongoing inflammation, and (3) the reduction of pulmonary macrophage clearance. The occurrence of these three factors has been documented individually or in various combinations in previous reports of chronic lung disease with fibrosis (5–7). Thus, the major endpoints of this study are the following: (1) time course and dose–response intensity of pulmonary inflammation and cytotoxicity, (2) airway and lung parenchymal cell proliferation, and (3) histopathological evaluation of lung tissue. This method utilizes both dose and time-course variations in order to determine the range of hazard potentials as well as particle clearance assessments. The multidisciplinary approach of this bioassay/screen is considered to be an important component in the accuracy of its predictions of pulmonary toxicity and for the investigation of clearance and inflammatory mechanisms. It is also expected that the results from the bioassay will provide valuable level-setting information for subchronic and chronic inhalation studies.

2. Materials

2.1. Intratracheal Instillation Exposure

1. Animals: Groups of male Crl:CD®(SD)IGS BR rats (Charles River Laboratories, Inc., Raleigh, NC). Five animals/particulate-type or control group/dose/time point. The rats should be approximately 8 weeks old at study start (mean weights between 240 and 255 g).

2. Particle suspensions in phosphate-buffered saline (PBS) solution (purchased from Sigma in a packet, Ca, Mg-free pH = 7.4, instillation volume = 0.5 mL): (a) particle of interest, (b) carbonyl iron powder (metallic iron, as a negative control, BASF Chemical Company (Wayne, NJ) at >99.5% purity), (c) crystalline silica (Min-U-Sil 5, α-quartz, as a positive control (5, 8, 9), US Silica Co. (Berkeley Springs, WV) at >99% purity).

3. Halothane anesthetic.

4. Disposable animal feeding needle (20 gauge – 1.5 in.) modified from a gavage needle.

2.2. Pulmonary Lavage and Cell Differentials

1. B-Euthanasia-D (sodium pentobarbital, 390 mg/mL) (Intervet/Schering-Plough Animal Health).

2. Scissors.

3. Forceps.

4. Tubing adapter.

5. Suture material.

6. Syringes.

7. PBS, pH = 7.4.

8. Trypan blue filtered using a 0.2-µm Nalgene syringe filter – 25 mm surfactant-free cellulose acetate filter. This dye is filtered once and prepared in a 1:1 ratio with lavagate containing cells.

9. Hemocytometer.

10. Microscope slides.

11. Diff-Quick staining solutions.

12. Optical microscope.

2.3. Biochemical Assays on Bronchoalveolar Lavage Fluid

1. Olympus AU640 Chemistry Analyzer.

2. Olympus® reagents for analyses of bronchoalveolar lavage (BAL) fluid lactate dehydrogenase (LDH) and alkaline phosphatase (AlkP).

3. Reagent kit based on Coomassie blue dye binding (QuanTtest, Quantimetrix, Hawthorne, CA) for measuring lavage fluid protein.

2.4. Pulmonary Cell Proliferation Studies and Histopathological Evaluations

1. 5-Bromo-2′deoxyuridine (BrdU): 100 mg/kg body weight.
2. PBS solution, pH = 7.4.
3. Scissors.
4. Forceps.
5. 10% Formalin.
6. Euthanasia-B solution (0.3–0.5 mL depending on body weight).
7. Anti-BrdU antibody with a 3-amino-9-ethyl carbazole (AEC) marker. 100 mg/kg body weight.
8. Hematoxylin.
9. 70% Ethanol.
10. Optical microscope.
11. Preservation jars.
12. Tissue monocassettes.
13. Paraffin.
14. Microtome.
15. Microscope slides.
16. H&E staining solutions.
17. Butterfly catheter (Abbott Labs, North Chicago, IL).

3. Methods

3.1. Intratracheal Instillation Exposure (See Ref. 10)

1. Prepare all particles in 0.5 mL PBS at a concentration of 1 or 5 mg/kg (see Note 1).
2. Anesthetize the rats with halothane in a jar and watch for slow respiration. Gauge level of anesthesia by assessing reflexes (see Note 2).
3. For the BAL studies and lung tissue studies, intratracheally instill groups of rats (five rats/group/dose/time point) with single doses of either 1 or 5 mg/kg of the particle type of interest, benchmark control particles [i.e., crystalline silica (α-quartz) (positive control) or carbonyl iron particles (negative control)], or PBS solution (vehicle control) (see Table 1 and Note 3). Use the same rats for the lung cell proliferation and histopathology studies, but different rats for the BAL studies.
4. Evaluate the lungs (see Subheadings 3.2–3.5) of sham and particle-exposed rats at 24 h, 1 week, 1 month, and 3 months postinstillation (after a recovery period) (see Note 4).

Table 1
Protocol for intratracheal instillation particle bioassay study

Exposure Groups

- PBS (vehicle control)

- Particles (1 and 5 mg/kg)

  o α-Quartz particles (positive control)

  o Carbonyl iron (negative control)

  o Other particles of interest

Instillation

Exposure

Post-instillation evaluation via BAL and lung tissue

| 24 h | 1 wk | 1 mo | 3 mo |

3.2. Pulmonary Lavage and Cell Differentials (See Figs. 1 and 2)

1. Euthanize the rats via intraperitoneal injections of Euthanasia-B at 0.1 mL/300 g of animal.

2. Lavage the lungs, trachea, and airways of particle-exposed and sham rats with PBS solution warmed to 37°C.

3. Carry out this BAL procedure 2–3 times or until 12 mL of fluid is collected from each animal.

4. Centrifuge lavaged fluids at $700 \times g$ (1,800 rpm) to collect the cells in a pellet.

5. Remove 2 mL of the supernatant, store it on ice for biochemical analysis (see Subheading 3.3), and decant the remaining supernatant (see Note 5).

6. Resuspend the cell pellet in 5 mL of cold (4°C) PBS.

7. Add filtered trypan blue to an equal volume of resuspended lavage fluid and quantify cell numbers and viabilities using a hemocytometer (see Note 6).

8. Stain cytocentrifuge preparations using Diff-Quick and perform cell differential counts using light microscopy with a minimum of 500 cells per slide. Identify, quantify, and categorize the cells according to the following cell-types: macrophages, neutrophils, lymphocytes, and eosinophils.

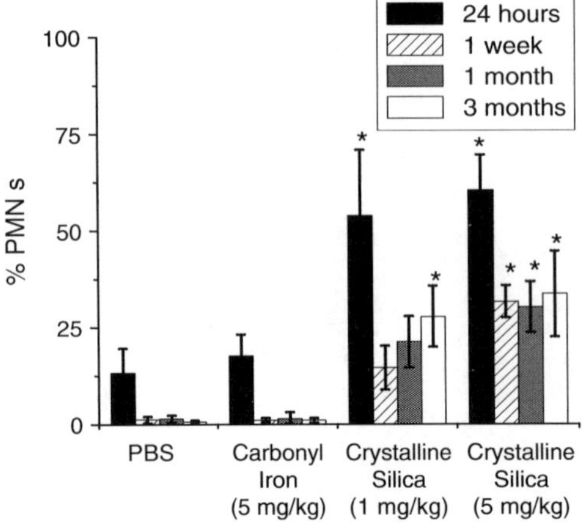
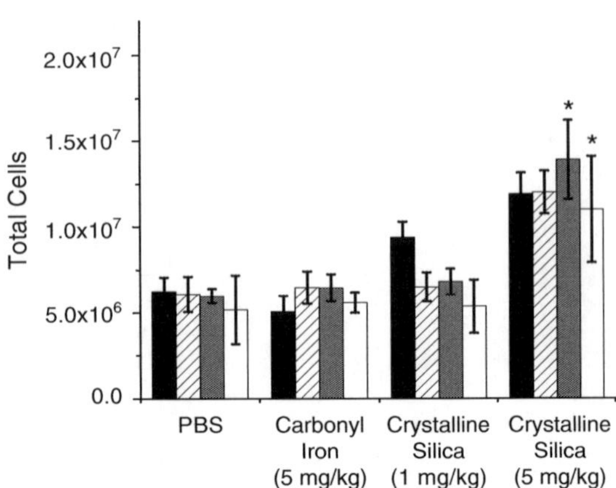

Fig. 1. *Left*: Pulmonary inflammation in sham and particle-exposed rats as evidenced by % neutrophils (PMN) in BAL fluids at 24 h, 1 week, 1 month, and 3 months postinstillation. Intratracheal instillation exposures of the carbonyl iron particles produced a short-term, pulmonary inflammatory response, as evidenced by an increase in the percentages/ numbers of BAL-recovered neutrophils, measured at the 24-h time point. However, the exposures to crystalline silica (α-quartz) particles (1 and 5 mg/kg) produced sustained pulmonary inflammatory responses, as measured through 3 months postexposure (*$p < 0.05$ vs. PBS controls). *Right*: Numbers of cells recovered in BAL fluids from sham and particle-exposed rats. The numbers of cells recovered by BAL from the lungs of high-dose crystalline silica (α-quartz) exposed (5 mg/kg) groups were higher than any of the other groups for all postinstillation time points, indicating pulmonary inflammation. In both figures, all values are given as means ± a standard deviation.

3.3. Biochemical Assays on Bronchoalveolar Lavage Fluid (See Fig. 3)

1. Perform all biochemical assays on BAL fluids at 30°C using a semiautomated clinical chemical analyzer.

2. Measure LDH and AlkP activity using commercially available reagent kits (see Note 7).

3. Measure lavage fluid protein using a commercially available reagent kit (see Note 8).

3.4. Pulmonary Cell Proliferation Studies and Histopathological Evaluations (See Fig. 4)

1. After the desired recovery period, intraperitoneally inject groups of sham and particle-exposed rats with BrdU (100 mg/ kg body weight in PBS); inject 5 mL/kg body weight. Euthanize the animals 6 h later with an intraperitoneal injection of Euthansia-B; this is referred to as a "6-h pulse." BrdU labels all dividing cells during the 6-h pulse period.

2. Following cessation of spontaneous respiration (within 1–3 min), fix the lungs of the rats either through the vasculature (vascular perfusion) or airway (intratracheal infusion) (11). In both cases, exsanguinate (bleed out) the animal to reduce artifacts, expose its trachea, and clamp it with a hemostat to prevent lung collapse. For intratracheal fixation, make a small incision below the clamp and secure a 19-gauge butterfly catheter into the trachea. Connect the catheter to a reservoir (containing a neutral buffered, 10% formalin fixative)

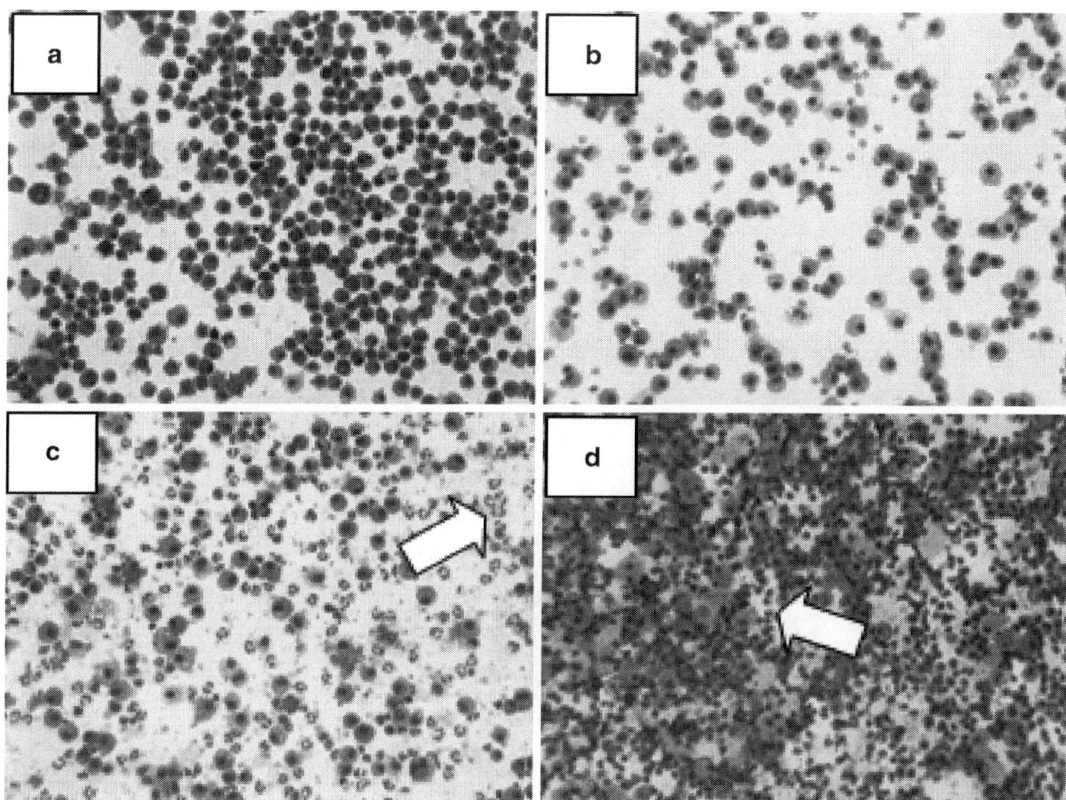

Fig. 2. Cytocentrifuge preparation of lavaged cells recovered from a rat exposed to 5 mg/kg of (**a**) carbonyl iron particles 1 week postinstillation exposure, (**b**) carbonyl iron particles 3 months postinstillation exposure, (**c**) crystalline silica (Min-U-Sil, α-quartz) particles 1 week postinstillation exposure, and (**d**) crystalline silica (Min-U-Sil, α-quartz) particles 3 months postinstillation exposure, demonstrating the sustainability of the α-quartz-induced pulmonary inflammatory responses. *Arrows* indicate neutrophil recruitment (magnification = 200×).

located 15 cm above the thorax of the animal and infuse the lungs at 21 cm H_2O.

3. After 15 min of fixation, carefully remove the heart and lungs together and immersion-fix them in 10% formalin.

4. In addition, remove a 1-cm piece of duodenum (which serves as a positive labeling tissue control) and store it in 10% formalin. The duodenum has a high cell proliferation rate and is utilized as a positive control tissue.

5. Subsequently, dehydrate the lung lobes, heart, and duodenum in 70% ethanol and then weigh the lungs (lung weight is a potential indicator of lung fibrotic responses). Section for histology.

6. For cell proliferation analyses, embed tissue sections from the right cranial, caudal, and left lobes of the lung, as well as duodenal sections, in paraffin using a tissue monocassette, cut them using a microtome and mount them on glass slides.

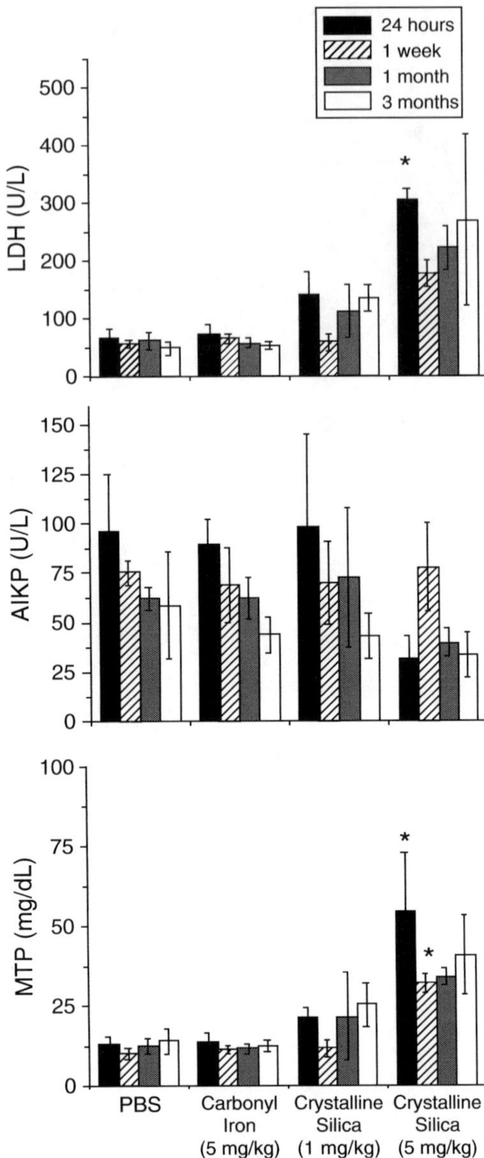

Fig. 3. BAL fluid (*top*) lactate dehydrogenase (LDH), (*middle*) alkaline phosphatase (AlkP), and (*bottom*) micrototal protein (MTP) values for sham and particle-instilled rats at 24 h, 1 week, 1 month, and 3 months postinstillation. Values given are means ± a standard deviation. Exposures to carbonyl iron did not produce any differences in the LDH, AlkP, or MTP values when compared to the PBS controls. Further, no significant increases in BAL fluid AlkP values are measured in any groups at any exposure time. Exposures to 1 or 5 mg/kg crystalline silica particles produced a sustained increase in BAL fluid LDH and MTP values vs. controls and the 5 mg/kg produced a decrease after week 1 through the 3-month postexposure period, demonstrating a sustained cytotoxic effect on the lungs ($*p < 0.05$ vs. PBS controls).

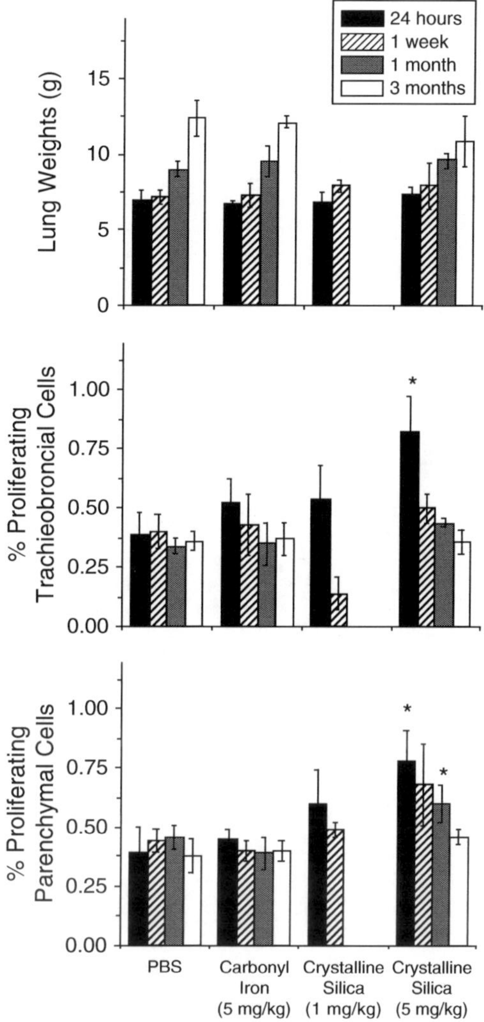

Fig. 4. *Top*: Lung weights, *middle*: tracheobronchial cell proliferation rates (% cells immunostained for BrdU), and *bottom*: lung parenchymal cell proliferation rates (% cells immunostained for BrdU) of sham and particle-instilled rats at 24 h, 1 week, 1 month, and 3 months postinstillation. Values given are means ± a standard deviation. Lung weights of rats increased with increasing postinstillation time; however, no difference in the lung weights was observed between the particle-instilled and sham groups at any postexposure time point. Significant increases in airway tracheobronchial cell proliferation indices were measured in high-dose α-quartz exposed rats at 24 h postinstillation, but these effects were not sustained. Significant increases in lung parenchymal cell proliferation indices were measured in rats exposed to high-dose α-quartz at 24 h and 1 month postinstillation ($*p < 0.05$ vs. PBS controls). Exposures to carbonyl iron particle did not produce any significant differences in cell proliferation indices compared to vehicle controls.

7. Stain the slides with an anti-BrdU antibody containing a AEC marker and counter-stain with aqueous hematoxylin.

8. For each treatment group, count immunostained nuclei in airways (i.e., terminal bronchiolar epithelial cells) or lung parenchyma (i.e., epithelia, interstitial cells, or macrophages) by light microscopy at 1,000× magnification (12, 13). Count a minimum of 1,000 cells/animal in both the terminal bronchiolar and alveolar regions of the lung.

9. For histopathological analyses, make sagittal sections of the left lung with a razor blade.

10. Dissect tissue blocks from upper, middle, and lower regions of the lung and prepare them for light microscopy (paraffin embedded, sectioned, and hematoxylin–eosin-stained).

11. Also, evaluate lungs for inflammatory reactions and lung fibrotic effects (see Note 9).

3.5. Statistical Analyses

1. Calculate a one-way analysis of variance (ANOVA) and Bartlett's test for each sampling time (see Note 10).

2. Compare the means of experimental values to their corresponding sham control values for each time point. Subsequently, normalize the data and represent it as a percentage of the sham control values for that experiment.

4. Notes

1. Ensure that the particles are evenly dispersed in the PBS vehicle otherwise a nonuniform exposure could result. Sonication can be used to disperse the nanoparticles if necessary.

2. Ensure that loss of animals does not occur during the instillation procedure due to an overdose of anesthesia.

3. When the experimenter incorporates a material (chemical or particle) that induces an inflammatory response (positive control), a material that induces little or no response in the animal (negative control), and the vehicle control (no particles), then accurate interpretation of data is more likely to be achieved.

4. Time-course experiments are needed to determine if the effects are transient or sustained; transient inflammation can occur at the 24-h postexposure time point.

5. All refrigerated lavaged samples containing enzymes and proteins are stable for a minimum period of 24 h.

6. Trypan blue is a stain that colors dead cells.

7. LDH is a cytoplasmic enzyme and an indicator of cell injury. AlkP activity is a measure of Type II alveolar epithelial cell secretory activity, and increased AlkP activity in BAL fluids is considered to be an indicator of Type II lung epithelial cell toxicity.

8. Increased concentrations of BAL fluid micrototal protein (MTP) generally are consistent with enhanced permeability of vascular proteins into the alveolar regions, indicating a breakdown in the integrity of the alveolar–capillary barrier. Normal variability for protein and LDH values among sham control samples for each time period average 17%.

9. Progressive lung tissue thickening is a prelude to the development of fibrosis.

10. When the F test from ANOVA is significant, the Dunnett's test is used to compare means from the control group and each of the groups exposed either to silica or to carbonyl iron. Significance is judged at the 5% probability level.

Acknowledgment

This work was supported by DuPont Haskell Global Centers for Health and Environmental Sciences.

References

1. Beck, B. D., Brain, J. D., and Bohannon, D. E. (1982) An in vivo hamster bioassay to assess the toxicity of particulate for the lungs. *Toxicol. Appl. Pharmacol.* **66**, 9–29.

2. Lindenschmidt, R. C., Driscoll, K. E., Perkins, M. A., Higgins, J. M., Maurer, J. K., and Belfiore, K. A. (1990) The comparison of a fibrogenic and two nonfibrogenic dusts by bronchoalveolar lavage. *Toxicol. Appl. Pharmacol.* **102**, 268–281.

3. Driscoll, K. E., Lindenschmidt, R. C., Maurer, J. K., Higgins, J. M., and Ridder, G. (1990) Pulmonary response to silica or titanium dioxide: Inflammatory cells, alveolar macrophage-derived cytokines, and histopathology. *Am. J. Respir. Cell Mol. Biol.* **2**, 381–390.

4. Warheit, D. B., Webb, T. R., Reed, K. L., and Sayes, C. M. (2007) Pulmonary toxicity study in rats with three forms of ultrafine-TiO_2 particles: Evidence for differential responses. *Toxicology* **230**, 90–104.

5. Lugano, E. M., Dauber, J. H., and Daniele, R. P. (1982) Acute experimental silicosis: Lung morphology, histology, and macrophage chemotaxin secretion. *Am. J. Pathol.* **109**, 27–36.

6. Bowden, D. H. (1987) Macrophages, dust and pulmonary disease. *Exp. Lung Res.* **12**, 89–107.

7. Reiser, K. M. and Last, J. A. (1986) Early cellular events in pulmonary fibrosis. *Exp. Lung. Res.* **10**, 311–355.

8. Morgan, A., Moores, S. R., Holmes, A., Evans, J. C., Evans, N. H., and Black, A. (1980) The effect of quartz, administered by intratracheal instillation, on the rat lung. 1. The cellular response. *Environ. Res.* **22**, 1–12.

9. Bowden, D. H. and Adamson, I. Y. R. (1984) The role of cell injury and the continuing inflammatory response in the generation of silicotic pulmonary fibrosis. *J. Pathol.* **144**, 149–161.

10. Warheit, D. B., Brock, W. J., Lee, K. P., Webb, T. R., and Reed, K. L. (2005) Comparative pulmonary toxicity inhalation and instillation and studies with different TiO$_2$ particle formulations: Impact of surface treatments on particle toxicity. *Toxicol. Sci.* **88**, 514–524.

11. Warheit, D. B., Chang, L. Y., Hill, L. H., Hook, G. E. R., Crapo, J. D., and Brody, A. R. (1984) Pulmonary macrophage accumulation and asbestos-induced lesions at sites of fiber deposition. *Am. Rev. Respir. Dis.* **129**, 301–310.

12. Warheit, D. B., Carakostas, M. C., Hartsky, M. A., and Hansen, J. F. (1991) Development of a short-term inhalation bioassay to assess pulmonary toxicity of inhaled particles: Comparisons of pulmonary responses to carbonyl iron and silica. *Toxicol. Appl. Pharmacol.* **107**, 350–368.

13. Warheit, D. B., Hansen, J. F., Yuen, I. S., Kelly, D. P., Snajdr, S., and Hartsky, M. A. (1997) Inhalation of high concentrations of low toxicity dusts in rats results in pulmonary and macrophage clearance impairments. *Toxicol. Appl. Pharmacol.* **145**, 10–22.

Chapter 21

Nanoparticle Therapeutics: FDA Approval, Clinical Trials, Regulatory Pathways, and Case Study

Aaron C. Eifler and C. Shad Thaxton

Abstract

The approval of drugs for human use by the US Food and Drug Administration (FDA) through the Center for Drug Evaluation and Research (CDER) is a time-consuming and expensive process, and approval rates are low (DiMasi et al., J Health Econ 22:151–185, 2003; Marchetti and Schellens, Br J Cancer 97:577–581, 2007). In general, the FDA drug approval process can be separated into preclinical, clinical, and postmarketing phases. At each step from the point of discovery through demonstration of safety and efficacy in humans, drug candidates are closely scrutinized. Advances in nanotechnology are being applied in the development of novel therapeutics that may address a number of shortcomings of conventional small molecule drugs and may facilitate the realization of personalized medicine (Ferrari, Curr Opin Chem Biol 9:343–346, 2005; Ferrari, Nat Rev Cancer 5:161–171, 2005; Ferrari and Downing, BioDrugs 19:203–210, 2005). Appealingly, nanoparticle drug candidates often represent multiplexed formulations (e.g., drug, targeting moiety, and nanoparticle scaffold material). By tailoring the chemistry and identity of variable nanoparticle constituents, it is possible to achieve targeted delivery, reduce side effects, and prepare formulations of unstable (e.g., siRNA) and/or highly toxic drugs (Ferrari, Curr Opin Chem Biol 9:343–346, 2005; Ferrari, Nat Rev Cancer 5:161–171, 2005; Ferrari and Downing, BioDrugs 19:203–210, 2005). With these benefits arise new challenges in all aspects of regulated drug development and testing.

This chapter distils the drug development and approval process with an emphasis on special considerations for nanotherapeutics. The chapter concludes with a case study focused on a nanoparticle therapeutic, CALAA-01, currently in human clinical trials, that embodies many of the potential benefits of nanoparticle therapeutics (Davis, Mol Pharm 6:659–668, 2009). By choosing CALAA-01, reference is made to the infancy of the therapeutic nanoparticle field; in 2008, CALAA-01 was the first targeted siRNA nanoparticle therapeutic administered to humans. Certainly, there will be many more that will follow the lead of CALAA-01 and each will have its own unique challenges; however, much can be learned from this drug in the context of nanotherapeutics and the evolving development and approval process as it applies to them.

Key words: Nanotherapeutics, FDA approval, Regulation, CALAA-01, Drugs

Sarah J. Hurst (ed.), *Biomedical Nanotechnology: Methods and Protocols*, Methods in Molecular Biology, vol. 726,
DOI 10.1007/978-1-61779-052-2_21, © Springer Science+Business Media, LLC 2011

1. The Drug Approval Process: An Overview

Drug development and approval by the US Food and Drug Administration (FDA) can be broadly categorized into three major phases (see Fig. 1) (1, 2). The *preclinical phase* includes the initial *discovery* of a candidate therapeutic, demonstration of *efficacy*, and investigation of *toxicity*. Following discovery, the major focus of preclinical development is animal testing to obtain evidence supporting drug safety and efficacy, and to determine appropriate dosing parameters. The data gathered during the preclinical phase is used to support an Investigational New Drug (IND) application filed with the FDA. Following IND approval, the candidate drug enters the *clinical phase* during which human trials are initiated and are subcategorized into Phases I, II, and III. If during the clinical phase, the drug is considered safe and efficacious, the manufacturer files a New Drug Application (NDA), also with the FDA. Upon NDA approval, the drug can then be marketed according to specific labeling and put into use by medical practitioners. As part of the NDA approval, the FDA may request further studies be done in what is referred to as the *postmarketing phase*, or Phase IV. Studies done after NDA approval are used to further confirm efficacy and/or safety, especially in groups that may not have been well studied in pre-NDA clinical trials. The entire FDA approval process is lengthy, labor-intensive, and stringent. It is estimated that it takes approximately 10–15 years to develop a new medicine at a cost of approximately $1 billion (3–5, 6). In addition, a number of drugs that are discovered and evaluated in the preclinical stages of drug development have an exceedingly high-attrition rate (5).

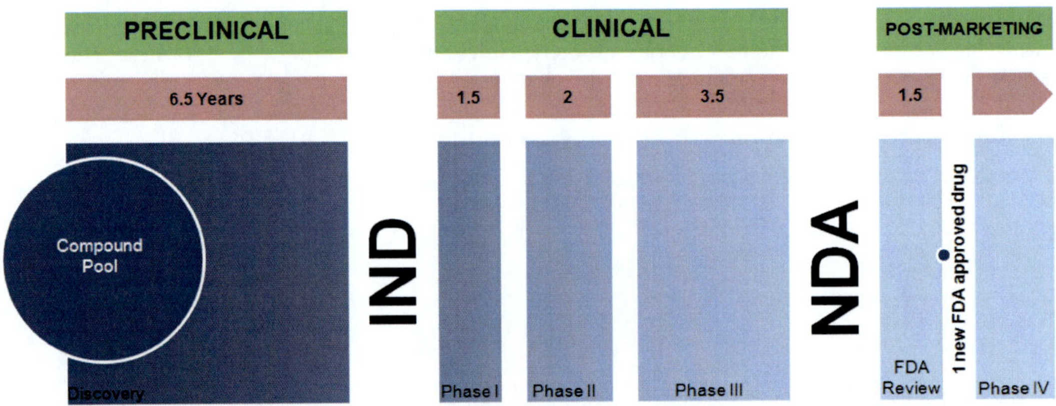

Fig. 1. The major phases of the FDA drug approval process.

1.1. The Preclinical Phase

The preclinical phase is the most diverse phase of drug development. New *small molecule* drug candidates are discovered through a number of different means. In some cases, candidates can be predicted in silico based upon known target and ligand interactions (e.g., crystal structure data) (7–10). Combinatorial approaches are also used to produce and screen libraries of small molecule drugs against known targets (11–13). Currently, high throughput methodologies are not commonplace in the fabrication and interrogation of nanoparticle therapeutics. In many cases, nanotherapeutics are built upon mature, well-characterized nanoparticle platforms whose syntheses and surface and/or internal modification are well understood and can easily be controlled. For instance, liposomes, metal nanoparticles (e.g., colloidal gold), metal oxides (e.g., superparamagnetic iron oxide nanoparticles), nanoshells (e.g., gold), quantum dots, self-assembling peptides/ proteins, fullerenes, and dendrimers are all mature platform technologies used to fabricate nanoparticle therapeutics (14).

In general, nanoparticle platforms require some degree of functionalization(s) to impart desired drug function (15–17). Nanotherapeutics are being built upon relatively few platform technologies, but with the potential for near endless platform modification(s) and chemical tailoring. Nanoparticle drugs often consist of more than one molecular component coupled to the nanoparticle surface or contained within (see Fig. 2). Systematic manipulation of individual, and often multiple, components of a nanoparticle platform can generate a substantial number of unique therapeutics. Owing to their high degree of tailorability, nanoparticle therapeutics can be rationally designed based upon preconceived notions of what the "ultimate" nanoparticle therapeutic

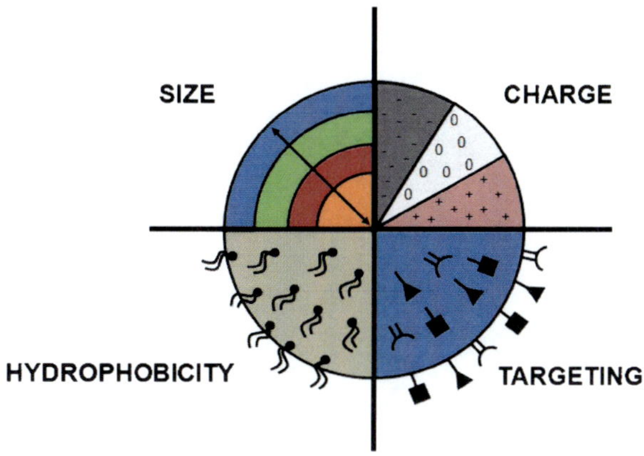

Fig. 2. The tailorability of nanoparticles allows for the manipulation of nearly every physical characteristic. Shown are four characteristics that can be manipulated to impart desired function, but that can also result in significant changes in in vivo drug behavior. Adapted by permission from the authors and from Macmillan Publishers Ltd, ref. 13. Copyright 2007.

represents from the standpoint of adaptability, ease of manufacture, tailorability, functionality, etc. (16–19). Nanoparticle therapeutics envisioned for human use must have robust methodologies for synthesis, characterization, quality control, and potential scale-up. These issues should be kept in mind during planning and early development phases to facilitate translation.

Realizing the need for robust methods to characterize the multitude of these nanoparticle entities prior to IND evaluation, the National Cancer Institute (NCI) created the Nanotechnology Characterization Laboratory (NCL) in 2004 (20). The NCL represents the desire of the NCI to fund and centralize the development of standardization protocols by which emerging nanotherapeutics and other nanomaterials can be preclinically evaluated with regard to efficacy and toxicity (20). The NCL has identified a number of physicochemical parameters that must be well controlled and understood in the early stages of nanotherapeutic development. Importantly, surface chemistry has a significant impact on in vitro and in vivo nanotherapeutic behavior (14, 20, 21). The systematic manipulation of surface chemistry is an ongoing effort, and a rich area for future research, in order to more accurately predict the behavior of nanoparticle therapeutics (20).

In developing a therapeutic for use in humans, significant in vitro and in vivo testing is required to assess drug safety and efficacy. According to the FDA, "During a new drug's early preclinical development, the sponsor's primary goal is to determine if the product is reasonably safe for initial use in humans, and if the compound exhibits pharmacological activity that justifies commercial development" (2). Addressing drug safety, preclinical testing focuses specifically on how the drug will interface with biological systems, and currently takes place on a drug-by-drug basis. A relatively comprehensive list of in vitro and in vivo testing currently performed on materials submitted to the NCL serves as a useful guide (20). Initial studies include data demonstrating efficacy in a number of cell and animal models appropriate to the disease under scrutiny. In cell culture, it is imperative to demonstrate the mechanism of action of the therapeutic and that the observed effect (e.g., apoptosis) is due to the therapeutically active component(s) of the nanoparticle drug and not the result of the nanoparticle platform devoid of drug component (i.e., vehicle). In order to get smooth transition from cell culture models to animal testing, it is important to pair cell culture models and controls with the same or comparable animal models; often, testing commences in mice or rat models. For nanoparticle therapeutics that contain nucleic acid-based therapeutic moieties (e.g., siRNA), testing should also assess potential off-target effects and any nucleic acid-mediated inflammatory response. Further, when possible, nucleic acid therapeutics (e.g., siRNAs) should target conserved sequences among different animal species that will be tested (18). It should also be kept in mind that most often the assay used for assessing a given response was not designed for a nanoparticle therapeutic (14). Thus, it is important to discern how the physicochemical properties of nanoparticle therapeutics (e.g., UV–Vis absorbance) impact assay readout and, potentially, generate false results (14).

Once a promising compound has been discovered, characterized, and demonstrated efficacy and safety in the treatment of a disease in cell culture and in an animal model(s), the ultimate goal of further preclinical testing is to amass sufficient evidence to support an IND submission to the FDA. Other data required to support an IND application includes toxicology, manufacturing information, and proposed clinical protocols (2). Importantly, such parameters are often assessed in multiple animal models (e.g., mouse, dog, nonhuman primate). Based on these data, the manufacturer must establish a dose range for initial administration to human subjects to be used in Phase I studies pending approval of the IND application.

1.2. Investigational New Drug Application

Before filing an IND application with the FDA, a manufacturer may choose to participate in the pre-IND Application Consultation Program (2). The consultation program is designed to foster early communication between the FDA and drug developers and to provide guidance regarding the information and data required for a successful IND application. Multiple consultation divisions exist and are organized by therapeutic class and organ system. Currently, there is no specific division for nanotherapeutics.

The IND application must contain information in three areas: animal pharmacology and toxicity, manufacturing information, and clinical protocols and investigator information (2). Animal pharmacology and toxicity information comes from the extensive preclinical testing of the drug that occurs after discovery and allows the FDA to determine whether the product is reasonably safe for testing in humans. This information also allows investigators to propose an initial dose to be tested in humans. Manufacturing information pertains to the composition, and stability of the drug as well as information on the manufacturer and the manufacturer's methods for quality control. The FDA uses this information to ensure that the company can adequately produce the drug in sufficient quantity and consistency across batches of the drug. The final piece of information is twofold; it concerns the protocols that will be used in the clinical phase of testing and the investigators who will oversee it. First, detailed protocols are scrutinized by the FDA to ensure that patients are not exposed to unnecessary risks and that proper informed consent will be obtained. This aspect is bolstered with a commitment for involvement by the Institutional Review Board (IRB) to review the study.

Second, information must be provided that demonstrates the qualifications of clinical investigators (generally physicians) who will oversee the administration of the compound. It must be shown that these physicians will be able to fulfill their clinical trial duties which often go above and beyond those of typical practice. After submission, the investigator must wait 30 days before initiating any clinical testing while the FDA reviews the application and ensures that research subjects will not be subjected to unreasonable risk. If approved, the potential drug is able to move into Phase I clinical testing in humans based on the study protocols proposed in the IND application.

Further, drugs are required to be the subject of an approved marketing application by federal law prior to their shipment across state lines (2). Exemption must be sought from this federal law for clinical trials designed to be conducted in multiple states. Technically, a successful IND application becomes the means by which drug developers are exempted from marketing application approval, and are able to conduct clinical drug trials across state lines. More practically, the approval of an IND application signifies the transition from preclinical development to clinical testing.

1.3. The Clinical Phase

1.3.1. Phase I

Phase I trials represent the first human dose of a drug whose IND application has just been approved (2). The main goal of Phase I testing is to assess dosing, acute toxicity, and drug excretion in humans. Typically, the candidate drug is administered to 20–100 healthy volunteers often with dose escalation. Healthy patients provide a baseline evaluation of initial dosing regimens derived from animal studies. In cases of severe or life-threatening illnesses, studies may enroll volunteers with the disease. For nanotherapeutics, as in the case of all therapeutics, testing for therapeutic-specific side effects identified in preclinical studies are particularly and formally scrutinized in Phase I trials. On average, Phase I studies can take from 6 months to one and a half years to complete. An estimated two thirds of Phase I compounds will move on to Phase II trials (1, 4).

1.3.2. Phase II

Phase II studies involve more patients (approximately 100–300) than Phase I studies and further evaluate the safety and efficacy of a potential drug in a group of patients who suffer from the disease being studied (2). Phase II studies also can be placebo-controlled in order to establish that the drug effectively treats the disease for which it is intended. To avoid unnecessarily exposing patients to a potentially harmful compound, the number of patients enrolled is based on appropriate power calculations, statistically restricting the study size to one where meaningful data can be expected with the fewest patient exposures. Phase II trials can take from 6 months to 2 years, however, recent evidence suggests that while the duration of Phase I and Phase III studies are staying relatively constant, the duration of Phase II studies has significantly increased (22). One reason for Phase II study prolongation is the desire to more thoroughly interrogate drug safety and efficacy in multiple patient groups, potentially, at multiple sites (22). This may increase the time that it takes to accrue patients. However, more data in Phase II provides increasing confidence to either halt evaluation or move forward to large and expensive Phase III trials (22).

1.3.3. Phase III	Phase III studies are large, randomized, placebo-controlled trials including typically 1,000–5,000 patient volunteers from hospitals, clinics, and/or physician offices across the country (1, 2). These studies are used to demonstrate further safety and efficacy and typically the investigational drug is directly compared to the gold standard treatment for a given indication, provided that one exists. Phase III studies can take from 1 to 10 years to complete. Even at this stage, after extensive testing is already completed, approximately 10% of medications fail in Phase III trials (1). Success in Phase III trials is a complex decision which largely rests upon drug safety and efficacy relative to the gold standard of care.
1.4. New Drug Application	Since 1938, every new drug has been the subject of an approved NDA before US commercialization (2). The NDA application is the vehicle through which drug sponsors formally propose that the FDA approve a new pharmaceutical for sale and marketing. The data gathered during all prior phases of drug development become part of the NDA. The documentation required in an NDA recounts the drug's entire history, including the drug identity, the results of animal studies, clinical trial outcomes, how the drug behaves in the body, and how it is manufactured, processed, and packaged. After appropriate deliberation, the FDA may request additional information by written request, issue a tentative approval under which minor deficiencies need to be corrected prior to final approval, or NDA approval may be granted and contain conditions that must be met after initial marketing, such as Phase IV studies (1, 2). Typically, an NDA takes from 6 months to 1 year for review (1).
1.5. The Postmarketing Phase	Postmarketing studies, also known as Phase IV studies, are undertaken after the NDA has been approved by the FDA (2). The impetus for postmarketing studies can come from a number of places including the FDA, practicing clinicians, or the manufacturer. As mentioned previously, the FDA may request a postmarketing study to examine its effects on a high-risk population, or a population that was not well represented in the Phase III trial. Clinicians may be interested in further study of the drug to assess

its efficacy or side effect profile compared to other treatments. Manufacturers may initiate Phase IV studies to assess long-term effects of the drug or to examine the drug's effectiveness in additional indications (23). Although not structured as a strict experiment-based study, postmarketing surveillance of drug administration continues throughout the lifetime of the drug. Physicians continue to report untoward side effects of the drug and manufacturers must submit periodic reports to the FDA describing any cases of adverse events.

As relatively new entities, nanoparticle therapeutics present the pharmaceutical industry with a vast array of opportunities. There is great anticipation surrounding the potential of nanotherapeutics and nanoparticle platform technologies to overcome current therapeutic barriers. This chapter heavily focuses on preclinical nanoparticle development due to the more robust literature available, and the paucity of nanotherapeutics that have reached the clinical stages of development. Progression of nanotherapeutics from preclinical to clinical phases of development is expected to be taken on a case-by-case basis with similar interrogation at each phase as would be applied with traditional therapeutics. Emphasized throughout this chapter is the need for robust nanotherapeutic physicochemical and biological characterization, and quality control so as to build a solid foundation for moving through preclinical and into clinical development. At the time of this writing, there are a number of nanoparticles in preclinical and early stage clinical development (14, 20) demonstrating that progress is being made. The approval of a new drug is a laborious and costly process and there is uncertainty due to a general lack of experience with this class of potential drugs. This is necessitating partnerships with academia, small start-up companies, and large pharmaceutical companies with expertise in a particular nanotherapeutic. Certainly, general knowledge of different nanoparticles and nanoparticle platforms will disseminate as individual materials and candidate nanotherapeutics find their way into development pipelines. Despite the unique challenges, the future is bright for the development of nanoparticle drugs, collaboration, and discovery. The hope is that the novel properties displayed by nanoparticle therapeutics will directly lead to advancements in the successful treatment of human disease.

2. Case Study

This case highlights CALAA-01, the first example of a targeted nanoparticle used to deliver a siRNA drug. At the time of this writing, CALAA-01 has progressed through the preclinical stages of drug development and is now in human clinical trials (18).

Work leading to the first targeted nanoparticle delivery of siRNA in humans, the drug CALAA-01, was begun by Mark Davis and colleagues at the California Institute of Technology in 1996 (18). The goal was to develop a multifunctional targeted cancer therapeutic which would enable the systemic administration of nucleic acids. Their self-stated design methodology was a systems approach to a multifunctional colloidal particle – a rationally derived nanoparticle therapeutic employed before the term "nanoparticle" was generally used. The initial drug schematic (see Fig. 3) presents the desired components of an envisioned therapeutic with (a) a cyclodextrin-containing polymer (CDP) core that spontaneously self-assembles with nucleic acids yielding small colloidal particles less than 100 nm in diameter, (b) a targeting ligand (R) providing for tumor cell specificity and uptake, and, (c) the appreciation of endosomal acidification as a mechanism for particle disassembly and endosomal escape of the therapeutic nucleic acid (18). This CDP-system was initially envisioned for use with plasmid DNA; however, as the association of the nucleic acid with CDP is based upon electrostatic interactions, it was appreciated that the approach could be a rather general one for candidate nucleic acid therapeutics. In addition to platform generality, the platform components were also chosen due to their amenability for scale-up and manufacture.

CALAA-01 is an embodiment of this initial vision. CALAA-01 ultimately evolved to include a number of key components which spontaneously assemble into therapeutic nanoparticles ~70 nm in diameter (see Fig. 4) (18). In addition to the CDP particle core

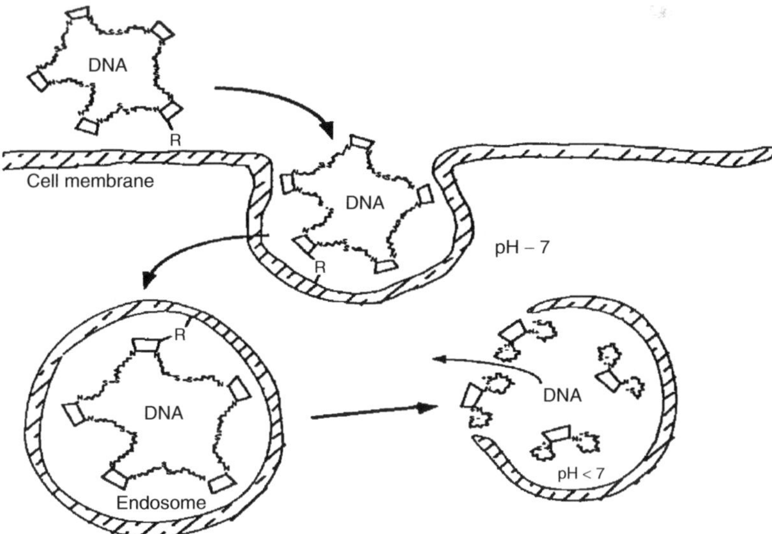

Fig. 3. Initial schematic of the delivery system that would become CALAA-01. Reprinted with permission from author and from ref. 15. Copyright 2009 American Chemical Society.

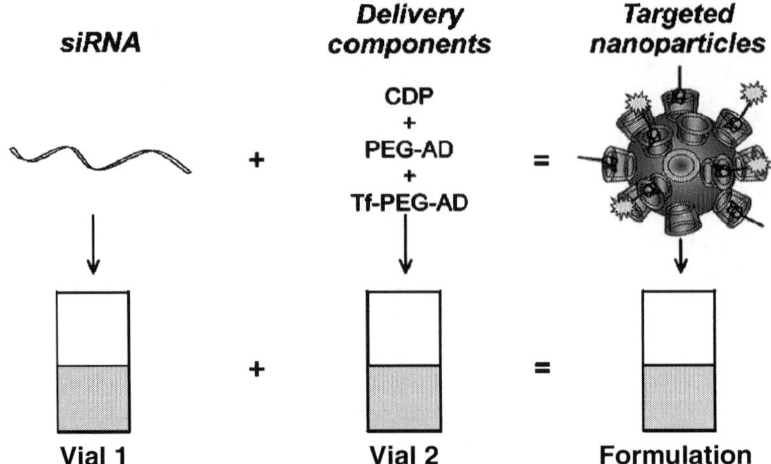

Fig. 4. Schematic of the two-vial formulation. The siRNA is contained in one vial, and the delivery components are contained in the other. Upon mixing, targeted nanoparticles form via self-assembly. Reprinted with permission from author and from ref. 15. Copyright 2009 American Chemical Society.

and the siRNA payload, the surface of the formed nanoparticles is decorated with (a) adamantane-polyethylene glycol (AD-PEG) for drug stabilization in biological matrices, (b) adamantane-PEG-transferrin (AD-PEG-Tf) for tumor-specific targeting and cellular uptake, and (c) imidazole residues to titrate the decrease in endosomal pH upon cellular uptake and promote the endosomal escape of the otherwise sequestered nucleic acid drug. Adamantane is a small molecule that binds tightly and forms an inclusion complex with cyclodextrin on the surface of the formed nanoparticles, thus, displaying PEG and Tf.

Shortly after the conjugate therapeutic was developed, investigators found that the nanoparticles could be formed by self-assembly through simultaneous component reconstitution and mixing (24, 25). This finding gave rise to a unique two-vial formulation strategy (see Fig. 4) which allows for the rapid and straight forward self-assembly of the nanoparticle delivery system components (CDP, AD-PEG, AD-PEG-Tf) with siRNA, at the point of care (25). This formulation provides for siRNA solvation immediately prior to reconstitution with the nanoparticle delivery system components. This is an advantage because siRNA is highly unstable when solvated, but, following the self-assembly process, is protected from nuclease degradation. In addition, this two-vial formulation allowed the molecular delivery system components of CALAA-01 to be separated from siRNA and tested separately for safety in animal models prior to introduction in humans (26). The composition of the formed nanoparticle therapeutic following mixing of the separated components has been well characterized from the standpoint of size and molecular composition (25).

The initial in vitro and in vivo demonstration of the CDP-based siRNA delivery system was published in 2005 using a disseminated murine model of Ewing's sarcoma (27). In these studies, the siRNA targeted the breakpoint of the EWS–FLI1 fusion gene, which is an oncogenic transcriptional activator, in TC71 cells positive for EWS–FLI1 and for the transferrin receptor (27). In addition to in vitro inhibition of their targeted gene product, TC71 cells transfected with firefly luciferase and injected into NOD/SCID mice served as a model system of metastatic Ewing's sarcoma where tumor dissemination and treatment efficacy could be assessed using bioluminescent imaging (27). In this murine model, investigators administered their targeted, CDP-based siRNA delivery particle and demonstrated antitumor effects and target-specific mRNA down regulation (27). Further studies provided evidence that the CDP-based delivery system does not illicit an innate immune response (27), and that active targeting to the transferrin receptor enhances tumor cell uptake (28). In the case of CALAA-01, siRNA targeting ribonucleotide reductase subunit 2 (RRM2) was identified (29). In addition to potency, siRNA targeting RRM2 demonstrates complete sequence homology in mouse, rat, monkey, and humans which allowed for a single siRNA to be used for conducting preliminary studies in all animal models. Targeting RRM2, Davis and colleagues confirmed effective protein knockdown with concomitant reduction in tumor cell growth potential in a subcutaneous mouse model of neuroblastoma (30).

Building upon the above experiments and drawing closer to the initiation of clinical drug testing, Davis and colleagues performed the first study showing that multidosing of siRNA, in the context of the CDP-based nanoparticles, could be done safely in a nonhuman primate (26). This study demonstrated that dosing parameters that were well tolerated were similar to those which had demonstrated antitumor efficacy in mouse models. Furthermore, reversible toxicity was observed in the form of mild renal impairment at high dose; however, extrapolations from mouse model efficacy studies suggested that the therapeutic dosing window would be large. With these promising animal results, a solid foundation was provided for moving the CDP-based siRNA therapeutic platform to the clinic.

In May of 2008, CALAA-01 became the first targeted delivery of siRNA in humans (18). Details of this study, and others focused on nanoparticle therapeutics, can be found at http://www.clinicaltrials.gov. In Phase I studies, patients with solid-organ tumors refractory to treatment were administered CALAA-01 to assess drug safety. Patients were administered CALAA-01 by way of intravenous infusion on days 1, 3, 8, and 10 of a 21-day cycle (18). Importantly, in a recent study by the Davis group, the authors demonstrate that CALAA-01 effectively targets RRM2 through an RNAi mechanism of action in tumor tissue taken from patients with melanoma after systemic administration of CALAA-01 (31). Certainly, the updated results of clinical trials with CALAA-01 are eagerly anticipated.

Many more nanoparticle therapeutics are on the horizon and will follow CALAA-01 into the clinical setting. By providing this brief case-study, certain elements of nanotherapeutic design and development are highlighted with the hope of providing a concrete example for referring back to the, more general, main text of this chapter. As a pioneering nanoparticle therapeutic, much can be learned from CALAA-01. For instance, scrupulous detail to the design, fabrication, characterization, and quality control of CALAA-01 provided a solid platform for moving the drug forward in preclinical and clinical trials. In addition, the choice of a species-generic but target-specific siRNA sequence targeting RRM2 provided direct access to multiple animal models, ultimately, for translation to humans. Finally, testing in a wide variety of preclinical animal models provided the data necessary to anticipate safe dosing parameters in humans, and dictated drug-specific safety monitoring protocols in humans. Building upon CALAA-01, it is anticipated that what is now an "infant" field will ultimately provide new therapeutics, based upon advances in nanotechnology, which will provide for significant improvements in human health.

Acknowledgments

C. S. T. wishes to thank the Zell Family for their generous support of his research through the Robert H. Lurie Comprehensive Cancer Center of Northwestern University, the National Science Foundation/Nanoscale Science and Engineering Center at Northwestern University for seed grant funding, the Howard Hughes Medical Institute for a Physician Scientist Early Career Award, the National Institutes of Health/NCI for funding through the Center of Cancer Nanotechnology Excellence, the International Institute for Nanotechnology at Northwestern University, the Institute for BioNanotechnology and Medicine at Northwestern University; and the Northwestern University Department of Urology, Feinberg School of Medicine. Furthermore, C. S. T. thanks all of the faculty, staff, postdocs, Urology residents, graduate students, and research technologists who dedicate their time and effort to the unwavering development of an ever expanding array of nanoparticle-based diagnostics and therapeutics in hopes of improving human health. A. C. E. wishes to thank the Howard Hughes Medical Institute for their Research Training Fellowship for Medical

Students and the Northwestern University Feinberg School of Medicine Departments of Radiology and Urology for their guidance and generous support.

References

1. Lipsky, M. S. and Sharp, L. K. (2001) From idea to market: the drug approval process. *J. Am. Board Fam. Pract.* **14**, 362–367.

2. http://www.fda.gov.

3. DiMasi, J. A., Hansen, R. W., and Grabowski, H. G. (2003) The price of innovation: new estimates of drug development costs. *J. Health Econ.* **22**, 151–185.

4. DiMasi, J. A. (2002) The value of improving the productivity of the drug development process: faster times and better decisions. *Pharmacoeconomics* **20** Suppl. 3, 1–10.

5. Kola, I., and Landis, J. (2004) Can the pharmaceutical industry reduce attrition rates? *Nat. Rev. Drug Discov.* **3**, 711–715.

6. Marchetti, S. and Schellens, J. H. (2007) The impact of FDA and EMEA guidelines on drug development in relation to Phase 0 trials. *Br. J. Cancer* **97**, 577–581.

7. Brooks, W. H., McCloskey, D. E., Daniel, K. G., Ealick, S. E., Secrist, III, J. A., Waud, W. R., et al. (2007) In silico chemical library screening and experimental validation of a novel 9-aminoacridine based lead-inhibitor of human S-adenosylmethionine decarboxylase. *J. Chem. Inf. Model.* **47**, 1897–1905.

8. Lee, P. H., Ayyampalayam, S. N., Carreira, L. A., Shalaeva, M., Bhattachar, S., Coselmon, R., et al. (2007) In silico prediction of ionization constants of drugs. *Mol. Pharm.* **4**, 498–512.

9. Shacham, S., Marantz, Y., Bar-Haim, S., Kalid, O., Warshaviak, D., Avisar, N., et al. (2004) PREDICT modeling and in-silico screening for G-protein coupled receptors. *Proteins* **57**, 51–86.

10. Sivanesan, D., Rajnarayanan, R. V., Doherty, J., and Pattabiraman, N. (2005) In-silico screening using flexible ligand binding pockets: a molecular dynamics-based approach. *J. Comput. Aided Mol. Des.* **19**, 213–228.

11. Aina, O. H., Liu, R., Sutcliffe, J. L., Marik, J., Pan, C. X., and Lam, K. S. (2007) From combinatorial chemistry to cancer-targeting peptides. *Mol. Pharm.* **4**, 631–651.

12. Batra, S., Srinivasan, T., Rastogi, S. K., and Kundu, B. (2002) Identification of enzyme inhibitors using combinatorial libraries. *Curr. Med. Chem.* **9**, 307–319.

13. Schnur, D. M. (2008) Recent trends in library design: 'rational design' revisited. *Curr. Opin. Drug Discov. Devel.* **11**, 375–380.

14. Dobrovolskaia, M. A. and McNeil, S. E. (2007) Immunological properties of engineered nanomaterials. *Nat. Nanotechnol.* **2**, 469–478.

15. Ferrari, M. (2005) Cancer nanotechnology: opportunities and challenges. *Nat. Rev. Cancer* **5**, 161–171.

16. Ferrari, M. (2005) Nanovector therapeutics. *Curr. Opin. Chem. Biol.* **9**, 343–346.

17. Ferrari, M. and Downing, G. (2005) Medical nanotechnology: shortening clinical trials and regulatory pathways? *BioDrugs* **19**, 203–210.

18. Davis, M. E. (2009) The first targeted delivery of siRNA in humans via a self-assembling, cyclodextrin polymer-based nanoparticle: from concept to clinic. *Mol. Pharm.* **6**, 659–668.

19. Davis, M. E., Pun, S. H., Bellocq, N. C., Reineke, T. M., Popielarski, S. R., Mishra, S., et al. (2004) Self-assembling nucleic acid delivery vehicles via linear, water-soluble, cyclodextrin-containing polymers. *Curr. Med. Chem.* **11**, 179–197.

20. http://ncl.cancer.gov/.

21. Dobrovolskaia, M. A., Germolec, D. R., and Weaver, J. L. (2009) Evaluation of nanoparticle immunotoxicity. *Nat. Nanotechnol.* **4**, 411–414.

22. Accenture (2007) The Pursuit of High Performance through Research and Development: Understanding Pharamceutical Research and Development Cost Drivers. http://www.phrma.org/files/attachments/AccentureR&D Report-2007.pdf.

23. Leonard, E. M. (1994) Quality assurance and the drug development process: an FDA perspective. *Qual. Assur.* **3**, 178–186.

24. Pun, S. H., Bellocq, N. C., Liu, A., Jensen, G., Machemer, T., Quijano, E., et al. (2004) Cyclodextrin-modified polyethylenimine polymers for gene delivery. *Bioconjug. Chem.* **15**, 831–840.

25. Bartlett, D. W. and Davis, M. E. (2007) Physicochemical and biological characterization of targeted, nucleic acid-containing nanoparticles. *Bioconjug. Chem.* **18**, 456–468.

26. Heidel, J. D., Yu, Z., Liu, J. Y., Rele, S. M., Liang, Y., Zeidan, R. K., et al. (2007) Administration in non-human primates of

escalating intravenous doses of targeted nano-particles containing ribonucleotide reductase subunit M2 siRNA. *Proc. Natl Acad. Sci. USA* **104**, 5715–5721.

27. Hu-Lieskovan, S., Heidel, J. D., Bartlett, D. W., Davis, M. E., and Triche, T. J. (2005) Sequence-specific knockdown of EWS–FLI1 by targeted, nonviral delivery of small interfering RNA inhibits tumor growth in a murine model of metastatic Ewing's sarcoma. *Cancer Res.* **65**, 8984–8992.

28. Bartlett, D. W., Su, H., Hildebrandt, I. J., Weber, W. A., and Davis, M. E. (2007) Impact of tumor-specific targeting on the biodistribution and efficacy of siRNA nano-particles measured by multimodality in vivo imaging. *Proc. Natl Acad. Sci. USA* **104**, 15549–15554.

29. Heidel, J. D., Liu, J. Y., Yen, Y., Zhou, B., Heale, B. S., Rossi, J. J., et al. (2007) Potent siRNA inhibitors of ribonucleotide reductase subunit RRM2 reduce cell proliferation in vitro and in vivo. *Clin. Cancer Res.* **13**, 2207–2215.

30. Bartlett, D. W. and Davis, M. E. (2008) Impact of tumor-specific targeting and dosing schedule on tumor growth inhibition after intravenous administration of siRNA-containing nanoparticles. *Biotechnol. Bioeng.* **99**, 975–985.

31. Davis, M. E., Zuckerman, J. E., Choi, C. H. J., Seligson, D., Tolcher, A., Alabi, C. A., et al. (2010) Evidence of RNAi in humans from systemically administered siRNA via targeted nanoparticles. *Nature* **464**, 1067–1070.

Chapter 22

Legislating the Laboratory? Promotion and Precaution in a Nanomaterials Company

Robin Phelps and Erik Fisher

Abstract

Legislation is a form of governance that directs attention and prescribes action. Within the domain of nanoscience, the US 21st Century Nanotechnology Research and Development Act contains mandates not only for rapid development for economic competitiveness but also for responsible implementation, which is required to take place by integrating societal considerations into research and development. This chapter investigates whether these two mandates tend more to coexist or compete with one another, both in the purview of nanoscience policy and in the venue of nanoscience practice. This chapter first reviews macrolevel analysis of the directives contained in the legislation. It then examines, drawing on an empirical case study, how these directives manifest at the microlevel of a nanoscience research and development laboratory.

Key words: Innovation, Integration, Nanotechnology, Precaution, Promotion, Responsible

1. Introduction

On December 3, 2003 the 21st Century Nanotechnology Research and Development Act (NRDA) was signed into law (1). This legislation established the National Nanotechnology Initiative (NNI) as the National Nanotechnology Program (NNP) and authorized multiyear federal funding for nanotechnology research and development. Since then, more than US\$6.5 billion of federal funding has been authorized over the 4-year period, from fiscal year (FY) 2005 to FY 2009, that the legislation has been in effect.

The genesis of the NNP was a series of informal meetings in 1996 of the federal agencies involved in nanotechnology research. In 1998, this informal group became a formal Interagency Working Group (IWG). Over the next year, the IWG issued

Sarah J. Hurst (ed.), *Biomedical Nanotechnology: Methods and Protocols*, Methods in Molecular Biology, vol. 726,
DOI 10.1007/978-1-61779-052-2_22, © Springer Science+Business Media, LLC 2011

three reports: *Nanostructure Science and Technology* (August 1999), *Nanotechnology Research Directions* (February 2000), and *National Nanotechnology Initiative* (July 2000). Combined, these reports provided a blueprint for the strategic intent of US investment in nanotechnology research and development. One foundational footing of the blueprint was rapid technological development and accelerated market deployment intended to keep the USA competitive in the international arena for economic and other gains projected to be realized from nanotechnology products. Another foundational footing was responsible implementation intended to proactively and adequately address public concern. Eventually, both of these foundations became incorporated into the US nanotechnology legislation. Since the policy foundations of "rapid" and "responsible" nanotechnology research and development appear, at least on the surface, to be contradictory (2), it remains unclear and uncertain whether tensions between the two foundational footings play themselves out in actual research and development contexts. In short, do they coexist, and perhaps even mutually reinforce one another? Or do they remain irreconcilable, competing for focus and attention?

This chapter examines how these two policy foundations manifest themselves in a US nanotechnology research and development laboratory. First, an overview of the tensions as defined in the various Program Activities of the Act provides context for the case study. We then provide a brief review of issues and concerns that have been stated and documented in preparation for the Act's reauthorization. This is followed by a description of the case study including the overarching research project it is situated in, the methodology and methods employed, the initial findings based on a limited analysis, and a discussion of those findings.

2. Discussion

2.1. US Nanotechnology Legislation (See Note 1)

The 21st Century Nanotechnology Research and Development Act (NRDA) defines eleven Program Activities which serve as guideposts for intent and implementation. A reproduction of Section 2(b) of the Act, listing the eleven Program Activities, is contained in Appendix 1. These Program Activities logically cluster into three groups, which we label here as technical, promotional, and precautionary.

Seven of the Program Activities pertain to the technical objectives of nanotechnology research. These consist of the methods and resources for the cultivation of nanotechnology as an interdisciplinary science. Though some have expressed skepticism about the interdisciplinary nature of nanotechnology research and development, case studies of biomedical nanotechnology research settings provide a measure of confirmation of such an interdisciplinarity (3–5).

The four remaining Program Activities address two areas that policy makers deemed crucial to the public business of nanotechnology: promotion of the economic outcomes and precaution regarding the societal dimensions.

2.1.1. Economic-Promotion Considerations

Three Program Activities focus nanotechnology research and development efforts on economic considerations that promote meeting global competitiveness and extensive projected opportunities for nanotechnology applications. These promotional considerations include "ensuring…global leadership," "advancing…productivity and industrial competitiveness," and "accelerating" nanotechnology deployment. These Activities represent a key policy objective behind the NNP: US domination in this new competitive global market. The economic prospect for nanotechnology is projected to be substantial. Lux Research, an international market research firm, projected that between 2006 and 2014 global revenues from nanotechnology-enabled products will grow from $50 billion to $2.6 trillion and will comprise 15% of projected global manufacturing output (6). Notably, nearly every industrialized and developing country has initiated national research programs in nanotechnology to capture a share of the projected economic and societal benefits.

Global competition for the prospective nanotechnology market had reportedly grown over the 5-year period before the NRDA's passage. Mihail Roco, chair of the National Science and Technology Council's Subcommittee on Nanoscale Science, Engineering, and Technology reported that at least 30 countries had created national nanotechnology programs and that international nanotechnology funding increased multiple times for a global investment of approximately US$3 billion (7, 8).

The NRDA specifically requires "accelerating the deployment and application of nanotechnology research and development in the private sector, including startup companies." This language seeks to position nanotechnology deployment on a well-established research-to-technology commercialization path within the US innovation system – a system consisting of academic and federal lab research, startup companies, venture capital firms, and other entrepreneurial supporting infrastructure.

Combined, these promotional activities drive a policy focused on rapid development not only to keep pace with international competition but also to capture the benefits as well as the pervasive impacts of nanotechnology, which have been deemed "crucial" for the country's future economic health (9).

2.1.2. Societal-Precaution Considerations

The single remaining Program Activity contained in the NRDA stands by itself as much for its content as for its intent. Program Activity (10) requires "ensuring that ethical, legal, environmental, and other appropriate societal concerns… are considered during the development of nanotechnology." While this requirement for responsible implementation can be seen as a counterpressure to that of rapid implementation, it is also possible to regard it as a notable recognition of the importance of social trust of institutions for commercial success.

The impetus for this legislative directive came from a number of concerns surrounding nanotechnology that had been expressed in public and political discourses prior to the legislation. Policy makers appeared eager to separate concerns they could regard as credible and convincing from others that could be regarded as too speculative or fictional. One prominent critical theme during this time cited potential harm from exposure to nanotechnology particles, suggesting its potential as the "next asbestos" (10). Additional concerns encompassed other potential health, safety, and environmental risks, and they extended to broad ethical and political questions, including the role of democratic governance in nanotechnology.

Citing a severe lack of governmental monitoring and regulation of nanotechnology, the nongovernmental organization ETC Group (11) called for a global, mandatory moratorium on nanotechnology research and product development to allow time for a closer examination of the potential negative impacts on environmental, health, and safety. The report criticized the "substantial equivalence" regulatory approach being implemented at the time for nanoscale materials, a policy that had been used previously to show the safety of genetically modified organisms (GMOs) without doing full toxicological analysis of GM crops. As applied to nanotechnology, substantial equivalence presumes that novel nanoparticles are similar enough to their larger-scale particles that they do not warrant new toxicology studies. One of the distinctive features of nanomaterials is that they have properties that are different from those of the analogous bulk material; substantial equivalence fails to take this feature into account. The suggested moratorium would remain in effect until scientific communities could come together to develop and adopt monitoring mechanisms and reporting procedures in a "precautionary principle" approach to regulatory governance.

Understanding societal implications and addressing societal concerns about nanotechnology was also a prominent topic within the US government prior to the legislation enactment. It was a frequent topic of discourse among the Government agencies involved directly or indirectly in nanotechnology. Also, the National Science Foundation (NSF) held a national conference in 2000 and issued a subsequent report in 2001 on the topic of societal implications of nanotechnology. More than 50 distinguished professionals and executives from government and national laboratories, academic institutions, and the private sector were among the conference participants and contributors. In April of 2003, the US Congress held a public hearing on the societal implications of nanotechnology signaling recognition among legislators that societal concerns were an important consideration that needed to be addressed publicly.

The NRDA contains provisions outlining how the sociotechnical integration is to be accomplished. General strategies encompass what can be termed both "wide" and "deep" integration, where "wide" consists of research into societal concerns and dissemination thereof, and "deep" consists of feeding research on societal concerns directly into the NNP including the nanoscience research and development itself. The interdisciplinary sociotechnical integration potentially allows research on societal considerations to shape the course and outcomes of nanotechnology research and development. As such, it envisions a new form of scientific research in which explicitly "societal" considerations manifestly influence the design and pursuit of scientific research and the technology it is meant to enable.

2.1.3. Coexisting or Competing Mandates

In total, the legislation is an acknowledgement that the success of any federal nanotechnology program will not occur solely based on best efforts to increase the pace of scientific discoveries and of technology developments. It is a recognition, at least rhetorically, that a broad range of legitimate societal concerns exist, some of which could manifest as health and environmental product-related issues, choice and governance issues, and distribution of benefits and burdens, to name a few examples. Any of these concerns, whether "real" or "perceived" (12), could influence public trust, and hence commercial success. On this view, socially acceptable outcomes and commercially robust products can be seen to result from a dual focus on economic and societal considerations of nanotechnology.

Yet efforts to attempt a dual approach that combines accelerated economic promotion with more deliberative precautionary methods could manifest as dueling pressures on laboratory researchers and administrators, who may be confronted with what appears to be a largely irreconcilable tension between these two policy objectives. Perhaps the key difference between the two objectives is in how societal concerns are factored in to nanotechnology development. In the traditional economic-promotion approach to R&D, societal concerns are to be corrected by mechanisms that are seen to be external to the laboratory, such as market forces and regulation. In contrast, the sociotechnical integration approach present in the NRDA would be an internal mechanism that encompasses and intentionally addresses societal concerns during R&D decisions.

This type of integrated approach represents a small but growing trend in US federal science and technology policy. Yet none of the other programs that have attempted to employ it are as explicit or as high level as the NRDA. The primary previous attempt was the Ethical, Legal, and Societal Implications (ELSI) program of the US Human Genome Project (HGP). The HGP mandate to examine and consider the ethical, legal, and social implications was thought by the program leaders to be both visionary and unique (13). The ELSI program, however, has been criticized for lack of integrative outcomes and in general for failing to fulfill its mandate (13, 14). The NNI has funded two Centers for Nanotechnology in Society, one at Arizona State University (CNS-ASU) and one at the University of California at Santa Barbara (CNS-UCSB). In particular, the CNS-ASU employs an integrative approach to research known as "Real-Time Technology Assessment" (15) and, more recently, has developed the strategic vision of "Anticipatory Governance" (16).

2.1.4. Nanotechnology Legislation Reauthorization

Since its authorization, there have been a number of reviews of the NRDA program performed – some as specified in the NRDA legislation, others independently and externally organized. In 2005, the President's Council of Advisors on Science and Technology (PCAST), an outside advisory board designated in NRDA legislation to provide biennial assessments of the NNI to Congress, acknowledged in its first report that current knowledge and data to assess the actual risks posed by nanotechnology products were incomplete (17). This point was reiterated by House Science and Technology Committee Chairman Bart Gordon in a press release issued after a 2005 committee hearing on the topic

> There seems to still be ample unanswered questions in this field, but what is clear is that commercialization of the technology is outpacing the development of science-based policies to assess and guard against adverse environmental, health and safety consequences. The horse is already out of the gate... Prudence suggests the need for urgency in having the science of health and environmental implications catch up to, or even better surpass, the pace of commercialization (18).

Later that same year, the Nanotechnology Environmental and Health Implications (NEHI) Working Group was established to provide an infrastructure for coordination with and between Federal agencies focusing on nanotechnology environmental, health, and safety research and programs. One year later, a comprehensive examination of the NNP was conducted by the National Research Council of the National Academies of Science (NAS) per their legislative directive to perform a triennial review. Their report noted that there was very little published research addressing the toxicological and environmental effects of engineered nanomaterials and that environmental, health, and safety issues were of "significant concern to and a topic of serious discussion by government agencies and commissions, nongovernmental organizations (NGOs), the research community, industry, insurers, the media, and the public" (19). According to the report, effective solutions required a balancing of promotion with that of precaution and recommended NEHI facilitate research and development in a full life-cycle analysis of the precautionary aspects. In February of 2008, NEHI released its report defining an environmental, health, and safety research strategy and calling for the six regulatory agencies in the NNI to work individually and jointly to implement the strategy (20). A subsequent 2008 National Academy of Science report delivered harsh criticism of the NEHI plan concluding that there was no strategy in place (21).

Reports, analysis, and testimony from nongovernmental sources contained similar conclusions and recommendations. A report by the Project for Emerging Nanotechnologies (22) argued that better and more aggressive oversight and new resources were needed to manage the potentially adverse effects of nanotechnology and promote its continued development. In its 2007 nanotechnology policy report, Greenpeace proclaimed that "no regulatory framework has been developed to address the emerging issues" (23). Richard Denison, a Senior Scientist at NGO Environmental Defense Fund and former analyst with the US Congress Office of Technology Assessment leveled a succinct summation of the criticisms:

> NNI and many of its member agencies are talking and writing a great deal about the need to address nanotechnology's risks as well as benefits...But there is a continuing disparity between NNI's words and actions (24).

Denison reiterated the NRC report call for "a balanced approach to addressing both the applications and implications of nanotechnology [as] the best hope for achieving the responsible introduction" of nanotechnology products. In his 2008 Senate committee testimony, Matthew Nordan, President of Lux Research, echoed this sentiment noting that the current ambiguity and the "glacial pace" of setting specific regulatory guidelines is becoming a gating factor for commercialization (25).

> On January 15, 2009, the US House of Representatives introduced the National Nanotechnology Initiative Amendments Act of 2009 (H.R. 544). In February of 2009, the legislation was passed by the House without amendment and forwarded to the Senate Commerce, Science and Transportation. The bill reauthorizes and makes incremental changes to several key provisions of the NRDA. One intention of the reauthorization bill, as passed in the House, was to better address environmental, health, and safety (EHS) issues associated with nanotechnology while continuing to encourage promotion of the commercialization of the technology for economic growth and competitiveness. As stated by House Science and Technology Committee Chairman Bart Gordon (26) in conjunction with the passage of the House bill in 2009:
>
> > It is important that potential downsides of the technology be addressed from the beginning in a straightforward and open way, both *to protect* the public health and to allay any concerns about the validity of the results. A thorough, transparent process that ensures the safety of new products will allow both the business community and the public *to benefit* from the development of these new technologies. (Emphasis added).

In summary, much of the reauthorization discourse has been directed toward developing a national governance framework with coordination amongst agencies and associated increase in funding to better address the precautionary aspects contained in the NRDA mandate to ensure consideration of the ethical, legal, environmental, and other appropriate societal concerns during nanotechnology development. The criticisms of the NNI's approach to responsible implementation suggest that this emphasis has received less attention and may be in direct competition to the emphasis on rapid implementation. It is also noteworthy that the reauthorization discourse focused on a more traditional top-down approach than that outlined in the NRDA's sociotechnical integration mandate. The next section describes a research project that investigates the possibility and utility of sociotechnical integration and then turns to a limited analysis of one of the case studies it has supported.

2.2. Socio-Technical Integration Research Project

The US legislative mandate for sociotechnical integration during nanotechnology R&D has opened up new opportunities to design and conduct experiments aimed at assessing the possibility and utility of sociotechnical integration to influence the direction of R&D. One such undertaking, the Socio-Technical Integration Research (STIR) project, is a three-year program that is administered by the Center for Nanotechnology in Society at Arizona State University. STIR is funded by the National Science Foundation (NSF #0849101) with the specific objective to assess and compare the varying pressures on and capacities for laboratory researchers to integrate broader societal considerations into their work. STIR places ten doctoral students into 20 laboratories in ten countries on three continents to conduct 20 "laboratory engagement studies," a cutting edge form of collaborative participant observation (27). The STIR method builds upon ethnographic qualitative research, a methodological paradigm pioneered by anthropologists and sociologists in the early twentieth century (28, 29). Ethnography uses extended, primarily participant observation, to examine the "shared patterns of behavior, beliefs, and language" of a "culture-sharing group" (29). Traditional laboratory studies employed an ethnographic method for studying science by examining the internal dynamics of scientific work through in situ observation "as it happens" and were pioneered by sociologist of science Bruno Latour (30–32).

The laboratory engagement study also transforms the "reflexive ethnography," which, in Woolgar's account, focuses on the reasoning practices used within the research laboratory to "generate awareness of reasoning practices as they are deployed in analysis" (32). Within STIR, the reflexive awareness is not only applied by the ethnographer to their own thinking about the phenomena they observe but also accomplished through an

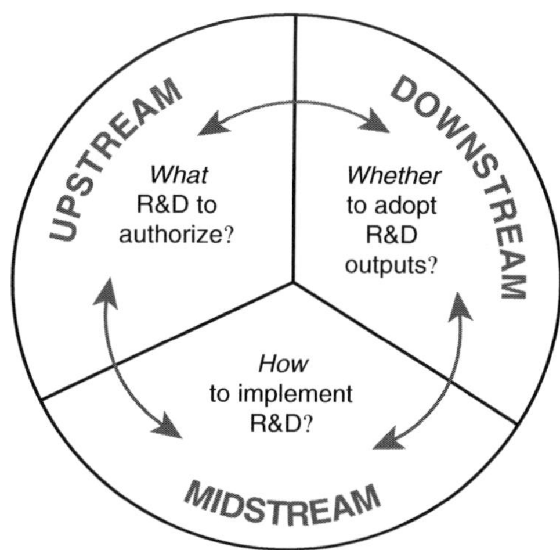

Fig. 1. Stages of science and technology governance (Adapted from STIR: Socio-Technical Integration Research Project Description, p. 6).

interdisciplinary collaboration between natural and social scientists. Thus, it is also a methodological practice of introducing ethnographic observations and findings into the laboratory research context itself – both for verification and so as to stimulate mutual learning and reflection by both parties to the sociotechnical collaboration (27).

STIR studies take place over a 4-month period and utilize the novel methodological approach of midstream modulation. As shown in Fig. 1, within laboratories research and development decisions are conceptually situated in the midstream of the science and technology governance process, occurring between upstream policy decisions and downstream regulatory and market activities. Midstream agents, including those who make basic research decisions, thus perform the functional role of implementing authorized research agendas. Research developments, which are measured by mapping the evolution of research decisions over time, are theorized to be modulated or incrementally shaped by a variety of institutional, social, and cognitive factors. Modulation at the midstream is posited to occur in three successive stages: de facto modulation, the factors that influence decisions; *reflexive* modulation, laboratory practitioners' awareness of these factors and of their own roles within larger social systems; and *deliberative* modulation, in which scientists consciously form decisions that are tempered by a reflexive awareness of these factors. Thus, midstream modulation provides a mechanism for evaluating and adjusting research decisions during the research process and constitutes a bottom-up approach for shaping research and development directions in light of relevant societal considerations – what has been termed "governance from within" (33). This unique application of reflexive ethnography to science and engineering research decisions serves to interact with the content of research decisions, thereby in theory lending visibility to both the promotional and precautionary influences on research decisions.

STIR engages in midstream modulation through interdisciplinary collaboration between social and natural scientists. Despite longstanding calls for such sociotechnical integration and collaborations (14), there have been very few laboratory engagement studies conducted using this approach. The NRDA mandate affords a renewed call for and recognition of the need for sociotechnical integration in nanotechnology development. Various types of relationships between social and natural science have recently been initiated in emerging technology research programs including nanotechnology, genetic engineering, and synthetic biology. Often the failed genetically modified crop debate is used as an example of the need and justification for including social scientists in these primarily public-funded research programs. In these relationships, Calvert and Martin (34) suggest two different roles for social scientists: the social scientist can perform the role of either a "contributor" or a "collaborator." A contributor is one who contributes to (at times as a representative of the "public"), facilitates the discussions of, and studies the ethical, legal, and social implications of research. In contrast, a collaborator is one who is involved with the research and interacts with the researchers in ways that can potentially shape the research agenda and influence the research direction. The collaborator role STIR scientist has shown some positive indications of success in initial laboratory engagements (27, 35). The next section offers findings drawn from one such engagement study.

2.3. STIR Case Study: Rocky Mountain Nanomaterials

The site for one STIR case study was a company, Rocky Mountain Nanomaterials (see Note 2), producing novel nanomaterials using a patented application technology. Nanomaterials can be considered a nanotechnology sector with numerous applications across the spectrum from biomedical, energy, and various technology and industrial markets. A report by market analyst firm Lux Research identifies nanomaterials at the beginning of the nanotechnology value chain (36). Thus, the nanomaterials sector represents a major portion of the economic potential for nanotechnology and is therefore posited to exhibit a number of influences for economic promotion. In addition, according to the Nanotechnology Industries Association (NIA), a UK-funded organization formed in 2005 to establish a framework for the safe, sustainable, and socially supportive development of nanotechnology, a complex and convoluted mixture of regional, national, trade, industry, and international voluntary and regulatory governance initiatives for nanomaterials exists (37). These disparate governance initiatives

1) Environmental toxicity and persistence

Material could be degraded rapidly or slowly [the product's intended life cycle needs to be taken into account

2) Human toxicity

Material could be total non-toxic through all routes of exposure, or conversely be highly toxic, a teratogen and possibly transmissible

3) Human exposure

Material might be used in a highly controlled treatment, or used in consumer or environmental quantities

4) In-vivo bio persistence

Material may accumulate and not be removed from body, organism, or organelles

5) Auto activity

Material may have no means of self recognition or environmental awareness, or may activate on response to, and in order to change, the environment

6) Mobility

Material may be permanently immobilized, it may become free as a result of intended use, designed to be free for purpose, transmissible, uncontrollable, and possibly even self-propelling

Fig. 2. Six dimensions of risk for Nanomaterials (see ref. 37).

need to address six dimensions of risk identified for nanomaterials (see Fig. 2). The nanomaterials sector is thus also posited as likely to exhibit evidence of influences for precaution, hence making it a viable source for examining the nature of interactions between the NRDA's promotional and precautionary mandates.

Rocky Mountain Nanomaterials is a university spin-out, a company which emanated from research conducted at a university in Colorado, which utilizes a patented process to create novel nanomaterials. University spin-outs have been shown to be important contributors in the emergence of nanomaterial applications (38). The company has been in business since 2002 and has a staff of four PhD scientists, one operations manager, and two founding executives. In addition, the laboratory has strategic partnerships with departments at the university where two additional founders of the company work.

This laboratory engagement study was conducted from April through August of 2009. The length of the study was predetermined and was consistent across the STIR laboratory sites. The primary empirical data collection methods were participant observation and semistructured and unstructured interviews. The researcher met individually each week with two scientists and participated in the weekly laboratory project review meetings. One scientist ("C4") received his PhD in Chemical Engineering from a university in the US Rocky Mountain region; the other scientist ("M1") received his PhD in Electro Chemistry from the same university. Interview responses and observations were recorded in a field notebook, and many of the individual interviews, with the scientist's permission, were digitally recorded. Documents, obtained with permission from the laboratory, and content from archival research form the remainder of the empirical data sources.

Interviews were guided by a protocol developed during the STIR pilot study (27). The model (see Fig. 3) consists of four distinct conceptual components intended to describe research decisions as well as to capture and make visible – to both the investigating STIR scientist and to the participating laboratory scientists – the de facto influences of societal considerations during research activities. The model was often utilized during the study to initiate the semistructured interviews with the scientists.

The remainder of this chapter presents the results of a limited study of a subset of the data generated in the Rocky Mountain

OPPORTUNITY Perceived state of affairs eliciting a response	**CONSIDERATIONS** Selection criteria that potentially mediate the response
ALTERNATIVES Perceived available courses of action	**OUTCOMES** Effects of selecting alternatives in light of considerations

Fig. 3. Decision protocol components (Adapted from STIR: Socio-Technical Integration Research Project Description, p. 7).

Nanomaterials case study. Rather than attempting to investigate the possibility and utility of sociotechnical integration, this particular study sought to develop broad classification categories which could characterize and relate the "nontechnical" or "societal" influence on technical research decisions.

2.3.1. Findings: Decision Influences

A preliminary analysis of a subset of the data – drawn from interviews, lab meetings, and informal conversations – was conducted using Conceptually Clustered Matrix display format (39). The data were placed into two major categories of influence: external and internal. Internal influences originate from the people and the policies within the company and indicate cultural norms that can guide decisions and behaviors. External influences originate from outside the company and indicate the institutional context of the innovation system, which can also guide decisions and behaviors. Within each category, the type of influence was categorized as either "technical" or "societal." From this grouping of empirical data, four distinct societal influences on the laboratory decisions emerged: economic, intellectual property, university relations, and environment, health, and safety (EHS). Of these, economic considerations had the greatest number of instances and dominated the external societal influences; however, it occurred only in a few instances in the case of internal societal influences. In contrast, university relations considerations were a much stronger internal societal influence but only occurred in a small number of instances as an external societal influence. That university relations were stronger internally is to be expected given the fact that the laboratory is a university spin-out and maintains ongoing ties to the university for research. Similarly, intellectual property was mentioned more often as an internal rather than as an external societal consideration. This may be due to the fact that intellectual property serves as a competitive advantage and as a barrier to entry into the market for others, thus having a significant potential economic impact.

EHS was the only consideration mentioned by all participants, and there was a near balance in the number of EHS considerations between internal and external societal influences. For example, a new opportunity required the use of hydrazine, an inorganic chemical compound. The researchers were aware of the potential negative and positive external societal considerations of hydrazine given its use in a range of applications from rocket fuel to pharmaceuticals to automotive airbags. They were aware that the US Occupational Safety and Health Administration (OSHA) was looking at toxicology, an internal EHS consideration. During the discussions of the opportunity, the question came up about the safe handling of hydrazine, an example of an internal EHS consideration. One of the participants agreed to make contact with the largest producer of hydrazine to find out the standard safe handling practices. In another example, one of the participants was looking at their current use of carbon in a nanomaterials application. He expressed an external EHS consideration and concern that in this specific application their existed the possibility of overheating potentially causing the nanomaterial to catch on fire because carbon, which was selected for this application because of its high conductivity, is combustible at high temperatures.

In addition to these four categories of societal influences, the conceptually clustered matrix also yielded a broader societal influence theme, which is here termed "green-nano." This theme appeared in both economic as well as EHS considerations. A majority of the external economic considerations were focused on so called "green energy," energy developed from renewable sources. This focus can more than likely be associated with the current Colorado and national priorities of becoming a "green economy" by both reducing the country's dependence upon foreign oil and reducing the carbon output during the production of energy from fossil fuels. It must be noted however that the only instance when "green-nano" was used in this latter sense was in the case of "carbon avoidance" as an internal EHS consideration. For the most part when "green-nano" was used as an economic consideration it was in reference to funding that could be obtained from government, venture capital, and/or strategic customers by pursuing green energy opportunities. (This finding is confirmed by one researcher's statement, "We can raise money with green" (see Note 3) and by another participant's statement that a carbon monetization mechanism called carbon credits could be a "cash cow" (see Note 4).) Notably, "green-nano" was employed as an EHS consideration both in pursuit of a green energy funding opportunity for the lab and in critically questioning the same opportunity. (The former inference is based on the potential funding source's emphasis on no carbon byproducts, while the latter is derived from one participant's statement, "How green is it when it uses nasty precursors?" (see Note 5)).

2.3.2. Analysis of Findings (See Note 6)

Analysis of the data subset produced four categories of "societal" influence on laboratory decisions in a nanomaterials laboratory: economic, intellectual property, university relations, and EHS considerations. Of these, both intellectual property and university relations emerged more in relation to economic justification (in cases of competitive differentiation and outsourcing partnership). Accordingly, these two categories fall primarily under economic promotion and appear less frequently under societal precaution. The analysis did find evidence of societal influences present in research decisions; however, economic considerations by far outweighed any other societal consideration. Thus, within the scope of the nanomaterials sector in which the Rocky Mountain

Nanomaterials company operates, promotion far outweighs precaution.

This limited analysis did discover some product-related societal influences concerning EHS; however, broader issues including those of power, choice, and distribution were usually not mentioned by research participants without prompting by the STIR scientist. It is intuitive that promotion considerations would be strongly influential given the fact that this laboratory is a startup company that uses a combination of market signals, customer opportunities, and government sources of funding to derive company growth and survival. It is also not unreasonable to anticipate that there would be a deficit of precautionary influences given the lack of clear downstream regulatory mechanisms governing risk, as previously detailed, and given a relatively unprecedented legislative directive for sociotechnical integration that in effect requires significant changes to the institutional settings governing nanoscale research and development. Such changes in the norms, values, and rules that shape organizational behavior regarding economic promotion and societal consideration would need to occur not only at the microlevel of laboratories but also at the mesolevel of institutional environments that constrain or encourage innovation (40–42). Institutions and organizations constitute primary elements of an established innovation system (43). Though innovation systems are evolutionary in many ways, they can also be slow to change and adapt in others (41, 42, 44). New technologies produce pressures on the institutions and organizations within a sector to change or adapt in response to new concerns such as the precautionary and promotional considerations of US legislation. Institutional and organizational response to these pressures is distinct within a sector and is based in part on the transformative capacity of the technology, whether it is endogenous or exogenous to the sector, and on the sectoral adaptability, the supportive or disruptive effects of the technology on the sector (45). The overall commercial success or failure of a technological regime can be a function in part of in what ways an innovation system retains old characteristics and in what ways it remains flexible and open to adaptation.

A thorough examination of the source and nature of the mesolevel institutions that shape the economic-promotion and societal-precaution influences would be a next reasonable step toward creating insight and understanding for policy and practice. Review of the initiatives of the US agencies involved in nanomaterials risk governance – the Federal Drug Administration, the Environmental Protection Agency, and the National Institute for Occupational Safety and Health – may provide insights into ways the nanomaterials sector may be responding to precautionary issues. A report from a UK pilot study (46) into how public policy might encourage the dual objective of promotion with precaution approach specifically for nanomaterial research and development could provide additional insight.

---

3. Notes

1. The material in this section draws heavily upon material in ref. 2.

2. The company's name has been changed for confidentiality purposes.

3. Participant interview, May 6, 2009, Laboratory site.

4. Participant interview, May 6, 2009, Laboratory site.

5. Project Review meeting, May 6, 2009, Laboratory site.

6. The limited analysis applied to the subset of data from the Rocky Mountain Nanomaterials case study did not indicate how best to characterize the relation between the two policy goals of promotion and precaution. Whether these two goals coexist or compete remains a broad question that requires more nuanced analysis. While the statements "we can raise money with green" would seem to indicate coexistence, the statement "how green is it?" implies competition. Similarly, this limited analysis did not seek to provide insight into the capacities of laboratories to engage in sociotechnical integration attempts or into how such capacities might be enhanced. We note that throughout this case study, broader societal dimensions of research decisions that were evident to the STIR scientist were not always indicated by the laboratory scientists. This was taken not to be due to intentional efforts by laboratory participants to ignore or negate these dimensions; rather, such dimensions simply did not appear to be in the de facto cognitive frame of decision alternatives of the company scientists. Once societal considerations were brought to the attention of a lab scientist by the STIR researcher through the use of the decision protocol, however, opportunities arose to discuss these dimensions further. Over time, these discussions extended from single decisions to more encompassing ones related to industry, market, and society. Other forms of analysis of the data set, including more narrative-based accounts, are therefore likely to produce more penetrating insights into the possibility and utility of sociotechnical integration mandated by the US legislation and investigated by projects like STIR. It is also worth noting that, in other settings, nanoscientists have been documented to be more concerned about some nanotechnology-associated risks than are members of the public (47).

Acknowledgments

The authors gratefully acknowledge the participants from the laboratory of this case study whose willingness to mutually share, discuss, and explore the content and context of laboratory decisions greatly increased the value of the experience. This material is based upon work supported by the US National Science Foundation under Grant No. #0849101.

Appendix 1. Program Activities of the National Nanotechnology Program Laid Out in Section 2(b) of the 21st Century Nanotechnology Research and Development Act

1. Developing a fundamental understanding of matter that enables control and manipulation at the nanoscale.

2. Providing grants to individual investigators and interdisciplinary teams of investigators.

3. Establishing a network of advanced technology user facilities and centers.

4. Establishing, on a merit-reviewed and competitive basis, interdisciplinary nanotechnology research centers, which shall

    (a) Interact and collaborate to foster the exchange of technical information and best practices.

    (b) Involve academic institutions or national laboratories and other partners, which may include States and industry.

    (c) Make use of existing expertise in nanotechnology in their regions and nationally.

    (d) Make use of ongoing research and development at the micrometer scale to support their work in nanotechnology.

    (e) To the greatest extent possible be established in geographically diverse locations, encourage the participation of Historically Black Colleges and Universities that are part B institutions as defined in section 322(2) of the Higher Education Act of 1965 (20 U.S.C. 1061(2)) and minority institutions [as defined in section 365(3) of that Act (2 U.S.C. 067k(3))], and include institutions located in States participating in the Experimental Program to Stimulate Competitive Research (EPSCoR).

5. Ensuring US global leadership in the development and application of nanotechnology.

6. Advancing the US productivity and industrial competitiveness through stable, consistent, and coordinated investments in long-term scientific and engineering research in nanotechnology.

7. Accelerating the deployment and application of nanotechnology research and development in the private sector, including startup companies.

8. Encouraging interdisciplinary research, and ensuring that processes for solicitation and evaluation of proposals under the program encourage interdisciplinary projects and collaborations.

9. Providing effective education and training for researchers and professionals skilled in the interdisciplinary perspectives necessary for nanotechnology so that a true interdisciplinary research culture for nanoscale science, engineering, and technology can emerge.

10. Ensuring that ethical, legal, environmental, and other appropriate societal concerns, including the potential use of nanotechnology in enhancing human intelligence and in developing artificial intelligence which exceeds human capacity, are considered during the development of nanotechnology by

   (a) Establishing a research program to identify ethical, legal, environmental, and other appropriate societal concerns related to nanotechnology, and ensuring that the results of such research are widely disseminated.

   (b) Requiring that interdisciplinary nanotechnology research centers established under paragraph (4) include activities that address societal, ethical, and environmental concerns.

   (c) Insofar as possible, integrating research on societal, ethical, and environmental concerns with nanotechnology research and development, and ensuring that advances in nanotechnology bring about improvements in quality of life for all Americans.

   (d) Providing, through the National Nanotechnology Coordination Office established in section 3, for public input and outreach to be integrated into the Program by the convening of regular and ongoing public discussions, through mechanisms such as citizens' panels, consensus conferences, and educational events, as appropriate.

11. Encouraging research on nanotechnology advances that utilize existing processes and technologies.

References

1. US Congress (2003) *21st Century Nanotechnology Research and Development Act of 2003*, Public Law no. 108–153.

2. Fisher, E. and Mahajan, R. L. (2006) Contradictory intent? US federal legislation on integrating societal concerns into nanotechnology research and development. *Sci. Public Policy* **33**, 5–16.

3. Meyer, M. and Persson, O. (1998) Nanotechnology-interdisciplinarity, patterns of collaboration and differences in application. *Scientometrics* **42**, 195–205.

4. Rafols, I. and Meyer, M. (2007) How cross-disciplinary is bionanotechnology? Explorations in the specialty of molecular motors. *Scientometrics* **70**, 633–650.

5. Rafols, I. and Meyer, M. (2010) Diversity and network coherence as indicators of interdisciplinarity: case studies in bionanoscience. *Scientometrics* **82**, 263–287.

6. Lux Research (2007) *The nanotech report.* 5th Ed., v. 1, New York, NY.

7. Roco, M. C. (2003) Government nanotechnology funding: an international outlook. National Science Foundation. 30 June. http://www.nsf.gov.

8. Roco, M. C. (2003) Broader societal issues of nanotechnology. *J. Nanopart. Res.* **5**, 181–189.

9. Hines, C. (2003) Nanotechnology: small miracles foreseen; some lawyers say trend will rival semiconductors and the Internet. *New York Law J.* **13**, 5 November.

10. Mnyusiwalla, A., Daar, A. S., and Singer, P. A. (2003) "Mind the gap": science and ethics in nanotechnology. *Nanotechnology* **14**, 9–13.

11. ETC Group (2003) Nanotech and the precautionary prince. Available at http://www.etcgroup.org.

12. Norden, M. (2008) Congressional testimony, Hearing before the Committee on Commerce, Science, and Transportation. US Senate. 24 April.

13. McCain, L. (2002) Informing technology policy decisions: the US Human Genome Project's ethical, legal, and social implications programs as a critical case. *Technol. Soc.* **24**, 111–132.

14. Fisher, E. (2005) Lessons learned from the ethical, legal and social implications program (ELSI): planning societal implications research for the national nanotechnology program. *Technol. Soc.* **27**, 321–328.

15. Guston, D. H. and Sarewitz, D. (2002) Real-time technology assessment. *Technol. Soc.* **24**, 93–109.

16. Barben, D., Fisher, E., Selin C., and Guston, D.H. (2008) Anticipatory Governance of Nanotechnology: Foresight, Engagement, and Integration. in: *The new handbook of science and technology studies* (Hackett, E. J., Amsterdamska, O., Lynch, M. E., and Wajcman, J., eds.) MIT, Cambridge, MA.

17. President's Council of Advisors on Science and Technology (2005) The national nanotechnology initiative at five years: assessment and recommendations of the National Nanotechnology Advisory Panel. May. Available at http://www.nano.gov.

18. US House of Representatives: Committee on Science and Technology (2005) Research is key to safety when tackling unknown aspects of nanotechnology. Press release 17 November.

19. National Research Council; Division on Engineering and Physical Sciences; Committee to Review the National Nanotechnology Initiative (2006) *A matter of size: triennial review of the national nanotechnology initiative.* p. 74, National Academics Press, Washington, DC.

20. NEHI – Nanotechnology Environmental Health Implications Working Group (2008) *National nanotechnology initiative strategy for nanotechnology-related environmental, health, and safety research.* National Nanotechnology Coordination Office, Arlington, VA.

21. National Research Council; Division on Engineering and Physical Sciences; Committee to Review the National Nanotechnology Initiative (2008) *Review of federal strategy for nanotechnology-related environmental health and safety research.* National Academies Press, Washington, DC.

22. Davies, J. C. (2006) Managing the effects of nanotechnology. Project on Emerging Nanotechnologies. http://www.nanotech-project.org.

23. Greenpeace (2007) Nanotechnology: policy and position paper. http://www.greenpeace.org.

24. Denison, R. (2007) Congressional testimony, Hearing before the Committee on Science and Technology, US House of Representatives, 110th Congress. 31 October.

25. Nordan, M. (2008) Congressional testimony, Hearing before the Committee on Commerce, Science, and Transportation, US Senate, 110th Congress. 24 April.

26. US House of Representatives: Committee on Science and Technology (2009) Bill introduced to ensure safety of nanotechnology, transparency of research. Press release 15 January.

27. Fisher, E. (2007) Ethnographic invention: probing the capacity of laboratory decisions. *NanoEthics* **1**, 155–165.

28. Sanday, P. R. (1983) The Ethnographic Paradigms. in: *Qualitative methodology* (Van Maanen, J., ed.) Sage, Newbury Park, CA.

29. Cresswell, J. W. (2007) *Qualitative inquiry and research design: choosing among five different approaches.* Sage, Thousand Oaks, CA.

30. Godfrey-Smith, P. (2003) *Theory and reality: an introduction to the philosophy of science.* The University of Chicago Press, Chicago, IL.

31. Latour, B. and Woolgar, S. (1979) *Laboratory life: the social construction of scientific facts.* Sage, Beverly Hills, CA.

32. Woolgar, S. (1982) Laboratory studies: a comment on the state of the art. *Soc. Stud. Sci.* **12**, 481–498.

33. Fisher, E., Mahajan, R. L., and Mitcham, C. (2006) Midstream modulation of technology: governance from within. *Bull. Sci. Technol. Soc.* **26**, 485–496.

34. Calvert, J. and Martin, P. (2009) The role of social scientists in synthetic biology. *EMBO Rep.* **10**, 201–204.

35. Schuurbiers, D. and Fisher, E. (2009) Lab-scale Intervention. *EMBO Rep.* **10**, 424–427.

36. Lux Research (2004) *Sizing nanotechnology's value chain.* New York, NY

37. NIA – Nanotechnology Industries Association (2007) Risk classification made easy. Presented at NanoEXPO 2007, May 16, in Nottingham, England.

38. Libaers, D., Meyer, M., and Geuna, A. (2006) The role of university spinout companies in an emerging technology: the case of nanotechnology. *J. Technol. Transf.* **31**, 443–450.

39. Miles, M. B. and Huberman, A. M. (1994) *Qualitative data analysis.* Sage, Thousand Oaks, CA.

40. Hollingsworth, J. R. (2000) Doing institutional analysis: implications for the study of innovations. *Rev. Int. Polit. Econ.* **7**, 595–644.

41. Nelson, R. R. (1994) The co-evolution of technology, industrial structure, and supporting institutions. *Ind. Corp. Change* **3**, 417–419.

42. Nelson, R. R. (1995) Co-evolution of industry structure, technology, and supporting institutions, and the making of comparative advantage. *Int. J. Econ. Bus.* **2**, 171–184.

43. Edquist, C. (1997) *Systems of innovation: technologies, institutions, and organizations.* Cambridge University Press, Cambridge, UK.

44. Nelson, R. R. and Winter, S. G. (1982) *An evolutionary theory of economic change.* Harvard University Press, Cambridge, MA.

45. Dolata, U. (2009) Technological innovations and sectoral change. Transformative capacity, adaptability, patterns of change: an analytical framework. *Res. Policy* **38**, 1066–1076.

46. Nightingale, P., Morgan, M., Rafols, I., and van Zwanenberg, P. (2008) *Nanomaterials innovation systems: their structure, dynamics and regulation. Report for the Royal Commission on Environmental Pollution.* Science Policy Research Unit, Brighton, UK.

47. Scheufele, D. A., Corley, E. A., Dunwoody, S., Shih, T.-J., Hillback, E. D., and Guston, D. H. (2007) Nanotechnology scientists worry about some risks more than the general public. *Nat. Nanotechnol.* **2**, 732–734.

Chapter 23

Navigating the Patent Landscapes for Nanotechnology: English Gardens or Tangled Grounds?

Douglas J. Sylvester and Diana M. Bowman

Abstract

The patent landscape, like a garden, can tell you much about its designers and users: their motivations, biases, and general interests. While both patent landscapes and gardens may appear to the casual observer as refined and ordered, an in-depth exploration of the terrain is likely to reveal unforeseen challenges including, for example, alien species, thickets, and trolls. As this chapter illustrates, patent landscapes are dynamic and have been forced to continually evolve in response to technological innovation. While emerging technologies such as biotechnology and information communication technology have challenged the traditional patent landscape, the overarching framework and design have largely remained intact. But will this always be the case? The aim of this chapter is to highlight how nanotechnology is challenging the existing structures and underlying foundation of the patent landscape and the implications thereof for the technology, industry, and public more generally. The chapter concludes by asking the question whether the current patent landscape will be able to withstand the ubiquitous nature of the technology, or whether nanotechnology will be a catalyst for governments and policy makers for overhauling the current landscape design.

Key words: Intellectual property, TRIPS, Patent thickets, Patent pools, Trolls, Technology innovation

1. Introduction

One is tempted to think of the patent landscape as a refined English garden. Views of gently rolling lawns spotted by outcroppings of majestic trees, a few revival buildings, and inundated by hundreds of floral and shrub varieties might leave the casual observer with the view that it is entirely organic and naturalistic. For those who look closer, however, one sees the tenders' efforts. The lack of straight lines, walls, or delineated beds masks the perfect visual delineation of the form – a form evolving over decades (if not centuries) and one that is largely in balance.

Sarah J. Hurst (ed.), *Biomedical Nanotechnology: Methods and Protocols*, Methods in Molecular Biology, vol. 726, DOI 10.1007/978-1-61779-052-2_23, © Springer Science+Business Media, LLC 2011

Organic contours hide the hundreds of small decisions that are continually being made to retain the appropriate balance. But those decisions are nevertheless made. Although such gardens may give the impression that they arise "just so," they hide the enormous and complicated efforts of their tenders to organize, weed, and design them.

Since the 1970s, patent gardens have been under continued attack, and their carefree style seems under threat as alien species invade these once tranquil spaces. The dual shocks of software (1) and much more importantly biotechnology (2) wreaked havoc on these once tranquil and seemingly unchanging spaces. Well-tended beds, ancient perennials, and majestic arbors were threatened by rapidly growing and unanticipated thickets and brambles (3–6). Rolling meadows were quickly dotted with pitfalls, and once-languid pools, (7) now choked with unforeseen infestations, threatened to become unsightly and unrecoverable bogs and quagmires (8). Worst of all, these gardens (and especially their beautiful marble bridges) were invaded by trolls (9, 10)!

In these dark days, the garden's tenders created new tools, brought in help from abroad (as discussed in Subheading 2.3) and, although forced to make certain concessions to these alien species (see, generally, the *Uruguay Round Agreements Act of 1994*), finally succeeded by the end of the millennia to return much of the garden to its former apparent placidity. Sure some ancient varietals were replaced by new and, for a time, foreign blossoms. A few hedges, long put up to keep out unwanted visitors, or at least to make the entry difficult, were, if not fully removed, trimmed a tad. Finally, it appears that furtive efforts were made to gather pitchforks and torches to drive off those pesky trolls – although some are obviously lurking underneath some of the murkier bridges (10). In the end, the threat, although not entirely gone, seemed largely under control and much of what we had always expected in our patent garden remained familiar and friendly.

However, new threats are looming on the edges of our serene plot. Tendrils of invasive species can be seen sprouting all over the garden, and destructive vines threaten the integrity of the garden's walls. In short, nanotechnology brings with it the potential to upset not only some aspects of the patent garden, but may also force a complete rethinking of its function and form (see Note 1). Biotechnology thickets may have grown over some beloved blooms, but nanotechnology's brambles threaten to take down the entire field (6, 8). Those seeking refuge in the cooling waters of the garden's pools, now clear after a decade of invasion, may find them once again choked to a vivid green. And, horror, the trolls appear to be breeding again! Our patent garden, so perfect in its form to handle the challenges of past patent revolutions, seems particularly unable to handle what nanotechnology may be bringing.

Leaving aside (thankfully we can only suppose) the metaphor of the garden, the increasing pace, complexity, and importance of technological revolutions has put real pressure on the patent system. Revered doctrines, designed for pre- and early industrial innovations, seem quaint (if not dangerous) for these times (6). Patent institutions, organized around silos of knowledge and focused on local inventorship, may not be able to stand in the face of massively complex and global innovations. Finally, the pace of patenting, both in terms of process and conceptual foundations, seems dangerously ill-suited to technological advances that have the ability to challenge existing national and international legal frameworks, including those relating to patents, in the blink of an eye. In numerous other publications, we have examined the many challenges nanotechnology poses to traditional regulatory structures related to environmental and human health and safety (see, for example, (11–15)). In this article, we examine the challenges that nanotechnology has and will pose for patent frameworks and institutions.

We already know that biotechnology radically shifted the patent landscape both in terms of patenting practices of researchers (16), and the scope and breadth of patentable subject matter (17–19). In just a few short years, biotechnology and rapidly advancing pharmaceutical patenting forced a fundamental rethinking of what was and should be patentable (19), as well as substantial hand-wringing about why we allow patenting of socially beneficial inventions at all (5). Although biotechnology, software, and pharmaceutical patenting may have spurred a substantial "rethinking [of] intellectual property rights" (19), much of the prior system remains in place. Patents are still, largely, national affairs (20) with massive bureaucratic costs that view patents as arising from traditional scientific disciplines (21, 22). In addition, patents are still issued without great oversight and on increasingly early stage technologies (23). These were, similarly, issues in the biotechnology and pharmaceutical revolutions of the 1980s – but the system was able to accommodate these issues without fundamentally altering the process and purpose of patenting.

Returning (apologies) to our metaphor – it may be time to decide whether the garden, as it has been known for centuries, should be dozed. The garden, whether we are conscious of it or not, picks winners and losers. It gives preference to those that bloom first – crowding out latecomers and preventing variations. Hedges and walls, meant to keep out unwelcome visitors and maintain tranquility, reduce hybridization, and competition and, arguably, reduce overall social utility. Finally, those darn trolls really need to be run out!

In this chapter, we explore the challenges that nanotechnology may pose for the traditional patent landscape. In particular, we wish to address more of the garden metaphors (although, mercifully, these are not our creations) that speak of potential thickets, bogs, brambles, quagmires, pitfalls, pools, and other problems that new technologies may pose for our little patent garden. To this end, this chapter proceeds in several sections. In Subheading 2.1, we set out some of the patent basics that have arisen over the course of the past few centuries. In so doing, our analysis focuses mainly on the USA, but we do include some important comparative and multinational aspects of the current landscape. In Subheading 2.2, we discuss some of the unique aspects of nanotechnology's development, current regulation, and potential that many have predicted will pose real problems for traditional patent systems. Further, we provide an overview of research into whether some of those predictions have started to come true. Subheading 2.3 sets out some of the remaining challenges that nanotechnology may pose for patent frameworks. Finally, in Subheading 2.4, we set forth some (admittedly tentative) thoughts about how we may avoid some of the potential problems with nanotechnology and how the patent system may need to reform itself to better accommodate technological invasions that will inevitably occur again.

2. Discussion

2.1. Patent Basics

For centuries, the patent system has been predicated on a few unchanging (and nearly universal) precepts. First, patents are national in scope (and often favor national inventors) (24, 25). Second, patents reward innovation by granting rights to inventors (24). Third, patents only apply to inventions and not discoveries (24). And, fourth, patents are ultimately intended to benefit society by encouraging technological innovation and must, therefore, seek the balance between encouraging invention and ensuring social benefits through access and use (20, 25). To achieve these various principles, patent systems around the world have created both institutions and doctrinal frameworks dedicated to ensuring their fulfillment.

These precepts were built up during periods of relatively low patenting and invention. They made sense in that era, but their relevance today is increasingly coming under question. First, the national scope of patents produces real inefficiencies in that inventors must seek approval in each nation, greatly increasing the cost on both inventors and societies to manage that system. In addition, in an era of increasing global research and patenting, it makes little sense to continue to favor one citizen inventor over another merely because of accident of birth.

As a result, there have been some tentative efforts to streamline this process. First, the United States Patent Office (USPTO) and Japanese Patent Office (along with many others) have begun to take some of the administrative burden off of inventors by allowing for streamlined patenting through individual treaty arrangements, or application of the Patent Cooperation Treaty. A discussion of how these procedures have streamlined the process is outside the scope of this chapter, but the real point is that it has not been wildly successful. It is still an extraordinarily complex and expensive process to engage in global patenting and calls for reform only continue to grow (26, 27).

A secondary, and closely related, framework for achieving patent's precepts is through doctrinal limitations on patentability. At their base, these rules seek to ensure that only those patents that add something to the world of knowledge in a given area are patentable (24). The doctrines of novelty, nonobviousness, and utility are all, in some respects, attempts to ensure that only those patents worthy of intellectual property protection are granted a bundle of legal rights (20, 28). Yet, given patent law's overall goal of providing social goods through appropriate incentivized innovation, a fundamental practical principle of nearly every patent system has been to "grant the patent and let the market figure it out!" In low-innovation periods, this approach makes perfect sense (23). In addition, it may even have net benefits in periods of explosive innovation in technological applications. Where it is deeply problematic is in times – and you guessed that the nanotechnology revolution may be one of those times – of immense patenting of basic research and fundamental research tools (6). Again, this is an issue we take up later in the chapter.

One area of genuine progress in patent law, at least in terms of overall efficiency and systemic fairness, has been in the area of doctrinal harmonization. In particular, the World Trade Organization's (WTO) *Trade-Related Intellectual Property Rights (TRIPS) Agreement* has provided an opportunity for tremendous progress in the harmonization of basic standards of patentability. Pursuant to Article 27(1) of the *TRIPS Agreement,*

> patents shall be available for any invention, whether products or processes, in all fields of technology provided they are new, involve an inventive step and are capable of industrial application.

While conditions for the granting of a patent and nature of the exclusive rights will vary between jurisdictions, Mandel (28) notes that these requirements, and therefore the bundle of rights granted, are "largely harmonized throughout the world." Unfortunately, as hinted above, substantive or doctrinal harmonization has not been accompanied with procedural efficiencies (25).

Despite these efforts at substantive harmonization, it is important to note that not all subject matter may be the subject of a patent grant and that, at its outer limits, the question of what is patentable is still an open discussion. As highlighted by Article 27 of the *TRIPS Agreement,* patents may be granted for "any inventions," but the Agreement does not define what constitutes an "invention." This area of potential disunity has been largely avoided as most nations have adopted permissive definitions of patentability (29), thereby ensuring continuation and global propagation of the "patent now, sue later" mentality noted above. To be clear, we do not view this practice as necessarily problematic. Indeed, as we note later in this chapter, an equally thorny problem is making patent grants too difficult and/or expensive to obtain – because they create disincentives to innovate, rob inventors of the fruits of their labor, or unduly delay beneficial applications. Nobody said tending the patent garden was easy.

As a result of the efforts of the WTO and, indeed, the USA through numerous bilateral agreements (30), there is wide consensus on what is *not patentable* as well as buy-in for the general principle that most things should be. According to Mandel (28),

> laws of nature, natural phenomena, abstract ideas, aesthetic creations, and information and data per se generally are not patent eligible. Almost everything else is.

Thus in most jurisdictions, "discoveries" or "products of nature" fall outside the traditional scope of patentable subject matter. Despite these similarities, there are some differences. Eisenberg (2), for example, has observed a "shifting landscape of discovery in genetics and genomics research" that presents moral and conceptual difficulties about what is or should be patentable. In particular, biotechnology (and, to a much lesser extent, software) forced national patent systems to reevaluate the scope of what they consider to be patentable subject matter. For example, in 1976, the Australian Patent Office adopted a fairly liberal approach to the patenting of living subject matter when it held that,

> living organisms were determined to be patentable provided they were not in a naturally occurring state and they had improved or altered useful properties, and not merely changed morphological characteristics which had no effect on the working of the organism (31).

While the USA has adopted a similarly liberal approach, the position of these two jurisdictions may be contrasted to that of, for example, the EU. Pursuant to the *European Parliament and Council Directive 98/44/EC of 6 July 1998 on the legal protection of biotechnological inventions,* the European Union places specific limitations on the patenting of, for example, plants and animal varieties (see, for example, Article 4).

Nanotechnology, and other emerging technologies such as synthetic biology, would appear to have the potential to further impact this landscape. Where biotechnology created fundamental challenges for many on the moral nature of what is patentable (32, 33), nanotechnology seems less controversial although the potential for ethical challenges remains. As a result, the real challenges for patenting of nanotechnology will flow from doctrinal conceptions of novelty and nonobviousness, and the fact that nanotechnology does not fall neatly into any one silo.

2.2. Nanotechnology Background

Nanotechnology is a field of technological effort that holds tremendous promise as well as potential peril (34, 35). Despite the concerns of many groups and governments, the regulatory response to nanotechnology has been, largely, one of research and development. While nanotechnology-specific safety or environmental regulation has been slow to develop (36), with the EU only recently adopting the first national or supranational legislative instrument containing nano-specific provisions (see Note 2), one area of legal action has not – the willingness of governments to fund research and development of nanotechnology. In the USA, for example, two (of the four) goals of the National Nanotechnology Initiative (NNI) are to "[a]dvance a world-class nanotechnology research and development program" and "[f] oster the transfer of new technologies into products for commercial and public benefit" (37). To that end, the US federal government has ponied up more than $US1.6 billion in 2010 alone to foster basic and applied research into the technology (38). Other countries have been equally quick to promote their nascent nanotechnology efforts with government funding (39, 40).

One consequence of this approach is that nanotechnology may be the most multijurisdictional and multinational technology to have emerged so far (41). The result of this is that nanotechnology patents are likely to have numerous inventors from more than one institution and, just as likely, from more than one country (42, 43).

For example, the atomic force microscope (AFM), one of the most basic research tools necessary to do almost any work in nanotechnology (44), was patented in 1988 (see below for a discussion of how patenting of basic research tools may be a problem) and awarded to IBM and, in particular, its Swiss research center (see Note 3). The initial patented invention was not multinational, however, a recent survey of patents arising out the original AFM patent shows that more than 3,000 patents now relate to (either in improvements, modifications, or processes) the original AFM patent (8). In writing this chapter, the authors conducted a rather unscientific survey of just a few dozen of those complementary patents and discovered related filings from Japan, Germany, the USA, China, Canada, France, and many others. Among these patents, a select few showed researchers from different jurisdictions collaborating on inventions (see Note 4).

In an even more unscientific study, we leafed through a few of the more than 2000 currently pending nano-patent applications and found numerous multinational collaborations on inventions including inventors from the USA collaborating with inventors from (1) India, (2) Great Britain, (3) the Netherlands, (4) Poland, (5) Belgium, and (6) Japan. A takeaway from this small sample is that the level of cross-country collaborations is growing. Perhaps more important, as collisions between patent holders will inevitably grow, so will calls for patent pools (see Subheading 2.4 below for a broader discussion of these) that will require inventors from numerous jurisdictions to collaborate on and share potential inventions. As inventors and inventions will become increasingly multinational, so too the pressure on national patent systems to increase efficiency and harmonize doctrines that discriminate against foreign inventors will also increase.

Massive government funding often means direct funding to universities for early stage research. In this way, nanotechnology is nothing new under the sun – governments have a long history of funding basic research into potential applied sciences. What *is* new, however, is that the outgrowth of this funding has come in an era of hyperpatenting on all points of the research curve. Indeed, this is one area that may separate nanotechnology from all other prior technological revolutions – every aspect of this technology may be patented (6).

All other technological innovations, from biotechnology to software, initially developed during a time when those who received federal funding (universities mainly) were unable to effectively commercialize or patent inventions (6). Software, arising in the 1960s and 1970s, not only arose in an era where basic research could not often be patented in the USA as a result of its federal funding, but also arose in a culture of publication over patenting. Biotechnology, although eventually reaching the hyperpatenting stage, was also developed in an era and culture that dampened much of the urge to patent. The only two historical parallels to nanotechnology are the patent wars of the radio and the airplane (6, 45, 46). In each of those cases, market-based attempts to override patent thickets failed. Government action was required in each case to break the patent logjam and allow research to transform to public goods.

Nanotechnology is global not only in its funding and commercial reach, but also in its ability to patent not only the applications of the technology but also the basic research tools necessary to conduct the research. The ability of inventors to patent basic research tools is not, doctrinally, novel (47). What is new is, as already noted, the willingness and desire of universities, the origin of most basic research in industrialized countries, to patent such tools and, more important, the erosion of traditional experimental use exceptions (48). Indeed, there are very few basic nanotechnology building blocks and research tools that are not, already, patented (6, 49, 50). There has been much hand-wringing about this fact – and real concern that nanotechnology's real potential will be swallowed up in a morass of litigation (51). This is an issue we explore in more detail in Subheadings 2.3 and 2.4.

Another characteristic, as described by Maynard (52), is its multidisciplinarity "which crosses established boundaries of scientific inquiry and agency jurisdiction." Based on the complexity and multidisciplinarity of most nanotechnology patents, many have been concerned that the patent offices of various governments will prove unable to handle, at least not appropriately, requests to patent these developments (53). Most concerning is, apparently, the fact that poor review at patent offices (6, 54) will result in overpatenting of inventions that do not meet minimum requirements of patentability in all areas (55). As one author put it, concerns over workload and expertise are prevalent at patent offices, but these concerns are "especially true [for nanotechnology] since the PTO's current staff lack the cutting edge knowledge to completely understand these inventions. This problem is exasperated [sic] by the fact that the PTO still faces an enormous backlog of nanotechnology patent applications" ((51); see also (56, 57)). As Tegart (51) noted, "Inventors need fast and unbureaucratic help to realize an idea with importance for the future." In the sections to come, we explore in more detail the problems hinted by Harris and attempt to assess the likelihood that these problems will increase.

Finally, as evidenced by their financial commitments, the potential of nanotechnology is obviously not lost on governments. Much of this potential is in public health and, in particular, in cancer research and other disease areas (34, 58). Although there is real concern that the commercial viability of nanotechnology applications may be hindered by patenting practices, a larger concern is that legal wrangling will slow the health benefits of nanotechnology. Slowing down economic growth may be a cost more easily borne by nations with strong patenting regimes – but as debates over pharmaceuticals have made clear, issues involving health applications of nanotechnology have the potential to rework patent systems in much more substantial ways.

2.3. Mapping the Current Nanotechnology Patent Landscape

Given the general background of nanotechnology patenting discussed above, the question that many have asked is whether the potential disaster many feared is starting to, or perhaps has, come to pass. The short answer is that it looks like things are not exactly going according to plan at the patent offices.

With Lux Research (50) having stated that "corporations, start-ups, and labs depend on patents to protect their nanotech innovation – and turn them into cash," the importance of securing patents for nanotechnology-based inventions – including both product and processes – is arguably best highlighted by reference to the increasing levels of patent activity within key patent offices such as the USPTO and the European Patent Office (EPO). Recent efforts to report on this activity, including performance data relating to jurisdictions, institutions, and individuals, have included work by, for example, Marinova and McAleer (59), Huang et al. (60–62), Bawa (63), Koppikar et al. (64), Heinze (65),

Lux Research (50, 53), and Chen and Roco (66). All have shown a deluge of nanotechnology patents.

In one of the first published studies examining longitudinal patent activity for nanoscale science and engineering activities within the USPTO, Huang et al. (61) reported rapid growth in patenting activity over the period 1976–2002. By using a "full text" keyword-based approach (see Note 5), and subsequent filtering process, the authors found that the USPTO processed approximately 8,600 "nano-based" patents over this period; patenting activity was found to be steep after 1997 and 2001. These periods of growth in patent activity occurred, as was observed by the authors, around periods of program growth and other institutional activities within, for example, the USA. It is perhaps unsurprising then that the authors (61) found that "the [nanoscale science and engineering] patents grew significantly faster than the USPTO database as a whole, especially beginning with 1997."

Other key findings reported by Huang et al. (61) included the diversity of countries and institutions involved in the patenting activity (albeit still dominated by the USA), and the strength of patenting activity within particular technological fields including chemicals, catalysts, and pharmaceuticals. The observed growth in patenting activity would appear to highlight the importance of patent law, and the protections therefore afforded to patentees under the legal framework; this is despite the costs associated with securing patent protection for an invention (see Note 6).

In a more recent study of patenting activity within the USPTO, Lux Research (50) similarly reported a "ramp-up" in the number of patents being issued by the national patent office, with steep growth continuing in the post-2003 period. According to their analysis,

> the number of nanotech patents issued ha[d] risen steadily from a base of 125 in 1985 to 4,995 today …Nanotech patents far outpace other areas of innovation, with a compound growth rate of 20% versus just 2% for patents overall (50).

Their analysis supported the findings of Huang et al. (61), with Lux Research noting that patentees were more likely to be from the USA than any other jurisdiction but with a growing percentage from other jurisdictions. In addition, patents were likely to be assigned to a patentee in the private sector than any other sector (for example, university, government and/or research organization). Along with significant growth in patent activity, the authors found that the average number of claims within each patent had also increased. In their words:

> inventors are authoring more sophisticated patents that cover more nuanced variations of the same theme in a single filing. The average nanotech patent issued in 2005 has 23.5 claims, compared to only 15.8 in 1985 (50).

As discussed below, this observed trend has significant implications for not only patent examiners who must be able to deal with the complexities associated with the applications, but also patent growth more generally.

In addition to overall patenting activity within the USPTO, Lux Research (50) also looked at patenting activity (applications and grants) for eight specific nanomaterials, each of which has the ability to be utilized across five different applications areas. They included the following: carbon nanotubes, metal nanoparticles, ceramic nanoparticles, dendrimers, quantum dots, fullerenes, and nanowires. The purpose of the report was to provide an in-depth analysis of patent density for each of these platform materials, to determine the breadth of the patent claims, and to identify areas of potential entanglement; vulnerability to potential challenges (conflict) and market potential of patents were also considered (50).

Based on this examination, the authors found that significant growth occurred in relation to patenting activity for all eight materials; ceramic nanoparticles and carbon nanotubes, which have broad applications across numerous fields, were found to have experienced particularly steep growth over the time period examined, resulting in high patent density (50). This, they suggested, had the potential to create an unfavorable patent environment for inventors/patentees in relation to, for example, carbon nanotubes within the electronics field and ceramic nanoparticles within personal health-care and cosmetics applications. However, the authors (50) went on to suggest that there may still be hidden opportunities in relation to these two materials, and that given their potential breadth as structural materials, "it [was] likely that these nanomaterials will emerge as battles worth fighting by 2008." Varying trends in patent filing and density, as well as future potential based on market opportunities, were observed for the other six materials (50).

The continued increase in patenting activity in key national patent offices suggests that industry, research organizations, universities, and other key bodies remain positive about the market opportunities and associated economic benefits for nanotechnology-based inventions. This is despite the costs associated with technological innovation and the increasingly vocal debates occurring within jurisdictions over potential, yet unquantified, risks associated with some facets of the technology. Yet, as the ETC Group (49) has sought to remind us, the successful granting of a patent, albeit for a nanotechnology-based invention or any other type, is not enough in itself to ensure the commercial success of that invention. Success or failure, as witnessed in the EU in relation to, for example, genetically modified foods is dependent on a far broader range of criteria, including consumer acceptance of the invention and/or technology (12, 67, 68).

While much of the literature relating to intellectual property rights and nanotechnology has canvassed the patent landscape and paints a detailed picture thereof, there is an increasing body of work that has focused on the potential challenges and barriers that the technology and its inventors may face in the coming years. Concerns have been expressed, for example, in relation to the breadth of patent claims for platform or structural materials, patent thickets, overlapping patent claims, and the institutional capacity of patent offices to assess nanotechnology-based applications. Many of these issues are not in themselves unique to nanotechnology; rather, as highlighted below, they are common to other emerging technologies, including synthetic biology and reflect many of the experiences of prior technologies.

However, nearly all current research into patenting activity for nanotechnology paints a gloomy picture of the patent landscape. National patent offices are overwhelmed, with the time for nanotechnology patents taking far longer than other technologies (50). Indeed, one study revealed that nanotechnology applications, the average time for a nanotechnology-based patent application to be processed and granted by the USPTO had increased from 33 months in 1985 to 47 months in 2006 (50). Whether the delays are caused by the sheer number of applications in relation to an understaffed patent office or if it is because of the immense complexity of patent applications is not yet clear (see Note 7). According to Halluin and Westin (69),

> because few individuals have an in-depth and complete knowledge of nanotechnology, patent examiners may not have the tools necessary to understand the complexities of the field. In fact, both the European and US Patent Offices have admitted that they do not fully understand nanotechnology.

Nevertheless, these delays have real consequences for the commercialization and development of beneficial nanotechnology applications, especially within the field of nanomedicine. Delays and costs at patenting obviously delay research and dissemination of key technologies and, just as obviously, deeply affect pricing. This challenge is not however unique to nanotechnology, with other sophisticated technologies likely to also challenge the capacity of patent offices.

In addition to these economic inefficiencies, the studies we have discussed above, as well as countless others, have shown that feared patent thickets, brambles, quagmires, and other natural-disaster-themed descriptions have apparently taken over the landscape (70–73). In basic research tools, fundamental materials, building-block structures, and numerous other crucially important aspects of nanotechnology, vast numbers of overlapping and broadly written patents, held by varied institutions and competitors, have already been issued. We have already seen, in the USA, a series of patent infringement lawsuits filed among competitors. Although no study has yet compared the level of infringement suit activity compared to prior technologies, there are many reasons to believe that nanotechnology's future may be threatened, or at least made less bright, by these looming controversies.

2.4. Moral and Ethical Implications of Nanotechnology Patenting

It is arguably not surprising, when considered against the backdrop of the patenting of human genes debate and associated concerns over the breadth of patents being granted on human genes, that this issue has also become a topic of debate in regards to the patenting of nanotechnology. This has been particularly the case regarding platform or structural materials, such as the eight considered by Lux Research (50) in their report. The concern here, as articulated by the ETC Group, is primarily in relation to the issues of concentrated ownership and therefore control over patents, and the subsequent implication of this in terms of economics, innovation, and access – especially for developing economics (49). Of course, such concerns are not new, nor unique to nanotechnology. However, it would appear that nanotechnology does create additional challenges here; the ETC Group (49) has suggested that, for example,

> breathtakingly broad nanotech patents are being granted that span multiple industrial sectors and include sweeping claims on entire classes of the Periodic Table.

They go on to suggest that (49),

> it's not just the opportunity to patent the most basic enabling tools, but the ability to patent the nanomaterials themselves, the products they are used in and the methods of making them.

The ETC Group's concern is that the dense concentration of patents, which are held by a small number of patent holders, combined with the ability to patent basic nanoscale materials, "could mean monopolizing the basic elements that make life possible" (49). The implications of this, it was argued, would be profound for developing countries with the ETC Group (49) stating that,

> researchers in the global South are likely to find that participation in the proprietary "nanotech revolution" is highly restricted by patent tollbooths, obliging them to pay royalties and licensing fees to gain access.

Given their concern over patent concentration and breadth of claims being made, the nongovernmental organization went on to highlight the emergence of so-called "patent thickets" within the nanotechnology patent landscape. This refers to, as explained by Shapiro (3), "an overlapping set of patent rights requiring that those seeking to commercialize new technology obtain licenses from multiple patentees." Patent thickets therefore have the ability to hinder technological innovation and therefore the commercialization of new technologies.

In their examination of patenting activity for five nanomaterials within the USPTO and the EPO (carbon nanotubes, inorganic nanostructures, quantum dots, dendrimers, and the Scanning Probe Microscopes), the ETC Group was able to paint a picture of the emerging patent thicket for some nanomaterials; this was illustrated primarily by reference to the number of patents relating to each of the applications currently held by different institutions, and the so-called patent density for each applications. By way of example, the ETC Group reported that the USPTO had issued 227 patents for carbon nanotubes between 1999 and 2004. While it found that the patents for the material were held by a number of different parties, across a range of different sectors, it nevertheless came to the conclusion that a patent thicket for the material had already occurred and that,

> a swarm of existing patents, whose claims are often broad, overlapping and conflicting, means that researchers hoping to develop new technology based on carbon nanotubes must first negotiate licenses from multiple patent owners (49).

The ETC Group is not the only commentator to have voiced its concern about the potential implications of overlapping patent claims and the emergence of "patent thickets" or "nano-thickets," with a number of commentators having expressed concern over the potential creation, and the implications thereof, for nanotechnology (45, 50, 53, 70–74).

Having observed the problems associated with patent thickets in other areas of technological innovation, including biotechnology and information and communication technologies, Clarkson and DeKorte (70) noted that patent thickets have the potential to give rise to a range of issues, including the unintentional infringement of patents and the subsequently liability created as a consequence of the said infringement, the problem of anticommons, the creation of barriers to entry, and the need for licensing. While the issues are not unique to any one area, they (70) noted that,

> the nanotechnology patent space experiences an even greater level of these problems because it is much more complicated than other technology areas.

In order to determine the extent of the growing patent thickets for nanotechnology, Clarkson and DeKorte (70) undertook a mapping exercise of patent space and density within the USPTO. The authors then used network analytic techniques as a way to "visualize" the growth at three time points – 2000, 2002, and 2004. By plotting individual patents and then references between patents, the authors demonstrated not only the growth in nanotechnology patents within the USPTO during that time period, but also the increasing interconnectivity – or network – between the patents. This visualization process enabled the authors to map potential patent thickets.

The increasing crowding of the nanotechnology patent space, and therefore emerging patent thicket in some areas, was found to support the findings of Lux Research (50, 53). Having determined that patent thickets were not just theoretical for nanotechnology, but were already occurring, the authors then took a somewhat pragmatic approach to the issue; if thickets cannot be avoided, what strategies can be employed to protect patents while also promoting innovation?

Traditionally cross-licensing arrangements – which Shapiro (3) has eloquently defined as "an agreement between two companies that grants each the right to practice the other's patents – are one way in which this may be achieved." However, while such arrangements are relatively straightforward when involving only two parties, Clarkson and DeKorte (70) note a number of limitations, including high transaction costs, which can make such arrangements prohibitive when more than two parties are involved in the contractual negotiations. In light of these limitations, Clarkson and DeKorte (70) proposed a second alternative for avoiding validity challenges and potential patent litigation: "patent pools." These contractual undertakings are, as summarized by Clark et al. (75) "an agreement between two or more patent owners to license one or more of their patents to one another or third parties." Such arrangements have been an important tool for providing parties with access to proprietary information for over a century. It has been suggested that the ability for parties to readily access patent information through such pool arrangements promote innovation within areas that may have otherwise become the subject of patent blocking and legal challenges – and at a lower transactional cost than cross-licensing (3, 70).

But establishing and relying on patent pools as a mechanism to access patent information would appear to be only one potential approach to addressing the challenges presented by the thickets. Another, arguably somewhat more radical, approach would be to promote an "open source" approach. The open source "movement" has been widely adopted in relation to software development (76) and to varying degrees within the field of medicine and drug development (77, 78). The movement's potential application to the nanotechnology patents landscape has therefore raised some level of discussion among commentators. One of the earliest contributions was from Bruns (79), who looked at the applicability of the open source movement to molecular nanotechnology. In his view, "open source approaches might offer advantages for faster, more reliable and more accessible research and development" (79). He advocated the adoption of an open source approach to the technology where public money had been used to generate the intellectual property in question. Bruns (79) argument was that such an approach would not only encourage innovation but also assist in diffusing the technology, and its associated benefits, to developing economies.

While the open source movement has continued to gain traction within, for example, the field of software development, "there is not yet an "open source nanotechnology" movement" (80).

This may in part be explained by the fact that software development is process based, where developers of nanotechnology are at this time largely focused on product generation. Moreover, with open source software, the primary 'cost' is the programmer's time, which they give freely to further develop and refine the code lines. The same cannot be said with the development of nanotechnology, which requires not only human resources but also infrastructure and consumables. Prisco (81) and Peterson (82) have however begun to further explore open source for nanotechnology. As such, we would suggest that the nanotechnology patent landscape is likely to significantly evolve over the short to medium term, with the open source movement just one way in which individuals and organizations attempt to circumnavigate the emerging patent thickets and promote technology innovation.

As any individual with a green thumb will know, tending to a garden – albeit a refined English garden or a small herb garden – requires constant care and attention. Any such garden is dynamic by its very nature and will evolve over time. A constant state of vigilance is needed to ward off pests and other challenges, and the more proactive, educated, and vigilant the gardener is, the better the outcome will be.

As this chapter has sought to highlight, there are many similarities between the needs and challenges of a garden and that of a patent landscape. As with our garden, the patent landscape has evolved and been refined over centuries in response to new species and the introduction of new technologies. Sometimes, the landscape has been better prepared to handle the attacks than others. Nanotechnology is the latest species to place a strain on the fundamental features of the landscape, pushing up against historical walls and threatening traditionally well-tended fields. This is due to a number of factors: its multidisciplinary character, its trans-jurisdictional nature, the ability for inventors to apply for and be granted patents not only to products but also to the basic building blocks, and claims which related to a diverse number of areas and/or applications. It is also in part due to the immense public and private sector interests in nanotechnology, and the rush to secure legal rights over their inventions.

But the question is: will nanotechnology be permitted to devastate that which has taken centuries to build up? Or will the gardeners – primarily national governments in this instance – see the emergence and growth of nanotechnology as an opportunity to reconsider the borders and features of the current landscape and revamp it accordingly? This would of course involve significant time and energy, but with other equally complex and multifaceted technologies already in the research and development pipeline, it would appear that policy makers need to "stop and smell the roses" to ensure that the economic and social benefits of the technology are released. Perhaps it is time to modernize the landscape to meet the needs of the current climate – a more global approach to patenting is one obvious option – and provide the gardeners with the tools that they need to do their job in a timely and efficient manner.

But those who use and enjoy the garden must also take some responsibility for its future to ensure that the benefits of the technology are realized. Rather than, for example, relying on costly and time-consuming litigation, beneficiaries of the patent system should be encouraged to explore arrangements such as cross-licensing and patent pools early on, or where appropriate, be encouraged to look to open source approaches. Governments can also play a role here by, for example, creating a framework which encourages and/or rewards these approaches.

3. Notes

1. In this chapter, we do not provide background or definitional sections on what we consider to be nanotechnology. For our views on this subject, see, for example, (11, 12, 83).

2. In November 2009, the European Parliament and Council adopted the final text for the Cosmetic Regulation (84).

3. See US Patent Number: 4 724 318.

4. Cites available upon request.

5. As noted by Huang et al., there were "seven basic keywords with several variations" (61).

6. For a more recent longitudinal study of nanotechnology-patenting activity within the USPTO, the EPO, and the Japan Patent Office (JPO). See also Chen and Roco (66).

7. Koppikar et al. (64) have suggested that the latter may be largely attributed to the interdisciplinary nature of the applications and the "many and diverse applications [are] often associated with a single nanotechnology invention."

References

1. Graham, S. and Mowery, D. C. (2003) Intellectual property protection in the U.S. software industry. In: *Patents in the Knowledge-Based Economy. Board on Science, Technology and Economic Policy (STEP)* (Cohen, W. and Merrill, D., eds.) The National Academies, Washington, DC. 219–258.

2. Eisenberg, R. S. (2002) How Can You Patent Genes? *Am. J. Bioeth.* **2**, 3–11.

3. Shapiro, C. (2000) Navigating the Patent Thicket: Cross Licenses, Patent Pools, and Standard Setting. *Innov. Pol. Econ.* **1**, 119–150.

4. Burk, D. L. and Lemley, M. A. (2003) Policy Levers in Patent Law. *Va. Law Rev.* **89**, 1575–1696.

5. Heller, M. A. and Eisenberg, R. S. (1998) Can Patents Deter Innovation? The Anticommons in Biomedical Research. *Science* **280**, 698–701.

6. Lemley, M. (2005) Patenting Nanotechnology. *Stanford Law Rev.* **58**, 601–630.

7. Bessen, J. Patent Thickets: Strategic Patenting of Complex Technologies. http://www.researchoninnovation.org/thicket.pdf. Accessed 26 December 2009.

8. D'Silva, J. (2009) Pools, Thickets and Open Source Nanotechnology. European intellectual property review, 31(6), 300–306.

9. Ratingen, J. (2006) Slaying the Troll: Litigation as an Effective Strategy Against Patent Threats. *Santa Clara Comput. High Technol. Law J.* **23**, 159–210.

10. Magliocca, G. N. (2007) Blackberries and Barnyards: Patent Trolls and the Perils of Innovation. *Notre Dame Law Rev.* **82**, 1809–1838.

11. Abbott, K. W., Sylvester, D. J., and Marchant, G. E. (2010) Transnational regulation of nanotechnology: Reality or romanticism? In: *International Handbook on Regulating Nanotechnologies* (Hodge, G. A., Bowman, D. M., and Maynard, A. D., eds.) Edward Elgar, Cheltenham. 525–544.

12. Sylvester, D. J., Abbott, K. W., and Marchant, G. E. (2009) Not Again! Public Perception, Regulation, and Nanotechnology. *Regul. Gov.* **3**, 165–185.

13. Marchant, G. E. and Sylvester, D. J. (2006) Transnational Models for Regulation of Nanotechnology. *J. Law Med. Ethics* **34**, 714–725.

14. Bowman, D. M. and van Calster, G. (2007) Does REACH Go Too Far? *Nat. Nanotechnol.* **1**, 525–526.

15. Bowman, D. M. and Hodge, G. A. (2009) Counting on Codes: An Examination of Transnational Codes as a Regulatory Governance Mechanism for Nanotechnologies. *Regul. Gov.* **3**, 145–164.

16. Rai, A. K. and Eisenberg, R. S. (2003) Bayh-Dole Reform and the Progress of Biomedicine. *Law Contemp. Probl.* **66**, 289–314.

17. Caulfield, T., Cook-Deegan, R. M., Kieff, F. S. and Walsh, J. P. (2006) Evidence and anecdotes: an analysis of human gene patenting controversies. *Nat. Biotechnol.* **24**, 1091–1094.

18. Klein, R. D. (2007) Gene patents and genetic testing in the United States. *Nat. Biotechnol.* **25**, 989–990.

19. Andrews, L. B. (2002) Genes and patent policy: rethinking intellectual property rights. *Nat. Rev. Genet.* **3**, 803–808.

20. Dinwoodie, G. B., Hennessey, W. O., and Perlmutter, S. (2001) *International Intellectual Property Law and Policy*. LexisNexis, Newark.

21. Eisenberg, R. S. (1989) Patents and the Progress of Science: Exclusive Rights and Experimental Use. *Univ. Chic. Law Rev.* **56**, 1017–1086.

22. Rai, A. K. (1999) Regulating Scientific Research: Intellectual Property Rights and the Norms of Science. *Northwest Univ. Law Rev.* **94**, 77–152.

23. Masur, J. S. (2008) Process as Purpose: Costly Screens, Value Asymmetries, and Examination at the Patent Office. http://ssrn.com/abstract=1105184. Accessed 26 December 2009.

24. United States Patent and Trademark Office (2005) General Information Concerning Patents. http://www.uspto.gov/patents/basics/index.html#patent. Accessed 15 December 2009.

25. Webber, P. M. (2003) Protecting Your Inventions: the Patent System. *Nat. Rev. Drug Discov.* **2**, 823–830.

26. Maskus, K. E. (2000) *Intellectual Property Rights in the Global Economy*. Institute for International Economics, Washington, DC.

27. Grossman, G. M. and Lai, E. C. (2004) International Protection of Intellectual Property. *Am. Econ. Rev.* **94**, 1635–1653.

28. Mandel, G. (2010) Regulating nanotechnology through Intellectual Property Rights. In: *International Handbook on Regulating Nanotechnologies* (Hodge, G. A., Bowman, D. M., and Maynard, A. D., eds.) Edward Elgar, Cheltenham. 388–405.

29. Caulfield, T., Gold, E. R., and Cho, M. K. (2000) Patenting Human Genetic Material: Refocusing the Debate. *Nat. Rev. Genet.* **1**, 27–231.

30. Abbott, F. M. (2006) Intellectual Property Provisions of Bilateral and Regional Trade Agreements in Light of U.S. Federal Law. http://www.unctad.org/en/docs/iteipc20064_en.pdf. Accessed 13 December 2009.

31. Australian Government (2008) *Patentable Subject Matter – Issues Paper*. Advisory Council on Intellectual Property, Canberra.

32. Drahos, P. (1999) Biotechnology Patents, Markets and Morality. *Eur. Intellect. Prop. Rev.* **21**, 441–449.

33. Bagley, M. A. (2003) Patent First, Ask Questions Later: Morality and Biotechnology in Patent Law. *William Mary Law Rev.* **45**, 469.

34. Royal Society and Royal Academy of Engineering (2004) *Nanoscience and Nanotechnologies: Opportunities and Uncertainties*. RS-RAE, London.

35. Maynard, A. D. (2007) Nanotechnology: The Next Big Thing, or Much Ado about Nothing? *Ann. Occup. Hyg.* **51**, 1–12.

36. Lux Research (2009) *Nanotech's Evolving Environmental, Health, and Safety Landscape: The Regulations Are Coming.* Lux Research, New York.

37. National Nanotechnology Initiative (undated) About the NNI – Home. http://www.nano.gov/html/about/home_about.html. Accessed 15 December 2009.

38. National Nanotechnology Initiative (2009) Funding. http://www.nano.gov/html/about/funding.html. Accessed 15 December 2009.

39. Roco, M. C. (2005) International perspectives on government nanotechnology funding in 2005. *J. Nanopart. Res.* **7**, 707–712

40. European Commission (2005) *Nanosciences and Nanotechnologies: An Action Plan for Europe 2005–2009.* European Parliament, Brussels.

41. Hullmann, A. and Meyer, M. (2003). Publications and Patents in Nanotechnology: an Overview of Previous Studies and the State of the Art. *Scientometrics* **58**, 507–527.

42. Zucker, L. G. and Darby, M. R. (2005). *Socioeconomic Impact of Nanoscale Science: Initial Results and Nanobank* (working paper 11181). NBER available at http://www.nber.org/papers/w11181.

43. Zucker, L. G., Darby, M., Furner, J., Lieu, R., and Ma, H. (2007). Minerva Unbound: Knowledge Stocks, Knowledge Flows, and New Knowledge Production. *Res. Policy* **36**, 850–863.

44. Binnig, G., Quate, C. F., and Gerber, C (1986) Atomic Force Microscope. *Phys. Rev. Lett.* **56**, 930–934.

45. Sabety, T. (2004) Nanotechnology Innovation and the Patent Thicket: Which IP Policies Promote Growth? *Albaby Law J. Sci. Technol.* **15**, 477–516.

46. Johnson, H. A. (2004) Wright Patent Wars and Early American Aviation. *J. Air Law* **69**, 21–64

47. Mueller, J. M. (2001) No Dilettante Affair: Rethinking the Experimental Use Exception to Patent Infringement for Biomedical Research Tools. *Wash. Law Rev.* **76**, 1–66.

48. Sylvester, D. J., Menkhus, E., and Granville, K. J. (2005) *Innovation Law Handbook.* Available at SSRN: http://ssrn.com/abstract=999451. Accessed 26 December 2009.

49. ETC Group (2005) *Nanotech's "Second Nature" Patents: Implications for the Global South.* ETC Group, Ottawa.

50. Lux Research (2006) *Nanotech Battles Worth Fighting.* Lux Research, New York.

51. Harris, D. L., Hermann, K., Bawa, R., Cleveland, J. T., and O'Neill, S. (2004) Strategies for Resolving Patent Disputes Over Nanoparticle Drug Delivery Systems. *Nanotechnol. Law Bus.* **1**, 372–390.

52. Maynard, A. D. (2006) *Nanotechnology: A Research Strategy for Addressing Risk.* Project on Emerging Nanotechnologies, Washington, DC.

53. Lux Research (2005) *The Nanotech Intellectual Property Landscape.* Lux Research, New York.

54. Thomas, J. R. (2001) Collusion and Collective Action in the Patent System: A Proposal for Patent Bounties. *Univ. Ill. Law Rev.* **1**, 305–353.

55. Tegart, G. (2004) Nanotechnology: the Technology for the Twenty-First Century. *Foresight* **6**, 364–370.

56. Zekos, G. I. (2006). Nanotechnology and Biotechnology Patents. *Int. J. Law Inf. Technol.* **14**, 310–369.

57. Bowman, D. (2007) Patently Obvious: Intellectual Property Rights and Nanotechnology. *Technol. Soc.* **29**, 307–315.

58. Ferrari, M. (2005). Cancer nanotechnology: opportunities and challenges. *Nat. Rev. Cancer* **5**, 161–171.

59. Marinova, D. and McAleer, M. (2003) Nanotechnology Strength Indicators: International Rankings Based on US Patents. *Nanotechnology* **14**, R1–R7.

60. Huang, Z., Hu, R., and Pray, C. (2003) Longitudinal Patent Analysis for Nanoscale Science and Engineering: Country, Institution and Technology Field. *J. Nanopart. Res.* **5**, 333–363.

61. Huang, Z., Chen, H., Chen, Z. K., and Roco, M. C. (2004) International Nanotechnology Development in 2003: Country, Institution, and Technology Field Analysis Based on USPTO Patent Database. *J. Nanopart. Res.* **6**, 325–354.

62. Huang, Z., Chen, H., Li, X., and Roco, M. C. (2006) Connecting NSF Funding to Patent Innovation in Nanotechnology (2001–2004). *J. Nanopart. Res.* **8**, 859–879.

63. Bawa, R. (2004) Nanotechnology Patenting in the US. *Nanotechnol. Law Bus.* **1**, 31–51.

64. Koppikar, V., Maebius, S. B., and Rutt, J. S. (2004) Current Trends in Nanotech Patents: A View From Inside the Patent Office. *Nanotechnol. Law Bus.* **1**, 24-30.

65. Heinze, T. (2004) Nanoscience and Nanotechnology in Europe: Analysis of Publications and Patent Applications including

Comparisons with the United States. *Nanotechnol. Law Bus.* **1**, 1–19.

66. Chen, H. and Roco, M. C. (2008) *Mapping Nanotechnology Innovations and Knowledge.* Springer, New York.

67. Bauer, M. W. and Gaskell, G. (eds.) (2002) *Biotechnology: The Making of a Global Controversy.* Cambridge University Press, London.

68. Jasanoff, S. (2005) *Designs on Nature: Science and Democracy in Europe and the United States.* Princeton University Press, Princeton.

69. Halluin, A. P. and Westin, L. P. (2004) *Nanotechnology: The Importance of Intellectual Property Rights in an Emerging Technology.* Howrey Simon Arnold & White, Washington, DC.

70. Clarkson, G. and DeKorte, D. (2006) The Problem of Patent Thickets in Convergent Technologies. *Ann. NY Acad. Sci.* **1093**, 180–200.

71. Lee, A. (2006) Examining the Viability of Patent Pools for the Growing Nanotechnology Patent Thicket. *Nanotechnol. Law Bus.* **3**, 317–328.

72. Harris, D. L. (2008) Carbon Nanotube Patent Thickets. In: *Nanotechnology & Society: Current and Emerging Ethical Issues* (Allhoff, F. and Lin, P., eds.) pp. 163–186. Springer, New York.

73. Miller, J., Serrato, R., Represas-Cardenas, J. M., and Kundahl, G. (2005) *The Handbook of Nanotechnology: Business, Policy, and Intellectual Property Law.* Wiley, New York.

74. Bastani, B. and Fernandez, D. (2004) *Intellectual Property Rights in Nanotechnology.* Fernandez & Associates, LLP, Menlo Park.

75. Clark, J., Piccolo, J., Stanton, B., and Tyson, K. (2000) *Patent Pools: A Solution to the Problem of Access in Biotechnology Patents?* USPTO, Washington, DC.

76. Lakhani, K. R. and von Hippel, E. (2003) How Open Source Software Works: "Free" User-to-User Assistance. *Res. Policy* **32**, 923–943.

77. Economist (2004) An Open-Source Shot in the Arm? *Economist,* 10 June, 17.

78. Munos, B. (2006). Can Open-Source R&D Reinvigorate Drug Research? *Nat. Rev. Drug Discov.* **5**, 723–729.

79. Bruns, B. (2001). Open Sourcing Nanotechnology Research and Development: Issues and Opportunities. *Nanotechnology* **12**, 198–210.

80. Kelty, C., Lounsbury, M., Yavuz, C. T., and Colvin, V. L. (undated) Towards Open Source Nanotechnology: Arsenic Removal and Alternative Models of Technology Transfer. http://opensourcenano.net/images/GRC-Poster2.pdf. Accessed 15 April 2009.

81. Prisco, G. (2006) Globalization and Open Source Nano Economy. *KurzweilAI.net.* http://www.kurzweilai.net/meme/frame.html?main=/articles/art0659.html. Accessed 12 April 2009.

82. Peterson, C. L. (2008) Citizen-Controlled Sensing: Using Open Source & Nanotechnology to Reduce Surveillance & Head Off Iraq-Style Wars. http://www.opensourcesensing.org/proposal.pdf. Accessed 12 April 2009.

83. Hodge, G. A., Bowman, D. M., and Ludlow, K. (2007) Introduction: big question for small technologies. In: *New Global Regulatory Frontiers in Regulation: The Age of Nanotechnology* (Hodge, G. A., Bowman, D. M., and Ludlow, K., eds.) pp. 3–26. Edward Elgar, Cheltenham.

84. EurActiv.com (2009) Germany opposed to 'nano' label for cosmetics, 24 November. http://www.euractiv.com/en/enterprise-jobs/germany-opposed-nano-label-cosmetics/article-187583. Accessed 15 December 2009.

Chapter 24

Scientific Entrepreneurship in the Materials and Life Science Industries

Jose Amado Dinglasan, Darren J. Anderson, and Keith Thomas

Abstract

Scientists constantly generate great ideas in the laboratory and, as most of us were meant to believe, we should publish or perish. After all, what use is a great scientific idea if it is not shared with the rest of the scientific community? What some scientists forget is that a good idea can be worth something – sometimes it can be worth a lot (of money)! What do you do if you believe that your idea has some commercial potential? How do you turn this idea into a business? This chapter gives the aspiring scientific entrepreneur some (hopefully) valuable advice on topics like choosing the right people for your management team, determining inventorship of the technology and ownership shares in the new company, protecting your intellectual property, and others; finally, it describes some of the various pitfalls you may encounter when commercializing an early stage technology and instructions on how to avoid them.

Key words: Entrepreneurship, Technology transfer, Nanomaterials, Nanotechnology, Start-up, Venture capital

1. Introduction

Why become an entrepreneur? Certainly, there are risks associated with becoming an entrepreneur. The odds of failure are high, and entrepreneurs typically face an emotional "roller-coaster" of good news and very bad news. This can be exacerbated by the fact that entrepreneurs take a very personal interest in their company. It is easy to get so personally invested that it can be difficult to keep an even emotional keel.

So why become an entrepreneur? When people speak about becoming an entrepreneur, they often lead with the bad side of entrepreneurship, much like we have done here. The simple truth is not everyone should become an entrepreneur. Entrepreneurship requires significant risk tolerance, emotional balance, and a keen

Sarah J. Hurst (ed.), *Biomedical Nanotechnology: Methods and Protocols*, Methods in Molecular Biology, vol. 726, DOI 10.1007/978-1-61779-052-2_24, © Springer Science+Business Media, LLC 2011

379

sense of vision, but for those who are well suited and who are willing to take the risks, it is an incredibly fulfilling career option for scientists.

We believe that entrepreneurs are, more than anything else, motivated by impact. Impact can be in the purely monetary sense – few other scientific careers offer the opportunity for a PhD to be a multimillionaire 5 years after graduation. More importantly, however, impact also has a huge nonfinancial dimension. In a company that is commercializing a new technology, one can have an opportunity to help address major world issues, to directly impact the success or failure of your venture, to build a team of employees from ground zero, and to influence the culture of a new organization. In addition to all of this, if you are a recent graduate from a doctoral, master's, or undergraduate program, it is an opportunity to create an exciting role for yourself in a growing company that suits your unique skills.

To introduce ourselves and to give you a little perspective on where we are coming from, two of the authors of this chapter (Anderson and Dinglasan) are scientific entrepreneurs and cofounders of Vive Nano. The third author (Thomas) is a serial entrepreneur who has been Vive Nano's President and CEO since the very early stages of the company. Vive Nano is a spin-off company from the University of Toronto that was founded in 2006. At the time of writing (2010), we have 16 employees, a pilot manufacturing facility in Toronto, and a range of issued patents and pending patent applications. Vive Nano is focused on using its proprietary process to provide simple small solutions to big issues. Specifically, our development activities are targeted in two areas with major global impact: new crop protection formulations with improved efficacy and decreased environmental impact, and improved nanoscale heterogeneous catalysts.

In this chapter, we provide an overview of our experiences in founding Vive Nano, including its history to date, some high-level but useful advice to aspiring entrepreneurs from academic institutions on topics like choosing the right people for your management team, determining inventorship of your technology and ownership shares in your new company, protecting your intellectual property, and others; finally, we give descriptions of some of the various pitfalls you may encounter when commercializing an early stage technology and instructions on how to avoid them.

This chapter is primarily aimed at graduate students and postdoctoral fellows that are thinking of starting a company based on research that they have performed at a university. The information may also be useful to professors, postdoctoral fellows or research associates in government laboratories, or researchers in private industry. There are many different types of companies that can be founded by researchers in these circumstances, including service businesses and companies that start small and grow slowly, but provide sufficient profits for the founding team to live well. We are focused in this chapter on companies that are targeting large problems in large markets that will (typically) require financing from external resources.

Of course, there is no one-size-fits-all solution to starting a company, and circumstances may vary greatly. One thing that an entrepreneur learns very early on is to take all advice with a grain of salt and come to their own conclusions. We suggest you do the same with this chapter.

2. Discussion

The idea of starting a company began when a group of us (including Anderson and Dinglasan) took part in a noncredit course offered by the University of Toronto called "Entrepreneurship 101." Most of us were at turning points in our careers and were trying to decide what to do after graduate school or a postdoctoral fellowship. This full-year seminar course was geared toward scientists/academic personnel who wanted to explore the idea of starting a business based on the technologies or ideas that they had worked on in the laboratory. Each of the course's lecturers was practicing in industry and had advanced scientific training. Lecture topics included intellectual property (IP), market research, project management, licensing, and others. This course gave us some insight into how to bring something that has commercial potential out of the laboratory and make it into a business.

At the end of the course, we were expected to make a business pitch based on a technology that had been developed as part of our research. The process of making the pitch made us think about the impact the technology that we were trying to commercialize could have on certain markets/sectors, the advantages (and disadvantages) of our technology over existing technologies, the different products we could potentially produce, and a possible plan for the business. Most importantly, it also taught us just how much we did not know about starting a company. Although we realized that a lot of additional groundwork would be needed to even start a sustainable business, we also convinced ourselves that the technology that we had developed was exciting enough to form the foundation of a new company.

We were also fortunate enough to belong to a research group that had prior start-up experience. Professor M. Cynthia Goh, in whose laboratory we were working, had previously spun out Axela Biosensors, a diagnostics company that was based on a technology developed in her laboratory. This company was about 5 years into its existence at that point, and had about 20 employees, and we had seen it grow from an idea in the laboratory to a well-financed, product-based company. This means that we knew it was possible to start a company from an idea developed in the laboratory and that we had access to people who could help advise us about the risks, opportunities, and surprises common to early stage companies. We hope to provide much of the same advice to the readers of this chapter.

2.1. The Idea

As scientists, we create and generate knowledge in the research laboratory. What do we do if we think this knowledge is potentially commercializable? For a start-up company, this knowledge is one of the two critical early stage assets of the company (the other is people, outlined in further detail in Subheading 2.2). This knowledge can take many forms, and if it can be owned by the company, it is known as intellectual property (IP). Intellectual property could be a new technology that it is possible to get patented – and thereby prevent anyone else from using – or simply "know-how," things that you know that are necessary for the knowledge to be applied, but may not be patentable. For most start-up companies, patentable IP is the most important type of IP and will need to be protected using patents at a very early stage.

Because of the importance of IP, it is important to clearly define at the very beginning who owns the IP and how it will be commercialized. Assuming you developed an invention or technology while working at a university, depending on the IP policies set up at that particular university, the university may own part, if not all, of your invention. This is also true in many government laboratories. Depending on the country in which you live, each university will have its own version of a technology transfer office that is responsible for handling ownership of IP developed within university premises. They can usually assist the inventors of the technology with commercializing the technology, if desired.

The first step once an invention or technology has been developed is to let the university know about the invention. This is normally done by filing an invention disclosure to the university's technology transfer office. The disclosure documents the circumstances under which the invention was created as well as provides the university with the information necessary to evaluate inventorship, patentability, and obligations to research sponsors outside the university. Depending on the policies at your institution, you and the other inventors may now have the option to commercialize the technology or invention yourself, or you may need to work with the technology transfer office to do so.

Usually, one of the first activities that is required is to file a patent on the invention that will help you prove that you own the invention or technology. Unless a technology has been protected in this way, it is often difficult to attract the financing that is necessary to build the business. It is only possible to get a patent on an invention if it has not yet been disclosed to the public. Public disclosure includes publishing a scientific paper or presenting a poster or a talk at a conference, so it is critical to make sure that you have filed any relevant patents before you give even a single presentation about your technology. Your technology transfer office can assist you with understanding this issue in more detail.

A discussion of drafting patents is beyond the scope of this chapter, but we do suggest that you recruit legal counsel (a patent agent or a lawyer) to assist you with this process. While expensive, IP is a critical asset for the company, and a mistake made at this stage could cause dramatic problems down the road.

It also is important to determine the inventors of the invention at an early stage. The rules of determining inventorship of an invention are quite different from the rules of determining authorship of a scientific paper. Inventors are the people who were responsible for the creative breakthrough that led to the invention. You, along with the colleagues that worked with you, must come into agreement about who contributed to the original idea and overall intellectual input to the invention in question. People that merely work under the direction of another are not considered coinventors unless they have made inventive or creative contributions to a concept that led to the invention. For you to be considered an inventor, you should be able to point to a specific idea that you contributed without which the invention could not exist. It is essential to settle any inventorship issues at this early stage because certain problems can arise later on if inventorship is not correctly attributed. Further, outstanding inventorship issues can leave your future patent open to challenges from other parties that may wish to invalidate it. From a practical point of view, it is just good sense to think about these issues early, while there is no money involved. Things become quite different once the invention is actually worth something. If necessary, your patent agent or patent lawyer can assist you with determining the inventors of an invention.

2.2. The Team

Now that you have a technology or an invention that you have decided to commercialize, it is time to start recruiting the people that will make your company a success. Before bringing on external people, we recommend you turn your attention to the founding group. In our experience, the start-up company has a better chance of success if there are multiple founders. This guarantees different points of view when the critical early stage decisions for the company are made and also increases the odds that people on the founding team have complementary skills. However, having multiple founders does complicate issues, for instance, in determining how much of the new company each founder will own.

For the sake of argument, let us say there are 5 individuals that decide they want to start a company together. Each of the individuals brings different contributions and skills to the table; some are inventors of the technology, some have business savvy and can help with initial financing, and some can provide mentorship or training to the others. In addition, usually (though not always) the university takes some stake in the start-up company by providing the background facilities necessary for the invention to be developed. In this scenario, each individual obviously deserves partial ownership of the company that is to be created based on the invention, but how do we now assign how much of the start-up company each individual should own? Do you divide the company equally? There is really no perfect way to do this. It is all really quite arbitrary at this point, but what is essential is that everyone comes into agreement on the amount of initial shares assigned to them. Each individual should negotiate with the group for his/her fair share. It is important to make sure that, in exchange for the shares in the company, the individuals involved have contributed to the company in a significant way. Ownership should not be handed out unless it has truly been earned. The goal is that everyone is comfortable that the company's initial ownership is fair and that everyone has been treated equitably. This is much easier to do now, when the company (and therefore the shares) is worth very little. Unfair ownership distribution or unearned ownership can cause major problems down the road, when the shares are actually worth something.

It is helpful not only to consider the amount of time each individual has already contributed to the company up to the point where the shares are being divided, but also the potential amount of time he or she will be spending with the company after the shares have been assigned. Those who are sticking around should be given some incentive (i.e., someone who is willing to stay on for a year or more should be allocated more shares than someone who is not going to be with the company next year). You could also set up a scheme to reward individuals for making significant contributions such as hitting certain milestones, developing new products, or finding funding, just to name a few examples. Regardless of the exact details of what you work out, it is essential to reward individuals with enough incentive to make them feel that what they have done and will do in the future is valuable to the company. We have found that employee stock option grants based on performance are one of the best ways to do this.

Once you raise any external financing, it is usually important to set up a shareholder's agreement that will govern share ownership, voting rights, and many other important issues that define how the company is run. At this time, the ownership of the company among the founding team is locked in, and some of the flexibility that existed initially disappears.

The founding team (and investors, depending on their interest) will need to recruit other people who can help the new company succeed. In addition to IP, the people involved in the new company are its greatest asset. In order to meet potential new hires, the most critical piece of advice that we can provide is to network. Just as scientific entrepreneurship is not for everyone, joining a new, growing company is not for the faint of heart! This means that people who are well suited to your new company can be difficult to find – you will need to search hard for the best people. One way to expand your network is by joining professional organizations and trade associations, as well as by attending scientific conferences and trade shows. Entrepreneurship programs run by your university or other nearby institutions are an excellent place to meet people with a keen interest in entrepreneurship. Usually, even if the speakers and attendees are not available to consider joining your team, they can connect you with good potential hires.

As one of the founding members of a technology-based company, you should realize that as smart as you are on the technical side of things, you probably do not know much about business. This will be particularly true if you are founding the company straight out of graduate school or a postdoctoral fellowship, though this point can apply just as much to professors! You will need other people – people with different backgrounds and expertise – to help create a successful company. The following is a brief overview of the people you will need in your team to build a successful start-up.

2.2.1. Management Team

You should preferably try to find people with previous start-up experience, like serial entrepreneurs who have gone through the ups and downs of starting their own companies and have previous experience or familiarity with the markets or industry you are focused on. These people will be directing and mentoring the people in the company (including you), so their vision and the way they interact with people is important. If you are lucky, you may be able to recruit experienced entrepreneurs who have been successful in past companies and may be willing to invest in your company, along with providing experienced management skills.

2.2.2. Scientific Team

Remember, your company will need people with a wide range of technical skills in addition to those of the members of the founding team. This means that you are not necessarily going to want to hire people with the same background as yourself, but will instead want to look for people with complementary skills. The specifics of what you need will, of course, depend on the nature of the technology you are commercializing, as well as the markets that you are targeting. Keep in mind that you have (likely) not worked in the industry that will use your products, so recruiting someone from that industry would be useful. Also, if you are a scientist, you may need engineering resources to scale-up your product or assist with other types of product development. Note that the number of people you hire on the management and scientific teams will obviously depend on your resources, and the type of people you will hire as employees #1 and #2 will likely be very different from employee #30.

2.2.3. Service Providers

In addition to the people inside your company who work for you, you will also need assistance from a wide range of other people as you and your new team establishes and grows the company. These people will include lawyers, accountants, bank account managers, Web site designers, and many others. Particularly for the more critical service providers, it is important to make sure that they have experience working with new companies in your industry. After all, you are unlikely to be experienced in these areas, and you want service providers that can keep you from making common early stage mistakes. These run the gamut from things as simple as missing payroll tax payments to accidentally publishing a new technology before filing for a patent – invalidating any chance you have to get a patent on that technology. Ideally, your service providers will also be able to provide other assistance, including introductions to financing sources, potential hires, and potential customers. The chances that they will be able to assist in this way are greatly increased if they have past experience working with companies such as yours.

As a new company, you are likely going to have limited resources that you can spend on instruments, so you will also need to find facilities and people that you can use for analytical services. Universities can be ideal for this purpose as their facilities can be less expensive to use than those in industry; it is also typically less expensive to use university facilities rather than purchasing the instrumentation yourself. Further, you can often access government grants to set up collaborations with universities. However, it is again important to make sure that the specific group you are working with at the university has experience working with companies like yours, as you are likely to be moving at a very fast pace and will need fast, effective turnaround of analytical services.

2.3. The Product and Sales

Products and sales are the most important reason for your company's existence. If no one wants to use your product, you have a product with no value. Ergo, you have a company with no value.

Your technology foundation will need to be made into products that are critical to the market, robust, and economically viable. If your technology is none of these, but it is interesting to you and your staff, it will keep you interested in your work, but ultimately it should stay as a university research project, and not be the foundation for a company. In addition, if you get too excited about the technology and forget about what your customers will want, you might as well stay in academia.

Product development and sales in start-ups are an interlinked and iterative process, largely driven by the need to fund growth. Typically, you will start by creating an "alpha" product internally and then fix problems in it to create a "beta" product that is then trialed with initial customers. This product is then modified again to create something that is ready for a broader market launch. The materials that we currently sell to researchers are always being improved – as we receive feedback; we add characterization details, tighten specifications, add new products, and decrease costs. This continuous improvement will lead to the best possible product – and it is critical to start that process as early as possible.

It is important to get your initial trial customers to help with product development, as they can provide: technical guidance, market access, customer validation that you are focused on the right problem, and reference customers, for other sales and importantly for financing your next development. In addition, these initial customers act as validation to investors to give you additional money, which you can then use to build the next product iteration.

It is also important to target the right size of initial customer. Everyone – you, your investors, and your family – will think it is fantastic if you land a large company as a customer. True, it will help you impress investors and enable them to dream of what would happen if your product is successful. But these "marquee" clients will be hard negotiators and the deals may not be that lucrative. We found that small-to-medium-sized companies were the most fertile ground for trial customers. These customers can make decisions more quickly, act as true partners as they are closer to your size and, in some cases, may still remember what it was like to start a new business. Though you should still work to sell to them, large company buyers may steer away from small company product development projects – unless the technology has great potential.

You may require additional staff to help build your sales. Always remember, though, that your first salesperson is yourself. You are inexpensive, know the technology, and should be able to synthesize customer feedback to develop and improve your product. But once you can no longer carry the burden alone, you should add a business development person, rather than salesperson. The distinction here is important. Our view was that a salesperson is very good at using his or her contacts to sell large quantities of a well-defined product when pointed in the right direction. A person specializing in business development, on the other hand, suits an earlier stage of product development.

These are people who can be trusted advisors for target customers, walk the halls to determine what their needs are, and then inform product development staff. In addition, a salesperson is largely commission driven, but an advisor typically is compensated with a higher base and lower commissions.

2.4. The Cash

There is no one who will give you money for free. You need to demonstrate value before any investor – besides your family – investor will think that there is a value in what you do. You will need to iteratively build your products and customers to demonstrate value to investors.

It is important that you constantly try to get money – and not worry about giving away too much when you get it. Some people say that you should think of your shares as if they could be worth $100 each in the future, but they will not get there unless you have money. Many start-ups go out of business simply because they run out of money, not due to a bad idea or product. One thing that you can count on is that things always take longer, and cost more, than you expect.

There are a range of sources for financing your venture, including:

- Friends and family – Fairly evident from the title, these are people who are close to you that will often invest because they trust in you. It is your responsibility to ensure that this trust is not misplaced and that they are suitable candidates for risky investments. It can be very uncomfortable at events with friends if you just lost their money. Frequently, arm's-length investors want to ensure that friends and family are invested, as it is a test of your resolve to make the company work. Your desire to not disappoint your friends and family then is a big driver to ensure your success.

- Angel investors – These are typically individuals or groups of individuals who have from ten thousand to several hundred thousand dollars to invest in early stage companies. They are the bridge to later financing and will typically invest under terms that give them special preferred rights.

- Government grant and regional development programs – These are typically grant or loan programs that we have found to be very helpful in providing growth capital without giving away large amounts of the company. They will vary from country to country, but most governments have various programs in place.

- Customers – This is the ideal source of financing as it provides market validation and reference customers for other sales while building your company and not diluting your ownership stake. Customers can sometimes creatively finance capital expenditure and may be eventual buyers of your company.

- Venture capital – This is institutional money that invests in high-growth companies. This can be very useful, and most people look at this as the "holy grail" because they hear stories of venture-backed companies like Google that have done fantastically well. But it is important to remember that the chances of getting venture funding are very low, and typically only 10% of those companies are really successful. In addition, only certain types of companies will be attractive to these investors, and you will largely be signing control of your company over to someone else.

In our case, we built Vive Nano largely through angel investors, customers, and grants.

As a scientist, you should not rely on your ability to learn about and raise financing on your own. It is critical to add financing capability at an early stage. While our company's initial management team was all technical, the next stage of the management team's development was in hiring individuals that could provide our company with the ability to finance itself. Many companies fail because they have outlasted their cash, so this was a critical development. So crucial, in fact, that at one point, 60% of our management team had experience in raising financing.

You should realize that the financing process is rarely fast. We have seen a typical "fast financing" take 6 months and a longer financing took 18 months. You will need to budget for this when considering your cash needs.

Lastly, once you get the money, manage it very carefully. One of us (Keith Thomas) used to be a banker and would quickly take loans away from companies that were "building temples to themselves." If you are focusing more on working in – or worse – building a lavish workspace, you have taken your mind off getting and keeping customers.

The Chinese philosopher Lao-tzu said that, "A journey of a thousand miles begins with a single step." The science you have created in the university laboratory is the first step but only the first step.

You may believe that you have invented the next world wonder. You might be right. But you will definitely need other people's help to turn your science into something that will have any benefit. Business people will tell you that it is a long road to turn your science into replicable technology, find a profitable application, build it into a saleable product, and package it for sale. This is all true. The entrepreneurial journey entails a lot of risks – risks of failure and risks of success. You just have to have it in your guts to take the risk. The rewards can be great, and taking these risks is necessary if you and your technology are going to change the world.

Good luck in your journey!

3. Notes

1. Consciously build a company culture. Culture can make or break your company. At our company, we focused on building a culture that is smart (smart in how we execute our work), open (open to new ideas and new challenges), and responsible (responsible in the application of what we do for the greater good).

2. Select employees who are a good fit for your company's culture. Our company required different resources at each stage of its development, but the cultural DNA of the people at all stages was common. Our people are aggressive, critical thinkers with the ability to network. To help foster a culture of critical thinking, we have a rule that at least two sets of eyes look at everything before it goes outside the company. This rule helps ensure consistent messaging, but more importantly makes sure that we test every document, research, or product before it is presented to clients.

3. Do not insist on doing it all by yourself and ensuring you get credit for it. It requires a range of partners to support the company. You should be thankful for anyone who helps build your company and share the credit with them. We were – and still are! – thankful for the university, government, and corporate partners who pushed us in the beginning to do better every day.

4. Do not only hire your friends. It is good to work with people with whom you are compatible and potentially even with your friends, but you may need to discipline or fire staff at some point and this is hard to do to close friends.

5. Do not hire the wrong start-up manager. Every company needs what is sometimes called "gray hair" or "adult supervision." Adding this person is critical and where we have seen a number of companies go wrong. Your company will require the experience that comes from having "done things" before to speed growth and avoid pitfalls, especially in cash and people management. There are a number of former executives from large companies who are looking to work in small companies, but it is important to not accept the first "gray hair" you see. A good start-up manager has experience in building start-ups and can provide access to their own money or network to get things going. In addition, that person will be your chief salesperson – to employees, customers, investors, funding agencies – until you hire a salesperson or business development manager later. If they cannot sell, they are not the right person.

6. Do not refuse to "let people go." Not everyone will be able to grow with the company, and this leads to hard decisions. It is often better for the company and for the individual to let that individual go after the job has surpassed their capabilities. Even retaining an employee in a reduced role and hiring someone to be their boss can lead to unnecessary tension.

7. Remember that there is a difference between how businesspeople and academics think. We have come to the understanding that this is one of the primary reasons why commercialization of university science is so tough. You have two partners working together, each with differing worldviews. The academic wants academic freedom – to research what they want and to enable their discovery to be widely known through publishing or conferences– while the businessperson wants the academic to perform dedicated research and to keep the discovery quiet to enable patenting. There are ways to ensure both goals are met, but they require understanding from both sides.

Acknowledgments

D. J. A. and J. A. D. thank Professor M. Cynthia Goh, who has been an excellent mentor for them both. The authors also thank their coworkers at Vive Nano, who keep the company growing.

Chapter 25

Applying the Marketing Mix (5 Ps) to Bionanotechnology

Michael S. Tomczyk

Abstract

This chapter, based on concepts developed for my book, NanoInnovation (Tomczyk, Nanoinnovation: What Every Manager Needs to Know, 2011), is one of the first attempts to evaluate nanotechnology in the context of the "marketing mix" – a conceptual challenge given that nanotechnology is not one product or even a set of products, but rather a technology that is incorporated in an expanding list exceeding a 1,000 products – encompassing materials, structures, processes, and devices. My purpose is to use this context to identify some of the critical issues and factors that will influence development of "nanotechnology markets" at this very early stage in the evolution of nanotechnology, and more specifically, bionanotechnology. As technological innovations continue to promote the market growth for nanotechnology, especially in the field of medicine and healthcare, sensemaking frameworks are needed to help decision makers keep pace with these evolving markets. One of the best frameworks is the "marketing mix" which has been used for decades to identify the controllable factors that decision makers can influence through marketing strategies. With so many game-changing innovations poised to move from nanotech research to commercialization, marketing issues are becoming increasingly important to decision makers in science/academia, business/venture development, and government/policymaking.

Key words: Bionanotechnology, Bioscience, Biotechnology, Innovation, Marketing, Nanobiotechnology, Nanoinnovation, Nanotechnology

---

1. Introduction

The promise of nanotechnology, especially in medicine and healthcare, is profound.

Each day, we hear how cancer researchers are using nanoparticles to tag, target, and destroy tumors. We read about nano-enabled labs-on-a-chip and portable devices that will affordably identify dozens of diseases, and bring sophisticated diagnostics to remote corners of the world. Science journals describe how nanoparticles are delivering drug molecules to diseased cells without harming healthy tissue, and delivering therapies *inside*

Sarah J. Hurst (ed.), *Biomedical Nanotechnology: Methods and Protocols*, Methods in Molecular Biology, vol. 726, DOI 10.1007/978-1-61779-052-2_25, © Springer Science+Business Media, LLC 2011

cells – activated by a pulse of light, a magnetic or electrical charge, or infrared heat. Researchers are striving to create a test that can read an entire human genome in less than an hour for $50 or $100. These are only a few examples.

Marketing nanotechnology is not like marketing a traditional product or service (1). Nanotechnology involves nanoscale materials, processes, structures, and devices, typically 1–100 nm in size. In context, a strand of DNA is about 2.5 nm wide. The diameter of the West Nile virus is about 50 nm (2).

The products that represent the nanotechnology market cover an enormous spectrum of applications and industries. They include imaging systems such as scanning tunneling microscopes that help medical researchers observe, understand, and manipulate biological processes to effect novel diagnostics and therapies. They include material forms that we have only known about for a couple of decades, such as carbon nanotubes, quantum dots, and nanowires, and variations such as nanoshells that can be used to deliver chemotherapy molecules. A key capability of nanotechnology is the capability to manipulate collections of atoms and molecules that exhibit quantum effects at the scale that make elements such as gold – inert in bulk form – surprisingly reactive at the nanoscale. These capabilities and phenomena are fueling emerging fields, such as "theranostics," which combine diagnostics and therapy.

Semiconductors are undoubtedly one of the most familiar nanotechnology products – nanoscale circuits have been present in computers and cell phones for decades. Intel has developed circuits as small as 22 nm, enabling the placement of 2.9 billion circuits on a chip the size of a small fingernail (3).

As exciting as these innovations and possibilities might be, they also introduce a variety of marketing challenges. Moving what works in a laboratory to what works in a treatment regimen can take years or decades to achieve – while patients and practitioners read the headlines and assume that a treatment or cure is imminent. For example, delivering a nanoparticle to one type of cell in a Petri dish is different from targeting a given amount of nanoparticles to a specific location in the human body, when the body includes more than 200 different types of cells.

Beyond the research challenges, nanotechnology faces a variety of image challenges associated with the fuzzy nature of health, safety, and environmental issues, many of which are unresolved. For example, until recently there were no globally accepted standards for nanomaterials or engineered nanoparticles. If you buy carbon nanotubes from two suppliers, the specifications can vary considerably. Studies linking nanoparticles to disease or environmental contamination are still being validated – which raises yet another marketing issue: should manufacturers advertise "nano-inside" or "nano-free" on their labels, or ignore nano-branding altogether?

To put these issues in context, I present some sensemaking frameworks and personal perspectives using a classic marketing framework known as the "marketing mix."

2. The Nanotechnology Marketing Mix

In traditional marketing, companies use a framework called the "5 Ps," based on factors that decision makers can control and incorporate into their business strategy. Traditionally, the 5 Ps include the following: Product, Price, Place, Promotion, and People. A version proposed by Dev and Schulz in 2005 (4) builds on this framework with an expanded next-gen mix called SIVA, an acronym that stands for: Solution, Information, Value, and Access – where Product becomes Solution, Price becomes Value, Place becomes Access, and Promotion becomes Information. Although Dev/Schultz did not include the fifth P (People) in their framework, I am including People and adding the dimension of "Community," which expands "SIVA" to "SIVAC." Communities of interest have become an important element in global marketing. In the case of bionanotechnology, communities of interest include scientists and research groups as well as patients and practitioners.

Each of the five parameters in the marketing mix poses some thought-provoking issues and challenges for decision makers. In Table 1, I have applied this expanded marketing mix to bionanotechnology and included some general examples for each parameter (1).

2.1. P1/Product (Solution): Nanotech Definitions, Maps, and Killer Applications

The first wave of products/solutions that emerged from nanotechnology included tools such as imaging systems, materials such as carbon nanotubes, and nano-sized catalysts, which are more efficient than comparable "bulk"-size catalysts. Nanotechnology products range from raw materials such as silver nanoparticles used in antimicrobial coatings (for example, in medical environments) to titanium dioxide nanoparticles used in sunscreens. Probably, one of the best-known nanostructures is the carbon nanotube (CNT), a tubular form of carbon molecule that is typically 1 or 2 nm in diameter. CNTs come in several varieties including single-walled and multiwalled nanotubes – most of these are used in applications that are embedded in industrial processes and hidden from view.

2.1.1. The "Products" of Nanotechnology

In nanotechnology, the "product" we are marketing is not always tangible; this makes the SIVAC term "solution" more appropriate. "Bionano" involves biological structures and processes as well as engineered materials and devices. These solutions range from nano-sized drugs, to therapies that use nanoparticles to deliver

Table 1
Applying the marketing mix to nanotechnology

Examples (5 Ps)	5 Ps	SIVAC	Examples (SIVAC)
(a) Labs-on-a-chip (b) Nanoshells for drug delivery (c) Nano-sized drugs for bioavailability	P1 Product	Solution	(a) $1 disposable disease test (b) A "silver bullet" cure for cancer (c) Longer lasting chemotherapy drugs with minimal side effects
Nanoparticles and nanoshells are being used to treat cancer; the first patients have been "cured" (experimentally)	P2 Promotion	Information	How cancer patients can enroll in clinical trials using nanotherapies; where clinical trials are underway
$1,000 personal genome (not currently available, but an industry goal)	P3 Price	Value	Affordable personal gene profile to identify susceptibility to disease (*especially inherited conditions*)
Medical research centers and research hospitals	P4 Placement	Access	Clinical trials, experimental research programs; foreign countries with these
Patients with diseases (*such as cancers*) that are most responsive to nano-enabled (a) diagnostics and (b) therapies	P5 People	Community	(a) High-risk patients who need to have their cancers detected as soon as possible (b) Patients who need nanotherapy to minimize chemotherapy side effects (these communities can share life-saving info online, worldwide)

drugs or destroy diseased cells, to digestible computer chips that can be embedded in a pill and monitored wirelessly.

While there are only a few examples of "nanodrugs" on the market today, dozens of nanodrugs and nano-enabled therapies are currently in clinical trials in the USA and overseas. Nanosizing a drug can make it more soluble (and thus more bioavailable), allow it to remain intact until it reaches the desired organ, or provide a "time-release" feature that keeps it working longer. One of the early successful uses of bionanotechnology was the ability to deliver water insoluble drugs to tumors. Early examples include nano-sized drugs coated with a liposome, most notably Doxil (a first generation drug approved in 1995) and Abraxane. Abraxane is a nano-sized formulation of the cancer drug paclitaxel that uses nanoparticles made from the human protein albumin to treat metastatic breast cancer. In 2009, Abraxane was approved for the treatment of pancreatic cancer. Abraxane minimizes toxic side effects and delivers a 50% higher dose in 30 min than standard paclitaxel, which originally needed to be administered over several hours (5).

In bionanotechnology, sometimes we are marketing a discovery – such as how to unwind a DNA molecule so that it can be maneuvered through a channel on a biochip. We may be marketing a phenomenon such as increased solubility that results when a drug is nano-sized, or a new capability that a researcher has demonstrated in a laboratory showing, for example, how wrapping a drug in a lipid makes it more bioavailable. The "product" may be a revelation that explains why a disease resists or responds to a particular treatment. While not always of immediate commercial value, such innovations do create value in the marketplace, through shared research, patents, product pipelines, and commercial applications.

Finding a comprehensive definition that encompasses imaging systems, nanomaterials, bionanostructures and processes, devices, and more is an important part of the marketing puzzle.

2.1.2. Defining the Nanotechnology "Product"

A core challenge in marketing nanotechnology (including "bionano") is getting the definitions right. The problem is that nanotechnology is not one discrete thing. It is a set of ideas, concepts, and metrics that describe not only what exists and can be observed and manipulated at the nanoscale, but also what is possible in the future as we increase our ability to engineer atoms and molecules.

Unfortunately, there is still no ubiquitous consensus definition for nanotechnology, per se. Nanotech means different things to different people and organizations. However, this is not as serious a consideration as we might imagine, since the definitions used are each tailored to the needs of the (mostly scientific) organizations that are doing the defining. The nuances are subtle, but important. Here are a few examples:

The National Nanotechnology Initiative (NNI) defines nanotechnology as: "the understanding and control of matter at dimensions between approximately 1 and 100 nm, where unique phenomena enable novel applications." The National Institutes of Health and the US patent office also use the NNI definition.

The National Science Foundation uses one of the longest and most comprehensive definitions for nanotechnology: "Research and technology development at the atomic, molecular or macromolecular levels, in the length scale of approximately 1–100 nm range, to provide a fundamental understanding of phenomena and materials at the nanoscale and to create and use structures, devices and systems that have novel properties and functions because of their small and/or intermediate size. The novel and differentiating properties and functions are developed at a critical length scale of matter typically under 100 nm. Nanotechnology research and development includes manipulation under control of the nanoscale structures and their integration into larger material components, systems, and architectures. Within these larger scale assemblies, the control and construction of their structures and components remains at the nanometer scale. In some particular cases, the critical length scale for novel properties and phenomena may be under 1 nm (e.g., manipulation of atoms at ~0.1 nm) or be larger than 100 nm (e.g., nanoparticle reinforced polymers have the unique feature at ~200–300 nm as a function of the local bridges or bonds between the nanoparticles and the polymer) (6)."

Most definitions of nanotechnology reference the nanoscale – from 1 to 100 nm. However, if we use this scale we exclude atoms, which are under 1 nm in diameter – and manipulation of atoms is an important function of nanotechnology. The European Commission addresses this by using the term "of the order of 100 nm" which includes some flexibility at both ends of the scale (7). Many bioscience applications – stem-cell scaffolds and clusters, use of nanoreagents in diagnostic tests, drug delivery systems, and convergent applications that combine semiconductors with diagnostics – may be greater than this scale; but most of these applications do involve nanoparticles, nanoscale structures, and processes.

Further, there is no consensus definition for bionanotechnology, which is often characterized as the intersection or convergence of biology and nanotechnology. This is also the juncture where novel applications for medicine are being developed. Nanotechnology structures and processes – atoms and molecules and their interactions, essentially – form the foundations of biology and chemistry. Many biological processes, such as self-replication of biological structures, are ingrained in Nature. Medical researchers are drawing inspiration from biology. Examples include self-replicating and self-healing systems, nanoscale structures that can be used as stem-cell scaffolds, and various types of biomarkers and drug delivery mechanisms. The term "biomimetics" describes the field of replicating natural structures and processes, sometimes called "biomimetic nanotechnology."

2.1.3. Mapping Bionanotechnology

The accompanying technology map (see Fig. 1) shows the complexity of bionanotechnology, with particular focus on human health care (1). This map is not intended to be comprehensive, but rather representative of the range of functions, applications, and innovations encompassed by bionanotechnology. This is a dynamic map, which is constantly changing. It shows the myriad domains encompassed by bionanotechnology, including some emerging sectors that are being created by the convergence of different streams of research such as theranostics (the convergence of diagnostics and therapeutics).

While these are technology sectors and not market segments, they offer a starting point for understanding the promising potential of bionanotechnology to address a wide variety of markets in the key areas of instruments and tools, diagnostics, and

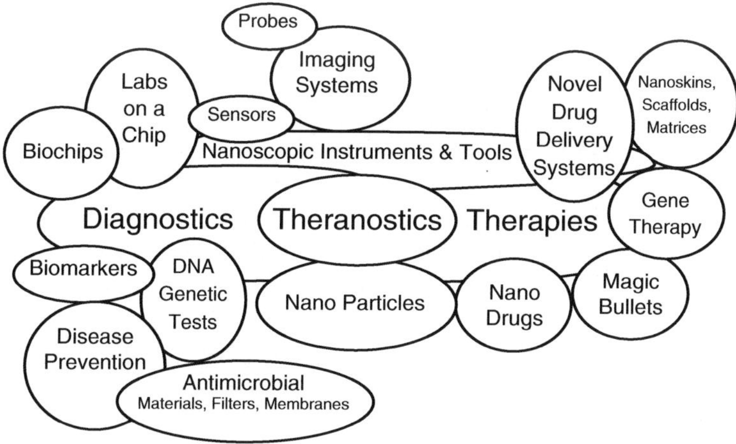

Fig. 1. A bionanotechnology technology map. Copyright © 2011 by Michael S. Tomczyk. All rights reserved.

therapies – including "killer applications" that are the bionano market incarnation of these emerging technologies (see Subheading 2.1.4).

Each of these circles represents not just commercial products, but also a set of very important and promising solutions across a broad spectrum of possibilities.

2.1.4. Killer Applications

In any market, the hottest products are those that address killer applications. A killer application (or "killer app") includes a use of technology that provides a core value, such as a computer operating system, which is where the term originated. In popular jargon, a killer app can also include an application that becomes so popular and widely diffused that it establishes itself as indispensable.

Many of the most important applications are under development – and in this sense, it can be argued that bionanotechnology is lagging the "nanotech revolution" in contrast to semiconductors, which are leading the charge. This is due in part to the long development and testing cycle for biomedical research in general, as well as the need to overcome technical issues that range from scaling up laboratory results to providing efficient methods for microencapsulation, genetic manipulation, etc.

In medicine, there are several holy grails that bionanotechnology is striving toward, the foremost being the quest to find a novel treatment or cure for cancer – one that destroys cancerous cells and tumors without damaging healthy cells and tissues. Many novel therapies that use nanoparticles are being researched and tested, including those which can be activated or triggered by electricity, heat, magnetism, and light. One experimental therapy involves using hollow gold nanoshells to deliver a drug to a specific disease site. Another approach involves concentrating gold nanoparticles in tumors and using near infrared radiation to heat the particles to kill the tumors (leaving the surrounding healthy tissue undamaged). Yet another involves tagging a drug with a semiconductor device similar to an RFID chip – made out of digestible materials – that tracks and transmits the location of the drug, or simply confirms that it was swallowed (an important consideration given that an alarming percentage of patients delay or stop taking their life-saving medications). In the area of diagnostics, researchers are investigating novel methods for detecting cancer cells before they form tumors.

Other "killer applications" that are being intensively pursued by research teams include: delivering disease-fighting molecules to specific cells, tissues, and organs that are difficult to treat with conventional therapies, delivering genetic materials using nanocarriers, providing controlled release of biological chemicals such as toxic chemotherapeutic agents, reducing the size (and cost) of reagents used in diagnostic tests, reducing the cost and efficacy of labs-on-a-chip, automating and cost-reducing personal genome sequencing, and using nanocapsules to target and deliver drugs to disease sites. Nanotechnology researchers are striving to revolutionize diagnostic testing by moving the field toward point-of-care testing. Scientists are engineering portable diagnostic systems that can be used in a doctor's office instead of having to be sent to a specialized laboratory. Portable diagnostic devices can also be used to quickly and inexpensively identify diseases in rural villages, and in remote regions of the world. These are just a few examples.

Entrepreneurs, companies, and venture capitalists may view these as "products" but they are much more than that. Most of these medical applications represent solutions that will lower diagnostic costs and greatly improve the delivery and effectiveness of health care. As these solutions begin to move from laboratories to commercial availability, there will be a need for a clear marketing message to position these innovations and show the relative benefits in relation to existing tried-and-true solutions such as traditional drugs, surgery, medical devices, and other modalities. It is not enough to simply announce that a new nanotherapy now exists. Practitioners, hospitals, and clinics will need to be shown that the "switching costs" justify adopting a nanotechnology solution, no matter how effective it might be. For example, why use a nano-sized version of a drug instead of the meta-sized version? One reason is that the nano-sized drug may be more soluble and bioavailable. Such distinctions are important and need to be included in product descriptions, to differentiate novel applications from existing alternatives.

2.2. P2/Price (Value): Trillions in Revenues, Thousands in Cost Savings

In marketing, "price" includes value throughout the supply chain. At the macro level, we can ask: what is the expected value of the nanotechnology market? The "value" of nanotechnology is significant, whether we calculate revenues represented by the entire nanotech sector, or cost/price savings for individual products.

Lux Research has forecasted that the US nanotechnology market will reach $2.5 trillion by 2015, after adjusting for the effects of the recession. However, as the Lux report observes, this estimate includes the value of finished goods such as expensive automobiles that incorporate nanomaterials, which is not a true indication of the value of the nanomaterials themselves. In 2000, the National Science Foundation predicted that the nanotechnology market will reach $1 trillion by 2015, with nanomedicine representing about $180 billion. Since then, other industry watchers including Cientifica and Lux Research (8, 9) have endorsed this prediction, with the caveat that these trillion dollar predictions include the value of nano-enabled products, which is much greater than the value of nanomaterials themselves. Lux Research has observed that the nanotechnology "market" is actually a nano-technology "value chain" that ranges from nanomaterials to nanointermediaries to nano-enabled products. By any measure, we are talking about a technology-driven, trillion dollar industry.

One of the benefits of nanotechnology is the ability to reduce the cost (and price) of expensive products and services. If we drill down to the price of individual nanotechnology solutions, we can easily gain a sense of the value proposition that lies ahead, even if only a few nanoinnovations achieve success.

For example, taking advantage of the larger surface area-to-volume ratio of nanoparticles enables the use of smaller numbers of particles in expensive diagnostic tests. This lowers the cost and can make the tests faster.

Many innovations (most notably nanotech "labs-on-a-chip") will lead to compact, disposable, "instant" diagnostic tests for a variety of diseases. The cost target for these tests is $1 per test. The combination of labs-on-a-chip with portable readers offers the potential to deliver quick, convenient tests to people in cities as well as in rural areas. The low cost can make diagnostics available in developing countries that otherwise could not afford this technology. Portable diagnostics may also make home tests more readily available, which could eventually change the paradigm for medical diagnostics in general.

Personal genomics is another emerging market that will benefit from lower costs enabled by nanotechnology. The first individual human genome sequenced in 2007 cost about $60 million. The second genome to be sequenced was that of DNA pioneer James Watson – 454 Life Sciences did Dr. Watson's profile in 2008, which took about 2 months and cost under $1 million. Currently, personal genome tests cost up to $200,000 depending on the level of detail that is required (10); with six billion base pairs of DNA in an individual genome, the processing power, time, and costs remain high. Several gene testing services offer targeted tests at lower costs, checking for specific traits and diseases, including testing for "point mutations." These profiles are less expensive – $400–$3,000 – however, they are not complete. They do offer a way to identify specific genes in at-risk populations where hereditary conditions may exist. The current objective, expressed by companies such as IBM and Invitrogen, is to lower the cost of a reasonably detailed personal genome analysis to $1,000. Several research teams are working to unwind and analyze DNA molecules – most notably the Philadelphia-based Princeton University spin-out BioNanomatrix, which has announced that it will beta-test its single-molecule DNA analyzer in 2010 (11). This line of research could result in targeted DNA tests at a price as low as $100–$200.

Of course, price and value are relative, when you consider that a bionanotech solution may be an early stage therapy or medical choice that can save a life – or thousands of lives. And in an era when it is imperative to reduce the cost of health care, nanotechnology innovations could play an important role in delivering these life-saving solutions at an affordable cost.

If bionanotechnology delivers extremely cost-effective solutions, it may be easier to market these products/technologies; on the other hand, it could be more difficult to market these solutions, if the profits are too low to recover research investments, or if the solutions (such as $1 laboratory tests) are commoditized.

There are also "switching costs" involved in transitioning from an existing technology to a bionanotechnology solution. It is conceivable to envision a time when there will be a dozen different ways to deliver drugs and treat diseases that use nano-technology and that are radically different than current proce-dures. Whether these are heated nanoparticles, or metal shells, or lipid envelopes, or boxes made from DNA, or magnetic materials such as iron oxide, or other innovations remains to be seen. The marketing question is: when these solutions do become available, will practitioners switch from surgery and drugs to nanotherapies, and if so, how soon?

2.3. P3/Promotion (Information): Nano-inside or Nano-free?

Two decades ago, medical solutions were provided by practitioners through hospitals and clinics, but patients were often adminis-tered solutions without fully understanding the exact purpose of a surgical technique, why a particular drug was prescribed, alter-native treatments, or salient details concerning their medical condition. Today in the era of ubiquitous communication, the same information available to medical researchers and practi-tioners is available online to the patients who are the "customers" of bionanotechnology. Unfortunately, some aspects of bionano-technology, including the safety of many nanomaterials, are still largely unknown, and these unknowns can color public percep-tion and acceptance.

Strategically, one of the decisions companies need to make is whether to promote the use of nanotechnology in their products, or not. This is a "nano-inside" or "nano-free" decision that could have a positive or negative impact. This marketing decision depends on many variables, from the value of a solution (e.g., a medical cure versus a cosmetic cream), the risk–reward tradeoffs, and perhaps most important, whether the public accepts and trusts nanotechnology in general or not.

2.3.1. Public Acceptance

So far, it appears that nanotechnology has enjoyed a free ride in terms of public acceptance. The term "nano" has not encountered the level of resistance that plagued genetically modified organisms (GMO) in the 1990s. This could be due to low public awareness, or to the fact that there has not been a public relations disaster or events to trigger a wave of public concern. For example, genetically modified foods were being introduced in Europe soon after the "mad cow disease" scare, which made the public especially sensitive about scientists and researchers tampering with their food supply.

Many nanoinnovations are marketed in the media as if they are already available, although most are years or decades away – these nanotech "breakthroughs" are trumpeted on the Internet and even in the most reputable science publications.

In November 2009, I did a quick search on Google for "nanotechnology breakthroughs" and generated 7.1 million hits. Does this represent marketing hype, or hope, or glimpses of realities to come?

Let us say that some of these seven million search items are redundant. Even at a rate of ten redundancies for each item, this equates to 700,000 breakthrough items online. So let us assume these search items span at least a decade, so divide the total by 10 and this yields 70,000 "breakthrough" items per year.

Given the amount of buzz around nanotechnology and anything that carries the name "nano" (including such products as cars (the Tata Nano) and media players (the iPod Nano) that may not actually include nanotechnology), promotional awareness of "nano" would seem to be fairly well established. Americans in particular should have a strong awareness of nanotechnology, given the amount of activity on the Internet – and the fact that there are more than 1,000 consumer products that use nanotechnology, according to the Project on Emerging Nanotechnologies at the Woodrow Wilson International Center for Scholars (http:// www.nanotechproject.org).

However, while we are supposedly being flooded with nano news, several sources report that the public at large, including college graduates, is largely *uninformed* when it comes to nanotechnology. According to one 2009 study, 68% of survey respondents reported having heard "just a little or nothing" about nanotechnology. Other studies have confirmed that public awareness of nanotechnology is low, including Andrew Maynard, chief science advisor for the Project on Emerging Technologies (12).

So, how do we reconcile this enormous number of nanotechnology breakthrough search items with survey results that show that a majority of Americans (and probably people in most countries) are largely unaware of nanotechnology? Is there a communication disconnect? Are only a minority of knowledgeable insiders reading these thousands of announcements, articles, news items, and blogs?

One interpretation of these results is that the plethora of nanotechnology news is only reaching an elite audience or that nanotechnology simply has not resonated with the public. This can also be explained by the fact that most nanotechnology products to-date have been hidden in industrial processes, such as coatings and catalysts, and are not designated on consumer products including foods that contain nanoparticles.

In nanobiotechnology, only a few dozen drugs have been approved, and the 100 or so in clinical trials are not widely publicized. Most companies do not advertise the fact that they have nanotechnologies in their products. You will not find a "nanotechnology" label on most sunscreens, for example. However, this is changing. In November 2009, the European Union announced a regulation requiring all cosmetics products to include the word "nano" in brackets after any ingredient that is less than 100 nm (13). This labeling requirement applies to all cosmetics marketed in the EU but could easily be extended to other product categories, from foods to medicines. This regulation could also provide a framework for new labeling standards in other countries including the USA.

Opponents of the new regulation have argued that this designation could be viewed by the public as a warning. The issue is, how will "nano-labeling" affect the marketing of consumer products, foods, and medicines? It depends on whether the pro-nano or antinano forces win the tug-of-war for public perception. Whether "truth in nano" laws and labels are beneficial, detrimental, or carry no effect remains to be seen.

Regardless of whether "nano" is perceived as a threat or a benefit, it is easy to envision a day very soon when nano-sized drugs, nanoscale therapies, and in vivo diagnostics or theranostics that involve nanomaterials will require doctors, nurses, and patients to sign consent forms. Tighter controls could be triggered by insurance companies providing risk protection, or by a specific "nano incident." A few nano incidents have already occurred, although they did not receive widespread attention. A major incident could, of course, damage public opinion and lead to laws and regulations that could restrict the use of nanotechnology, including biotechnology therapies.

2.3.2. The Implications of a Nanoincident

There have been a few notable examples of PR brushfires involving "nano" although they did not turn into firestorms. In 2006, an aerosol product called Magic Nano, a household glass and ceramic tile sealant sold in an aerosol can, was blamed for over 90 customer reports of respiratory distress in Germany.

The product was promptly recalled. However, it was not confirmed that the aerosol particles were in fact nano-sized. A side issue that this case raised is that a product such as a solution in an aerosol can may not be nano in its present form, but it may become nano when you push the button on the top of the can.

Another example of negative publicity involves the use of silver nanoparticles used in Samsung washing machines, refrigerators, and other appliances under the trademarked brand, "Silver Nano™" also marketed as "SilverCare™" and as the "Silver Nano Health System™." Samsung claims that this silver nanotechnology sterilized against over 650 types of bacteria by releasing up to 400 billion silver ions into fabrics to create a sterilizing protection up to 30 days after washing (14). It is feared that these particles will find their way into waterways and kill fish and other aquatic life – which is a real concern given that ionic silver can be toxic to aquatic life. While metallic silver is not water soluble, if you remove one electron from an atom of metallic silver you create silver ions which are not only water soluble, but toxic at certain levels. Samsung's Silver technology includes a patented release mechanism for silver ions that claims to kill 99% of bacteria in cold water. However, the wastewater from washing machines eventually enters streams and groundwater, which is potentially troubling since ionic silver can be toxic to friendly bacteria as well as larger aquatic species. In 2007 the US Environmental Protection Agency (EPA) determined that the silver ions in Samsung's washers were subject to the Federal Insecticide, Fungicide, and Rodenticide Act (FIFRA) and needed to be regulated as a pesticide (15).

Time will tell if nanotechnology dodges these and other negative PR bullets. A very real concern is that one significant "toxic nano" event could taint the entire field of nanotechnology. Labeling the entire area of science and technology across industries with the descriptive term "nanotechnology" runs the risk that if one incident occurs in any one sector, this could be extrapolated to the entire field of "nano."

The most significant potential hazard – and potential image problem – associated with nanotechnology has been the comparison of carbon nanotubes to asbestos fibers. Carbon nanotubes have a thread-like structure that is similar to asbestos, and studies have demonstrated an "asbestos-like pathogenicity" in mice (16). The question is that is this a problem for humans, and if so, is it limited to nanotubes of a certain length – for example, longer nanotubes seem to be more hazardous than shorter tubes. The jury is still out on this issue, yet it points to the need to address the handling and disposal of nanotubes and the specifications for CNTs used in products. On the other hand, it can be argued that nanomaterials have been used for decades in many industries. In bionanotechnology, clinical trials are in place to determine safety, and at the other end of the life cycle, there are procedures for handling and processing "biohazard" materials so that nanotech waste residues should be able to be properly disposed of.

There are some other important marketing issues evoked by these examples. First, labeling the entire field of nanotechnology under one umbrella term runs the risk that one negative incident could taint the entire field. I believe that the term "nanoscale technology" is more appropriate and that "nanotechnology" or "nanotech" – while effective as a buzzword and currently friendly in the marketplace – groups the entire sector across all industries under one label. If one fatal incident occurs that is associated with a certain type of nanoparticle or nanomaterial, it is possible that headlines could create a backlash against nanotechnology in general.

Another aspect that companies in particular need to be aware of is the possibility that some of their products such as paints and coatings may contain nanoparticles and if so, they could wake up one morning to discover that their existing products – which may be a 100-years-old – are suddenly reclassified as "nano" and forced to meet stringent and costly requirements for packaging, handling, labeling, and insurance.

How companies and institutions deal with these image issues will determine whether nanotechnology, including bionanotechnology, will be accepted as a customer-friendly technology like microelectronics, or as a "customer-beware" technology such as genetically modified organisms (GMO).

Informing the public about the benefits as well as the potential (including unknown) risks of nanotechnology – while guarding against misleading the public and encouraging fear-mongering – is a major marketing challenge that scientists as well as marketers need to address proactively.

2.4. P4/Place (Access): Diffusing Nanotechnology

The "place" of nanotechnology is not a traditional retail outlet such as a pharmacy, or even the Internet. Most nanotechnology products such as carbon nanotubes are only available from a few reliable sources worldwide. Most bionanotech solutions are being researched in multimillion dollar laboratories in large corporations, research hospitals, and government-sponsored programs. Aside from nano-sized drugs, the most promising emerging innovations, the really radical solutions, are still being developed in research laboratories or are only available in early stage clinical trials. "Placing" these solutions in the health-care market requires the solution to run a gauntlet of animal tests and clinical trials before it is commercialized, and even then a new therapy needs to compete with existing accepted therapies that range from surgical procedures to drugs.

One marketing question we can ask is: how will bionanotechnology solutions be diffused, once there are more commercial applications available? Manufacturers need expensive imaging systems – or nanoimaging services – to produce nanoparticles, fabricate nanoscale biochips and other devices, and perform quality control on products that use nanotechnology. In medicine, nano-imaging is needed to diagnose the impact of nanoscale therapies and to monitor the amount of nanoparticles that linger in the body. The cost of these tools can be a limiting factor to the development and diffusion of nanotechnology products and solutions. While the public may not worry about this issue, it is possible that some solutions may take longer to reach the market, or to become diffused, because companies developing nanosolutions may not be able to afford the tools. The high cost of imaging systems also may limit the ability of schools to train the next generation of nanotechnology scientists.

2.4.1. The Role of Imaging in Bionano Diffusion

Imaging systems play a critical role in the adoption and availability of bionanotechnology solutions. Scanning tunneling microscopes and other "nanoscopes" cost as much as $50,000–$500,000 depending on the accessories such as probes and sensors that are included. The resource Web site, Nanotechnology Now, lists about 200 companies under "Nanotechnology Tool Makers and Service Providers (17)." If there is one parameter that could expedite the diffusion of nanotechnology research worldwide, it is the availability of lower cost imaging systems.

Until the tools of nanotechnology become more affordable and accessible, access to "bionano" will be limited to research and educational laboratories. This does not mean that we should despair. Many analogous examples exist of innovations that required expensive infrastructures to create the market, including MRI imaging systems, medical implants such as pacemakers and stents, and high-definition television and cell phone technologies. We can learn from these examples as bionano therapies, diagnostic/theranostic tests, and medical devices become available.

2.4.2. The Impact of Health and Safety Regulations

Wherever a nanotechnology solution is developed, tested, manufactured, delivered, and/or used, there are important safety and environmental considerations that need to be considered. Safety issues in particular can impact not just market acceptance, but commercial viability and access. Regulations by government agencies have the power to impact public access in many ways, from policies that restrict government funding for research (e.g., stem cells) to regulations that require additional tests and studies as part of a clinical trial.

While access to nanotechnology in general is currently not regulated in most markets, globally – in most countries, nanotechnology falls under existing health and safety regulations. In the USA, the FDA – which does not regulate "technology" per se – has been studying how to regulate the "combination products" represented by nanobiotechnology (drugs, medical devices, and biological products) (18).

The European Union enacted the REACH program (Registration, Evaluation, Authorisation, and Registration of Chemicals) in 2007. REACH requires companies to manage the risks associated with chemicals (including nanomaterials) and to provide safety information on process chemicals, mixtures of chemicals, chemicals used to prepare or manufacture products, and even chemicals that might be released from products. Manufacturers and importers are required to gather information on the properties of their chemicals, so they may be handled safely, and to register the information in a central database run by the newly formed European Chemicals Agency (19).

Most industrial organizations are still wrestling with nanotech standards. For example, while nanoscale features have been used in semiconductors and electronic devices for more than a decade, organizations such as IEEE, ISO, and IEC are still a few years away from establishing standards for the use of nanomaterials in electronics and other sectors.

We have not seen a heated public debate over the safety of bionanotechnology because there are so few solutions available at this early stage in the development of this emerging market. Bionano is lagging the consumer market in terms of applications, which provides some breathing room – however, this "market lag" suggests a scenario where we could see bionano being subjected to the policies and regulations developed to address "nonbio" applications of nanotechnology.

For example, if carbon nanotubes or "silver nanoparticles" are shown to be hazardous to human health or to the environment, it is reasonable to expect that bionano applications in hospitals, pharmacies, and diagnostic laboratories will face more stringent controls than those that currently exist. On balance, reasonable standards of safety and efficacy should prevail, and these are precepts that already exist.

For the foreseeable future, it is conceivable that we will not be able to advertise or describe many applications involving nanotechnologies as "safe" or "healthy" or "environmentally friendly" with any degree of certainty until more research is available. In the meantime, the ability of marketing professionals to guarantee that there will only be positive effects and no negatives or liabilities will remain in limbo until generally accepted standards, rules, and policies are put into place. These policies will range from safe handling during processing of nanomaterials to administration and safety in the delivery of bionano therapies and to disposal of devices and materials that contain nanomaterials. In this area, the bionano community will benefit from existing procedures that already exist for the handling of biohazardous materials.

2.5. P5/People (Community): Creating a Global Nanomarket

As a "marketplace," the nanobiotechnology market acts more like a community – actually, a network of several communities – than a traditional consumer market. On the technology side are scientists and research teams who are networked through scientific conferences, webinars, online publications – and the companies and institutions that provide commercial solutions. On the "customer" side are patients and practitioners who can receive information about nanotechnology discoveries, as well as how they relate to orphan diseases, medical therapies, charitable services, and experimental research programs. Policymakers, legislators, and regulators determine how solutions are provided and regulated.

Influencing these communities are networks of media experts, pundits, bloggers, and commentators. Nanoscale drugs and therapies already on the market as well as those in clinical trials and research laboratories are actively discussed and tracked by advocacy groups and patient networks.

The key to reaching this vast networked market with bio-nanotechnology solutions will be to demonstrate how a novel bionano therapy is more efficacious and cost-effective than existing treatments. This is especially true for therapies that are not drugs or surgery but something entirely different that does not fall neatly into either category.

The "market demand" for bionanotech solutions is strong. Worldwide deaths from cancer alone are expected to increase from 7.4 million in 1974 to 12 million people by 2030, according to the World Health Organization. Bionanotechnologists are working on a wide variety of novel therapies for cancer. In most countries including the USA, people are living longer. They need therapies that can provide better quality of life in their 70s, 80s, and 90s. People in remote villages need nano-enabled portable devices and test kits to help detect and treat diseases.

Innovations enabled by bionanotechnology offer the promise to shrink the size and cost of biochips, to identify, track, and treat disease at the cellular level, and to target diseases in organs that were previously inaccessible. Nanotechnology scaffolds are already enabling stem cells to grow into blood vessels and bladders; other solutions on the horizon that hold promise include inexpensive DNA tests, artificial synapses, and much more.

As they become available, some of these solutions will be easier to communicate than others. A surgeon may one day tell a patient: "We can use this drug to kill your tumors but it may kill some healthy tissue as well, you will lose your hair, and you could die from the chemicals…OR…we can inject nanoparticles into your system that will migrate to the tumor cells, then we'll use radiation or magnetic resonance to heat those particles so they kill your tumor with less destruction of healthy cells…and you can go home 2 h after the procedure." Of course, communicating the implications of a DNA test result, or a genetic defect revealed by a nano-enabled labs-on-a-chip, will be more complicated and involve more sophisticated messaging.

Perhaps the best people to explain bionanotechnology to the beneficiaries of these innovations are the patients themselves. In an era when health care reform is a vigorous public issue, large communities of interest are already paying increased attention to the promise of bionanotechnology. Patients and practitioners maintain vigorous communities of interest that share life-saving, life-enhancing medical information through blogs and forums. As bionanotechnology solutions move from laboratories to hospitals and clinics, "viral marketing" (an ironic term) will communicate which of these solutions are most efficacious, affordable, and available.

3. Summary

At this early stage in the evolution of nanobiotechnology, scientists need to remain aware of the impact that the marketing mix can have on the funding, development, adoption, perception, and ultimate success of the solutions they are developing. In many areas, what happens outside of the laboratory will play a vital role in what happens to bionanotechnology solutions in the market-place. Many of the marketing considerations mentioned in this discussion will impact the translation of bionanotechnology solutions from laboratories to hospitals and clinics.

References

1. Tomczyk, M. S. (2011) *Nanoinnovation: What Every Manager Needs to Know.* Wiley-VCH, Weinheim, Germany.
2. Boutin, C. (October 9, 2003) Purdue team solves structure of West Nile virus. *Purdue News.*
3. Intel Press Release (September 22, 2009) Moore's Law Marches on at Intel.
4. Dev, C. S. and Schulz, D. E. (2005) In the mix: a customer-focused approach can bring the current marketing mix into the 21st century. *Mark. Manage.* **14**, 18–24.
5. McGee, P. (2006) Delivering on nano's promise. *Drug Discov. Dev.* **9**, 12–18.
6. National Science Foundation (December 2009) Nanotechnology definition NSF website, http://www.nsf.gov/crssprgm/nano/reports/omb_nifty50.jsp
7. Williams, D. (May/June 2008) Defining Nanotechnology. *Medical Device Technology.*
8. Cientifica (April 2007) Halfway to the Trillion Dollar Market? A Critical Review of the Diffusion of Nanotechnologies.
9. Lux Research (2009) Nanomaterials State of the Market Q1 2009: Cleantech's Dollar Investments, Penny Returns.
10. McGowan, K. (June 22, 2009) Your Genome, Now Available for a (Relative) Discount. *Discover.*
11. Karow, J. (October 20, 2009) BioNanomatrix Plans Beta-Launch of Single-Molecule DNA Analyzer for Q2 of 2010. *Genomeweb.*
12. Hart Research Associates (September 22, 2009) Nanotechnology, Synthetic Biology, & Public Opinion. Commissioned by the Project on Emerging Technologies, Woodrow Wilson International Center for Scholars.
13. Stafford, N. (2010) New nano rule for EU cosmetics. *Chem. World* **7**, 6.
14. Samsung (March 29, 2005) Samsung Silver Nano Health System™ Gives Free Play to Its 'Silver' Magic. *Samsung Press Release.*
15. Bolles, D. (December 3, 2007) The Irony of Nanotechnology's Promise. *Water & Wastewater News.*

16. Poland, C. A., Duffin, R., Kinloch, I., Maynard, A., Wallace, W. A. H., Seaton, A., et al. (2008) Carbon nanotubes introduced into the abdominal cavity of mice show asbestos-like pathogenicity in a pilot study. *Nat. Nanotechnol.* **3**, 423–428.

17. Nanotechnology Now (June 27, 2009) *Nanotechnology Tool Makers & Service Providers.* http://www.nanotech-now.com/tool-makers.htm

18. US Food & Drug Administration (November 7, 2009) *FDA Regulation of Nanotechnology Products: FDA and NanoPharmaceuticals.* http://www.nanopharmaceuticals.org/FDA.html

19. Bergeson, L. L. (May 23, 2007) REACH and Nano. *Nanotechnology Law Blog.*

Chapter 26

Managing the "Known Unknowns": Theranostic Cancer Nanomedicine and Informed Consent

Fabrice Jotterand and Archie A. Alexander

Abstract

The potential clinical applications and the economic benefits of theranostics represent a tremendous incentive to push research and development forward. However, we should also carefully examine the possible downsides. In this chapter, we address the issue of how theranostics might challenge our current concept of informed consent, especially the disclosure of information concerning diagnosis and treatment options to human subjects. We argue that our lack of data concerning long-term effects and risks of nanoparticles on human health and the environment could undermine the process when it comes to weighing the risks against the benefits. Our lack of an agreed upon framework for risk management in nanomedicine may require us to adopt an "upstream" approach that emphasizes communication and transparency among all relevant stakeholders to help them make informed choices that enable safety or progress.

Key words: Informed consent, Nanomedicine, Cancer, Theranostics, Ethics, Policy, Risk management

1. Introduction

Nanotechnology enables its users to control matter and exploit novel phenomena and properties at the nanoscale. Advances in nanotechnology provide an opportunity to develop innovative interventions in various areas of medicine such as cancer treatment. The heterogeneous nature of cancer tumors and the need to target specific cells constitute major challenges nanomedicine could overcome contrary to conventional drug therapies. Indeed, some nanoscale particles or devices smaller than 50 nm can penetrate cells while nanodevices smaller than 20 nm have the potential to move through the blood stream. These type of capabilities may allow for better selectivity of drugs toward cancer cells, which reduces toxicity, increases efficacy of chemotherapy, and leads to better dosing of medications (1). Theranostics takes advantage of

Sarah J. Hurst (ed.), *Biomedical Nanotechnology: Methods and Protocols*, Methods in Molecular Biology, vol. 726,
DOI 10.1007/978-1-61779-052-2_26, © Springer Science+Business Media, LLC 2011

413

novel nanotechnology platforms by incorporating therapy, imaging diagnostics, and cell targeting into one system with multifunctional capabilities. Theranostic nanomedicine exploits the multifunctional capabilities of nanoscale devices and the molecular knowledge of the human body to diagnose, treat, and possibly, prevent diseases (2).

Current research and development in theranostic cancer nanomedicine focus on combining three major areas of nanomedicine to engineer innovative multifunctional devices that are capable of (1) targeted drug delivery, (2) diagnostic imaging, and (3) genetic testing. This combination approach may offer clinicians an opportunity to diagnosis and assess disease, deliver a targeted therapy, and monitor therapeutic responses simultaneously (3, 4). Some believe that theranostics will become the future standard of care (5), because it enables clinicians to tailor a specific treatment regimen to the biomarkers expressed by a single patient. This may prove beneficial in cancer treatment, where high levels of variations in molecular markers or extreme heterogeneity of the disease occur (6).

Researchers at UT Southwestern Medical Center and the University of Texas at Dallas are working on the development of multifunctional nanomedicine platforms for cancer diagnosis and therapy. They utilize a "bottom-up" strategy to create highly integrated architectures with multiple functions for tumor targeting, imaging ultra-sensitivity, and controlled drug delivery. New molecular targets of cancer are exploited to achieve greater imaging specificity of molecular probes and higher therapeutic indices of drugs. Other researchers focus on theranostic agents that merge both diagnostic and therapeutic capabilities in one platform. There are currently few viable approaches to theranostics, and of those available, most combine existing fluorescence or MRI imaging approaches with traditional chemotherapeutic drugs in the same molecular complex. For therapy, single wall nanotubes (SWNTs) efficiently convert absorbed near infrared (NIR) light into heat, and the thermal ablation of model tumor cells has been demonstrated. Not only is the therapeutic potential for nanomedicines promising, but also analysts predict a bright economic future for this market.

Market analysts believe that theranostic applications are worthwhile because they see nanoscale structures as a transformative technology that will revolutionize our health care system (7). Theranostic companies (imaging source) are partnering with pharmaceutical companies to merge their diagnostic and therapeutic products into combinations (dual-use products) that may improve our clinical outcomes. The upside for these partnerships may be greater than the more traditional standalone pharmaceutical firms, because their dual-use products may lead to reductions in our health care costs. Some financial analysts predict that the global market for theranostics may be in the billions of dollars by 2014 (8), up to $40 billion (9). In fact, several major pharmaceutical companies are seeking mergers with diagnostic imaging companies to take advantage of these opportunities within the US market.

While the potential clinical applications and the economic benefits of theranostics represent a tremendous incentive to push research and development forward, we should also carefully examine possible downsides. In this chapter, we address the issue of how theranostics might challenge our current concept of informed consent, especially the disclosure of information concerning diagnosis and treatment options to human subjects. We argue that our lack of data concerning long-term effects and risks of nanoparticles on human health and the environment could undermine the process when it comes to weighing the risks against the benefits. Our lack of an agreed upon nanospecific framework for risk management in nanomedicine may require us to adopt an "upstream" approach that emphasizes communication and transparency among all relevant stakeholders to help them make informed choices that enable safety or progress or possibly undermine them, depending on how the public and private sector handle its content and delivery.

2. Implications of Theranostics in the Clinical Setting

The development of new therapeutic applications of nanomaterials in cancer research may offer cancer victims new hope but they may also create ethical, legal, and policy challenges for its stakeholders and their institutions. One major reason for concern lies with the ability of "multicomponent nanomedicine with modular designs" to personalize therapy. Personalization of therapy may lead to "adaptive targeting" where a course of therapy is altered in real-time as a response to the adaptive resistance of cancers (3). Hence, the possibility of "adaptive targeting" raises concern for the validity of the informed consent process that must account for a real-time change of the therapeutic options (in this case dosage) without a formal consent procedure to account for each new change or alteration. Not only will "adaptive targeting" raise concern for the validity of informed consent, but it also may reshape our concepts of the patient–physician relationship (10). The use of nanomaterials in combination with diagnostic/therapeutic application may produce an "automation of medical expertise," where treatment may change without any exchange of information between patients and their physicians, unless, of course, there is a feedback mechanism for real-time assessment (11). But even then, it is unclear whether patients will have sufficient time to determine the best course of action especially if therapy involves a higher level of toxicity requiring them to reassess their risks and benefits. In other words, the distinction between the process of communicating a diagnosis and the individual's assessment of treatment options may become blurred. Patients may be forced to make therapeutic decisions without knowing all of relevant information necessary to make an informed decision. In an ideal world we may believe that researchers and clinicians foster the best interests of human subjects and patients alike, and thus, a change in therapy options would be immaterial. However, because of the complexity of our medico-legal system, the consenting process has transitioned from an ethical principle to a legal doctrine with legal consequences. Ultimately, theranostics not only poses new ethical challenges to the process of informed consent, but likewise it also creates challenges to existing legal and institutional policies.

Any ethical, legal, or policy challenges raised by the transformative nature of theranostic nanomedicine, especially those impacting the informed consent process, may be further complicated by our lack of risk data and analysis (12–15). Even so, forecasters predict more, not less, research and development with nanoscale materials and structures, especially those utilized in nanomedicines and health care (7–9). To protect these developing markets and reap the health benefits related to nanomedicines, all stakeholders must assume responsibility for insuring the benefits of these emerging technologies outweigh their risks or hazards. Given that the general public remains mostly ignorant of nanotechnology, clinical research, and development of nanomedicines may continue unabated until our first health disaster or nanomedicine scare goes public (16). Once this happens, stakeholders may not be able to tackle risk issues after the fact or control public perceptions to preserve their use or markets.

Some commentators warn that the end result for all nanotechnologies will mirror those Europe experienced for genetically modified organisms (GMOs) or nuclear power and other emerging technologies. A mistrustful public may simply shun all nanotechnologies to stifle further research and development. Reality is we had our first nanoscare and -recall in 1999 when the German manufacturer (Kleinmann GmbH; Sonnenbuehl, Germany) recalled its aerosol dirt-repellant spray called Magic-Nano after it reportedly sickened nearly 100 of its users (17). Fortunately for Magic-Nano and its company, authorities discovered that the "nano" was not responsible for the adverse events, because it lost its "nanocharacteristics" during manufacturing. Nevertheless, authorities simply had no choice but to hunker down and do damage control while nongovernmental organizations (NGOs) called for more toxicology studies and regulations. Although the company survived its scare, many NGOs and interested stakeholders remain worried about their risks and some express concern for further research and dissemination of nanotechnology into the market (17). Perhaps, the better approach is to focus on achieving more transparency for informed decision-making through the informed consent (13–16).

3. Informed Consent

During the last few decades, advances in biomedical research have provided better diagnostic tools and increased the number of therapeutic options. Theranostics harvest the benefits in both of these areas. The hope is to increase objectivity in diagnosis to create algorithms for a well-defined understanding of the condition of the patient and optimization of treatment. However, treatment cannot rest on objective information alone. It requires subjective data gained from the perspective of the patient and the subjective judgments of the physician (18). Indeed, the subjective dimension stems from the necessity to let the patient decide a particular treatment option and assess what is morally acceptable for that individual in relation to potential risks and benefits. In other words, even though technology will provide tremendous accuracy in diagnosis and patient-specific treatments (e.g., pharmacogenomics and proteomics), it is highly unlikely that the biomedical sciences will ever eliminate the role of subjective data in patient care. Because medicine deals with human beings who are masters of their own bodies and destinies, informed consent constitutes an ethical, and by extension, a legal and policy imperative.

The doctrine of informed consent fosters the autonomy of individuals and protects them from abuse. It allows individuals to decide what should be done with their bodies, where no one should be compelled to act against their will (19). It also provides guidance and boundaries within the patient–physician relationship to encourage communication and trust. The process of informed consent facilitates the meaningful exchange of information between the physician or researcher and the patient seeking treatment or subject participating in research, respectively (20). These discussions should be more than mere recitals of checklists of information whether they pertain to a choice of treatment or a decision to participate in a research protocol (21). Both processes have oral and written components, but their scope or information content may vary, depending on whether the consent process focuses on treatment or clinical research. If it is treatment, then discussions generally include the diagnosis, nature and purpose of treatment, material risks and outcomes commonly known or expected for the particular patient, disclosure of all feasible alternatives and risks, prognosis, if recommendations declined, prognosis, if recommendations accepted, and disclosure of conflicts of interest. Conversely, the clinical research consent process may involve identifying the study as research and its purpose, describing the reasonably foreseeable risks and benefits, disclosing appropriate alternatives, handling of confidentiality and research-related injuries, supplying contact information, and obtaining a statement of voluntariness (20). And unlike the process for treatment, this may require an entirely new written consent form, which must be reviewed and approved by an Institutional Review Board (IRB) before it ever reaches a participant. Success of the process in both treatment and clinical research requires participants to address all components and meaningfully exchange information to respect ethical norms, or it fails, subjecting parties to potential legal liability. Not only is the processes of informed consent essential for good care, but also its successful completion keeps all parties out of our judicial system and avoids wasting time and judicial resources litigating matters that could have been avoided by respecting ethical norms.

Based on a previous analysis (22), we will reiterate our acceptance of the seven element model of informed consent developed by Beauchamp and Childress (23) for the analysis of nanotechnology. Some scholars have promoted a three-element model (decision-making capacity, voluntariness, and informed understanding) (24) while usually the following five elements are recognized: (1) competence, (2) disclosure, (3) understanding, (4) voluntariness, and (5) consent (23). However, the complexity of clinical trials warrants a more refined approach as proposed by Beauchamp and Childress who recognize seven distinctive elements of informed consent, organized in three main groups:

1. *Threshold elements (preconditions)*: (a) Competence (to understand and decide) – the ability to make a rational decision; (b) voluntariness (in deciding) – absence of coercion.

2. *Information elements*: (a) Disclosure (of material information) – the patient/research subject must be fully informed; (b) recommendation (of a plan) – the patient/research subject must be provided specific recommendations concerning his/her medical condition and/or the research procedure; (c) understanding (of 2a and 2b) – the patient/research subject must be able to process information and understand it.

3. *Consent elements*: (a) Decision (in favor of a plan) – the patient/ research subject's ability to make a choice; (b) authorization (of the chosen plan) – the patient/research subject must consent to the treatment and/or experimental procedure.

The use of theranostics nanomedicine for cancer therapy calls into question the validity of informed consent from an ethical standpoint. As already stated, individuals might need to decide about, and consent to, therapeutic options before a diagnosis is provided. In order to avoid compromising the ethical standards of informed consent, it is paramount for us to determine whether we should rethink the consent process when the potential risks or side effects are unknown. Specifically, there are concerns about the disclosure of information (2a) and recommendation (2b) with regards to the condition of the individual. How we adequately inform an individual about his or her medical condition and the potential risks of a treatment option is even more challenging in cancer therapy because adaptive resistance of cancer cells leads to "adaptive targeting" and real-time alterations of therapy (3). How individuals will make decisions about, and consent to, potential treatment options will depend on the type of information provided to them and a set of personal considerations when confronted with two or more potential treatment outcomes (risk–benefit analysis). What does or should the individual understand about the risks associated with theranostics? What mechanism(s) should a clinician or clinical researcher choose to frame the discussion of scope of informed consent? Should there be a centralizing theme if one component is therapeutic while another component is nontherapeutic? To address these questions adequately requires a better understanding of the potential risks and effects associated with the use of nanoparticles and nanodevices in therapy and diagnosis. Unfortunately, at this point, a definitive answer is not possible due to our lack of data concerning long-term effects.

4. Theranostics and the "Known Unknowns"

No one disputes our need for more information on the risks and benefits of nanoscale materials including those utilized in theranostics. Unfortunately, we are (1) dealing with materials that have novel properties which we may not fully comprehend, (2) lacking the necessary and sufficient toxicology studies we need to define its hazards, exposures, and life cycles, (3) moving products from our laboratories to our markets at a pace that exceeds the ability of our existing ethical norms and regulatory frameworks to adapt and respond, (4) lacking long-term studies concerning the majority of the nanomaterials we produce, and (5) hoping our benefits from nanomaterials, especially those in medicine and cancer therapy, will outweigh our risks (12–15). The resulting risk issues for the nanoscale materials in nanomedicines may qualify as what former Secretary of Defense Donald Rumsfeld calls the "known unknowns" where the existence of possible outcomes is recognized, but their actual occurrence is uncertain or unknown (13).

Not only will our failure to address the "known unknowns" of its risks impact current research and development of new products sooner rather than later, but also it may dissuade future participants from participating in clinical trials. Potential participants may choose not to participate because clinical researchers may not have the information necessary to communicate and explain potential risks. Unfortunately, the negative experiences with a commercial product such as Magic-Nano or those at Hospital of Pennsylvania ("HUP") following the death of one of its gene-therapy participants may foretell of things to come if we do not begin assessing and managing risks now (25).

In 1999, investigators at HUP were conducting a gene-therapy trial when one of their participants experienced a tragic death. After a thorough investigation by the FDA, HUP ceased all clinical trials with similar agents and followed by some rather troubling revelations about their clinical trials process with this agent (25). Officials at HUP admitted under intense pressure from the local press that the parents of the deceased did not know about the risks of the vector in research animals. Worse still, the parents did not know that some participants experienced flu-like symptoms and their son did not actually qualify for the protocol. Once this information came to light, the parents claimed they would have never agreed to enroll their son and participate. The bitter lessons learned by everyone were communication and full disclosure of the risks or hazards are essential. But in the case of all nanoscale materials, including those used in nanomedicines, these lessons may be lost because much of the information physicians, patients, and participants will need for a meaningful exchange during informed consent remains unknown (15).

Unfortunately, knowing the risks and balancing them against their potential benefits may be easier said than actually done. Clearly, most stakeholders from the basic scientists to policymakers recognize that we lack information on the health, safety, environmental, ethical, legal, and social issues associated with nanomaterials (26–30). The problem is our "known unknowns" related to risk identification, assessment, management, and communication, or risk governance, are daunting at best (13, 31, 32). Notwithstanding our "known unknowns" about their risks, we also have local, national, and international regulatory gaps. Almost every nation is researching and developing nanotechnologies without drafting nanospecific laws or regulations because they believe their existing regulatory schemes are adequate (26–29, 31, 32). In fact, most nations, especially the developing ones, are shying away from agreeing to any formalized, transnational approach to the regulation of nanomaterials because they fear falling further behind more powerful nations in the race to corner the market on nanoproducts including those in nanomedicine. Simply put, developing nations do not want to miss out on the economic benefits of this technology. To regulate or not to regulate may ultimately depend on what we learn about our "known unknowns" on risk and how we assess, manage, or govern them.

5. Defining Risk, Assessment, and Management

The crucial question then is "what is risk?" Briefly stated risk represents the likelihood of an adverse event occurring that produces consequences (31, 33). To adequately manage risk and reduce adverse events, individuals must identify, assess, manage, and communicate risk information. In short, individuals must know the "known unknowns" related to risk, and the existence of a framework helps to manage them. In a review of multiple risk management frameworks over the past decade, Jardine and her co-workers found that agencies responsible for dealing with risk may do so using a variety of approaches, but all of them generally require: (1) *risk assessment* (describing and estimating adverse outcomes using hazard identification, dose–response assessments, exposure assessments, and risk characterization), (2) *risk management* (using a process that identifies, evaluates, selects, and implements a set of actions based on science and designed to cost-effectively reduce risk while recognizing the social, cultural, ethical, political, and legal contexts), and (3) *risk communication* (facilitating interactive exchanges of information between stakeholders and essential to effective management) (33). They also recognized seven key elements or principles essential to comprehensive risk management programs dealing with human health, ecological, and occupational risks (33). These seven key elements include (1) problem formulation, (2) stakeholders ("interested and affected parties"), (3) risk communication among stakeholders, (4) quantitative risk assessment components, (5) iteration and evaluation, (6) informed decision-making among stakeholders, and (7) flexibility. Of these elements, the problem formulation stage may be the most critical element because stakeholders who fail to identify the right problem usually spend their time, manpower, and capital solving it to arrive at wrong or unworkable solutions. More importantly, they found a failure to incorporate any of these elements will likely generate conflicts and problems among stakeholders that make risk management more, not less problematic (33). Jardine and her co-workers also noted these seven principles should be supported by ten ethical principles in risk management decision-making: (1) beneficence, (2) fairness, (3) equity, (4) utility (adequate risk management), (5) honesty, (6) caution when uncertain ("better safe than sorry"), (7) respect of autonomy, (8) repeatability, (9) realization that risk cannot be eliminated ("life is not risk free"), and (10) Golden Rule. Several of the ethical principles supporting sound risk decision-making may also serve as the underpinnings for informed consent (15, 25).

Although elements, principles, and practices may prove useful with more traditional problems, nanotechnologies may raise new challenges with their "known unknowns" (13, 14, 31, 34, 35). Given some commentators already question our ability to identify the risks or hazards associated with all nanomaterials (13–16, 34), then it also seems stakeholders will likely target the wrong problem without a sound problem formulation stage. Our need for discussions about risk management may be rising as reports of potential toxicities for nanomaterials continually surface in the academic literature and lay press (25, 31, 32, 35). Depending on the particle and the animal model, investigators are identifying particles that may affect our lungs (36), renal cells (37), and other organ systems (10). Although most of these studies and reports focus on animal models and cell cultures, a recent report from China raises the specter for human involvement (38). Apparently, a group of Chinese workers developed symptoms and pulmonary findings similar to other pneumoconiosis such as asbestosis following their exposure to nanoparticles over several months (38). Unlike previous reports in animal models, the materials incriminated in this report here were not carbon nanotubes (CNTs), but nanoparticles. Because it represents a single report and likely has other confounds, it may generate more questions than it answers. It does, however, emphasize our need to know more about our "known unknowns."

Some commentators believe the inadequacies of our traditional risk management principles may further undermine our public trust and confidence when it comes to managing the real or imagined risks of nanomaterials (39–43). Marchant and his colleagues consistently question whether our current risk management principles and programs are up to the task of managing nanotechnology (39). They point out that identifying and quantifying the health, safety, and environmental risks may prove difficult at best, because we lack both knowledge and experience with its risk. They believe traditional risk management principles such as (1) acceptable risk (utilizing risk assessment to identify and reduce risks to acceptable levels), (2) cost–benefit analysis (weighing the costs and benefits of proposed risk management options), (3) best available technology (reducing risks to lowest level technologically or economically possible), and (4) the precautionary principles (applying a "safe is better than sorry" approach to control risk) are unworkable for nanotechnologies. For example, the acceptable risk approach may not be applicable because it favors risk reduction over benefit attainment, while the cost–benefit analysis suffers from the uncertainties about the "known unknowns" of nanotechnology. In the case of the best available technology approach, both the risks and benefits of nanotechnology may be simply ignored and this could lead to policymakers to make arbitrary decisions. As for the precautionary principle, it appears to be the least favored approach by several commentators including Marchant for a variety of reasons including its multiplicity of versions, low level of risk required to invoke it, and ability to stop work and hinder progress (32, 39). More importantly, all of these risk management principles suffer from the lack of risk information that the public and its policymakers require to effectively manage risk. To better manage the "known unknowns," some commentators and NGOs suggest we adopt risk assessment and management schemes specifically designed for nanomaterials (32, 39, 44–46). Currently, there exist various attempts to construct risk management frameworks, some of which we outline in the next section.

6. Exploring Risk Management Frameworks

One regulatory framework for nanotechnology developed by Bowman and Hodge covers six regulatory frontiers by creating an "enforcement pyramid" that will allow regulators to use a variety of enforcement options (45, 46). Their "enforcement pyramid" addresses major legal areas related to nanotechnology that include intellectual property, privacy, product safety, occupational health and safety, international law, and environmental law. They believe such an approach will allow multiple stakeholders to utilize current regulations to help manage nanomaterials and their risks. Even so, some commentators question whether this framework may be too static and likely inapplicable to nations with poorly developed legal systems (38). One alternative to the enforcement pyramid is the *Nanorisk Framework* which is also a nanotechnology-based paradigm developed by Environment Defense (ED) and DuPont to regulate a specific industry (44). This framework uses a stepwise approach that focuses on risk identification, emphasizes safety through the life cycle of a given nanomaterial, fosters transparency, and tracks success of risk management schemes. Although it focuses on an industry, this scheme may be applicable to nanomedicines, because it focuses on safety, stakeholder participation, and monitoring that fosters trust.

Unlike these frameworks, the International Risk Governance Council (IRGC) envisions a framework that provides both governance and risk governance of nanomaterials. Risk governance incorporates the totality of circumstances, institutions, processes, and stakeholders related to these activities that lead to risk-related decisions or actions within the context of the risk situation (32). Here, governance differs from the traditional concepts of legislative control or government that mandates behavior through "command

and control measures." Governance is generally preferable to "command and control" government initiatives when dealing with emerging technologies such as nanotechnology. The IRGC framework acknowledges the inherent "riskiness" of existing and future nanotechnologies and looks for a flexible process to lead stakeholders toward prudent risk decisions. Renn and Rocco in their paper discussing the IRGC framework on risk governance see the uncertainty or ambiguity of risks of nanomaterials as becoming greater or more problematic as nanostructures in later generations increase in their sophistication. To help all stakeholders cope with these eventualities, the IRGC proposes a risk management framework specifically designed for all nanotechnologies, including those utilized in nanomedicine.

The IRGC framework categorizes risk-related knowledge as (1) simple (clear causes and effects), (2) complex (multiple causal events and effects), (3) uncertain (knowledge lacking), and (4) ambiguous (variability of norms and interpretations), depending on the generation and complexity and the generation of the nanostructure (32). Perhaps, preassessment is the most critical phase because problems are identified and framed into Frame 1 for passive, first-generation nanomaterials or Frame II for increasingly complex nanomaterials in generations two through four. Once risk-related knowledge is accrued then the risk process moves from preassessment (risk assessment and risk concern) to risk tolerability assessment (risk characterization and evaluation) to risk management and its risk decision-making and implementation phases. The goal of this complex nanospecific framework is to allow stakeholders to identify problems sooner rather than later while addressing their potential social, legal, and policy impacts. Each step of the process incorporates stakeholders and utilizes communication and transparency which are the same key elements and principles identified by Jardine and her co-workers. Although this framework is nanospecific, it is highly complex unlike the framework suggested by Marchant and his co-workers or other commentators (39).

Compared to IRGC risk management framework, Marchant and his co-workers have crafted a less complex framework that incorporates many of the aspects of the aforementioned frameworks as well as concepts from Ayres and Braithwaite to build and to create a dynamic regulatory pyramid (39). Rather than viewing persuasion, soft law, self-regulation, and command and control regulations as static events, they extend them through time. Their approach also focuses on self-regulation and stakeholder generated norms and policies. Each stage of their framework looks to involve stakeholders who are touched by nanotechnologies. Their process begins with information gathering, assessment, and information dissemination (immediate) stage followed by a stakeholder-based self-regulation and norms (short term) stage that moves to an enforced self-regulation (long term) stage. Each level incorporates more supervision while also increasing transparency and participation among stakeholders. The key to their approach is its incrementalism that allows nanotechnology to advance while keeping vigilant to risks concerns. It also allows regulators to use a tit-for-tat regulatory strategy to get compliance without unduly hindering progress. Such a framework may allow all nanotechnologies to advance while allowing everyone to reap its benefits.

Because most, if not all, of these risk management frameworks and principles depend on our "known unknowns" about nanomaterial risks, we will always have uncertainties and ambiguities that will need defining. In fact, many "risky" theranostic nanomedicine products reside within either the first (invasive and non-invasive diagnostics and quantum dots) or second generation (targeted drugs) of nanomaterials. They have "known unknowns" of risk. Even with formalized risk governance frameworks, stakeholders will continually be forced to weigh and balance the "known unknowns" of nanomaterial risks against any real or imagined benefit. As noted, our "known unknowns" may become more uncertain and ambiguous as nanotechnologies become more complex and diverse in their uses.

If Marchant and other commentators are correct about the deficiencies in our existing risk management principles (39), then anyone using them for risk calculus may be misled. The lack of risk information on nanomaterials will likely impact IRB members who must identify, access, and manage risks related to theranostic cancer nanomedicines in clinical trials (15, 25). In fact, several commentators have already questioned the approval process the "nanoparticle albumin-bound" or nab-technology (Abraxis Biosciences) anticancer agent Abraxane for market and clinical use (47). Although many hail the approval process for this agent as a triumph for abbreviated drug trials (48), others question the validity of the process since the toxicology on the long-term effects of the agent are lacking (49). Of the nearly 81,478 clinical trials identified by the U.S. National Institute of Health (NIH) on their web site *only 1,023* or less than 2% of all trials specify nanoparticles in their protocols and over 60% of them utilize the drug Abraxane or nab-paclitaxel (50). The question remains whether the use of risk–benefit data from "old" or preexisting macromolecular counterparts is appropriate for nanomaterials with novel properties (27, 51). Reality is there may be long-term effects that are not accounted for by more traditional risk–benefit models for macromolecular forms. More importantly, the companies and regulatory agencies responsible for assessing risk and safety may not know the proper benchmarks or safety thresholds for comparison with these novel nanomaterials (47).

If our traditional risk management principles and benchmarks are inapplicable to nanoparticles because of their "known unknowns," then IRB members may have to rethink their use of existing IRB risk management strategies such as the component analysis (52) or the "net risks test" to evaluate the risks and benefits of protocols with nanoparticles (53). Certainly, both methods require some knowledge of the risks and benefits of the therapies and procedures ("interventions") to make their comparisons meaningful. And although the former looks at clinical equipoise to guide therapeutic research while the latter does not, both must compare proposed research interventions with existing therapies and procedures. If both assessment schemes rely on knowing the risks and benefits of nanomaterials, then both assessment schemes could be flawed by the "known unknowns" and our lack of knowledge of toxicological studies on their long-term effects (25, 47). This knowledge gap may be compounded by IRB members who may choose to fill the gaps by drawing on their preexisting knowledge and assumptions about the macromolecular counterparts to evaluate risks, or benefits, or both (15, 51). It may be the IRB equivalent of the computer axiom that "garbage in equals garbage out," where calculations based on bad information simply produce more misinformation. Not only may IRB members find their handling risk-related matters for granting protocols problematic, but they may also face challenges with crafting informed consent documents that convey risk information meaningfully to potential candidates for clinical trials (15, 25). In this case, IRB members may be unable to help investigators draft informed consent documents that reasonably convey the risks and benefits to participants or second (research team members) and third parties (public).

Worse still, some allege that both clinical trial participants, and ultimately patients, may be misled by researchers into participating in these protocols, because participants may be assuming they are participating in a "bioequivalency trial" when they are actually not. This situation may be further complicated by the need for companies to keep their proprietary information on their nanoparticle-based formulations confidential. In some cases, as in the case of drugs such as Vioxx, companies and researchers may not reveal information on safety concerns until it is too late (47, 54). If this occurs, then everyone, including participants and patients, may be receiving information that is either deficient or inaccurate which could open the door to ethical and legal challenges to informed consent (25, 47). If this lack of information becomes the basis for a claim of deception, then everyone may find themselves at risk for a legal challenge to the process of informed consent (20, 55).

7. Managing Our "Known Unknowns": Transparency and Communication

The current lack of an adequate framework for risk management requires careful communication on the part of the scientific and medical community with various stakeholders to insure a successful development of theranostics nanomedicine (56). The economic stakes and the potential therapeutic benefits demand a strategic plan that will allow the adequate assessment, management, and communication of risks. In this chapter, we favored a risk management framework that emphasizes self-regulation, responsibility, transparency, and the involvement of all relevant stakeholders. This bottom-up approach does not guarantee a unanimous adherence to the aforementioned principles but it allows a rapid flow of information between stakeholders, the involvement of all relevant parties and the continuation of research and development despite our "known unknowns." Until we get an agreed upon risk management framework, this approach minimizes the potential of stigmatizing the technology (32, 39).

That being said, the "known unknowns" related to nanotechnology may make it almost impossible for stakeholders to avoid the public reaching premature opinions on its risks and benefits (47). The propensity of the public to make risk–benefit decisions based on mental shortcuts or "heuristics" makes nanotechnology vulnerable to interpretational errors (39). This means the public may be exposed to an untoward event related to nanotechnology and form an opinion based on their initial perception of the risks. Once formed, it may be difficult or impossible to change. Reality is imagery and social reinforcement may give certain groups, such as NGOs or the media, real leverage in framing public perceptions of nanotechnology. For these reasons, all stakeholders must openly discuss the "known unknowns" related to nanomaterials and engage the public to build public confidence and trust before rather than after some adverse event taints the debate.

Maybe the best strategy to avoid public uproar and misinformation is for all stakeholders to begin engaging in discussions about the "known unknowns" before some adverse event occurs. We need to encourage "upstream ethical and policy reflections" at various level of discourse that includes the public, the scientific, and medical community as well as various political and economic stakeholders. Ultimately we need to develop risk management frameworks that will help us identify the appropriate problems, engage parties, foster discussions, and develop management schemes to minimize risks and maximize our benefits.

References

1. Abel, J. "Nanomedicine and Cancer" Available online: http://www.chem.usu.edu/~tapaskar/Joe-Seminar-Nano-Medicine.pdf.

2. European Science Foundation Nanomedicine Report (2005) Available online: http://www.esf.org/publication/214/Nanomedicine.pdf.

3. Sumer, B. and Gao, J. (2008) Theranostic nanomedicine for cancer. *Nanomedicine* **3**, 137-140.

4. Muthu, M. and Singh, S. (2009) Targeted nanomedicine: effective treatment modalities for cancer, AIDS and brain disorders. *Nanomedicine* **4**, 105–118.

5. Tassinari, O., Caiazzo R. J., Ehrlich J. R., and Liu, B.C-S. (2008) Identifying autoantigens as theranostic targets: antigen arrays and immunoproteonics approaches. *Curr. Opin. Mol. Ther.* **10**, 107–115.

6. Del Vecchio, S., Zannetti, A., Fonti, R., Pace, L., and Salvatore, M. (2007) Nuclear imaging in cancer theranostics. *Q. J. Nucl. Med. Mol. Imag.* **51**, 152–163.

7. Ramachandran, G. (June 26, 2009) Theranostics: an evolving field catering to the unmet needs of the medical world. http://www.frost.com/prod/servlet/market-insight-print.pag?docid=170968864. Accessed 28 October 2009.

8. Visogain. (May 29, 2009) *In vitro* diagnostics: market analysis 2009-2024. http://www.prlog.org/10246394-in-vitro-diagnostics-market-analysis-20092024.pdf. Accessed 28 October 2009.

9. Blair, E. D. (2008) Assessing the value-adding impact of diagnostic-type tests on drug development and marketing. *Mol. Diag. Ther.* **12**, 331–337.

10. Jotterand, F. (2007) Nanomedicine: how it could reshape clinical practice. *Nanomedicine* **2**, 401–405.

11. 3rd Horizon Scanning Workshop (2007) Theranostics – Ethical, legal and social aspects. Muenster, 6th-8th June 2007.

12. Girod, J. and Klien, A. R. (2009) Predictive health technologies: personalized medicine and toxic exposures. *Hous. J. Health L. Policy* **9**, 163–178.

13. Wilson, R. F. (2006) Symposium article: Nanotechnology: the challenge of regulating known unknowns. *J. Law Med. Ethics* **34**, 704–713.

14. Rakhlin, M. (2008) Regulating nanotechnology: a private-public insurance solution. *Duke Law Tech. Rev.* **2008**, 2–45.

15. Resnick, B. R. and Tinkle, S. S. (2007) Ethical issues in clinical trials involving nanomedicine. *Contemp. Clin. Trials.* **28**, 433–441.

16. Bawa, R. and Johnson, S. (2007) The ethical Dimensions of nanomedicine. *Med. Clin. N. Ann.* **91**, 881–887.

17. Wolinski, H. (2006) Nanoregulation. *EMBO Rep.* **7**, 858–861.

18. Katz, J. (2009) Informed consent – Must it remain a fairy tale? in: *Ethical Issues in Modern Medicine (7th edition)* (Steinbock, B. Arras, J. D., and London, A. J., eds.) McGraw Hill, New York.

19. Furrow, B. R., Greaney, T. L., Johnson, S. H., Jost, T. S., and Schwartz, R. L. (2000) *Health Law (2nd edition)*. Hornbook series, West Group, West Publishing Co.

20. Grim, D. A. (2007) Informed consent for all! No exceptions. *New Mexico L. Rev.* **37**, 39–83.

21. Etchells, E. (1999) Informed consent in surgical trials. *World J. Surg.* **23**, 1215–1220.

22. Jotterand, F., McClintock, S. M., Alexander, A. A., and Husain, M. M. (2010) Ethics and informed consent of Vagus Nerve Stimulation (VNS) for patients with Treatment-Resistant Depression (TRD). *NeuroEthics* **3**, 13–22.

23. Beauchamp, T. L. and Childress, J. F. (2009). *Principles of Biomedical Ethics (6th edition)*. Oxford University Press, Oxford.

24. Brock, D. (1987). Informed consent. in: *Health Care Ethics* (Van De Veer, D. and Regan T., eds.) Temple University Press, Philadelphia.

25. Virdi, J. (2008) Bridging the knowledge gap: examining potential limits in nanomedicine. *Spon. Gen.* **2**, 25–44.

26. Breggin, L., Falkner, R., Jaspers, N., Pendergrass, J., and Porter, R. (2009) Securing the promise of nanotechnologies. Chatham House, London, England: Regulating Nanotechnologies in the EU and US. http://www.lse.ac.uk/internationalRelations/centresandunits/regulatingnanotechnologies/nanopdfs/REPORT.pdf. Assessed 29 October 2009.

27. Scientific Committee on Emerging and Newly Identified Health Risks (SCENIHR) (2009) Risk Assessment of Products of Nanotechnologies. http://ec.europa.eu/health/ph_risk/committees04_scenihr_o_023.pdf. Accessed 28 October 2009.

28. Nanotechnology: a report of the U.S. Food and Drug Administration. Rockville, Maryland: Nanotechnology Task Force (2007). http://www.fda.gov/downloads/ScienceResearch/SpecialTopics/Nanotechnology/ucm110856.pdf. Accessed 10 October 2009.

29. Pelley, J. and Saner, M. (2009) International approaches to the regulatory governance of nanotechnology. Ottawa, Ontario: Regulatory Governance Initiative. http://www.carleton.ca/

regulation/publications/Nanotechnology_Regulation_Paper_April2009.pdf. Assessed 20 September 2009.

30. Greenpeace: policy and position paper on Nanotechnology (2002) http://www.greenpeace.org/raw/content/denmark/press/apporter-or-dokumenter/nanotechnology-policy-positi.pdf. Accessed on 1 October 2009.

31. Tyshenko, M. G. and Krewski, D. (2008) A risk management framework for the regulation of nanomaterials. *Int. J. Nanotechnol.* **5**, 143–160.

32. Renn, O. and Roco, M.C. (2006) Nanotechnology and the need for risk governance. *J. Nanopart. Res.* **8**, 153–191.

33. Jardine, C. G., Hrudey, S. E., Shortreed, J. H., Craig, L., Krewski, D., Furgal, C., et al. (2003) Risk management frameworks for human health and environmental risks. *J. Toxicol. Environ. Health, Part B* **6**, 570–718.

34. Maynard, A. D. (2006) Nanotechnology: assessing the risks. *Nanotoday* **1**, 22–32.

35. Patra, D., Ejnavarzala, H. and Basu, P. K. (2009) Nanoscience and nanotechnology: ethical, legal, social and environmental issues. *Curr. Sci.* **96**, 651–657.

36. Orberdorster, G., Oberdorster, E., and Oberdorster, J. (2005) Nanotechnology: an emerging discipline evolving from studies of ultrafine particles. *Environ. Health Perspect.* **113**, 623–839.

37. Azou, B. L., Jorly, J., On, D., Sellier, E., Moisan, F., Fleury-Feith, J., et al. (2008) *In vitro* effects of nanoparticles on renal cells. *Part. Fibre Tox.* **5**, 22.

38. Song, Y., Li, X. and Du, X. (2009) Exposure to nanoparticles is related to pleural effusion, pulmonary fibrosis and granuloma. *Eur. Respir. J.* **34**, 559–567.

39. Marchant, G. E., Sylvester, D. J., and Abbott, K. W. (2008) Risk management principles for nanotechnology. *Nanoethics* **2**, 43–60.

40. Linkov, I., Steevens, J., Adlakha-Hutcheon, G., Bennett, E., Chappell, M., Colvin, V., et al. (2009) Emerging methods and tools for environment risk assessment, decision-making, and policy for nanomaterials: summary of NATO Advance Research Workshop. *J. Nanopart. Res.* **11**, 513–527.

41. Linkov, I., Satterstrom, F. K., Steevens, J., Ferguson, E., and Pleus, R. C. (2007) Multicriteria analysis environmental risk assessment for nanomaterials. *J. Nanopart. Res.* **9**, 543–554.

42. Farber, D. and Lakhtakia, A. (2009) Scenario planning and nanotechnologies futures. *Eur. J. Phys.* **30**, S3-S15.

43. Kuzma, J., Paradise, J., Ramachandran, G., Kim, J.-A., Kokotovich, A., and Wolk, S. M. (2008) An integrated approach to oversight assessment for emerging technologies. *Risk Anal.* **28**, 1197–1219.

44. Nano Risk Framework, Washington D.C. and Wilmington, Delaware: Environmental Defense – DuPont Nano Partnership (2007) http://www.edf.org/documents/6496_Nano%20Risk%20Framework.pdf. Accessed on 20 October 2009.

45. Bowman, D. M. and Hodge, G. A., (2007) A small matter of regulation: an international review of nanotechnology regulation. *Columbia Sci. Tech. L. Rev.* **8**, 1–36.

46. Bowman, D. M. and Hodge, G. A. (2006) Nanotechnology: mapping the wild regulatory frontier. *Futures* **38**, 1060–1073.

47. Vines, T. and Faunce, T. (2009) Assessing the safety and cost-effectiveness of early nanodrugs. *J. Law Med.* **16**, 822–45.

48. Till, M. C., Simkin, M. M. and Maebius, S. (2005) Nanotech meets the FDA: a success story about the nanoparticle drugs approved by the FDA. *Nanotechnology Law & Bus.* **2.2**, 163–167.

49. Faunce, T. A. (2008) Toxicological and public good considerations for the regulation of nanomaterial-containing medical products. *Expert Opin. Drug Saf.* **7**, 103–106.

50. U.S. National Institutes of Health, Nanoparticle Studies. [Clinical Trials.gov] http://www.clinicaltrails.gov/ct2/results?term=nano. Accessed on Nov. 10, 2009.

51. Beasley, J. C., Kramer, V. L., and Priest, S. H. (2008) Expert opinion on nanotechnology: risks, benefits, and regulation. *J. Nanopart. Res.* **10**, 549-558.

52. Weijer, C. and Miller, P. B. (2004) When are research risks reasonable in relation to anticipated benefits? *Nat. Med.* **10**, 570–573.

53. Wendler, D. and Miller, F. G. (2004) Assessing research risks systematically: the net risks test. *J. Med. Ethics* **33**, 481-486.

54. Avom, J. (2006) Dangerous deception – hiding the evidence of adverse drug effects. *NEJM* **355**, 2170–2171.

55. King, S. K. and Moulton, B. W. (2006) Rethinking informed consent: the case for shared medical decision making. *Am. J. Law Med.* **32**, 429–493.

56. Plows, A. and Reinsborough, M. (2008) Nanobiotechnology and Ethics: Converging Civil Society Discourses. in: *Emerging Conceptual, Ethical and Policy Issues in Bionanotechnology* (Jotterand, F., ed.) Dordrecht, Springer.

INDEX

Sarah J. Hurst (ed.), *Biomedical Nanotechnology: Methods and Protocols*, Methods in Molecular Biology, vol. 726,
DOI 10.1007/978-1-61779-052-2, © Springer Science+Business Media, LLC 2011